Maritime Situational Awareness

for

Tyrians

Maritime Security & Safety Organizations, Agencies, Networks and Systems

Joachim Beckh

Maritime Situational Awareness
- MSA -
for Tyrians

Entities and Systems in Maritime Situational Awareness

FIRST EDITION

This information was prepared and is provided as an account of personal work experience, hopefully avoiding the phenomenon of „*expertise induced amnesia*". However, for a dynamic topic like MSA a comprehensive coverage is hardly achievable, therefore submissions of correction or additional information are most welcome.

The author makes no warranty, expresses or implies, nor assumes any legal liability or responsibility for the accuracy, completeness, or any third party's use of the information or the links provided, nor of the use of any information, apparatus, product, or process disclosed. Reference herein to any specific commercial product, process, or service by trade name, trademark, manufacturer, or otherwise, does not necessarily constitute or imply its endorsement, recommendation, or favoring by the author.

The views and opinions of author expressed herein do not necessarily state or reflect those of any entity named or referred to. External information see listed reference or link.

Copyright © 2019 Joachim Beckh
All rights reserved.
ISBN-10: 1542959683
ISBN-13: 9781542959681

The Cantos

"And by the beach-run, Tyro,

Twisted arms of the sea-god,

Lithe sinews of water, gripping her, cross-hold,

And the blue-gray glass of the wave tents them,

Glare azure of water, cold-welter, close cover."[1]

"... thick smoke, purple, rising

bright flame now on the altar

the crystal funnel of air

out of Erebus, the delivered,

Tyro, Alcmene, free now, ascending"[2]

---

[1] Ezra Pound refers to Tyro in The Cantos. In Canto 2 he takes up her rape by Poseidon.
[2] In a later Canto (74) Pound connects her to Alcmene, imprisoned in the world of the dead, but in a later paradise vision he sees her *"ascending"*.

MSA for Tyrians Content

PREFACE 15

Maritime Picture (MP) and Recognized Maritime Picture (RMP) 30
Maritime Operations, Maritime Terrorism, Piracy 49

MARITIME ENTITIES 55

Africa 65
African Maritime Safety and Security Agency (AMSSA) 66
Arctic Council 68
Association of Southeast Asian Nations (ASEAN) 71
Australia, Canada, New Zealand, the United Kingdom and the United States (AUSCANNZUKUS) 74
Baltic and International Maritime Council (BIMCO) 76
Balkan & Black Sea Security Forum 77
Baltic Sea Region Border Control Cooperation (BSRBCC) 78
Benelux 81
Belgium, the Netherlands, and Luxembourg *81*
Netherlands Coast Guard Center 82
Belgium Coast Guard Center 84
Brazil (BRA) 85
Sistema de Informações Sobre o Tráfego Marítimo (SISTRAM) *86*
Canada (CAN) 88
Polar Epsilon Project *89*
Polar Epsilon 2 Project 90
China (CHN) 92
BeiDou Navigation Satellite System 93
Maritime Safety Administration of the People's Republic of China (CMSA) 95
Cooperation Council for the Arab States of the Gulf (GCC) 97
Council of the Baltic Sea States (CBSS) 101
Djibouti Code of Conduct 102
Economic Community of Central African States (ECCAS) 104
Centre Régional De Securité Maritime de l'Afrique Centrale (CRESMAC) *106*
Multinational Center of Coordination (CMC) in Duoala 108
Economic Community of West African States (ECOWAS) 109
Regional Coordination Center for the Maritime Security of West Africa (CRESMAO) *110*

Center for Information and Communication (CINFOCOM) 111
Maritime Trade Information Sharing Center - Gulf of Guinea (MTISC-GOG) 113
Inter-Regional Coordination Center (ICC/CIC) 115
European Union (EU) 116
Permanent Structured Cooperation (PESCO) 119
European Union Maritime Security Strategy (EUMSS) 123
Integrated Maritime Policy (IMP) 125
Common Information Sharing Environment (CISE) 130
The Seven Maritime CISE User Communities 134
Cooperation Project (CoopP), CISE Handbook, CISE Incubator, and CISE Data Model 136
European Coast Guard Functions Forum (ECGFF) 139
European Coast Guard Functions Academy Network Projects 142
European Coordination Centre for Accident and Incident Reporting Systems (ECCAIRS) 144
Critical Maritime Routes Indian Ocean (CRIMARIO) 148
Indian Ocean Regional Information Sharing (IORIS) 149
European Police Office (EUROPOL) 152
European Union Patrol Network (EPN) 155
European Union SECRET (EU SECRET) Network 156
European Union Naval Force SOMALIA (EU NAVFOR SOMALIA) 157
European Union Naval Force Mediterranean (EUNAVFOR MED) 158
European Union Maritime Safety Agency (EMSA) 159
CleanSeaNet (CSN) 164
Electronic Quality Ship Information System (EQUASIS) 166
Integrated Maritime Data Environment (IMDatE) 167
SafeSeaNet (SSN) 169
Thetis 175
Fisheries Areas Network (FARNET) 176
Fishery Data Exchange System (FIDES) 178
FRONTEX 179
European Border Surveillance System (EUROSUR) 183
Galileo 185
Maritime Security (MASE) Program /Projects 188
Piracy, Maritime Awareness & Risks (PMAR) 189
Piracy Incident Reporting and Information Exchange System (PIRATES) 201
Smartfish Program 202
Vessel Detection System (VDS) in the European Union 208
Vessel Monitoring System (VMS) 209
Vessel Traffic Monitoring (VTM) 211
Vessel Traffic Monitoring System (VTMS) 212

- Vessel Traffic Service (VTS) 213
- Calais-Dover Reporting System (CALDOVEREP) 214
- Container Tracking System (CTS) 215
- FIVE POWER DEFENSE ARRANGEMENTS (FPDA) 216
- FRANCE (FRA) 218
- SYSTÈME D'INFORMATION ET DE COMMANDEMENT 21 (SIC 21) 219
- GERMANY (DEU) 222
- KIEL INTERNATIONAL SEA POWER SYMPOSIUM (KISS) 223
- NATIONAL MARITIME CONFERENCE (NMK) 224
- GERMAN MARITIME FORCES STAFF (DEUMAR FOR) 225
- Community-Of-Interest specific Maritime Situational Awareness (COI Specific MSA) 226
- MARITIME SAFETY AND SECURITY CENTER (MSSC) 228
- System Maritime Verkehrstechnik (SMV) 232
- Integrated Sea Surveillance System 232
- GULF OF GUINEA COMMISSION (GGC) 233
- HELSINKI COMMISSION (HELCOM) 236
- HELSINKI COMMISSION AIS SYSTEM (HELCOM AIS SYSTEM) 238
- INDIA (IN) 239
- INDIAN OCEAN NAVAL SYMPOSIUM (IONS) 240
- INFORMATION MANAGEMENT AND ANALYSIS CENTER (IMAC) 242
- National Command Control Communication and Intelligence System (NC3I) 243
- INDIAN REGIONAL NAVIGATION SATELLITE SYSTEM (IRNSS) 244
- INDIAN OCEAN COMMISSION (IOC) 246
- REGIONAL INTEGRATION SUPPORT PROGRAM (RISP) 249
- INDIAN OCEAN RIM ASSOCIATION (IORA) 251
- INTERGOVERNMENTAL AUTHORITY ON DEVELOPMENT (IGAD) 253
- INTERNATIONAL ASSOCIATION OF LIGHTHOUSE AUTHORITIES (IALA) 258
- INTERNATIONAL ASSOCIATION OF LIGHTHOUSE AUTHORITIES NETWORK (IALA-NET) ... 260
- INTERNATIONAL ASSOCIATION OF PORTS AND HARBORS (IAPH) 261
- INTERNATIONAL CHAMBER OF SHIPPING (ICS) 262
- INTERNATIONAL CARGO HANDLING COORDINATION ASSOCIATION (ICHCA) 263
- INTERNATIONAL FISHERY ORGANIZATIONS (IFO) 264
- FISHINFONETWORK (FIN) 267
- INTERNATIONAL HYDROGRAPHIC ORGANIZATION (IHO) 269
- INTERNATIONAL MARITIME BUREAU-PIRACY REPORTING CENTER (IMB-PRC) 272
- INTERNATIONAL MOBILE SATELLITE ORGANIZATION (IMSO) 273
- INTERNATIONAL POLICE (INTERPOL) 276
- INFORMATION EXCHANGE SYSTEM FOR INTERPOL 24/7 (INTERPOL-I24/7) 279
- West African Police Information System (WAPIS) 279

Entry	Page
INTERNATIONAL TELECOMMUNICATION UNION (ITU)	280
MARITIME MOBILE ACCESS AND RETRIEVAL SYSTEM (MARS)	281
ITALY (ITA)	282
TRANS-REGIONAL MARITIME NETWORK (T-RMN)	283
VIRTUAL-REGIONAL MARITIME TRAFFIC CENTER (V-RMTC)	283
Virtual Regional Traffic Centre Wider Mediterranean Community (V-RMTC WMC)	284
Virtual Maritime Traffic Center 5+5 (V-RMTC 5+5)	285
Virtual Maritime Traffic Center 8+6 (V-RMTC 8+6)	285
Virtual Maritime Traffic Center Lebanon (V-RMTC Lebanon)	286
Service-oriented infrastructure for MARitime Traffic tracking (SMART)	287
SMART FENIX	289
JAPAN (JPN)	290
QUASI-ZENITH SATELLITE SYSTEM (QZSS)	292
COSPAS-SARSAT	294
LLOYDS OF LONDON	297
LLOYD'S REGISTER GROUP LIMITED (LR)	298
Lloyd's Register Fairplay	298
MDA WatchKeeper	299
MARITIME ANALYSIS OPERATIONS CENTER – NARCOTICS (MAOC – N)	300
CENTRE DE COORDINATION POUR LA LUTTE ANTI-DROGUE EN MEDITERRANÉE (CECLAD-M)	301
WEB ENABLED TEMPORAL ANALYSIS SYSTEM (WEBTAS)	302
MARITIME MULTI-NATIONAL IP INTEROPERABILITY (M2I2)	303
GLOBAL COUNTER-TERRORISM FORCE (GCTF)	304
COMBINED MARITIME FORCES (CMF)	305
COMBINED TASK FORCE 150	307
COMBINED TASK FORCE 151	309
COMBINED TASK FORCE 152	311
COMBINED ENTERPRISE REGIONAL INFORMATION EXCHANGE SYSTEM (CENTRIXS)	312
Atlantic, Barents, Baltic and Arctic Regions (ABBA)	314
MARITIME ORGANIZATION FOR WEST AND CENTRAL AFRICA (MOWCA)	315
MARITIME SECURITY CENTER HORN OF AFRICA (MSCHOA)	317
MERCURY	318
SHIP SECURITY REPORTING SYSTEM (SSRS)	319
MARITIME SURVEILLANCE (MARSUR) NETWORKING	320
Maritime Surveillance (MARSUR) Capability	324
MEDITERRANEAN COAST GUARD FUNCTIONS FORUM (MEDCGFF)	331
NORTH AFRICAN PORT MANAGEMENT ASSOCIATION (NAPMA)	333
NORTH ATLANTIC COAST GUARD FORUM (NACGF)	334
NORTH ATLANTIC TREATY ORGANIZATION (NATO)	335

Active Endeavour 342
Allied Command Operations (ACO) 343
   Maritime Command (MARCOM) 343
Allied Command Transformation (ATO) 344
Allied Worldwide Navigational Information System (AWNIS) 346
   Planning Board for Ocean Shipping (PBOS) 347
NATO Centers of Excellence (COE) 348
   Analysis and Simulation Center for Air Operations (CASPOA) 351
   Center of Excellence Defense Against Terrorism (COE DAT) 351
   Center of Excellence for Operations in Confined and Shallow Waters (COE CSW) 352
   Center of Excellence for Cold Weather Operations (COE CWO) 354
   Civil-Military Cooperation Center of Excellence (CIMIC COE) 354
   Combined Joint Operations from the Sea Center of Excellence (CJOS COE) 355
   Command and Control Center of Excellence (C2 COE) 357
   Cooperative Cyber Defense Center of Excellence (CCDCOE) 358
   Counter-Improvised Explosive Devices Center of Excellence (C-IED COE) 359
   Counter Intelligence Center of Excellence (CI COE) 360
   Crisis Management and Disaster Response Center of Excellence (CMDR COE) 360
   Energy Security Center of Excellence (ENSEC COE) 360
   Explosive Ordnance Disposal Center of Excellence (EOD COE) 361
   Human Intelligence Center of Excellence (HUMINT COE) 361
   Joint Air Power Competence Center (JAPCC) 361
   Joint Chemical Biological Radiological Nuclear – Defence Center of Excellence (JCBRN COE) 362
   Maritime Security Center of Excellence (MARSEC COE) 363
   Military Engineering Center of Excellence (MILENG COE) 365
   Military Medicine Center of Excellence (MILMED COE) 366
   Military Police Center of Excellence (MP COE) 367
   Modeling and Simulation Center of Excellence (M&S COE) 367
   Mountain Warfare Center of Excellence (MW COE) 368
   Naval Mine Warfare Center of Excellence (NMW COE) 368
   Stability Policing Center of Excellence (SP COE) 369
   Strategic Communications Center of Excellence (STRATCOM COE) 370
NATO Science and Technology Organization (STO) 371
NATO SECRET Wide Area Network (NSWAN) 373
NATO Shipping Center (NSC) 374
Naval Cooperation and Guidance for Shipping (NCAGS) 376
Planning Board for Ocean Shipping (PBOS) 377
Standing NATO Maritime Groups (SNMG1 and SNMG2) 378
Tidepedia 380
   Technology for Information, Decision and Execution Superiority (TIDE) 388
   TIDE Sprints 390

Maritime Information Services Conference (MISC) 398
BASELINE FOR RAPID INTERACTIVE TRANSFORMATIONAL EXPERIMENTATION (BRITE) 401
BI-STRATEGIC COMMAND AUTOMATED INFORMATION SYSTEM (BI-SC AIS) 404
BATTLEFIELD INFORMATION COLLECTION AND EXPLOITATION SYSTEM (BICES) 406
MARITIME COMMAND AND CONTROL SYSTEM (MCCIS) 407
TRITION 409
NORTH PACIFIC COAST GUARD FORUM (NPCGF) 411
PACIFIC ISLANDS FORUM (PIF) 412
PAN-AFRICAN ASSOCIATION FOR PORT COOPERATION (PAPC) 415
PORT MANAGEMENT ASSOCIATION OF EASTERN AND SOUTHERN AFRICA (PMAESA) 417
PORT MANAGEMENT ASSOCIATION OF WEST AND CENTRAL AFRICA (PMAWCA) 418
UNION OF PORT ADMINISTRATIONS OF NORTHERN AFRICA (UAPNA) 419
PORT STATE CONTROL (PSC) 420
RED OPERATIVA DE COOPERACIÓN REGIONAL DE AUTORIDADES MARITIMAS DE LAS AMERICAS (ROCRAM) 421
ORGANIZATION OF AMERICAN STATES (OAS) 422
INTER-AMERICAN COMMITTEE AGAINST TERRORISM (CICTE) 428
REGIONAL COOPERATION AGREEMENT ON COMBATING PIRACY AN ARMED ROBBERY AGAINST SHIPS IN ASIA (ReCAAP)
 431
INFORMATION NETWORK SYSTEM (IFN) 433
RUSSIA (RUS) 434
GLOBALNAYA NAVIGATSIONNAYA SPUTNIKOVAYA SISTEMA (GLOSNASS) 438
SCANDINAVIA 440
NORWAY (NOR), DENMARK (DNK), SWEDEN (SWE), FINLAND (FIN) AND ICELAND (ISL) 440
BARENTS WATCH 445
C2SöC 447
CRIADS 447
MDA-TOOL 448
NORWEGIAN COMMAND AND CONTROL INFORMATION SYSTEM (NORCCIS) 453
MARIA GEO DEVELOPMENT TOOLKIT (MARIA GDK) 453
SEALION 455
SITAWARE SUITE 459
SJÖBASIS 461
SEA SURVEILLANCE COOPERATION BALTIC SEA (SUCBAS) 464
Sea Surveillance Cooperation Baltic Sea (SUCBAS) Interface Technology 468
SEA SURVEILLANCE COOPERATION FINLAND SWEDEN (SUCFIS) 469
Sea Surveillance Information System (SSIS) 470
SEYCHELLES (SYC) 471
REGIONAL FUSION AND LAW ENFORCEMENT CENTER FOR SAFETY AND SECURITY AT SEA (REFLECS3), ... 471

- Singapore (SGP) 472
  - Information Fusion Center (IFC) 472
    - Malacca Straits Patrol Info-System (MSP IS) 474
    - Open & Analysed Shipping Info-System (OASIS) 474
    - Regional Maritime Information Exchange (ReMIX) System 475
    - Sense-making, Analysis and Research Tool (SMART) 475
    - Surveillance Picture (Surpic) 476
- South Africa (ZAF) 477
  - South African Maritime Safety Authority (SAMSA) 477
  - Southern Africa Development Community (SADC) 478
- Spain (ESP) 480
- Turkey (TUR) 481
  - Operation Black Sea Harmony (OBSH) 481
- United Kingdom of Great Britain and Northern Ireland (GBR) 482
  - United Kingdom Maritime Trade Operations (UKMTO) 483
  - National Maritime Information Centre (NMIC) 485
  - INMARSAT 488
- United Nations (UN) 495
  - International Maritime Organization (IMO) 496
    - Automated Information System (AIS) 500
    - Satellite-based AIS (S-AIS) 507
    - Electronic Chart Display and Information System (ECDIS) 510
    - Global Integrated Shipping Information System (GISIS) 512
    - Global Maritime Distress and Safety System (GMDSS) 514
    - Gulf of Finland Reporting (GOFREP) 519
    - International Ship and Port Facility Security Code (ISPS Code) 520
    - Navigational Telex (Navtex) 522
    - Long Range Identification and Tracking (LRIT) 523
    - The 100 Series Rules 526
    - West European Tanker Reporting System (WETREP) 527
- United States of America (USA) 528
  - African Partnership Station (APS) 529
  - Customs-Trade Partnership against Terrorism (C-TPAT) 531
  - Global Command and Control System (GCCS) 534
    - Global Command and Control System - Army (GCCS-A) 534
    - Global Command and Control System - Joint (GCCS-J) 534
    - Global Command and Control System – Maritime (GCCS-M) 535
  - Global Maritime Partnership (GMP) 537
  - Global Positioning System (GPS) 541
  - International Seapower Symposium (ISS) 544

Joint Interagency Task Force (JIATF) 545
Linked Operations-Intelligence Centers Europe (LOCE) 549
Maritime Domain Awareness Executive Steering Committee (MDA ESC) 554
Maritime Information Sharing Environment (MISE) 555
Maritime Liaison Office (MARLO) Bahrain 558
Maritime Tactical Command and Control (MTC2) 559
National Information Exchange Model (NIEM) 560
National Maritime Intelligence-Integration Office (NMIO) 561
National Maritime Law Enforcement Academy (NMLEA) 563
North American Aerospace Defense Command (NORAD) 564
Office of Global Maritime Situational Awareness (OGMSA) 567
Office of Naval Intelligence (ONI) 570
Surface Picture (SURPIC) 571
United States Department of Transportation (US DOT) 573
Maritime Safety and Security System (MSSIS) 577
SeaVision 578
Western Pacific Naval Symposium (WPNS) 579
World Port Source (WPS) 580

SYSTEMATOGENESIS 581

Concept, Development & Experimentation (CD&E) to Production 581
Basic Knowledge on Terms, Abbreviations and Definitions 592
Architecture 595
Concept 598
Demonstration 598
Experiment 598
Project 600
Prototype 600
Standardization 601
Communications and Information Systems (CIS) to C5ISR 603
System-of-Systems (SOS) 605
Fusion & Correlation 610
Fusion 610
Correlation 611
Fusion & Correlation Models 612
Interoperability 613
Technical Interoperability 614
Operational Interoperability 614

- Procedural Interoperability ... 615
- *INFORMATION EXCHANGE REQUIREMENT (IER)* ... 616
- *NETWORK TOPOLOGY* ... 617
- *OPERATING CAPABILITY* ... 625
- IOC (Initial Operating Capability or Initial Operational Capability) ... 625
- FOC (Full Operational Capability) ... 626
- *POLICY* ... 627
- *PROCEDURE* ... 628
- Standard Operating Procedure (SOP) ... 628
- *SAFETY* ... 629
- *SECURITY* ... 630
- National Security ... 631
- Security Modes ... 632
- Compartmented Information ... 633
- Information Security (INFOSEC) / Information Technology Security (IT Security) ... 633
- Information Security ... 634
- Information Technology Security (IT-Security) ... 634
- Information Assurance (IA) ... 635

PREFACE

In Greek mythology Tyro was the daughter of King Salmoneus who had two brothers, Athamas and Sisyphus. Tyro obeyed her father's status and married Cretheus, with whom she had three children, Aeson, Pheres, and Amythaon. When she fell deeply in love with the river god Enipeus, he unfortunately turned down her approaches. Poseidon on the other hand, was filled with lust for Tyro, disguised himself as Enipeus and got lucky. Obviously Tyro had a crash for the men from the seas, called Sailors.

Understandably Tyro was in the follow-on disappointed about being tricked into love-making, getting pregnant, and then just being dropped. She took revenge and later disposed Poseidon's children, the two twin boys Pelias and Neleus, on a desert mountain and left them to die. However, the boys were found by a herdsman who raised them as his own. When the boys reached adulthood, they started searching for their mother and found Tyro, who was at the time being mistreated by her stepmother, Sidero.

King Salmoneus had married Sidero when Alkidike, his former wife and the mother of Tyro, had died. Sidero now tried to hide from the brothers in a temple of Hera, but Pelias found and killed her. This desecration caused in turn the undying hate of Pelias by the goddess Hera and her glorious patronage of Jason and the Argonauts in their long and famous quest for the Golden Fleece.

Pelias' half-brother Aeson, the son of Tyro and Cretheus, was hence the father of Jason. Being the eldest son of Tyro, Jason's father, Aeson, should have inherited the kingdom of Iolcus, a seaport in Thessaly (today north-eastern Greece), which Cretheus had founded. The story of Jason and the Argonauts talks about another great journey by water to a faraway land eastwards of Greece.

Soon after the killing of Sidero, Tyro married Sisyphus, the brother of King Salmoneus and her paternal uncle, who was famed as the craftiest of all men. Now King, Sisyphus founded the city of Ephyra[3] and promoted navigation and commerce, but he was avaricious and deceitful. He killed travelers and guests, a violation of Xenia, which fell under Zeus' domain. Maintain an iron-fisted rule by killing possible opponents is common to all ages, but now Sisyphus also tried to outsmart Zeus and the other gods.

Sisyphus even tried to talk Tyro into one of his plots to kill her own father Salmoneus. Eventually Tyro slayed her own children she bore from Sisyphus, when she discovered that his plan was to use them to dethrone her father. Sisyphus was punished for his self-aggrandizing craftiness and deceitfulness by being forced to roll an immense boulder up a hill, only to watch Zeus kicking it to roll back down each time again, consigning him to an eternity of endless trying in useless efforts with unending frustration.

Týros is also the Greek name of a city at the coast of the Mediterranean, located about 80 km south of Beirut in the South Governorate of Lebanon. The name of the city in local language means "*rock*" after the rocky formation on which the town was originally built. The adjective for the Lebanese name "*Tyre*" is Tyrian, as an ancient Phoenician city and the legendary birthplace of Europa and Dido (Elissa). The inhabitants were the Tyrians and their merchants were among the first who ventured to navigate the Mediterranean waters.

Tyre originally consisted of two distinct urban centers, where Tyre was on an island just off shore, and the associated settlement of Ushu was located on the adjacent mainland. This provided protection from threats on the mainland and is possibly one of the earliest maritime anti-access/area-denial operations. Alexander the Great later connected the island to the mainland by constructing a causeway during his siege of the city, conquered it and demolished the old city just to reuse its cut stones.

The original island city had also two harbors, one on the south side and the other on the north side of the island. These two harbors enabled Tyre to gain the maritime prominence that it had for a long time; the harbor on the north side of the island was, in fact, one of the best harbors on the eastern end of the Mediterranean. The harbor on the south side has silted over, but the harbor on the north side is still in use today.

In ancient times, the island-city of Tyre was heavily fortified and the mainland settlement, originally called Ushu (later called Palaetyrus, meaning "*Old Tyre*" by the ancient Greeks), which was actually more like a line of suburbs rather than one city. Ushu was used primarily as a source of water and timber for the main island city of Tyre.

The maritime records of Josephus show, that the two cities fought against each other on occasion, but because they both benefited from the island city's wealth from maritime trade and the mainland area's source of timber, water and burial grounds they most of the time supported one another.

---

[3] supposed as the original name of Corinth

In "*Maritime Situational Awareness for Tyrians*" using the plural of the word "*Tyrian*" stands for the novice in MSA. Situational Awareness (SA) has repeating processes to establish knowledge, it is a truly Sisyphus work and, like the innumerable characters and chronicles of the Greek mythology, it can be very confusing for any Tyros, aka new member, in the maritime community. Like the Greek history and the ancestry of the gods, Situational Awareness is complex and sometimes confusing, to a great extent it is psychology, supported and based on smart technology.

Awareness is the state or ability to perceive, to feel, or to be conscious of events, objects or sensory patterns. On this level of consciousness, sense data can be confirmed by an observer without necessarily implying an understanding of its meaning and importance in a context. More broadly, it is the state or quality of being aware of something.

Awareness is a relative concept, sometimes defined as a "*unit of knowledge*". Only when people are aware of threats, dangers, and circumstance, they will act naturally on them with their profession. If they are not aware, panic might rule and the advantage of possible action through utilization of professional power will be lost. Therefore we have to make sure others are able to recognize there is a problem and possible ways to a solution.

Situational Awareness is in general non-dividable as each area possibly interacts at any given time with any or all other divided types of awareness, which involves the planning, conduct, and direct interagency information exchange in detection and monitoring of activities. It requires flexible response operations in order to detect, monitor, disrupt and deter the cultivation, production and transportation of targeted products and/or outcomes. Situational Awareness is often divided into different, but interacting and linked, categories.

<u>Selection of Situational Awareness categories used</u>:

- Individual Situational Awareness (ISA)
- Land Situational Awareness (LSA)
- Air Situational Awareness (ASA)
- Maritime Situational Awareness (MSA) / Maritime Domain Awareness (MDA)
- Space Situational Awareness (SSA)
- Joint and/or Combined Situational Awareness (J/CSA)
- Enhanced Situation Awareness in Sea, Air and Land (ESASAL)
- Global Maritime Situational Awareness (GMSA)
- Foreign Situational Awareness (FSA, Friend/Foe)
- … and more specialized or sub-categories of SA

Land, Air, Maritime, Space and Cyber can be seen as the core situational awareness categories. The Individual or Own Situational Awareness (ISA) is a more pin-pointed, individual or personal, situational view of i.e. a single ship's captain, tank commander or airplane pilot. Maritime Situational Awareness (MSA) and Maritime Domain Awareness (MDA) are quite similar in their meaning and have mainly a different origin.

While Space Situational Awareness (SSA) is obvious, the Joint and/or Combined Situational Awareness (J/CSA, „*Jigsaw*") is more like a puzzle of pieces from a jigsaw to be set together, while the Enhanced Situation Awareness in Sea, Air and Land (ESASAL) is seen as a combination of already established situational awareness levels combined. The Global Maritime Situational Awareness (GMSA) has an even wider and comprehensve approach, while Foreign Situational Awareness (FSA, Friend/Foe) sets the focus further into Blue and Red Force tracking and less into a division between Land, Air, and/or Maritime, which makes it another possible combination or category of the "*situational awareness prime numbers*". Eventually we need a Global Joint and/or Combined Situational Awareness (GJ/CSA, spoken „*Gee Jigsaw*"[4]). Basically Situational Awareness is an omnivore where every spectator has his own reality and his own point-of-view.

Situational Awareness (SA) & Point-Of-View (POV)

Individual stakeholder's reality results in different POV

Each situational awareness category can be argued depending on the domain reality of the stakeholder and the individual point-of-view, it may exist in various security levels/classifications, due to the level of sensitivity of the information.

While Situational Awareness definitions may be required in concept, development, law and politics, they are less beneficial or supportive for the work of the end-users, operators

---

[4] … God Bless you!

and technicians, due to the interconnection between the categories in the daily operational workflow and when establishing knowledge for Situational Awareness in general.

An analysis of an operational environment must consider the four traditional pillars of air, land, maritime and space, and the fairly new cyber[5] domain. Exploiting the information environment underpins the achievement of influence in and across these domains without possibly defined clear borderlines of categorization at the beginning. Prior to collecting information on an objective or event, the evaluator cannot know, in which category and responsible domain the final analyzation will eventually set the event case. In the beginning of an investigation case the event shall not be ditched in any domain or category bucket; the approach must be always universal, over traditional domains of air, land, maritime, space, and cyber Situational Awareness (SA).

Certainly, offensive and defensive cyber capability will offer specific advantages to competitors, disrupting targeted networks and systems or countering any offensive cyber operation. Information and infrastructure security challenges are significant, with cyber-attacks anticipated to grow in scope, frequency and impact, they also play a role in Situational Awareness. Innovation needs to question all operational routines, any systems or organization/cooperation that underpin core competencies and can involve large-scale changes, requiring a shift in doctrine, structure and technology in order to achieve better knowledge for an improved awareness.

The content of this book may concentrate out of personal expertise on maritime affairs; however, the generic understanding of its value and meaning in all fields has to be stressed and is most important. Terms and definitions also exist in a galore array, the content of *"Maritime Situational Awareness for Tyrians"* is basically divided and categorize into:

- Preface
- Maritime Entities
- Concepts, Demonstrations & Experimentations and Systematogenesis
- Terms, Abbreviations and Definitions – Basic Knowledge

Maritime Situational Awareness (MSA) started mainly as a US-terminology and in the International Maritime Organization (IMO), was later introduced in NATO, EU, and became a success story all over the world; ... unfortunately with altered definitions. MSA is today sometimes hard to distinguish from the later introduced US-definition of Maritime Domain Awareness (MDA). Later nations and stakeholders started to use the abbreviations MSA and MDA interchangeably, i.e. in the EU the generic term Maritime Surveillance (MARSUR) fits roughly equivalent the definition of MSA.

No matter which term we chose to define Maritime Situational Awareness, the strategy,

---

[5] Cyberspace may be defined as an operating environment consisting of the interdependent network of digital technology infrastructures (including platforms, the Internet, telecommunications networks, computer systems, their processors and controllers.

concept, down to the daily operational work deals with very similar content, multiple cases in a wide array of civilian, military, non-military, governmental and non-governmental, activities associated with the maritime environment.

Maritime Situational Awareness (MSA) is a similar operational process of evaluation information as centuries ago, simply the individual point-of-view is today much broader, there are more entities and stakeholders involved, and the supporting tools got complex and quite sophisticated.

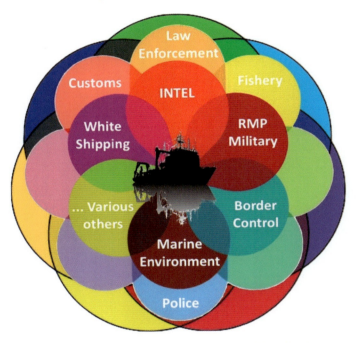

Individual Stakeholder Point-Of-Views on identical event case

The National Strategy for Maritime Security (NSMS) of the United States of America dates back to an U.S. Government interagency and international maritime security initiative from 2004 and defines the objectives and elements in the Maritime Domain Awareness (MDA) on … "*the ability to know, so that pre-emptive or interdiction actions may be taken as early as possible*". The US Maritime Domain is defined as all areas and things of, on, under, relating to, adjacent to, or bordering on a sea, ocean, or other navigable waterway, including all maritime-related activities, infrastructure, people, cargo, and vessels and other conveyances. MDA is defined as the effective understanding of anything associated with the maritime domain that could impact the security, safety, economy, or environment.

In NATO the concept of Maritime Situation Awareness (MSA) is close to the US MDA aims:

"The purpose of Maritime Domain Awareness (MDA) is to facilitate timely, accurate decision-making. MDA does not direct actions, but enables them to be done more quickly and with precision. MDA is achieved by collecting, analyzing and disseminating data, information and intelligence to decision makers, and applying functional and operational knowledge in the context of known and potential threats. A United States Government MDA capability that is integrated, interoperable, and efficient, coupled with continually improving knowledge is required to meet today's mission requirements.

The aim is to persistently monitor Vessels and craft, Cargo, Vessel crews and passengers, all identified areas of interest, access and maintain data on vessels, facilities, and infrastructure in the global maritime domain, to collect, fuse, analyze, and disseminate information to decision makers to facilitate effective understanding. Access, develop and maintain data on MDA-related mission performance." It is *"the effective understanding of anything associated with the global maritime domain that could impact the security, safety, economy, or environment of the United States."*.

AWARENESS contributes significantly to any task of early warning, monitoring and protection and requires normally a longer period of surveillance to reveal changes in a given situation independently from any specific area.

Maritime Situational Awareness (MSA) can be defined as *"The understanding of military and non-military events, activities and circumstances within and associated with the maritime environment that are relevant for current and future operations and exercises where the Maritime Environment (ME) is the oceans, seas, bays, estuaries, waterways, coastal regions and ports"* [6] and therefore an important concept for many nations and organizations.

The maritime domain is defined by the U.S. Maritime Security Policy as *"All areas and things of, on, under, relating to, adjacent to, or bordering on a sea, ocean, or other navigable waterway, including all maritime-related activities, infrastructure, people, cargo, and vessels and other conveyances"*[7].

Global Maritime Situational Awareness (GMSA) is defined in the U.S. National Concept of Operations for Maritime Domain Awareness, December 2007, as *"the comprehensive fusion of data from every agency and by every nation to improve knowledge of the maritime domain"* [8].

It might be undisputed that not one country, department, or agency holds all of the authorities and capabilities to establish an effective Maritime Domain Awareness by itself. By combining separate parts of information from agencies at the federal, state, local, and

---

[6] http://yadda.icm.edu.pl/baztech/element/bwmeta1.element.baztech-article-BWM4-0021-0048/c/
[7] The US National Security Presidential The maritime domain is defined by the U.S. Maritime Security Policy1 Directive-41/Homeland Security Presidential Directive-13.
[8] https://en.wikipedia.org/wiki/Global_Maritime_Situational_Awareness

social/tribal level around the world with information from the maritime industry and other non-governmental organizations, it is possible to keep track of the status of every cooperative ocean-bound and sea-bound vessel.

GMSA results from combining intelligence given by other regions of the world into a complete picture for identifying trends and detecting anomalies for the Global Commons[9]. However, by their purpose and nature, non-cooperative objects will always create a grey zone in any situational pictures.

Maritime Domain Awareness (MDA)[10] is defined by the International Maritime Organization as the effective understanding of anything associated with the maritime domain that could impact the security, safety, economy, or environment. The maritime domain is defined as all areas and things of, on, under, relating to, adjacent to, or bordering on a sea, ocean, or other navigable waterway, including all maritime-related activities, infrastructure, people, cargo, and vessels and other conveyances. The mentioning of the environment sets a broad approach.

The extent of future sea-level rise remains uncertain, sea-level rise generally is anticipated to have a range of economic, social, and environmental effects. Higher sea levels increase permanent or temporary coastal land inundation, change shoreline dynamics and coastal erosion, and increase saltwater intrusion and hydrodynamic changes to coastal freshwater aquifers. These impacts are anticipated to result in more nuisance flooding and impeded drainage, more loss of lands to inundation, more shifts in habitat types (e.g., freshwater wetlands converting to brackish wetlands), less freshwater supply, and potential water-quality impairment. These changes may affect coastal species as well as coastal developments and their amenities.

Global Sea Level (GSL) is the average height of the Earth's oceans, as measured by satellite altimetry relative to a calculated reference ellipsoid. These global measurements, available since the first satellite ocean altimeters were placed in orbit in 1992, are combined so that height of the world's oceans can be averaged into one number. The GSL value is significant because it allows scientists to measure trends, namely GSL rise, without having to consider whether the land surface is moving up or down along the coastline.

Policymakers are interested in sea-level rise because of the risk to coastal populations and infrastructure and the consequences for coastal species and ecosystems. From 1901 to 2010, global sea levels rose an estimated 187 millimeters, averaging a 1.7 mm rise annually. Estimates are that the annual rate rose to 3.2 mm from 1992 to 2010. Since 1900, expanding oceans due to warming ocean water and melting glaciers and ice sheets have been the

---

[9] Global Commons is a term typically used to describe international, supranational, and global resource domains in which common-pool resources are found. Global commons include the earth's shared natural resources, such as the high oceans, the atmosphere, outer space and the Antarctic in particular. Cyberspace may also meet the definition of a global commons.
[10] https://en.wikipedia.org/wiki/Maritime_domain_awareness

main drivers of global sea-level rise. Oceans have warmed due to a combination of natural variability and the influence of greenhouse gas emissions on atmospheric temperatures. Similarly, glaciers and ice sheets since 1900 have been melting due to both natural variability and greenhouse gas emissions.

Sea-level rise is directly connected to the climate change, a change in the statistical distribution of weather patterns when that change lasts for an extended period of time (i.e., decades to millions of years). Climate change may refer to a change in average weather conditions, or in the time variation of weather around longer-term average conditions (i.e., more or fewer extreme weather events). Climate change is caused by factors such as biotic processes, variations in solar radiation received by Earth, plate tectonics, and volcanic eruptions. Certain human activities have also been identified as significant causes of recent climate change, often referred to as global warming.

"… more than at any time in our history, our species needs to work together. We face awesome environmental challenges:

climate change, food production, overpopulation, the decimation of other species, epidemic disease, acidification of the oceans,"[11]

Environmental issues and Maritime Situational Awareness (MSA) have reached the agenda of G7 and many other high level political forums and conferences, and just like in Cyber Defense, suddenly experts are found in world-wide abundance. "Maritime Situational Awareness (MSA) for Tyrians" is therefore aiming at the new navigators, who follow a pursuit attaining proficiency, the beginner or novice, who is trying to research on Maritime subjects, to get an overview and links for further information - truly a Sisyphus's work - but also for the professional Subject Matter Expert (SME) in Maritime Safety & Security as food-for-thought and discussion background.

"When facts are few, experts are many."[12]

Maritime Safety (MS), being part of Maritime Situational Awareness, can be described as the combination of preventive and responsive measures intended to protect the maritime domain against threats and intentional unlawful acts, and limit the effect of, accidental or natural danger, harm, and damage to environment, risk or loss.

In some languages the word for Safety & Security is identical and similar to Situational Awareness the different definitions for Safety & Security, required i.e. for concepts and developments, laws and politics, are seen less beneficial or supportive on for the actual operational work of the operators and technicians. When it comes to establishing knowledge in the daily operational workflow each MSA category is in direct relation to others.

---

[11] Prof. Stephen Hawking on Donald Trump and Brexit: *"We are at most dangerous moment in humanity"*, http://www.independent.co.uk/news/science/stephen-hawking-donald-trump-brexit-dangerous-moment-humanity-a7452406.html
[12] Donald R. Gannon

The ability to utilize the Maritime Environment is vital due to the dependencies of the global economy and the strategic security interests of all nations. Loss of access to significant global trade routes, connecting nations, people, markets and manufacturers around the world, rapidly impacts all nations; and economy development has impacts on the natural environment.

National doctrines are normally well documented and represented within the international context and not subject to this essay. Nations as single entity all have their own national strategy, their national organizations, agencies, networks and systems and therefore also have a different need to adjust themselves independently in the overarching concepts driving in example in UN, NATO, EU.

Situational Awareness on the other hand requires a holistic approach, a common concept for Command, Control, and Information (C2I) up to Command, Control, Communications, Computers, Combat Systems, Intelligence, Surveillance and Reconnaissance (C5ISR) systems, in order to efficiently coordinate operations. Information gathered by domestic, national and/or international stakeholders/partners from all participating, or collected from non-cooperative entities, need to be fused and disseminated to achieve more and better knowledge.

Any information, in any given situation, will need to be combined and compiled with technology in systems and networks used by the Maritime Entities in order to produce a COMMON Maritime Picture (CMP). Common is here in the understanding of the relevant information being shared, not all information collected in one system or by one entity. Maritime Entities can enhance the individual national awareness by sharing services depending on national and international agreements, laws and regulations with priority given to locally and nationally available track history, notifications and alerts.

The concept of a COMMON Picture does not imply that all information needs to be displayed on a single screen or information layer, especially when a national or regional picture information is concerned. It is by far easier to enable the exchange of information from different domains, sources, and stakeholders, which leads to an overall benefit of quantity as well as quality of available information for each participant. Each stakeholder will always have sensible data which might never be exchanged for national security reasons. Therefore, the enabling of the most relevant data for exchange is the prime key for Maritime Situational Awareness.

"The COMMON Maritime Picture (CMP) is seen as the sum of maritime surveillance systems enabled to share relevant information electronically."[13]

Due to asymmetric threats in the 21st Century, the realm of Maritime Situational Awareness (MSA) transcends both the military and non-military realms and involves interoperability and/or collaboration with a wide variety of non-traditional data sources, organizations

---

[13] Joachim Beckh

and actors. MSA includes all maritime information collected in an overall Maritime Common Picture (MCP). The picture is described as common, because it contains basically the same or similar track data, looks alike in the military and the civilian environment, since no evaluation of the contained information has been conducted at the lowest collection instance. MSA involves interoperability and/or collaboration with a wide variety of non-traditional data sources, organizations and actors in the military and civilian and non-governmental realms.

Because countries subsist in an environment where internal and external threats to security are both common and ever present, the effectiveness of their coercive arms becomes the ultimate measure of power. Military capabilities allow countries to defend themselves against all adversaries, foreign and domestic, while simultaneously enabling their state managers to pursue whatever interests they wish, if necessary over and against the preferences of other competing entities. Therefore, the military capability for information exchange is the ultimate yardstick of any national, or any organization in awareness.

NATO and its members have been the by far biggest collective Maritime Picture entity in the last decades, while collectively, with Europe being the world's second largest military power. However, it the European Union i.e. still ranges far behind the United States and suffers from inefficiency in spending due to national sovereignty causing duplications, lack of interoperability and technological gaps. Defense budgets in Europe have been shrinking in recent years, while other global actors (i.e. the United States, China, Russia and Saudi Arabia) have been upgrading their defense sectors on an unprecedented scale. Without a sustained investment in defense, any industry is on risks losing the technological ability to build the next generation of critical defense capabilities. Ultimately, this will affect the strategic autonomy and the ability for power balance and as security provider.

The proliferation of anti-access and area denial capabilities, proven successful in historic examples like by Alexander the Great, enable today a wide range of potential adversaries to contest access to, and freedom of movement within, operational areas. Adversaries will likely deter opposing powers by raising the potential cost of action, to exclude adversary armed forces from the aimed theaters or to limit their effective employment. This can be done either directly or by exploiting vulnerabilities of wider capabilities. In the broadest sense, anti-access may involve political and economic exclusion, which could translate in modern times into refusal for basing, staging, transit, port facilities or overflight rights. Under more hostile circumstances like war, lethal anti-access systems will include more sophisticated and technological advanced long-range weapons. One needs to see it coming, to be able to prevent it.

"AWARENESS is achieved by gaining knowledge through any observation by any type of entity, in any geographical area, by any means of information, by any person, in order to detect, identify and track any activities and/or situation of interest."[14]

---

[14] Joachim Beckh

The distinction between war and peace is sometimes blurred and adversaries, both state and non-state, threaten on various fields the stability of the rules-based international order that underpins the long-term security of all nations. The range, geographic spread and capabilities of potential adversaries make a distinction between home and overseas operations obsolete and ask for new approaches. Tactical, operational and strategic success will escape a military that continues to embrace only traditional views of conflict in a persistent and multi-faceted state-on-state competition.

Competition has changed and challenged the decision-making process which uses a broad range of tools with both attributable and non-attributable methods to apply pressure below traditional military response thresholds. Recognizing and responding effectively to hybrid-warfare[15] is increasingly important.

"Military power expresses and implements the power of the state in a variety of ways within and beyond the state borders, and is also one of the instruments with which political power is originally created and made permanent."[16]

MSA Capability enables the users to consult, manage and process maritime information from both commercial and military sources. MSA is therefore a core capability which seeks to deliver the required information in the maritime environment in order to achieve a common understanding of the maritime situation and environment. The maritime capability environment comprises the oceans, seas, bays, estuaries, waterways, coastal regions and ports and considerable resources for merchant shipping data are available ranging from commercially or open accessible databases to transient records of vessels passing through the world's ports, which is posted on port websites.

The World Port Index (WPI) contains i.e. a tabular listing of ports throughout the world (approximately 64,000 entries) describing their location, characteristics, known facilities, and available services. To be used in a Maritime Picture (MP) compilation of the information from sources like WPI require an automated data extraction, sometimes referred to as *"Port Scrapping"*. Similar procedures apply to databases with hydrographic, meteorological and oceanographic information.

The maritime information contained in off-line or national/organizational protected databases is a greater challenge in MSA, since authorized users manage (modify, delete, export, backup, archive) all data in their internal storage. Furthermore laws and regulations are often preventing easy access and rapid exchange of information.

In the current global security environment, there are a bewildering variety of terms, definitions and acronyms referring to various types of picture building efforts. Compiling an, at first step local maritime awareness picture, in conjunction with other entities, is an ongoing process of evaluating maritime activities in the seven oceans, coastal waters, rivers

---

[15] Hybrid warfare is a form of warfare combining conventional and unconventional military and non- military actions to achieve a specific goal. (definition is currently awaiting NATO agreement).
[16] Peter Paret, "Military Power", The Journal of Military History, Vol. 53, No. 3 (July 1989), p. 240.

and includes subsurface, air, land and even space awareness in parts; a plot compiled to depict maritime activity (at least in the naval context) referred to as a Recognized Maritime Picture (RMP).

The term "*RECOGNIZED*" is used to indicate that the picture has been evaluated prior to its dissemination. In other words, rather than having operators and their systems simply pass data back and forth, there is a central authority to whom data is forwarded for compilation, evaluation and dissemination as a *recognized* picture – an expert's evaluation of what is happening in a given area. Furthermore a Recognized Maritime Picture can be a singularly recognized picture or the awareness of multiple systems with recognized pictures compiled.

Multiple event cases in different Recognized Maritime Picture (RMP) compilations

A maritime picture is the one produced for an ocean area that encompasses waters well past the 200 nautical mile Exclusive Economic Zone, involving information and cases with far reach into the national domain of other countries. Although the arbitrary area of responsibility is nationally assigned for management purposes and according to international rules and regulations, the various databases supporting the picture actually always contain data far beyond national responsible areas and even for the entire globe, which is required to track incoming or outgoing objects and provide event predictions for a Maritime Situational Awareness.

A maritime picture is built from all available data sources that can be accessed regarding maritime traffic in the area of concern. Normally, a data source will provide position and identity information regarding a given vessel and, increasingly, amplifying data such as the vessel's owner, cargo and other background information for information sharing. Assembled into the picture, all this data provides an awareness of the volume, location and nature of shipping activity and provides a background for deeper analysis of trends and vulnerabilities. Data sources are identified through actively seeking them out and then collecting and assembling the data in a format suitable for exchange between numerous partners. Resources are not available for comprehensive surveillance of any large area of responsibility if only national or organizational assets are used.

Therefore, any picture should always be compiled in cooperation as many other agencies possible, that possess data on maritime activities. With regards to these external data sources, it is clear that the picture is built on a shared effort between many partners.

"The importance of integrated, all-source analysis cannot be overstated. Without it, it is not possible to "connect the dots". No one component holds all the relevant information."[17]

Connecting all the "*dots*", connecting all information, in technical network in which devices are linked with many redundant interconnections between various network nodes is called a mesh topology or a mesh network. In a true mesh topology every node has a connection to every other node in the network.

A mesh network (or simply meshnet) is a local network topology in which the infrastructure nodes (i.e. bridges, switches and other infrastructure devices) connect directly, dynamically and non-hierarchically to as many other nodes as possible and cooperate with one another to efficiently route data from/to clients. This lack of dependency on one node allows for every node to participate in the relay of information. Mesh networks dynamically self-organize and self-configure, which can reduce installation overhead. The ability to self-configure enables dynamic distribution of workloads, particularly in the event that a few nodes should fail. This in turn contributes to fault-tolerance and reduced maintenance costs.

Mesh topology may be contrasted with conventional star/tree local network topologies in which the bridges/switches are directly linked to only a small subset of other bridges/switches, and the links between these infrastructure neighbors are hierarchical. While star-and-tree topologies are very well established, highly standardized and vendor-neutral, vendors of mesh network devices have not yet all agreed on common standards, and interoperability between devices from different vendors is not yet assured.

There are two types of mesh topologies, the full mesh and the partial mesh, each one with the respective advantages and disadvantages. Partial mesh topology is less expensive to implement and yields less redundancy than full mesh topology. With partial mesh, some

---

[17] National Commission on Terrorist Attacks upon the United States. pg. 408, Kean, T. H., & Hamilton, L. (2004). The 9/11

nodes are organized in a full mesh scheme but others are only connected to one or two in the network. Partial mesh topology is commonly found in peripheral networks connected to a full meshed backbone.

Full Mesh Network Topology Formula

Number of connections (c) linking all nodes (n) in a full mesh network c = n(n−1)/2

In a fully connected and federated network topology, meaning all nodes are interconnected in an solely decentralized architecture, the full mesh network, the number of links between the nodes increases differently with each new member. The simplest fully connected network is a two-node network. A fully connected network doesn't need to use packet switching or broadcasting, but the number of connections (c) grows quadratically with the number of nodes (n): c = n(n−1)/2. The Full Mesh Network Topology[18] does not trip and affect other nodes in the network, however, the quadratically growing number of connections makes it impractical for large networks/nodes (aka partners).

Maritime Situational Awareness therefore deals with many variables in order to develop a combined product, a Maritime Picture (MP), and at the highest level of complexity a Recognized Maritime Picture (RMP).

---

[18] https://en.wikipedia.org/wiki/Network_topology#Fully_connected_network

Maritime Picture (MP) and Recognized Maritime Picture (RMP)

Situational Awareness is an omnivore. The holistic and comprehensive MSA approach includes a wide array of architectural elements, civilian organizations with open source and low classification as well as the military, with up to highly classified intelligence information on locations of ships and events, in order to provide a Recognized Maritime Picture (RMP).

The WP cannot fulfil the classic requirements of the military RMP in its original definition, however the abbreviation is today used in military and civilian context, but with different meanings. The civilian *"White Shipping"* includes sensitive data from Maritime Safety & Security, General Law Enforcement, Maritime Environment, Fishery and Border Control, and Customs, forming in general the White Picture (WP) as the collection of civilian maritime tracks and vessels and other information available. This WP is sometimes incorrectly equalized to an RMP, which is the reason to distinguished it in the book as Civilian RMP (CRMP).

As synonym for the word picture is the image as a Case File (CF), an electronic envelop that contains all information needed to reproduce a live working copy of an event and situation. Recognizing is to acknowledge the existence of something, realize or discover the nature of this event, and treat it as worthy of consideration or validation. Recognition therefore involves a human action at some point in the process. In Situational Awareness are various Recognized Pictures related:

- Recognized Air Picture (RAP)
- Recognized Ground Picture (RGP)
- Recognized Maritime Picture (RMP)
- Recognized Logistics Picture (RLP)
- Recognized Civil Picture (RCP)
- Recognized Civil-Military Cooperation Picture (RCIMICP)
- Recognized CIS Picture (RCISP)
- Recognized Electromagnetic Picture (REMP)
- Recognized Engineer Picture (RENGP)
- Recognized Environmental Picture (REP)
- Recognized Intelligence Picture (RIP) …

The classical Recognized Maritime Picture (RMP) is a managed geographic presentation of, processed all source contact and information data known at a given time, of surface, sub-surface, amphibious and maritime air activities in the maritime operating environment. The comprehensive MSA profile including architectural elements such as classified intelligence and locations of warships produced in the Maritime Military Environment is here referred to as the Military Recognized Maritime Picture (MRMP).

The NATO Common Operational Picture (NCOP or NATO COP) derived out of the importance of clearly defining geographic areas of responsibility (e.g. Co-ordination Areas

(CA), Areas of Responsibility (AOR) and Joint Operating Areas (JOA)) and component (e.g. maritime, air, special forces and ground) AORs or areas of exclusive use for one of the above components or subcomponents thereof down to unit size is an essential condition for the proper management of the various databases comprising the NATO COP.

Ensuring that a single Manager can handle the responsibility for a given geographic area and component establishes clear responsibility for management and reporting, and significantly reduces the potential for ambiguous, circular or dual-reporting. Two or more separate reports on the same track/force element is, besides multiple feeds of overlapping surveillance areas, one of the most common causes for disturbance in the work of the operators.

The exception of single responsibility may be in the case of a CJTF, which may have reporting responsibilities for all track data (maritime, air and ground) within a designated Joint Operational Area (JOA). Reporting responsibilities within CAs/AORs are established by the NATO ACO. Generally, the various component pictures are managed as Recognized Pictures (RP) as follows:

- The Recognized Air Picture (RAP) is provided by the regional CAOCs, or a CJTF in a designated JOA and will include data from units as well as plans and orders, i.e. Air Control Orders and Air Traffic Orders;

- Recognized Ground Picture (RGP) which would likely be provided by a Land Component Commander (LCC) or CJTF in a designated JOA;

- The Recognized Maritime Picture (RMP) is provided by the Maritime Component Commander (MCC) or a CJTF in a designated JOA.

The JOA may rely on one of the two land based RMP Managers. Additionally, a sea-based CJTF may be assigned RMP management responsibilities within a designated JOA. Additional command information and functional enhancements (e.g. WSM, environmental, plans) provided by various commanders contributing to the NATO COP (NCOP).

The NCOP tool was in the past tested in the NATO Coalition Warrior Interoperability eXploration, eXperimentation, eXamination, eXercise (CWIX) with the Interim Geo-Spatial Intelligence Tool (iGeoSIT) and COP IM/LM to trial partners requiring these capabilities to provide situational Awareness for network centric sharing of geospatial data and operational overlays by the combination of geo maps, satellite imagery such as town plans and aerial photography along with daily incidents, minefields, bridges, tunnels, medical facilities, airports, real-time vehicle tracking and any other overlays, contributing to the Situation Awareness.

The Joint Common Operational Picture (JCOP) tool was also an interim operational capability in the a version of JCOP compliant with common NATO interfaces (MIP BL2 and NATO Vector Graphics - NVG) as information management tool to define the COP, consuming information and producing the COP to be distributed to various viewers and C2 System.

The interim NATO capability for Joint Situational Awareness is the Joint Common Operational Picture (JCOP) capability filling the short and medium term (was intended as 2007-2010) time frame in the existing gap of joint situational awareness at the NATO levels of command.

NATO therefore stablished the ACO C2 IPT (Allied Command Operations - Command and Control - Integrated Project Team) with the propose to present solutions to bridge this deficiency in short and medium term while staying in-line with longer term capability acquisition projects (like NSIP NCOP). In the summer of 2006 the NATO C3 Agency (NC3A) was tasked by the ACO C2 IPT to propose c technical solution for an interim JCOP capability.

In MSA the NATO Recognized Maritime Picture (RMP) is a managed geographic presentation of processed all source and information data known at a given time, of surface, sub-surface, amphibious and maritime air activities in the maritime operating environment. This includes contact, geographic, environmental and other operational information. The NATO RMP is compiled in accordance with operational directives and tasking to support decision makers in the conduct of Command and Control of maritime forces and operations. The NATO RMP is an operational picture, not a tactical plot. It is not intended to be used to directly support weapon systems, target acquisition or engagements.

The RMP is the Recognized Maritime Picture that directly supports the command and control of maritime forces by providing commanders at each level of command with information relevant, but not limited to:

- Force Employment
- Operational Success or Failure
- Cueing Surveillance Assets
- Rules of Engagement
- Future logistics and re-supply planning
- Selection of submarine, surface, and maritime patrol areas, and
- Planning for tactical air support of maritime operations missions.

In comparison the Recognized Ground Picture (RGP) is described as an electronically-produced, integrated geographic presentation of the ground picture for a defined area. The RGP is compiled from various sources providing force element information, and also includes functional enhancements (e.g. environmental, intelligence) provided by/to the RGP management site. The level of granularity for force element data is determined by the Commander responsible for RGP management within a given AOR. RGP are likely to be provided by a Land Component Commander (LCC) or CJTF in a designated JOA.

In comparison the NATO Recognized Air Picture (RAP) as an electronically produced display compiled from radar and other sources covering a three-dimensional volume of airspace in which all detected air contacts have been evaluated and assigned an identification designation and track number. The NATO RAP is provided by the regional CAOCs, or a CJTF in a designated JOA and will include data from units as well as plans and orders, i.e. Air Control Orders and Air Traffic Orders.

Due to the requirements of a RMP any contact information entered into the established picture follows certain rules:

- AIS contact information is permitted only if validated by other means than AIS.
- All contacts within the NATO RMP must be identified and verified (e.g. unknown radar contacts or raw AIS are prohibited).

The actual terms in some Handbooks of RMP Managers are:

- *"Recognized"* - indicates that the plot is an evaluated and validated interpretation of available information by an assigned Manager for a given ocean area
- *"Maritime"* - indicates a focus on ocean areas for the benefit of maritime commanders and
- *"Picture"* - the presentation of the evaluated and validated data to the decision makers.

The Military Maritime Picture (MMP) is part of naval history and core decision element in all Navies. The MMP is the overall collection of all available information from the generic national Military Picture (MP), possibly containing information from joint systems (Air Force, Army, Ministry, a. o.), including the own Military Recognized Maritime Picture (MRMP), and the non-combatant elements from the White Picture (WP). The Military Maritime Picture (MMP) can therefore be seen as the broadest view and consists of all kinds of maritime data available and is provided by compiling and presenting the information related to both combatant and non-combatant vessels for Maritime Military Forces. The term MSA Picture (MSAP) is mostly avoided in this context as it could easily be confused with the Military Maritime Picture (MMP), the White Picture (WP or Civilian RMP (CRMP)) or any other described category.

Similar to all the above are all other national, international, coalition, organizational military and civilian definitions of creating an image of the current reality from the individual point-of-view; a picture, an instant capture of a situation for possible analyzation of the past events or future predictions.

Due to the security aspect any Military RMP (MRMP) and the Military Maritime Picture (MMP) the definition is always more detailed and in nature of a higher quality and therefore classification than any White Picture (WP) in the military or civilian environment. The MRMP includes tracking in the sub-surface environment as well as a local Recognized Air Picture (RAP) while a WP does not; while a MSA WP might. The information contained in a high classification MMP and MRMP is exchanged within military encrypted systems and networks (CLASSIFIED DOMAIN). The White Picture (WP) is in general unclassified or on the lowest classification level and can be exchanged in unclassified/low classified systems and networks (UNCLASS DOMAIN) and to the MMP and MRMP.

While the Recognized Maritime Picture (RMP), the Military Maritime Picture (MMP), and the Military Picture (MP) have well defined definitions within the respective military communities the term *"White Shipping"* and the White Picture (WP) refer in general to commercial shipping including passenger, cargo, tanker, ferries, container vessels, fishing vessels, leisure crafts and other non-military vessels. White Shipping is the basis of the sharing information on all manner of commercial vessels regardless of their displacement/tonnage, length or other specifications in the establishing of a White Picture (WP) within the military communities.

The current notion of White Shipping - and therefore a WP - tends to be limited to such vessels that exceed 300 tons displacement per the IMO AIS requirement. However, the current IMO requirement (effective 31 Dec 2004) for AIS which requires AIS to be fitted aboard all ships of 300 gross tonnage and upwards engaged on international voyages, cargo ships of 500 gross tonnage and upwards not engaged on international voyages and all passenger ships irrespective of size is expected to be changed.

Currently, among the littoral states, only Singapore has mandated that ships below 300 gross tonnages be fitted with a transponder system to monitor their positions within Singapore's waters. The United States of America and other littoral states tend to impose the same mandatory fittings within their territorial waters which is a logical development since i.e. recreational vessels, platforms for piracy and illegal migration are ranging normally far below the 300 gross tonnage.

In most cases when the term RMP is used by civilian authorities, it is derived from the military terminology and imposed on the White Picture (WP) collected from civilian systems and/or sensors. However without any or only few validated tracks this *"Civilian RMP"* (CRMP) constitutes a simplistic Maritime Picture (MP) and should be clearly differentiated from the military and recognized information collection for a naval or Military RMP (MRMP).

The classic RMP is generally an electronically produced display compiled from active and passive – mostly military - sensors covering a three-dimensional cube in which all detected maritime contacts have been evaluated against threat parameters and assigned a recognition category and track number. The RMP consists of all contacts in the maritime environment, both surface and subsurface, commercial, military and government platforms and vessels. The evaluation and validation of inputs and the dissemination of updates in the RMP needs to be highly automated and conducted continuously using automated rules. The ambiguity resolution of data using fusion techniques are also an ongoing activity, but may require operator assessments to deal with complex cases. The RMP compilation task also includes a need to conduct detailed analysis, such as route assessments, that involve significant amount of time to complete.

In the current global security environment, there are unfortunately a bewildering variety of terms, definitions and acronyms referring to various types of picture building efforts. However, the fundamental goal does not change - to support awareness through building a picture.

Compiling a - at first step local - maritime awareness picture by navies, in conjunction with many other agencies, is an ongoing process of evaluating maritime activities in the seven oceans, coastal waters, rivers and includes the sub-surface, air, land and even space awareness in parts; a plot compiled to depict maritime activity (at least in the naval context) referred to as a Recognized Maritime Picture (RMP). The term *"recognized"* is used to indicate that the picture has been evaluated prior to its dissemination. In other words, rather than having stations simply pass data between themselves, there is a central authority to whom data is forwarded for compilation, evaluation and dissemination as a *recognized* picture – an expert's operational evaluation of what is happening in a given area.

The maritime picture is the one produced for an ocean area that encompasses Atlantic waters well past the 200 nautical mile Exclusive Economic Zone, involving information and case with far reach into the national domain of other countries. Although the arbitrary area of responsibility is nationally assigned for management purposes and according to international rules and regulations, the various databases supporting the picture actually contain data for the entire globe.

A maritime picture is further built from all data sources that can be accessed regarding maritime traffic in the area of concern. Normally, a data source will provide position and identity information regarding a given vessel and, increasingly, amplifying data such as the vessel's owner, cargo and other background information is included. Assembled into the picture, all this data provides an awareness of the volume, location and nature of shipping activity and provides a background for deeper analysis of trends and vulnerabilities. Data sources are identified through actively seeking them out and then collecting and assembling the data in a format suitable for exchange between numerous partners. Resources are not available for comprehensive surveillance of any large area of responsibility if only national or organizational assets are used.

All the above requires a maritime picture to be compiled in cooperation with various stakeholders that possess own data on maritime activity. In regards to these external data sources a maritime picture is a shared effort between many partners.

"The importance of integrated, all-source analysis cannot be overstated. Without it, it is not possible to "connect the dots". No one component holds all the relevant information."[19]

The ability to adapt in effective assessments processes at the strategic, operational, technical and tactical levels is vital in Maritime Operations. Adaptability is here not solely adjusting to new external conditions or responding to adversaries. Adapting to changing complex environments, rather than seeking to control them, is today fundamental.

Military Maritime Operations are a set of military activities that are conducted by maritime air, surface, sub-surface and amphibious forces to attain and maintain a desired degree of control of the surface, sub-surface, and air above the sea, influence events ashore,

---

[19] National Commission on Terrorist Attacks upon the United States, Kean, T. H., & Hamilton, L. (2004), pg. 408.

and, as required, support land, air and space operations.

Non-combatant Maritime Operations are a set of civilian activities of the stakeholders in Maritime Safety and Security, General Law Enforcement, Maritime Environment, Fishery and Border Control, and Customs or any non-military actor.

A Military Maritime Operation is therefore a military action or the carrying out of a strategic, tactical, service, training, or administrative military mission; the process of carrying on combat, including movement, supply, attack, defense and maneuvers needed to gain the objectives of any battle or campaign. Military Maritime Operations include any actions performed by forces on, under, or over the sea to gain or exploit command of the sea, sea control or sea denial, and/or to project power from the sea. Any Maritime Operation requires visual images, compiled in a Maritime Picture, build on track information deriving from multiple sources as the base foundation decision making.

MSA Services enable users to collect, process, present and distribute information that support Maritime Operations being a set of military activities conducted by maritime air, surface, sub-surface and amphibious forces to attain and maintain a desired degree of control of the surface, sub-surface, and air above the sea, influence events ashore, and, as required, support land, air and space operations. The future of forces will increasingly need to integrate information and physical activity across multiple domains as part of a full spectrum approach alongside allies and partners.

Track sources can be manifold and include:

- Active and Passive Sensors
- Stationary and Mobile Terrestrial Sensors
- Stationary and Guided Satellite Sensors

A TRACK in any RMP is basically defined as the representation of a moving object in terms of its position, course, speed and general characteristics. A RECOGNIZED TRACK requires its verified identification and until today this process ends always with the manual validation by a human operator when producing the Military RMP. The reported tracks in a specific area are evaluated by an assigned RMP Manager. Recognized tracks can then be associated to Vessels in a database. Due to the manifold information collected and the verification process the database and system access needs to be protected against adversaries in a classified environment.

The number of commercial vessels in the world exceeded 50,000 in 2017, just in the USA the total of recreational boating vessels registered in 2015 reached 11.87 million. The number of maritime objects, each with an individual track history, is therefore in a three digit million digit range. The establishing, maintaining and exchange of Maritime Information is a quality and quantity technology challenge.

Operational users access, process and disseminate information manually from external sources (e.g. commercially available databases, Internet sites) and authorized users manually initiate new tracks, update or delete them. All manual, simulated or exercise tracks require also a defined manual update rate and shall not be correlated with Live Tracks and not be associated with vessels as this would cause conflicts in any Maritime Picture and even mutually interfere in a RMP track representation.

Initiation of automatic and manual tracks and managing their life cycle may be different due to the nature of the different track sources, update rate and time late. Automatic tracks need to be updated by their source at certain intervals and retransmitted to the RMP. If they are not updated for a given time periods then they are declared as "Lost".

AIS tracks i.e. are reported at intervals between two seconds and three minutes, the default Long Range Identification and Tracking (LRIT) updates rate is six hours, which can be increased to 15 minutes for a specified period of time. LRIT reports have by nature a positive identity/validation, however, when not verified by a human operator, they are often considered as manual or at least non-validated tracks in respect to a RMP, as they are never declared as Lost and the LRIT-Vessel Database will be updated automatically, while AIS tracks tend to get *"Lost"*. Due to the number of tracks and different sources any Maritime Picture requires electronically filtering, validation, evaluation and correlation processes.

When an object is reported or vessel data is received from a source, the identification of the track is normally used as the unique key to process the track and locate it in a track database. The attributes of the identification are normally automatically derived from the track attributes such as country, MMSI number or other values with a high valued quality based on the confidence level of the source. If the identification of the track is not found in the track database, then a new track is initiated.

If a track report or similar track data is found already existing in the database a correlation process needs to be initiated and defined criteria for a correlation satisfied. The same maritime vessel may be reported by more than one source, multiple tracks and track data is the usual case. Track information received from multiple sources but referring to the same track need at first a correlation process preserving all existing data. If the attributes of an automatic track are modified by the authorized user, then subsequent updates to the same track shall not change its attributes, because the track identification (track number) plays a vital role when forwarding and exchanging tracks system externally as well as in correlation of the tracks in the Maritime Picture.

However, updates to correlated tracks may cause a reevaluation of the correlation. The correlated tracks should then be uncorrelated if the defined condition in the correlation criteria is met. This process is i.e. hardly possible with regularly fused track information. Fusion merges the available data into on track file and the originator/source is therefore lost. MSA requires an Adaptive Track Fusion approach due to the multi-sensor and multi-source environment.

A track can be a reference object and/or be in relation to reference object. Reference objects are geographical or moving points, or lines or areas having an operational or even tactical meaning for the operational users. Points, lines and areas are defined in support of the Maritime Common Picture (MCP) to mark points, lines or areas of special interest. Some of these tracks move with an independent course and speed. Certain points and areas can be slaved to a moving track and its position needs to be updated along with the updates to the track.

MSA is far more than vessel tracking or a simple track validation. The increasing volume, variety and velocity of information transmission fuels the complexity of contemporary and future threats, crisis and conflicts. MSA requires rapid connection of new audiences and more tightly binds strategy to tactics. MSA-Strategy is therefore increasingly sensitive to tactical actions and the opinions of local, regional and global audiences.

MSA requires the exchange of maritime data, singular national/organizational maintained RMPs (regional RMP) which at the far end contribute to a virtual global RMP (VRMP). Virtual RMP in the sense as the information cannot be viewed on a single system or graphical user interface, however relevant information exchange is ensured. Any Regional RMP (RRMP) contains a part of the global collection of recognized maritime objects.

The number of global tracks, their evaluation and the amount of available information requires automated and manual anomaly detection functions in any RMP. Server farms need to ensure the data collection and storage for anomaly detection and historical data in relation with the respective alert, warning and notification.

Today's mainly hierarchical networks play an essential part of a maritime network centric information environment, but are not well suited to address the dynamic information sharing and safeguarding. The few existing federated networks are designed to combine existing networks are genuine network centric examples for sharing in a common information environment.

The security aspect requires often a *system high* approach to manage information asset protection within the network. This approach provides all information assets with only one level of protection, which is the highest level of protection afforded by the network. In other words, the network safeguards are unable to discriminate the protection requirement for each information asset within the network, thereby forcing the safeguard to assume that all information assets require the same high level of protection. The system high approach presents two challenges to the sharing and protection of information assets in a coalition environment.

The general and often inflexible protection rules of information security impede information sharing by preventing legitimate users to access and transfer information between assets in a coalition or alliance.

Over-classification of information assets leads to a reduction in sharing capacity; supporting the initial claim that the requirements for sharing and protecting information often strangle one another while network centric security favors protecting over sharing of information. This makes the information exchange difficult by design.

The sharing of information from different assets between coalition or alliances domain requires a sensibility evaluation of the asset before leaving and entering partner's network.

When exchanging information from open or unclassified sources to classified systems, the common exchange method today the "*Diode*", a limited one-way Information Exchange Gateway (IEG), which is stove-piping the information from the lower to the next higher classified domain without return option. The assessment processes, typically performed by exchange systems with a gateway including "*Gate Keeper*" (GK) functionalities, are cumbersome, and still too often require the human in the loop ("*Air-Gap*"-Media)[20].

International rules and standards for the handling of unclassified and classified information, the data exchange, a common labeling, and the required procedures, do not exist or are just evolving. Only when these prerequisites are fulfilled, the technology can evolve for international accredited security systems and exchange mechanisms, to connect different classified national and organisational environments.

In coalition environments, information sharing conditions express a desired state for a set of information sharing attributes, such as the assets sensitivity (classification), coalition partner identity and credentials, the network's security accreditation and operational conditions.

The operational conditioning distinguishes itself from other information sharing attributes. The value assigned to information asset sensitivity, a user credential and a network accreditation is pre-determined and authoritative, and changes to those attributes shall occur under a controlled process and environment.

Network safeguard and security assessment of the "*Gate Keeper*" tools lack the ability to adapt to changing operational conditions. The required security measurements weight against the possible open and flexible rules required. These challenges often lead to operators having to resort to standard commercial means of information sharing.

The next step in cross-domain, classification-independent information exchange will be reached with information labeling, information binding and the respective trust models. This dynamic coalitions problems highlight the intricacies, challenges and complexities of

---

[20] Mostly downloading the information on a USB-Stick or similar Transfer Media.

information sharing and safeguarding in a coalition environment. In short, in a coalition, the desired balance between information sharing and safeguarding is not constant but rather dynamic following the mission requirements.

The aim must be a more dynamic information protection concept referred to as a Data-Centric Security (DCS). The concept calls for the introduction of safeguards at the data access point, which protects an information asset at a level commensurate with applicable information sharing agreements. Information Exchange Gateways including Gate Keeper functionalities need to transform into a Network Information Area (NIA) in order to safeguard the security assessment from network security to coalition information sharing conditions, which need a high degree of automatization and flexibility for rules management.

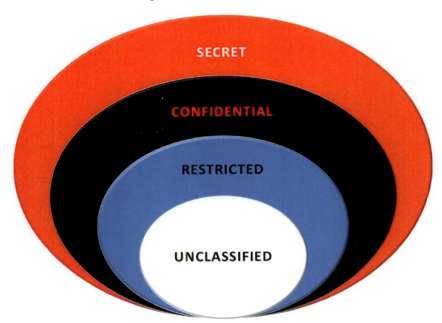

Multiple Layers of Labelled Information in System High up to SECRET

The introduction of a DCS safeguard in an information domain brings the required granularity for the safeguard to evaluate and authorize each information transaction request independently. The granularity and the resulting protection simplifies the movement of information in a coalition and alliances environment.

Each information label has needs its own layer, a unclassified system may contain only unclassified data, while a system in high mode, i.e. secret, contains information from unclassified, restricted, classified up to secret. The Labelling is the preconditioning for the information exchange with different classifications.

The use of the term "*Labelling*" is often intended to highlight the fact that the "*Label*" is a description applied from the outside, rather than something intrinsic to the labelled thing. Labelling is necessary for secure communication over different domains, environments and classifications.

Describing someone or something in a word or short phrase is "*marking*". Extensible Markup Language (XML) is a markup language that defines a set of rules for encoding documents in a format that is both human-readable and machine-readable and a key element for secure labelling and has proven its value in this field.

The Dynamic Information Protection concept rests on the integration of three capability pillars into existing and future CIS. The capability pillars are:

- Users Credential. This capability pillar delivers users (nation, organization or individual) identity and credentials (need to know) information to assist in determining whether or not information sharing conditions are satisfied.

- Data Classification. This capability pillar delivers information about information assets sensitivity which also assist in determining whether or not information sharing conditions are satisfied.

- Information sharing conditions enforcement point. This capability pillars delivers an automated and real-time mediation service for all IT transactions (move, store, process) that assesses a user need to know against the data classification and other information sharing conditions, and controls IT transaction requests. The challenge is the borderline of different information domains under different organisations and nations with different security and classification definitions and regulations, different protection and sensitivity needs.

Sensitive Information is data which is not classified but at the same time is not free for public release. Therefore the access to sensitive data must be restricted to identified users, because its misuse might adversely affect personal rights, international, national, or commercial laws, rules, values, assets or could bear potential to endanger national security. However, disclosure of sensitive information has no legal consequences in comparison to disclosure of classified information.

Information Sensitivity deals with the control of access to information or knowledge that might result in loss of an advantage or level of security if disclosed to others who might have low or unknown trust ability or undesirable intentions. Loss, misuse, modification or unauthorized access to sensitive information can adversely affect the privacy or welfare of an individual, trade secrets of a business or even the security, internal and foreign affairs of a nation depending on the level of sensitivity and nature of the information.

A classification policy is fundamental to protecting an organization's information assets, and sets up categories for governing the release of sensitive information in a Security Policy. The management must assign and/or identify the owner of any information and – if possible – assign a label or ID. Any Information Owner may also delegate the responsibility of protecting the data to his custodian or designee.

The minimum classification in any organization should distinguish between internal and public information. When in need of more detailed classification categories, they have to be clearly identified and their value, importance, the possible harm for the organization if mishandled, defined. Classified Information is sensitive information to which access is restricted by law or regulation to particular user groups.

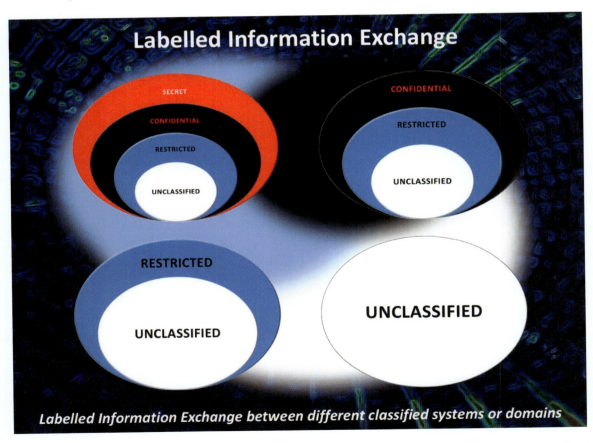

Labelled Information Exchange between different classified systems or domains

A formal security clearance is required to handle classified documents or access classified data. The clearance process requires a satisfactory background investigation. There are typically several levels of sensitivity, with differing clearance requirements. This sort of hierarchical system of secrecy is used by virtually every national government. The act of assigning the level of sensitivity to data is called data classification. Although classification systems vary from country to country, most have corresponding levels, i.e.:

Security Classifications:

1. Top Secret (TS): the highest level of classification of material on a national level. Such material would cause "*exceptionally grave damage*" to national security if made publicly available.);

2. Secret (S): material would cause "*grave damage*" to national security if it were publicly available;

3. Confidential (C): material would cause "*damage*" or be "*prejudicial*" to national security if publicly available;

4. Restricted (R): material would cause "*undesirable effects*" if publicly available. Some countries do not have such a classification.

Unclassified (N): technically not a classification level, but is used for government documents that do not have a classification listed above. Such documents can sometimes be viewed by those without security clearance.

The European Union / EU Commission uses five levels for Classified Information (SECURITY REGULATIONS OF THE COUNCIL OF THE EUROPEAN UNION) equivalent to the NATO Classifications:

EUROPEAN UNION	NATO
EU TOP SECRET	COSMIC TOP SECRET (CTS)
No equivalent	FOCAL TOP SECRET (FTS)
EU SECRET	NATO SECRET (NS)
EU CONFIDENTIAL	NATO CONFIDENTIAL (NC)
EU RESTRICTED	NATO RESTRICTED (NR)
EU CONCIL / COMMISSION	No equivalent

The Organization for Joint Armament Co-operation (OCCAR) as a European defense organization uses instead the three levels OCCAR SECRET, OCCAR CONFIDENTIAL, OCCAR RESTRICTED.

Cooperation environments are different from organizations due to their agreements and status by law and international regulations. The Sea Surveillance Cooperation Baltic Sea (SUCBAS) and Maritime Surveillance (MARSUR) have only the unclassified networks, even the member states have agreements for classified document exchange up to secret, there is no common security classification and the only protection is to use a copyright protection notice.

In a coalition environment, the information sharing conditions enforcement point would be at the digital border between a nation and the coalition information environment such as the Federated Mission Networking (FMN) in NATO. In this environment a common classification scheme, the NATO Classification, will be available.

FMN aims towards a common information environment with common standards and procedures to express coalition information sharing conditions and data classification. This provides a clear delineation between national and coalition information environments, which addresses the coalition information sharing requirement, while ensuring that each nation retains access and control over its national information assets and environment.

Classification Stovepipe

NATIONAL SECRET
NATO SECRET
EU SECRET
MISSION SECRET
NATIONAL CONFIDENTIAL
NATO CONFIDENTIAL
EU CONFIDENTIAL
MISSION CONFIDENTIAL
NATIONAL RESTRICTED
NATO RESTRICTED
EU RESTRICTED
MISSION RESTRICTED
PUBLIC

ONE WAY — ONE WAY

Information is only distributed to higher level of classifications but never back.

NATO has i.e. introduced the Standardization Agreements (STANAG) 4774 (Labeling, ratified), 4778 (Binding, draft), and has workshops for the STANAG on Trust Models. Once the information security in NATO, hopefully parallel in the EU and nationally in their Member States, has the procedures and documents in place, a Network Information Area (NIA) will be able to reach NATO, EU and national accreditation.

Network Interconnection Areas (NIA) will only allow the automated electronic exchange based on rules defined, meaning on the identified classification of the data, which requires a label bound to the information, and the indication, where the data originates from (Source), and where it is supposed to be received. For an electronic global information exchange IEG/NIA and the Labeling need to be commonly agreed and the used systems and environment accredited prior to operational use.

Labelled information is exchanged through all levels of classifications.

The benefits are no limited to the maritime world, the increasing focus on the automatic exchange of information and the blockchain technology over the recent years will continue to have a significant impact i.e. on financial institutions.

Best known as the immutable database that runs underneath cryptocurrencies like Bitcoin and Ethereum, blockchain technology is poised to play a critical role in every industry imaginable as businesses seek ways to cash in on the distributed ledger technology's promise of enabling a "*trustless*" consensus to validate transactions.

Financial transactions are typically guaranteed by a trusted third party (such as PayPal) and blockchain can be used to automate that process, reducing overall costs by cutting out the middleman with autonomous smart contracts acting as trusted intermediaries between parties on the network.

Blockchain is expected to be so influential over the coming years that some technologists foresee it ushering in a new type of Internet, one that stores and authenticates information about every asset, device and individual, opening the door to a range of new technological capabilities.

Besides simply being the backbone of cryptocurrency exchanges, the most powerful uses of blockchain technology are yet to emerge. It's envisioned by many to become a decentralized, real-time global distributed digital ledger of things for everything from tracking food supplies to managing identities.

The combination of all the above will provide the base foundation for a controlled information exchange between different national and international security domains. The first step is to label the content of all information, the second step is to bind it to the required security classification. Both steps combined will enable the NIA to filter/block information for the exchange from the highest classified systems down to the non-classified internet. So much for the technical aspect.

Audience always means interaction between humans, the art of communication, and the most important factor for best results in use of technical capabilities. The surest way to maximize operator happiness is to minimize the daily work load; decision-making is tiring. A great deal of research has found that humans have a limited amount of mental energy to devote to making choices. The second premise is that humans falsely believe they are in full control of their life and work by making own choices. As long as we can make the right choices, the thinking goes well, we'll put ourselves on a path toward life satisfaction, when conflict arrives, unhappiness and stress follows.

The truth is, decision-making is fraught with biases that cloud our judgment. People misremember bad experiences as good, and vice versa; they let their emotions turn a rational choice into an irrational one; and they use social cues, even subconsciously, to make choices they'd otherwise avoid.

That last factor can be harnessed for good. When two people interact in a company, organization or cooperation, their brain waves will begin to look nearly identical. One study of on movie audiences found the most engaging trailers all produced similar patterns in people's brains. The people we work with actually have an impact on your engagement with reality beyond what we can explain; one of the effects is we think and act alike.

Conclusion is that if people want to maximize happiness and/or success and their minimize stress, they should build a life, organization/cooperation that requires fewer decisions by surrounding themselves with people who embody the traits they prefer; have a common interest. Instead making two smaller decisions make one larger one.

Most important are the people we choose for cooperation. It is the team building factor in international cooperation and organizations. More facts and thinking you can find in the book *"Darwin's Religion - The Art of Communication Survival in the Information Jungle"*[21].

Military actions need to be effective, it must therefore be conducted within the context of an effective strategy and a supporting narrative that gives meaning to tactical actions. This is important, because the pervasive nature of information means that success is significantly influenced by the extent to which competing narratives[22] influence[23], or fail to connect with, the audiences. Influence will only be achieved with a clear focus on audiences and effects, and by integrating and synchronizing kinetic and non-kinetic activities conducted across the physical and virtual domains to try to achieve those effects.

---

[21] Darwin's Religion Paperback and Kindle Edition on Amazon: http://www.amazon.de

[22] Narratives can be spoken or written accounts of events. Narratives can dominate collective thought, and once ingrained can be very hard to shift. Moreover, narratives can be formed by imagination, myth and stories rather than fact, especially over time.

[23] Influence is defined as: the capacity to have an effect on the character, or behavior of someone or something or the effect itself.

A holistic approach to Situational Awareness therefore covers a very broad range of sources and requires complex processes of many stakeholders. Any Duty Officer in Situational Awareness requires the access to all information relevant to establishing an valid picture by evaluating all available data.

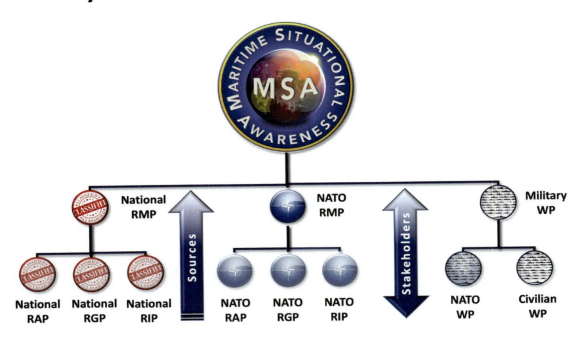

Maritime Picture compilation of a DO MSA for enhancing Situational Awareness

The Duty Officer MSA enhances awareness from the National Recognized Maritime Picture (RMP) with the information exchange from National classified and NATO classified Recognized Air (RAP), Ground (RGP) and Intelligence Picture (RIP), as well as from National Military White Picture (MWP), Civilian White Picture (CWP) and NATO White Picture (NWP), which can be extended to various other MSA sources and stakeholders.

The Duty Officer responsible for Ground/Air/Intel and all others enhance their awareness from the respective RGP/RAP/RMP/RIP and other sources available with an information flow arranged according to their mission priority and tasking.

Maritime Operations, Maritime Terrorism, Piracy

The classical military warfare area of maritime operations in conflicts like war or military/political tensions with the *"good, old"* RMP was always overlapping with Maritime Terrorism and Piracy. In today's complex heterogeneous world, military operations cover a wide variety of missions making it difficult to pre-design and pre-integrate Information Technology (IT) solutions.

Flexible and innovative IT capabilities are required to counter any potential threat in ever-changing global environments. Beside military conflicts and operations, Maritime Terrorism and Piracy have always been around in human history in some extend, however the global threats have taken a different quality in the latest years again.

The Council for Security Cooperation in the Asia Pacific (CSCAP) Working Group has offered an extensive definition for maritime terrorism: *"…the undertaking of terrorist acts and activities within the maritime environment, using or against vessels or fixed platforms at sea or in port, or against any one of their passengers or personnel, against coastal facilities or settlements, including tourist resorts, port areas and port towns or cities."*

This definition, however, does not define what terrorism is and whether it would only include maritime attacks against civilian (merchant) vessels or also attacks against military crafts. Maritime terrorism is motivated by political goals beyond the immediate act of attacking a maritime target. One definition of maritime terrorism could therefore be the use or threat of violence against a ship, civilian as well as military, its passengers or sailors, cargo, a port facility, or if the purpose is solely a platform for political ends. The definition can be expanded to include the use of the maritime transportation systems to smuggle terrorists or terrorist materials into the targeted area.

Piracy, in contradistinction, according to article 101 of the 1982 United Nations Convention on the Law of the Sea (UNCLOS) is defined as:

1. *"… any illegal acts of violence or detention, or any act of depredation, committed for private ends by the crew or the passengers of a private ship or a private aircraft, and directed: (i) on the high seas, against another ship or aircraft, or against persons or property on board such ship or aircraft; (ii) against a ship, aircraft, persons or property in a place outside the jurisdiction of any State;*

2. *any act of voluntary participation in the operation of a ship or of an aircraft with knowledge of facts making it a pirate ship or aircraft;*

3. *any act inciting or of intentionally facilitating an act described in sub-paragraph (a) or (b)."*

The UNCLOS definition of piracy developed into international law and the International Maritime Organization (IMO) recognized and accepted this definition. Thus, according to international law, any illegal acts of violence and detention which are committed within

State's territorial waters are not defined as piracy.

Established by the International Chamber of Commerce (ICC) in 1981, the International Maritime Bureau (IMB) came into existence with the backing of the IMO, the world's foremost agency for exchanging and collecting information on maritime crime. However, according to the IMO, it is estimated that piracy incidents are likely under-reported by a factor of two (meaning, they assume that for each attack that was announced, there were two additional attacks that were not announced). Moreover, it is likely that the statistics are subject to distortion as many smaller attacks go unreported. This originates from two factors:

1. the increase in insurance premiums often outweigh the value of the claim for smaller attacks; and ...

2. reporting a piracy attack is often time-consuming can lead to a delay of several days. Keeping in mind the running sunk costs of an idle ship, in many, especially smaller cases, it is cheaper not to report the incident.

While this wider definition allows the IMB to produce a more comprehensive picture about maritime crime, its definition is not recognized by international law.

According to the International Maritime Bureau (IMB), nearby all illegal acts in Southeast Asia occur within territorial waters and thus would not fall under the definition of piracy. Technically, if an attack occurs within the territorial jurisdiction of a state, the event is only classified as piracy if that nation's penal code criminalizes it as such. Moreover, the IMO defines any unlawful act of violence or detention or any act of depredation at anchor, off ports or when underway through a coastal State's territorial waters as armed robbery against ships.

In order to overcome the distinctions between high seas and territorial waters, the IMB defined piracy as: "... *an act of boarding (or attempted boarding) with the intent to commit theft or any other crime and with the intent or capability to use force in furtherance of that act.*"

Al-Qaida and other groups are threatening and operating land and air space, but have as well maritime activities and cyberwar potential. In 2013 Al-Qaida declared cyberwar on NATO, in 2009 the special Mobile Team Al-Ansar went operational and grew into today's Al-Ansar Media Battalion.

The organization Al-Qaida Arabian Peninsula (AQAP) was founded in January 2009 from sub-groups in Yemen and Saudi-Arabia forming at the time one of the most dangerous threats in the region. AQAP has an operational commando with clear leadership under the former Colonel of the Egyptian Special Forces Saif al-Adel. In the after-match of the collapse in Libya the organization came in possession of thousands of MANPADS Strela-2 ground-to-air missiles supplied through the regional Islamic Maghreb AQIM and the Libyan-Islamic Fighter Group (LIFG).

Activities of Al-Qaida are supported by other local groups with several ten thousand fighters in Yemen, Somalia and Mali. Logistics, support and replenishment is in these areas mainly conducted from the sea. The MANPADS were transported either directly over land from Libya or with fishing boats from Libya to the Sinai-Peninsula. The MANPADS Strela-2 are infra-red heat-seeking ground-to-air missiles without electronic counter measures and can be easily deviated. AQAP gave therefore instructions on how to deactivate the infrared heat-seeking and conduct a manual aiming.

During the Arabic Spring these forces were able to overrun regular forces, i.e. the 25. Mechanical Brigade in Yemen, capturing several coastal cities that could eventually only reconquered with the lead of American Special Forces, all together five Brigades with 25.000 soldiers, and Saudi-Arabian fighter and drones in close air support. After the coastal areas, especially Shaqra, was lost to AQAP the support continues over land and via smuggler boats from East-Africa.

In 2010 the regional group Al-Shabab join the AQAP with about 14.000 uniformed, well trained and paid fighters, since 2012 all Al-Qaida basic training is conducted in the camps of Al-Shabab. The main financial support came from custom taxes of the ports of Bosaso, Merca, Kismayo, and Shaqra was estimated in 2011 at an annual of 35-50 million US$. On 28. September 2012 Kismayo got under control of military troops of AMISOM, however the total control over the situation was not achieved yet, activities stretching over the areas of Puntland, Somaliland and Galmudug. Since the support is mainly established over these ports an increase of maritime activities of Al-Qaida was to be expected. Between 2009 and 2012 maritime operations were conducted in eight Al-Qaida organizations.

The Abu Sayyaf Group (ASG) could be viewed as the maritime component of Al-Qaida and was established in 1991. ASG joined Al-Qaida around 2007 with 300 - 3000 fighters of the Tausog people. The Tausog being mainly families of fisherman form a potential supply line for covered activities, like the attack on 24 February 2004 against the SUPERFERRY with 116 passengers killed. The operation was conducted together with the Rajah Solaiman Movement (RSM), founded in 2001 and consisting mainly of converted Christian believers that converted to Islam. The Abdullah Azzam Brigade is recruited mainly from Palestinians out of refugee camps in Lebanon and joined Al-Qaida in 2009 in Levante.

The Islamic State of Iraq and the Levant (ISIL), also known as the Islamic State of Iraq and Syria (ISIS), Islamic State (IS) and by its Arabic language acronym Daesh is a Salafi jihadist militant group and former unrecognized proto-state that follows a fundamentalist, Wahhabi, and heterodox doctrine of Sunni Islam. ISIL gained global prominence in early 2014 when it drove Iraqi government forces out of key cities in its Western Iraq offensive, followed by its capture of Mosul and the Sinjar massacre.

This group has been designated a terrorist organization by the United Nations and many individual countries. ISIL is widely known for its videos of beheadings and other types of executions of both soldiers and civilians, including journalists and aid workers, and its destruction of cultural heritage sites. The United Nations holds ISIL responsible for human rights abuses and war crimes and Amnesty International has charged the group with ethnic cleansing on a historic scale in northern Iraq.

ISIL originated as Jama'at al-Tawhid wal-Jihad in 1999, which pledged allegiance to al-Qaeda and participated in the Iraqi insurgency following the 2003 invasion of Iraq by Western forces. The group proclaimed itself a worldwide caliphate and began referring to itself as the Islamic State or IS in June 2014. As a caliphate, it claims religious, political and military authority over all Muslims worldwide. Its adoption of the name Islamic State and its idea of a caliphate have been widely criticized, with the United Nations, various governments and mainstream Muslim groups rejecting its statehood.

In Syria, the group conducted ground attacks on both government forces and opposition factions and by December 2015 it held a large area in western Iraq and eastern Syria containing an estimated 2.8 to 8 million people, where it enforced its heterodox interpretation of sharia law. ISIL is now believed to be operational in 18 countries across the world, including Afghanistan and Pakistan, with "*aspiring branches*" in Mali, Egypt, Somalia, Bangladesh, Indonesia and the Philippines. As of 2015, ISIL was estimated to have an annual budget of more than US$1 billion and a force of more than 30,000 fighters.

In July 2017, the group lost control of its largest city, Mosul, to the Iraqi army. Following this major defeat, ISIL continued to lose territory to the various states and other military forces allied against it, until it controlled no meaningful territory by November 2017.

Incidents in the waters off the Philippines revealed a new pattern. According to media reports, the strike on a Vietnamese cargo ship wasn't meant to cause death and injury, as it was aimed at capturing hostages. The attack were conducted in a manner similar to recent attacks by ISIS-affiliated groups in Southeast Asia. Militants are increasingly prone to using hostages as bargaining chips to extract concessions out of regional governments. Abu Sayyaf, one of the most significant terror organizations in Southeast Asia to have pledged allegiance to the Islamic State (IS), is said to hold many people, including many Malaysians, Indonesians, and Vietnamese hostage.

West of the Malacca, Indian observers worry also over the possibility of a similar terror tactic at sea. After the 26/11 attacks, India's maritime agencies have been on high alert looking for signs of another terrorist infiltration into Indian waters. Abandoned Pakistani fishing boats in the Rann of Kutch region caused speculation on a Pakistan-based terror groups to cross-over into Indian Territory. Some of these, Indian analysts surmise, could well have been connected to the Islamic State.

An IS-inspired maritime terror attack on the Indian seas might seem far-fetched, but isn't beyond the realm of conception. India's security managers are being forced to confront the possibility of a terrorist attack on Indian ports on the west coast, and even cruise ships on the high seas. As IS-motivated attacks in Asian waters rise, analysts say the number of reported violations of India's territorial seas by smaller boats and crafts continues to be high.

Militant cadres could possibly use new methods to strike maritime military facilities. One such tactic could be the targeting of maritime infrastructure, in particular naval operational and residential complexes, and logistical hubs. In December 2015, Australian police had arrested two young men in Sydney for planning an attack on a maritime facility. The men are said to have been radicalized and were attempting to carry out strikes on multiple targets in the Woolloomooloo naval base in Sydney, home to Australia's principle naval assets. A few days earlier, there was a bomb blast at a mosque in a naval base in Dhaka, Bangladesh that killed two people and injured many more.

Since the early 2000s, when a series of terror attacks riled the waters of Asia, regional maritime forces have been on high alert. The most famous of those was the attack on the USS COLE on October 12, 2000 that killed 17 sailors. That was a follow-up to a failed attack on the USS SULLIVAN that's supposed to have hardened the resolve of al-Qaeda to launch a successful attack on a US warship. It was carried out by a small boat in harbor and considered by far the biggest breach of security on a commissioned US warship. It was shortly followed by the attack on the MV LIMBURG in 2002, which left the supertanker blazing off the Yemeni coast.

In July 2015, radicalized elements of the PN colluded with al-Qaeda to give effect to a diabolical plan to forcibly take-over two Pakistani warships in Karachi Harbor. The plan - foiled only on the nick of time - was to use the hijacked ships to carry out attacks on US and Indian naval warships. The Indian Navy ramped up security measures in the Arabian Sea after that incident, instituting special security procedures to deal with a threat of terror related violence at sea.

The most high-profile attack on a naval facility was the strike on PNS Mehran, a premier Pakistani naval base at Karachi, where a group of well-trained militants carried out a full-fledged military assault, killing scores of people and inflicting severe damage on Pakistani military assets.

Analysts surmise a future terror hit on a port facility could well involve a lone wolf. Over the years, terrorists have shown themselves to be remarkably enterprising in planning attacks and it's entirely possible for radicalized individuals to carry out a covert strike. Cargo containers arriving from ships, some say, offer the perfect opportunity to terrorists to carry out an attack on a port facility. A jihadi inside a container could detonate a vast quality of explosive or a low-grade nuclear device. In premier container and trans-shipment ports across the Indian Ocean and Pacific, port authorities follow a layered risk-based approach to security where specific kinds of cargo is comprehensively scanned.

The passenger ship DON RAMON was attacked in 2005, the tanker M STAR 2010 in the Strait of Hormuz, and the COSCO ASIA in 2013 transiting the Suez Canal. The potentially controlled maritime terrorism risks face possible under-responded regular and common security issues such as: corruption, piracy, stowaways, cargo crime, contraband smuggling and people trafficking. 1,754 incidents of maritime crime have been recorded since January 2013, in which 1031 ships were successfully boarded affecting an estimated 20,000 seafarers. These security issues hurt the industry, the cost of stowaway incidents each year are estimated to be in the region of 15 million dollars. In 2013 the World Bank estimated the global economic cost of piracy off the coast of Somalia at 18 billion dollars.

Regional maritime forces have a hard stand to kept pace with the fast evolving threats and rapidly adapting terrorist groups, as only a global common approach can deal with a global common threat.

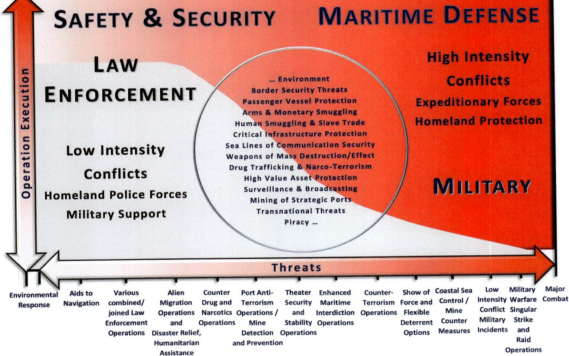

MARITIME ENTITIES

Based on the concept of an entity in Wikipedia, a MARITIME ENTITY (ME) can be described as any state, a group of states, an organization, any group or an group in an operation, acting alone or together, with an agreed framework of rules and procedures, designed to ensure security within the Maritime Environment. Any defined Maritime Stakeholder, an Institution, Regime o. a. that exists in itself, actually or potentially, concretely or abstractly, physically or not, being any type of organization, cooperation, or informal networking existence is therefore a Maritime Entity.

The Orwellian slogan *"Communication is the Foundation of Democracy"* suits well in Situational Awareness (SA), as any awareness is first about people and communication. Humans and their groupings, i.e. in nations, organizations, agencies, or in a cooperation need to be addressed as AUDIENCE or STAKEHOLDER.

An AUDIENCE is the result of specific situations and circumstances that organize individuals into a stakeholder group. Audience is an individual or group that witnesses an event or information conveyed through social audio-visual or printed media. Audiences need to be defined based on the strategic communication and the aims of the intended results and are therefore not predefined. However; individuals can by nature at the same time belong to multiple stakeholder groups, thus making stakeholder management even more complex.

However, some organizations are less detailed, i.e. NATO describes audiences simply as individuals, groups, or populations: *"A specified group or persons to whom NATO public affairs efforts are directed"*[24].

TARGET AUDIENCE describes any individual or group selected for influence, where diplomacy, with the totality of measures and means to inform, communicate and to cooperate with a broad range of target audiences in Maritime Entities world-wide, has the aim to raise the level of awareness and understanding, promoting policies and activities, thereby fostering support for the own plans and policies and developing trust and confidence in them.

When diplomacy fails, the deterrent threat or coercive use of force are the principal means by which military power influences people or changes the course of events. However, the power of a forceful and skilled narrative amplified by social media operations (PSYOPS) offer an significant advantage to adaptable and agile actors in SA.

STAKEHOLDERS are defined as potentially influencing individuals or groups, being influenced by an Entity - here a Maritime Entity - either positively or negatively. Stakeholder

---

[24] for information activities in MC 457/1, Military Decision, 19 Sep 07

groups might need to be prioritized due to their relevance for the achievement of organizational goals.

An ACTOR can be any individual, mainly the decision-maker and leader. Possible opinion leaders, opinion formers, like journalists, editors, and media publishers, are stakeholders for the target audience. But any actor can also be defined as a group of the population, or parts of, as defined by a region, ethnicity, religion, and activity. The groups can be formed of active individuals, and/or by the opinion lead followers. Organizations like i.e. government agencies and governmental or international organizations, nongovernmental organizations, private volunteer organizations, regional and international enterprises, even company from global players to local businesses can form individual's groups and therefore an acting entity; an Actor.

Population growth and urbanization caused mega cities and urban areas to be more connected and created a physically, culturally and institutionally complex terrain. An increasingly complex and ambiguous tapestry of actors with shifting allegiances for any situational awareness.

Safety and security actors rely on a fair number of people and organizations as information resources, even though they are not generally seen as members of the community. MSA depends on crucial cornerstones, like the people, their organization, the policy, the business process, technology and system/s, and well-defined Information Exchange Requirement (IER). Basic questions to analyze the overall readiness of stakeholder to share information are:

- An Authority with the Will to share Information
- Regional, multi-agency, international organizational structures for information sharing
- Policy foundation for operational information sharing
- Business procedures to enable situational awareness
- Technology in use to enable information sharing for situational awareness
- Caveats and Challenges

Government reports, academic literature, and primary resource documents collected, participation in communities, attending and participating in local, regional, national or international events, conferences, conducting informal interviews with individuals, achieve nothing without the people's will to participate, dedicate themselves and share their knowledge.

The importance of the overall willingness to share information cannot be underestimated, especially when it comes to incident-driven information exchange.

Incidents happening are always made very visible and often highly publicized by the media, and if one will be explicitly blamed, it might be to cover another. Situational Awareness is a core concept of incident-driven information sharing. Awareness on number of incidents, crises, and disasters that are avoided on a day-to-day basis as the result of the successful collaboration of the numerous stakeholders is difficult to achieve in any community and is sometimes made less public, receives less attention.

Different functions of a role in the communities, often using the same name/term with possibly different definitions, lead to differences in understanding and practice overall. Operators and Technicians who play a role based on their prior experience working at another maritime community organization provide for a cross-community experience and serve as a foundation for good relationships and fluent information sharing with their former organization.

Information ownership affects access and sharing. Security clearances are means of including individuals in or excluding them from information exchanges. The national security classification governs all national security information and is normally under the Ministry of Interior in nations or a special security office in organizations/agencies.

The process for obtaining a clearance under this system and for gaining access to classified information within the system is needs to be formal and well-controlled. Access to classified information is authorized by an individual clearance level and by the need to know. An information technology system has to account for the specific information attributes assigned to the piece of information and for the individual who wishes to access that information.

The barriers attributed to classification counter often efforts to develop a shared operational picture, which often requires integrating information drawn from classified and unclassified sources. Other instances concern difficulty in processing or handling the movement of classified or sensitive information, such as the need for specialized equipment under security agreements.

Many instances represent cases where one or more parties want to share or receive a particular piece of information but are prevented from doing so because they lack the clearance needed to be a recipient of that information. IT-Security, network administrator and international cooperation in general do not mix well, therefore sensitive/classified information is often informally shared, most of the time inappropriately according to the formal national security regulation, simply to get important everyday security work in MSA done, as tasked by the superiors.

Trust is the information sharing success factor leading to a better awareness.

The critical importance of experience and the development of trusted relationships are today still the key elements driving the information sharing behavior in any Situational Awareness, not formal procedures, authority, laws or regulations.

An international formal agreement for exchange in the highest form is a Memorandum of Understanding (MOU) or sometimes a Memorandum of Agreement (MOA), however in most cases these official documents need high governmental involvement and have a long process in negotiating and staffing. In most cooperation entities turn to an Operational or a Technical Agreement (OA/TA) instead, which can be staffed quicker and needs a lower level of ministry/organization signature/authorization. These documents are the rooftop for the following Terms of References (TOR) and the required operational and technical documentation.

Aligning information technology to mission requirements and its workflow while guided by policy agreements needs international standardization in security classifications and respective operational and technical exchange mechanisms. Technology solutions designed to impose uniform information sharing are ineffective tools in MSA.

Technology solutions for cross-domain and multiple security classification information exchange are the future solutions. Technology concepts and developments for federated future network architectures are available, however an abundance of various stakeholders and projects lacking an overall approach. Even NATO or EU accredited solutions would simply function only within the respective organization or unity, the logical MSA solution requires a worldwide accredited security solution for information exchange, hence from the United Nations, as the participants in a future mission are impossible to be predicted.

Improving the utility of information systems demands a high degree of awareness on how the community operates and might be changing in future developments. Few of the difficulties associated with information sharing can be addressed through purely technological solutions. Particular information sharing challenges account to a purely technological issue in roughly 10% of all cases.

Among these technological issues are bandwidth capacity, system errors, interoperability, data standardization, system and hardware functionality, user adoption, and the dependency on specialized, but scarce, experienced personnel to operate complex systems and equipment. A policy of acquiring and purchasing one-size-fits-all technological solutions appears therefore as ill-advised in Situational Awareness. Most promising and successful seems the civilian world of decentralized sharing mechanisms[25] linking existing systems and their individual capabilities into networks for information sharing. This concept was transformed for the civilian and military federated, decentralized approaches fulfilling region-specific operational requirements.

Regional and international cooperation require international and cross-domain interfacing standards to connect homogeneous national and organizational systems. Interagency dependencies and use cases need to be aligned and harmonized with international and/or cross-domain dependencies and their use cases in order to provide an functional workflow and to enable information sharing for a comprehensive situational awareness.

An overall MSA conceptual framework is required, taking into account standards, specifi-

---

[25] Here federated media sharing networks like Gnutella (https://en.wikipedia.org/wiki/Gnutella) o. a.

cations, governance, considering and involving the single users up to the main stakeholder. However, size matters. The economist Mancur Olson described the advantages of the smaller number of entities in a group over a certain vast number of members in an organization. The critical number is of course dependent on many aspects and the complexity and will need to be analyzed carefully. It is a key for fast development and the ability to adapt to changes and challenges.

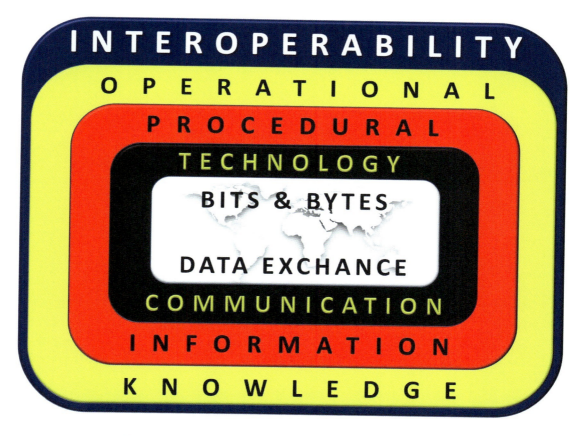

"There is a critical number of participants depending on the complexity of the aims and objectives in any organization."[26]

Organization behave slow and become at large hardly manageable, because communication difficulties increase. A discussion below the number of ten entities is difficult; above ten any consensus is hardly possible, and reaching a number of thirty it becomes nearly impossible. Examples are the Sea Surveillance Cooperation Baltic Sea (SUCBAS) and Maritime Surveillance Networking (MARSUR Networking), both having very similar decentralized/federated environment, but a different speed development, mainly due to their numbers of participants and the difference in homogeneous operational environ-

---

[26] Joachim Beckh

ments. The United Nations with 193 members requires obviously the most intensive negotiating and the smallest common nominators for any agreement. To be able to reach consensus requires a common language and agreed definitions in terms for functional communication.

A MESSAGE is any thought or idea expressed briefly in plain, coded, or secret language, prepared in a suitable form for transmission by any means of communication. Media is any personnel, i.e. in management, owners, financiers, stakeholders, publishers, editors, journalists, employees using messages in order to influence target audiences.

COMMUNICATION is in general the imparting or exchanging of information by speaking, writing, or using some other medium and denotes the exchange of information between individuals through a system of signs, symbols, or behavior. Communication systems are the technical means of sending or receiving information, the providing tool for a messaging capability.

CAPABILITY defines in general the resources giving a certain ability to undertake a particular kind of action. An actor's capacity for action is dependent upon his physical capabilities and their utility in a particular situation. Information activities will seek to affect those capabilities, such as communications infrastructure and propaganda facilities that enable actors to understand a situation and apply their will. A technical Capability is a technical resource in respect of systems.

INFORMATION in general is the unprocessed and mostly unstructured data, send and received in a message, which may be used in the production of knowledge and intelligence. Information Management is the, most of the time, futile attempt to bring order into to chaos of information overload in today's information age by means of technology, plans, concepts and strategies.

INFORMATION SYSTEM is an assembly of equipment, methods and procedures organized to accomplish information processing functions. However; the necessary personnel organized to accomplish information processing functions may also be referred to an organization itself, i.e. stakeholders, groups or any other human entity forming such.

Information critical to operations comprises both quantitative and qualitative data, but structured data and the formal systems that hold it are mostly lacking flexibility to adapt to fast changing operational requirements. Security clearances create further difficulties in information sharing, while many elements of information that are critical to safety and security at the regional level are very dynamic, diverse, nuanced, informal, and trust-based. These elements cannot be captured in current structured data formats at the enterprise level.

Technology requires a strategy in overcoming challenges to the community information sharing, but technology is often perceived simply as an asset, but not a complete solution under a strategy and concept. Technology is often not adopted because it introduces considerable overhead tasks, when adopted often perceived as an additional challenge and dismissed without a full understanding of its potential for the problem solving; often it is seen as part of the challenge.

More information and an abundance of communication is at this point not simply better situational awareness; the key is Quality of Information (QoI). Mass of information does not provide the instant solution on prior unsolved challenges, i.e. receiving millions of unvalidated vessel tracks will not automatically provide for a better Recognized Maritime Picture (RMP); rather the opposite is the case. It is the question of interoperability of human cultures in all fields of procedures, operations, and technology in order to ensure:

- the right source
- to send the right information
- at the right time
- at the right place
- in the right format
- for the right user
- in the right amount
- to the right recipient

The Maritime Community includes an vast amount of organization or individuals with active roles in the wide maritime domain, but at times it may include unknown and outside actors on a borderline that might be extremely difficult to demarcate. The Global Maritime Community of Interest (GMCOI) includes, among other interests, federal, state, and local departments, agencies, inter-departmental actors with responsibilities in the maritime domain. Because certain risks and interests are common to government, business, and citizen alike, GMCOI also includes public, private and commercial stakeholders, as well as foreign governments or international stakeholders.

"*Communities - either mission focused communities of interest, or professionally or technically focused communities of practice - provide a way to build coalitions and deepen relationships to mutual benefit. With our journey to accelerate responsible information sharing, the key is to bring together mission-focused and functional communities, and together to drive secure and trusted collaboration.*"[27]

Safety and security in the maritime environment, as in air, land space or other, is one of the most critical components of national resilience and well-being of citizens. Networking is aimed at creating knowledge by a continuous, adaptive and linked information sharing activity. These Networks can be of a social environment of any organization or group simply in exchanging messages, i.e. in discussion forums in a meeting or in technical form of the social network of the online blog for such information exchange forums or existing information systems with tracking and accounting databases. Important is the quality of information, as i.e. the International Maritime Organization (IMO) found in a study of seven

---

[27] Kshemendra Paul, Information Sharing Environment Program Manager, June 5, 2014

participating countries[28], that 32% of the 25,284 inspected containers had false cargo declarations.

All maritime related activities involve a large set of stakeholders from many sectors and levels of government. The resilience and well-being of nations depend upon the ability of these diverse stakeholders to network in support of the effective collaboration. Maritime operations are a large, diverse, and interconnected system depend upon the organization of many individual operators and agencies into a community that supports effective and efficient information sharing and coordination.

To achieve influence and effect in today's ever changing maritime operating environment has become more complex and competitive, and yet it is increasingly central to future success. It is a multifaceted task to provide credible civilian and military options to maintain the freedom of action and political utility. Pragmatic responses with a strong capability base are required. Prevention is easier from a position of strength and it demands the credibility of a robust force confident to fight and adapt in contact when deterrence fails, which it sometimes will.

However; for most community members, security is not the primary focus of their day-to-day work. Community members, particularly those in the private sector, are often unaware of how their day-to-day operations relate to security. Activities related to safety and security are driven by an awareness of risk, i.e., they are driven by the need to prepare for possible incidents.

The information sharing environment built during day-to-day operations is a critical component of emergency response and management in MSA. The work of aligning the resources and building the capability to respond to incidents occurs outside the context of an incident. Successful information sharing requires expertise in the community itself.

Motivation is the key for unidirectional information flow, to break instances where one partner takes information, but gives little in return. Today there is an galore of networks with their existing the technology to share information, still commonly there is challenge concerning partners not providing information, often for security reasons, but sometimes due to internal structures and processes or even cultural differences.

Entities in the maritime community require motivation by competition for profit, resources, and credit, but they commonly worry about the possible impacts of sharing too much information even in Maritime Situational Awareness (MSA). The entities involved in commerce including the ports, terminal operators, shipping lines, fishing vessels, tug boats, trucking companies, rail, and many others that operate in a competitive environment which could cause severe financial damages when business processes, information, or

---

[28] Belgium, Canada, Chile, Italy, the Republic of Korea, Sweden, and the United States. International Maritime Organization (2006). Sub-Committee on Dangerous Goods, Solid Cargoes, and Containers. Report to the Maritime Safety Committee. (DSC 11/19). 11th Session. Agenda Item 19.

security vulnerabilities are exposed or hampered.

However, this motivation through competition for profit can scarcely be achieved in i.e. governmental or military communities, which have valid different requirements and points-of-view as well as threats in safety and security. To harmonize the military/governmental and civilian/commercial MSA interests in safety and security is a challenge, as already the expert terminology often uses the same or similar terms with quite different meanings and definitions in each environment. Most beneficial would be an agreed common "Maritime Language", identifying identical abbreviations and terms.

Maritime Situational Awareness (MSA) and Interoperability

MARITIME ORGANIZATIONS & AGENCIES include any form of non-technical networking, while Maritime Technology intends to cover the most relevant industrial developments and technological system networks, which in most cases result from Maritime Projects & Programs.

TECHNOLOGY is in general the collection of techniques, skills, methods and processes used in the production of goods or services or in the accomplishment of objectives, such as scientific investigation. Technology can be the knowledge of techniques, processes, and the like, or it can be embedded in machines or systems which can be operated without detailed knowledge of their workings.

MARITIME TECHNOLOGY is in consequence used in this book to name technology used in the Maritime Environment, but most of the time any technology is not limited to the maritime use. The technology needs to be interoperable in order to enable the operators tasks for a common and validated picture in all situational awareness cases. The term Maritime Technology is often used to distinguish and/or describe any technology in systems and network used by Maritime Entities in order to produce a Maritime Picture.

The definition of a SYSTEM in Wikipedia defines the term as a set of interacting or interdependent component parts forming a complex/intricate whole. Every system is delineated by its spatial and temporal boundaries, surrounded and influenced by its environment, described by its structure and purpose and expressed in its functioning. The term SYSTEM should therefore not be limited to only technology, as any entity is more likely a complex system of within itself.

System may also refer to a set of rules that governs structure and/or behavior. Alternatively, and usually in the context of complex social systems, the term is used to describe the set of rules that govern structure and/or behavior. Here the generic term system refers to an entity and/or a technical system or even network entity. Maritime Entities are described aside their use of technical SYSTEMS or NETWORK.

Prior to any systems or network development there needs to be a concept, the development of a prototype and the experimentation. Concepts, Developments and Experimentation (CD&E) are listed separate from the Maritime Entities, sometime the name of an Entity is used for the technical system, sometimes the later system and/or network may be called differently or the name may alter in the realization process.

Generic terminology and the diverse understanding of similar definitions pointed at the content of Maritime Entities, Concepts, Demonstrations & Experimentations and *"Systematogenesis"* to be covered simultaneously in *"Maritime Situational Awareness for Tyrians"*. Each information given is in a constant flux and should be double checked on its actual status in the internet.

Nations/Countries are by definition a Maritime Entities, priority lays within geographical position, coastal area, economic and military capabilities, however their individual situation and requirements are not subject to this publication. The countries listed are used as national Maritime Entity Identifier to simplify and distinguish the examples of organizations or systems. In other cases an international organization / cooperation or system might be listed singular and independent.

Africa

Africa is the world's second largest and second most-populous continent (behind Asia in both categories). At about 30.3 million km^2 (11.7 million square miles) including adjacent islands, it covers 6% of Earth's total surface area and 20% of its land area. With 1.2 billion people as of 2016, it accounts for about 16% of the world's human population.

The continent is surrounded by the Mediterranean Sea to the north, the Isthmus of Suez and the Red Sea to the northeast, the Indian Ocean to the southeast and the Atlantic Ocean to the west. The continent includes Madagascar and various archipelagos and is therefore by geographical position, coastal area, and economic capability one of the larger Maritime Entities.

It contains 54 fully recognized sovereign states (countries), nine territories and two de facto independent states with limited or no recognition. The majority of the continent and its countries are in the Northern Hemisphere, with a substantial portion and number of countries in the Southern Hemisphere.

The African Union (AU) is a 55-member federation consisting of all of Africa's states. The union was officially established on 9 July 2002 as a successor to the Organisation of African Unity (OAU). In July 2004, the African Union's Pan-African Parliament (PAP) was relocated to Midrand, in South Africa, but the African Commission on Human and Peoples' Rights remained in Addis Ababa. There is a policy in effect to decentralize the African Federation's institutions so that they are shared by all the states.

Although it has abundant natural resources, Africa remains the world's poorest and most underdeveloped continent, the result of a variety of causes that may include corrupt governments that have often committed serious human rights violations, failed central planning, high levels of illiteracy, lack of access to foreign capital, and frequent tribal and military conflict (ranging from guerrilla warfare to genocide). According to the United Nations' Human Development Report in 2003, the bottom 24 ranked nations (151st to 175th) were all African.

National initiatives and projects are few as the geographical and financial setting requires a combined approach in most case. For this reasons most of the projects are inter-governmental, international and in relation with the European Union or selected cooperation with European or other countries. These Maritime Entities are listed under the related cooperation organization or project.

In most African countries there is a lack of important arterial rivers or lakes, which requires extensive water conservation and control measures, growth in water usage is outpacing supply, there is pollution of rivers from agricultural runoff and urban discharge, air pollution resulting in acid rain in the big cities including soil erosion and desertification.

African Maritime Safety and Security Agency (AMSSA)

The African Maritime Safety and Security Agency (AMSSA)[29] provides a scientific and intelligence platform for the African Member States and other Maritime Stakeholders on matters relating to the safe, secure and clean movement of maritime transport, and the prevention of the loss of human lives at sea.

The development of the African Maritime Safety and Security Agency (AMSSA) was first introduced during the proceedings of the 3rd Assembly of the International Cargo Handling Coordination Association (ICHCA)[30] Canaries-Africa Regional Charter Ports conference held 10-11 December 2007 in Accra, Ghana. Aim was to strengthen the African port logistics, trade, the maritime economy and the overall capacities of the African integrated maritime transport and supply chain.

The discussion relating to Maritime Safety and Security related to the socio-economic and environmental benefits of developing a maritime safety orientated intelligent support framework. The purpose was to provide new scientific research, harmonized statistical data, and the development of management models in relation to the economic impacts for the port and shipping industries of maritime safety and security issues. The development of a North West Africa Maritime Safety Agency (NWAMSA) was considered as a process that would optimize various activities.

The conference proceeding found that effective operations of African port activities are vital for the well-being of communities, and that the 'turn around' for incoming vessels is an important factor relating to the economic viability of the sector. It was agreed that overall concerns relating to maritime safety and security affairs were impinging upon the operational capacity of the African maritime sector, and it was noted that such issues are aggravating the overall maritime management process.

The ICHCA Canaries-African ports and shipping industries representatives' consensus was that the development of methodologies to respond to Maritime Safety and in particular the irregular immigration phenomena, is a priority for the maritime transport sector. The change of title occurred after the printing of the first report when representatives of the African maritime community noted that the tools and mechanisms developed through the organization would provide benefits for the whole of Africa due to the international perspective of maritime it was thus justified that the NWAMSA should eventually be the organization AMSSA, becoming a legal entity during 2010.

AMSSA is to respond directly to the Africa - EU Joint Strategy and to help take the Africa-EU relationship to a new, strategic level with a strengthened political partnership and enhanced cooperation at all levels; this with specific regards to Maritime Safety, Security and

---

[29] http://www.amssa.net/
[30] https://ichca.com

Marine Environmental Protection. The proposed AMSSA partnership will be based on a Euro-African consensus on values, common interests and common strategic objectives.

The AMSSA partnership and its further development is guided by the fundamental principles of the unity of Africa, the interdependence between Africa and Europe, ownership and joint responsibility, and respect for human rights, democratic principles and the rule of law, as well as the right to development. This will be managed through the development/implementation of a relevant institutional architecture, which will allow and promote intensive exchange and dialogue on all issues of common concern.

In the light of both EU-African partnership it will also commit itself to enhance the coherence and effectiveness of existing agreements, policies and instruments. The backdrop for the AMSSA proposition is the implementation of the New African Maritime Transport Charter and the response to the associated resolutions agreed during the 2nd African Union Conference of Ministers Responsible for Maritime Transport held in Durban, South Africa – October 2009, and the (2007) European Integrated Maritime Policy, (in particular the International Dimension) and the associated action plan.

AMSSA will establish and coordinate the formation of collaborative projects, of African Maritime Safety, Security and for Marine Environmental Protection and their associated networks. These integrated sub-projects will promote the African Maritime Transport Charter and will aim to develop actions, services and initiatives by progressing projects from concept through the embryonic phases to the implementation stage by building capacity.

Arctic Council

The Arctic Council[31] is a high-level intergovernmental forum to promote cooperation, coordination and interaction among the Arctic States. While Antarctica is a land area and governed by a designated treaty, the Arctic Ocean is an ocean surrounded by national states. Although parts of the Arctic Ocean are covered by ice, the Law of the Sea applies in full in this area as in the other oceans of the earth.

The first step towards the formation of the Council occurred in 1991 when the eight Arctic countries signed the Arctic Environmental Protection Strategy (AEPS). The 1996 Ottawa Declaration established the Arctic Council as a forum for promoting cooperation, coordination, and interaction among the Arctic States, with the involvement of the Arctic Indigenous communities and other Arctic inhabitants on issues such as sustainable development and environmental protection. The Arctic Council has conducted studies on climate change, oil and gas, and Arctic shipping.

The Arctic Council is the only circumpolar political forum that involves all the Arctic states and provides for the active participation of the Indigenous Peoples of the Arctic. There are eight member states of the Arctic Council: Canada, Denmark (including Greenland and the Faroe Islands), Finland, Iceland, Norway, Sweden, the Russian Federation and the United States of America.

The chairmanship of the Arctic Council rotates every two years among the eight member states. Canada had the chairmanship from May 2013-2015, followed by the United States in 2015-2017 and Finland 2017-2019. The chairmanship is assisted by Arctic Council Secretariat staff, responsible for arranging and hosting the Arctic Council meetings during its period. Each member state country has appointed a Senior Arctic Official (SAO), who manages its interest in the Arctic Council. Senior Arctic Official meetings are held once in the spring and once in the fall.

Ministerial meetings, gathering Foreign Ministers from all 8 states, are held biannually in the country holding the chairmanship and result in the signing of the Arctic Council declarations, setting the priorities of the Arctic Council for the coming period.

At the 7th Ministerial meeting in Nuuk, Greenland, on 12. May 2011, the Arctic Council adopted the first legally binding Agreement on Cooperation in Aeronautical and Maritime Search and Rescue in the Arctic (SAR agreement), which enhances cooperation on search and rescue in the Arctic region. At the 8. Ministerial meeting in Kiruna, Sweden on 15. May 2013, Ministers signed the second legally binding under the auspices of the Arctic Council. The Agreement on Cooperation on Marine Oil Pollution Preparedness and Response in the Arctic will substantially improve procedures for combatting oil spills in the Arctic. During the meeting six new non-Arctic states were also approved as accredited

---

[31] https://arctic-council.org

observers to the Arctic Council. Further, the Standing Arctic Council Secretariat officially became operational in Tromsø, Norway on 1. June 2013.

There are six Permanent Participants (PPs) organizations in the Arctic Council that participate actively and are fully consulted in all deliberations and activities of the Council. The organizations representing the Arctic Indigenous Peoples are:

- Aleut International Association (AIA),
- Arctic Athabaskan Council (AAC),
- Gwich'in Council International (GCI),
- Inuit Circumpolar International (ICC),
- Russian Association of Indigenous Peoples of the North,
- Siberia and Far East (RAIPON) and Saami Council (SC).

There are six Arctic Council working groups that engage in scientific-oriented studies on issues concerning the Arctic environment and its inhabitants. The working groups focus on issues such as: monitoring, assessing and preventing pollution in the Arctic, climate change, biodiversity conservation and sustainable use, emergency preparedness and prevention; as well as, living conditions of Arctic residents, studies on oil and gas and on Arctic shipping. The scientific reports provide advice and recommendations to the Arctic Council.

The work of the Council is primarily carried out in six Working Groups.

- The Arctic Contaminants Action Program (ACAP) acts as a strengthening and supporting mechanism to encourage national actions to reduce emissions and other releases of pollutants.
- The Arctic Monitoring and Assessment Program (AMAP) monitors the Arctic environment, ecosystems and human populations, and provides scientific advice to support governments as they tackle pollution and adverse effects of climate change.
- The Conservation of Arctic Flora and Fauna Working Group (CAFF) addresses the conservation of Arctic biodiversity, working to ensure the sustainability of the Arctic's living resources.
- The Emergency Prevention, Preparedness and Response Working Group (EPPR) works to protect the Arctic environment from the threat or impact of an accidental release of pollutants or radionuclides.
- The Protection of the Arctic Marine Environment (PAME) Working Group is the focal point of the Arctic Council's activities related to the protection and sustainable use of the Arctic marine environment.
- The Sustainable Development Working Group (SDWG) works to advance sustainable development in the Arctic and to improve the conditions of Arctic communities as a whole.

The Council has also provided a forum for the negotiation of three important legally binding agreements among the eight Arctic States. The first, the Agreement on Cooperation on Aeronautical and Maritime Search and Rescue in the Arctic, was signed in Nuuk, Greenland, at the 2011 Ministerial Meeting. The second, the Agreement on Cooperation on Marine Oil Pollution Preparedness and Response in the Arctic, was signed in Kiruna, Sweden, at the 2013 Ministerial meeting. The third, the Agreement on Enhancing International Arctic Scientific Cooperation, was signed in Fairbanks, Alaska at the 2017 Ministerial meeting.

Under certain conditions, observer status in the Arctic Council is open to a) non-arctic states (currently 12); b) inter-governmental and inter-parliamentary organizations, global and regional (currently 9); and c) non-governmental organizations (currently 11). The Council may also establish Task Forces or Expert Groups to carry out specific work. The Task Forces operating during the Chairmanship of Finland (2017-2019) are:

- Task Force on Arctic Marine Cooperation (TFAMC)
- Task Force on Improved Connectivity in the Arctic (TFICA)

During the 2017-2019 Finnish Chairmanship there is also one Expert Group operating:

- Expert Group in support of implementation of the Framework for Action on Black Carbon and Methane (EGBCM)

Association of Southeast Asian Nations (ASEAN)

The Association of Southeast Asian Nations (ASEAN)[32] was established on 8 August 1967 in Bangkok, Thailand, with the signing of the ASEAN Declaration (Bangkok Declaration)[33] from 8. August 1967 by the Founding Fathers of ASEAN, namely Indonesia, Malaysia, Philippines, Singapore and Thailand. Brunei Darussalam then joined on 7. January 1984, Vietnam on 28. July 1995, Laos and Myanmar on 23. July 1997, and Cambodia on 30 April 1999, making up what is today the ten Member States of ASEAN: Brunei, Cambodia, Indonesia, Laos, Malaysia, Myanmar, Philippines, Singapore, Thailand, Vietnam.

The ASEAN Declaration states the aims and purposes of ASEAN as:

- To accelerate the economic growth, social progress and cultural development in the region through joint endeavors in the spirit of equality and partnership in order to strengthen the foundation for a prosperous and peaceful community of Southeast Asian Nations;
- To promote regional peace and stability through abiding respect for justice and the rule of law in the relationship among countries of the region and adherence to the principles of the United Nations Charter;
- To promote active collaboration and mutual assistance on matters of common interest in the economic, social, cultural, technical, scientific and administrative fields;
- To provide assistance to each other in the form of training and research facilities in the educational, professional, technical and administrative spheres;
- To collaborate more effectively for the greater utilization of their agriculture and industries, the expansion of their trade, including the study of the problems of international commodity trade, the improvement of their transportation and communications facilities and the raising of the living standards of their peoples;
- To promote Southeast Asian studies; and
- To maintain close and beneficial cooperation with existing international and regional organizations with similar aims and purposes, and explore all avenues for even closer cooperation among themselves.

In their relations with one another, the ASEAN Member States have adopted the following fundamental principles, as contained in the Treaty of Amity and Cooperation in Southeast Asia (TAC) of 1976:

---

[32] http://asean.org
[33] http://asean.org/the-asean-declaration-bangkok-declaration-bangkok-8-august-1967/

- Mutual respect for the independence, sovereignty, equality, territorial integrity, and national identity of all nations;
- The right of every State to lead its national existence free from external interference, subversion or coercion;
- Non-interference in the internal affairs of one another;
- Settlement of differences or disputes by peaceful manner;
- Renunciation of the threat or use of force; and
- Effective cooperation among themselves.

The ASEAN Vision 2020, adopted by the ASEAN Leaders on the 30th Anniversary of ASEAN, agreed on a shared vision of ASEAN as a concert of Southeast Asian nations, outward looking, living in peace, stability and prosperity, bonded together in partnership in dynamic development and in a community of caring societies.

At the 9th ASEAN Summit in 2003, the ASEAN Leaders resolved that an ASEAN Community shall be established and at the 12th ASEAN Summit in January 2007, the Leaders affirmed their strong commitment to accelerate the establishment of an ASEAN Community by 2015 and signed the Cebu Declaration on the Acceleration of the Establishment of an ASEAN Community by 2015.

The ASEAN Community is comprised of three pillars, namely the:

- ASEAN Political-Security Community
- ASEAN Socio-Cultural Community
- ASEAN Economic Community

As some sort of fourth community the ASEAN Defense Minister's Meeting (ADMM) was established, which was enlarged to the ASEAN + Australia, China, India, Japan, New Zealand, South Korea, Russia, USA (ADMM Plus) underlining the importance of ACEAN as the most relevant association in the Asian region and abroad.

Countries also meet in the ASEAN Regional Forum (ARF) and the East Asia Summit (EAS). Each pillar has its own Blueprint, and, together with the Initiative for ASEAN Integration (IAI) Strategic Framework and IAI Work Plan Phase II (2009-2015), they form the Roadmap for an ASEAN Community 2009-2015.

The ASEAN Charter serves as a firm foundation in achieving the ASEAN Community by providing legal status and institutional framework for ASEAN. It also codifies ASEAN norms, rules and values; sets clear targets for ASEAN; and presents accountability and compliance.

The ASEAN Charter entered into force on 15 December 2008. A gathering of the ASEAN Foreign Ministers was held at the ASEAN Secretariat in Jakarta to mark this very historic occasion for ASEAN.

With the entry into force of the ASEAN Charter, ASEAN will henceforth operate under a new legal framework and establish a number of new organs to boost its community-building process.

In effect, the ASEAN Charter has become a legally binding agreement among the 10 ASEAN Member States.

Australia, Canada, New Zealand, the United Kingdom and the United States

(AUSCANNZUKUS)

AUSCANNZUKUS[34] is the abbreviation for the naval Command, Control, Communications and Computers (C4) interoperability organization involving the Anglosphere nations of Australia, Canada, New Zealand, the United Kingdom of Great Britain and Northern Ireland and the United States of America. It is also used as security caveat in the UKUSA Community, known as *"Five Eyes"*.

Early in World War II communications interoperability between Allied forces was poor, forces where thrown together according to their availability and readiness. During March 1941 the first high-level proposals to formally structure combined operations between the United States and the United Kingdom were considered. These discussions were the genesis of the current Combined Communications Electronics Board (CCEB).

The origins of the AUSCANNZUKUS organization arose from dialogue between Admiral Arleigh Burke, USN, and Admiral Lord Louis Mountbatten, RN, in the 1960s. Their intention was to align naval communications policies and prevent, or at least limit, any barriers to increase interoperability, with the imminent introduction of sophisticated new communications equipment. Their initiative matured to the current five-nation organization in 1980 when New Zealand became a full member in AUSCANNZUKUS.

The organization's remit has expanded over the years, and its mission now includes fostering knowledge sharing and C4 interoperability between the navies of the five nations in order to increase operational effectiveness. The organization's vision and mission, Objectives, Strategies and Guiding Principles, and Structure are presented on the AUSCANNZUKUS Information Portal.

The current AUSCANNZUKUS organization consists of the Supervisory Board, C4 Committee, and various other subordinate groups.

AUSCANNZUKUS liaises closely with Washington based management groups of the Combined Communications Electronics Board (CCEB), Multinational Interoperability Council (MIC), American, British, Canadian & Australian Armies (ABCA Armies), Air and Space Interoperability Council (ASIC (Air Force)) and The Technical Cooperation Program (TTCP).

The AUSCANNZUKUS Information Portal provided in the past access for AUSCANNZUKUS members to the documents created supporting continued Coalition Interoperability at sea, but is for security reasons no longer available on the web.

---

[34] https://en.wikipedia.org/wiki/AUSCANNZUKUS

Strategy is to:

- Establish C4 policy and standards.
- Identify C4 interoperability requirements and risks.
- Identify, develop and utilize new technologies.
- Exchange information on national C4 capabilities, plans, and projects.
- To improve national awareness of AUSCANNZUKUS.
- Leverage exercises, experiments and demonstrations to deliver capability.
- Inform and influence multinational defense fora.

Objectives:

- To achieve internal sharing and understanding of Maritime C4 knowledge.
- To produce products and processes to achieve Maritime C4 interoperability.
- To increase external sharing and understanding of AUSCANNZUKUS.

Guiding Principles are:

- The focus of all activities is to be on the requirements of the naval warfighter.
- All knowledge-sharing initiatives are to aim at providing innovative options, which are affordable to all AUSCANNZUKUS navies.
- All relevant information is to be shared with appropriate joint and combined organizations.

Baltic and International Maritime Council (BIMCO)

The Baltic and International Maritime Council (BIMCO)[35] was founded under the title of *"The Baltic and White Sea Conference"* in Copenhagen in 1905 was one of the first organization to see the benefit in joining forces with other countries to secure better deals and standard agreements in shipping.

The sailing ship, previously the carrier for cargoes, was being replaced by the faster and more cost effective steamship in the early 20th century. More and more steamship owners were met with increasing competition with freight rates and something needed to happen to stop a number of owners going out of business.

Thomas Cairns, of Cairns, Noble & Co from Newcastle Upon Tyne and John Hansen of CK Hansen, Copenhagen realized that owners working together - despite being in competition - would be the answer to some of the safety and security challenges. As the new season was approaching for the timber trade, Cairns sent Hansen a telegram in January 1905 asking him whether owners in the Baltic states might want to agree on timber trade rates.

The result was a gathering from 16.-18. February 1905 in Copenhagen of ship owners from across Scandinavia, Great Britain, The Netherlands, Germany, Belgium, Finland, France and Latvia. A minimum freight rate was set for the coming year and agreed by all the delegates. In addition to agreeing a firm rate for freight at the gathering, a number of delegates requested establishing uniform charter party terms to safeguard profit for the owners. The fixed freight rate pricing would never happen but the standardization of charter parties and other shipping documents would go on to benefit the maritime industry for over the next 100 years and is still going strong today.

As the organization grew and became more international it was renamed into Baltic and International Maritime Council (BIMCO). It operates a number of committees that make decisions about the future of BIMCO and direction of its business. BIMCO is today the world's largest international shipping association, with more than 2,200 members globally. They provide a wide range of services to their global members including ship owners, operators, managers, brokers and agents. BIMCO is also recognized worldwide for the clarity, consistency and certainty of its standard maritime contracts.

The aim of BIMCO is to produce flexible commercial agreements that are fair to both parties. They work with industry experts to produce modern contracts tailored to specific trades and activities. The recognized contracts are widely used and this familiarity provides greater certainty of the likely commercial outcome - helping their members manage contractual risk.

---

[35] https://www.bimco.org

Balkan & Black Sea Security Forum

<u>Balkans & Black Sea Cooperation Forum</u>[36] is an independent, nonprofit, nongovernmental economic conference, fostering cooperation, business relations and sustainable growth across CEEC & SEE regions and beyond.

True international outlook, substantial content and cross-sectoral agenda for leading business- & policy makers, melting top class audience from the public, business and the diplomatic world about connecting the dots and expanding your world.

Forums' multidisciplinary agenda focuses on UN SDGs 2030, Digital Economies, Maritime/Transports, Agriculture, Environment, Energy, Education, and Innovation, Tourism/Culture and Women in leadership, being layers influencing & affecting the real world.

Balkans and the Black Sea, a strategic, although sensitive region, requires multilevel approach about enhancing mutual understanding towards deepening economic relations and cross-border cooperation.

At the crossroads of Europe, Middle East and Central Asia, the region is in the center of gravity for investments and geopolitical influence of world's superpowers during a time of larger region-wide and global change.

At the cross section of public and business sectors, Balkans & Black Sea Cooperation Forum aiming at identifying, strengthening & promoting economic relations, business opportunities, cross-border cooperation and sustainable growth & development across the Balkans, the Black Sea region and beyond.

Balkans & Black Sea Cooperation Forum made to be for long term. Furthermore willing to elevating people *from* and stakeholders *in* the region focusing on debating and influencing hot topics of geopolitical importance and economic potential, being also a unique channel about building and refreshing your network of economic diplomacy and doing business.

---

[36] http://balkansblackseaforum.org

Baltic Sea Region Border Control Cooperation (BSRBCC)

The <u>Baltic Sea Region Border Control Cooperation (BSRBCC)</u>[37] was established at the 5th Ministerial Session of the Council of the Baltic Sea States (CBSS) in 1996 at the level of the heads of the relevant national authorities in the fields of Border/Frontier/Coast Guards and covering the needs for a Baltic Sea Coast Guard Forum (BCGF), which would then be established as the BSRBCC.

The BSRBCC focuses on security-related issues concerning border control in the Baltic Sea Region. Among the most important issues on its agenda are the development of practical forms of cooperation, simplifying communication routines between the parties (e.g. the development of the data transmission system COASTNET[38]), and the cooperating with FRONTEX.

The BSRBCC is a flexible regional tool for daily inter-agency interaction in the field of environmental protection and to combat cross-border crime in the Baltic Sea region, with a certain maritime focus. Cooperation Partners are Police, Border Guards, Coast Guards and Customs Authorities.

The foundation of a cooperation of law enforcement agencies working at the borders in the Baltic Sea Region was initiated by Finland's Foreign Secretary, Mrs. Tarja Halonen, during a conference of the Council of the Baltic Sea in Kalmar / Sweden in 1996. Consequently the establishment of the BSRBCC (Baltic Sea region Border control Cooperation) was decided in Turku/Finland in the context of the first meeting of the heads of the Border Guard Services of the -10- Baltic Sea States in 1997. Member states are: Estonia, Denmark, Finland, Germany, Latvia, Lithuania, Norway, Poland, Russia and Sweden. Iceland holds an observer status.

The field of operations of the BSRBCC covers extensively the entire field of cross - border criminality and carries out environmental protection in the maritime area. National coordination centers (NCC) which are connected 24/7 have been built up in all member states. Moreover, the BSRBCC is the frame for carrying out operations, further education and regular meetings, also under observation and participation of Frontex.

According to Frontex, the BSRBCC is regarded as an example for other cooperation forms. Strategic Partners are the Council of the Baltic Sea States, FRONTEX, and the European Association of Airport and Seaport Police.

The field of operations of the BSRBCC covers extensively the entire field of cross - border criminality and carries out environmental protection in the maritime area.

---

[37] http://www.bsrbcc.org
[38] CoastNet was not established under this name and the term is not to be confused with Australia's Premium Internet Service Provider (https://coastnet.net.au)

The meeting of the BSRBCC Border Guard Chiefs continues to take place annually. The meeting of chiefs serves as the supreme body for the cooperation initiative, and it approves the operational strategy. An operational and economic mandate for cooperation is created through the meeting of chiefs.

The presidency of the BSRBCC lasts one year. Every change of presidency is implemented with the support of the previous presidency nation, which in particular should support the Secretariat and the ICC. A qualitative continuation of the cooperation will thus be ensured in connection with any change of presidency.

In order to grant continuity in the BSRBCC, the composition of the BSRBCC- Secretariat follows a revolving system, this means the Secretariat is composed of representatives of the previous BSRCCC-Presidency (2016/Latvia) the current BSRBCC-Presidency (2017/Norway) and the following BSRBCC-Presidency (2018/Sweden). The BSRBCC-Secretariat serves as a tool for the presidency state and coordinates any interfacing with other cooperation bodies (Frontex, Baltic Sea Task Force, Council of the Baltic Sea States etc.). The Secretariat is the essential representative for the cooperation initiative in cooperation with other bodies.

The Baltic Border Committee (BBC) forms the operative supreme body of cooperation. The BBC evaluates the interaction with other cooperation forms, while the Secretariat supports the presidency country in the planning and coordinating of operative activities on demand. The BBC is a cooperation network that is formed of individuals who represent national points of contact. It is tasked with preparing and implementing the cooperation strategy. The BBC is authorized to implement the strategy by putting operative practices into action between countries in the Baltic Sea Region.

All information about joint operations is to be linked to the BBC, whose core task is to network joint operations, drawing primarily on existing structures, and creating new ones if required, whilst taking local needs into account. Through the BBC, national representatives integrate national action and local cooperation into wider cooperation as necessary. The BBC convenes at the initiative of the presidency state or the Secretariat, and whenever necessary at the initiative of a national representative.

Drawing on the information exchange network that has been formed, the BBC is responsible for coordinating national information about joint operational sectors covering the entire area, thus enabling functional analyses and threat assessments in the operational sectors. Furthermore, the BBC is responsible for collecting and communicating information about local joint operations. The BBC's national representative has a key role in coordinating bilateral and multilateral cooperation.

The National Coordination Centre (NCC) is a central national communication and operative contact point for and to the respective authorities of the Baltic Sea Region. By using the NCC network, the partners exchange operational information in real time. The NCC network is on duty 24/7 and able to communicate via a secure internet platform, the so-called COASTNET system. Each participating state will bring the resources of the NCC up to a level that enables daily cooperation.

The NCC conveys the information required in joint operations from the national level to the ICC and vice versa. Additionally, the role of the NCC is to identify solutions to concrete problems in the field of action for cooperation at a national level, either in its own organization or that of other competent authorities.

The presidency state runs an International Coordination Centre (ICC), in replacement of or in addition to the National Coordination Centre (NCC). The ICC coordinates the information exchange with respect to the obligations of the presidency state.

Benelux

Belgium, the Netherlands, and Luxembourg

The Benelux Union is a politico-economic union of three neighboring states Belgium, the Netherlands, and Luxembourg in western Europe. The name Benelux is formed from joining the first two or three letters of each country's name and was first used to name the customs agreement that initiated the union (signed in 1944). It is now used more generally to refer to the geographic, economic and cultural grouping of the three countries.

The main institutions of the Union are the Committee of Ministers, the Benelux Parliament, the Council of the Union and the Secretariat-General, while the Benelux Office for Intellectual Property and the Benelux Court of Justice cover the same territory but are not part of the Union.

The Benelux General Secretariat is located in Brussels. It is the central administrative pillar of the Benelux Union. It handles the secretariat of the Committee of Ministers, the Council of Economic Union and the various committees and working parties.

Belgium and the Netherlands have a close cooperation in Maritime Situational Awareness along their coast and shipping lines and are traditional maritime power by geographical position, coastal area, and economic capability. The Netherlands are bordering the North Sea, between Belgium and Germany and are located at mouths of three major European rivers (Rhine, Maas or Meuse, and Schelde). About a quarter of the country lies below sea level, and only about half of the land exceeds one meter above sea level.

Luxembourg is landlocked between the Netherlands and Belgium and the only grand duchy in the world.

The Netherlands are in international environmental agreements party to Air Pollution, Air Pollution-Nitrogen Oxides, Air Pollution-Persistent Organic Pollutants, Air Pollution-Sulfur 85, Air Pollution-Sulfur 94, Air Pollution-Volatile Organic Compounds, Antarctic-Environmental Protocol, Antarctic-Marine Living Resources, Antarctic Treaty, Biodiversity, Climate Change, Climate Change-Kyoto Protocol, Desertification, Endangered Species, Environmental Modification, Hazardous Wastes, Law of the Sea, Marine Dumping, Marine Life Conservation, Ozone Layer Protection, Ship Pollution, Tropical Timber 83, Tropical Timber 94, Wetlands, Whaling.

Netherlands Coast Guard Center

The Netherland Coastguard Center[39] is the operational part of the Coastguard. It is accommodated on the Navy Base in Den Helder. The Ministry of Defense is responsible for the operational command of the Coastguard. It functions as a Communication and Co-ordination Centre and is officially appointed as the Dutch Maritime and Aeronautical Rescue Co-ordination Centre (JRCC). The Director Coastguard is an officer of The Royal Netherlands Navy.

The Communication and Co-ordination Centre is internally also called the Front Office. In a Back Office situated in the building of the Coastguard Center various Law Enforcement Services deliver personnel for this office. In case of incidents the various databases are consulted and the necessary information is combined to perform integral appearance in case of incidents and offenses.

The Netherlands Coastguard is an independent civil organization with own tasks, competences and responsibilities. The NL Coastguard has three main goals:

- A responsible use of the North Sea;
- To provide services that contribute to safety and security at sea;
- Upholding (inter)national laws and duties.

The Netherland Coastguard is controlled by the Coastguard Fourmanship (KW4) that can be seen as a daily management. Chairman of the KW4 is the Director North Sea of the Ministry of Infrastructure and the Environment (the coordinating Ministry for Coastguard affairs). Other members are the Chairman of the Law Enforcement Committee North Sea, the Director Planning and Control of the Royal Netherlands Navy (operational administrator of the Netherlands Coastguard) and the Director Netherlands Coastguard.

Besides that there is a Coastguard Council. The members of this Council are Directors General of the participating departments and it acts as a front office to the Council of Ministers. This Council gives its approval to the yearly Activity Plan and Budget (APB) and the annual Coastguard report.

The Coastguard makes permanent use of the lifeboats of the Royal Netherlands Sea Rescue Organization (KNRM) and we co-operate closely with a number of Traffic Centers at the large harbors (Rotterdam, Amsterdam, Flushing, Scheveningen and Den Helder) and manned lighthouses at Schiermonnikoog, Terschelling and Ouddorp.

The Royal Netherlands Meteorological Institute supplies the weather forecasts and storm warnings for the North Sea. The Coastguard Centre takes care of the transmission of these messages by radio and NAVTEX.

The Netherlands Coastguard carries out 15 tasks for six ministries (seven provision of service

---

[39] https://www.kustwacht.nl/en

tasks and eight law enforcement task:

Ministry of Infrastructure and Water Management

- Directorate-General for Mobility and Transport
- Directorate-General Public works and Water Management

Ministry of Defense

- Royal Netherlands Navy
- Royal Netherlands Marechaussee (Military Police)
- Royal Netherlands Airforce

Ministry of Justice and Security

- Public Prosecution Service
- National Police Force
- National Crisis Centre

Ministry of Finance

- Customs

Ministry of Economic Affairs and Climate Policy

- State Supervision of Mines

Ministry of Agriculture, Nature and Food Quality

- The Food and Consumer Product Safety Authority

From various Government Departments personnel is assigned to the Coastguard:

- Directorate-General Public works and Water Management
- Maritime Police
- Royal Netherlands Navy
- Royal Netherlands Airforce
- Customs
- Royal Netherlands Military Police
- The Food and Consumer Product Safety Authority
- Human Environment and Transport

Belgium Coast Guard Center

The Belgium Coast Guard Center[40] is made up of two centers complementing each other: the Maritime Rescue and Coordination Centre (MRCC) and the Maritime Security Centre Belgium (MIK). The MRCC can be compared to the fire and ambulance brigades on the mainland, while the MIK can be seen as the police and custom services at sea. The MIK takes charge of security at sea and ensures that legislation at sea is not violated. The MRCC is in charge of safety at sea and coordinates rescue operations.

The coast guard center, assisted by the vessels, the helicopters and the surveillance aircraft of the coast guard partners, operates as the eyes and the ears guarding the North Sea. Radar images, nautical charts and meteorological data enable the coast guard center's operators to watch over the North Sea and to prevent accidents.

The MIK has operators of the Navy, the Maritime and River police and Customs working closely together to ensure that legislation is respected at sea as well. The MIK's operators track down illegal activities. These can range from terrorist attacks over human and drug trafficking to forbidden fishery practices and illegal oil spillages. Surveillance by authorized coast guard partners is not a luxury but an absolute necessity.

Organizing emergency response and relief in moments of crisis is the MRCC's responsibility. When a disaster at sea enfolds, the governor of West-Flanders province activates the North Sea Contingency Plan. Together with a specialist committee, the governor coordinates the emergency relief out of the MRCC's crisis room. Both the MRCC and the MIK have a direct radio connection to the government's emergency response center and all units on and above the North Sea offering assistance.

The Coast Guard is made up of an operational branch and an administrative branch. There are three administrative bodies working in close cooperation: the policy-making body, the consultation body and the Coast Guard secretariat.

The Coast Guard and the Coast Guard center can make use of the different coastguard partners, such as Vloot (Fleet), the Search and Rescue (Vrijwillige Blankenbergse Zeereddingsdienst, VBZR) and the Ship Support Technics BVBA, the Seaking-Helicopters of the Ministry of Defence, the surveillance aircraft etc. pp. which forms an operational part of the Coast Guard.

In 2006, regulation concerning emergency planning was renewed and a new Royal Decree was published in the Belgian Government Gazette (16 February 2006). This Decree stipulates that all emergency plans have to have a well-defined structure. Emergency plans provide different phases according to the seriousness of the incident. The organization, coordination and tasks of the different disciplines are described for every phase. In every emergency plan, the tasks of the rescue services are divided amongst five groups.

---

[40] https://kwgc.be/en/content/coast-guard-centre

Brazil (BRA)

Brazil is by geographical position, coastal area, economic and military capability, one of the larger Maritime Entities. Brazil is slightly smaller than the USA, is the largest country in South America and in the Southern Hemisphere. It shares common boundaries with every South American country except Chile and Ecuador. Brazil shares Iguazu Falls, the world's largest waterfalls system, with Argentina and it includes the Arquipelago de Fernando de Noronha, Atol das Rocas, Ilha da Trindade, Ilhas Martin Vaz, and Penedos de Sao Pedro e Sao Paulo.

Brazil is in international environmental agreements party to Antarctic-Environmental Protocol, Antarctic-Marine Living Resources, Antarctic Seals, Antarctic Treaty, Biodiversity, Climate Change, Climate Change-Kyoto Protocol, Desertification, Endangered Species, Environmental Modification, Hazardous Wastes, Law of the Sea, Marine Dumping, Ozone Layer Protection, Ship Pollution, Tropical Timber 83, Tropical Timber 94, Wetlands, Whaling.

Sistema de Informações Sobre o Tráfego Marítimo (SISTRAM)

As a member of SOLAS Convention, Brazil is committed to assist vessels in emergency situation at SAR area under Brazilian responsibility. In order to control the movement of vessels in this area the navy developed an electronic system called <u>Sistema de Informações Sobre o Tráfego Marítimo (SISTRAM)</u>[41] ("*Maritme Traffic information System*"), under the responsibility of the Maritime Traffic Naval Control Command (COMCONTRAM).

Vessels sailing under foreign flags, not chartered to Brazilian owners, are invited to report routes and eventual occurrences while sailing in the Brazilian jurisdictional waters (200 miles from the coast) and obligated to report routes and any eventual navigation facts while sailing in the Brazilian Territory (12 miles from the coast). Vessels sailing with the Brazilian flag and the ones chartered by Brazilian shipowners are obligated to report routes and navigation facts anywhere in the world.

SISTRAM enables the monitoring of domestic and foreign merchant vessels and is divided into two modules, the Data Management Module and Data Module Graphic Presentation. The Data Management Module permits the registration of ships and their respective companies, managing their messages and different changes on the trips made. SISTRAM is similar to the USA's AMVER system and to other systems in the world. These systems allow quick identification of nearby vessels which may be capable of assisting the ship in distress. In addition, these systems can help render urgent medical assistance.

The greater the number of vessels participating in the program, the greater its effectiveness and reliability, thus enhancing the security of those same vessels. Therefore, all merchant ships are invited to participate in the "*SISTRAM-Network*". Brazilian ships participation is mandatory and foreign vessels is voluntary when outside the Brazilian territorial waters, however, whenever is sailing across territorial waters is also mandatory. This participation involves transmission of standard type messages about their planned voyage. These transmissions are free when are made through the Brazilian Coastal Radio Stations Network (RENEC).

The purpose of SISTRAM is to improve SAR efforts within the Brazilian maritime area. This is accomplished by gathering navigational information from participating vessels. This information is then used during a SAR effort to route nearby vessels to the scene. The ability to quickly divert nearby vessels to the scene provides faster response than can be provided from shore and increases the safety of life at sea. Benefits of participating in SISTRAM are:

- Rapid start of SAR operations.
- Designation of nearby ships for quick assistance.
- Urgent medical assistance for vessels without a doctor.

---

[41] http://www.sistram.mar.mil.br

ships of all countries are invited to participate in SISTRAM. To participate, the ship should send her sailing plan inside Brazilian SAR AREA, even if she is just crossing it bound for another countries' port.

This is accomplished by sending a Sailing Plan to COMCONTRAM (Comando do Controle Naval do Tráfego Marítimo) for each trip. The invitation implies in a voluntary participation for foreign merchant ships while not sailing in Brazilian territorial waters (12NM). Inside those limits the participation is required.

All Brazilian merchant ships are required to participate wherever they are, even in foreign waters with some basic types of messages like Sailing Plan, Position Report, Deviation Report and a Final Report.

Canada (CAN)

Canada is by geographical position, coastal area, economic and military capability. one of the larger Maritime Entities. Canada is the second-largest country in world after Russia, the world's largest country that borders only one country, and largest in the Americas and is bordering the North Atlantic Ocean on the east, North Pacific Ocean on the west, and the Arctic Ocean on the north, north of the conterminous USA. It has a strategic location between Russia and US via north polar route. Canada has more fresh water than any other country and almost 9% of Canadian territory is water; Canada has at least 2 million and possibly over 3 million lakes - that is more than all other countries combined.

Canada is in international environmental agreements party to Air Pollution, Air Pollution-Nitrogen Oxides, Air Pollution-Persistent Organic Pollutants, Air Pollution-Sulfur 85, Air Pollution-Sulfur 94, Antarctic-Environmental Protocol, Antarctic-Marine Living Resources, Antarctic Seals, Antarctic Treaty, Biodiversity, Climate Change, Desertification, Endangered Species, Environmental Modification, Hazardous Wastes, Law of the Sea, Marine Dumping, Ozone Layer Protection, Ship Pollution, Tropical Timber 83, Tropical Timber 94, Wetlands.

Polar Epsilon Project

Polar Epsilon[42] is a major Canadian Forces project to provide enhanced all-weather day and night surveillance capabilities utilizing imagery from the RADARSAT-2 earth observation satellite. The project includes two new ground stations, one at each coast. Data is primarily used to support military operations, but can also be accessed by other departments or agencies.

The Polar Epsilon Project provides enhanced surveillance capabilities for the Department of National Defense/Canadian Forces (DND/CF), which will improve their ability to act quickly in the event of a crisis at home and overseas.

The project was initially developed to address the need of the CF to improve surveillance capabilities over the Arctic and other large areas of responsibility. Polar Epsilon is a space-based wide area surveillance and support capability that is owned and run by DND. The project, valued at approximately $64.5 million, was approved on May 30, 2005.

The Polar Epsilon Project involves using information from RADARSAT-2 to produce imagery for military commanders in their areas of responsibility during the conduct of operations. This includes the surveillance of Canada's Arctic region, including its ocean approaches, the detection and tracking of foreign vessels, and support to CF operations globally. The Project delivered its Arctic surveillance capability to Canada Command on June 17, 2010, marking another significant step toward strengthening Canada's sovereignty and security in the Arctic. Polar Epsilon's capability to enhance CF situational awareness is due to its ability to provide all-weather day/night surveillance in areas where other sensors are limited or unable to operate.

Polar Epsilon has completed its definition phases and is implementing all capabilities including Arctic Surveillance, Environmental Sensing and Maritime Surveillance. These enhanced capabilities will help Canada exercise our sovereignty in the North and protect our environmental heritage.

The implementation phase of Polar Epsilon began in March 2009, and included the design and construction of two new RADARSAT-2 ground stations, one on the east coast in Masstown, Nova Scotia, and the other on the west coast in Aldergrove, British Columbia. The definition phase for the ground stations was completed by MacDonald Dettwiler and Associates (MDA).

The ground stations will be wholly owned and operated by the Government of Canada and was expected to be operational by late 2010. Completion of the Polar Epsilon project was expected by March 2011 and provided for the Polar Epsilon system.

The advantage of Polar Epsilon is that its imagery can be used for precise cueing and

---

[42] https://en.wikipedia.org/wiki/Polar_Epsilon

location of activities, which allows for a more efficient and cost-effective use of other Canadian military assets, such as patrol aircraft and ships. Polar Epsilon can also be used to survey for oil or water pollution, aircraft or satellite crash sites. The project however, does not have the capability to detect ballistic missiles, nor can it track small vessels or individuals. The data provided by Polar Epsilon is used primarily to support military operations, but will provide significant information to several departments and agencies to support their daily operations.

RADARSAT-2, the satellite from which Polar Epsilon draws its information, is a world leading commercially available radar satellite. The Government of Canada, through the Canadian Space Agency, has invested $445 million in the RADARSAT-2 program. RADARSAT-2 is commercially owned and operated by MDA, a Canadian company headquartered in Richmond, British Columbia. RADARSAT-2 was launched in mid-December 2007.

Polar Epsilon 2 Project

The Polar Epsilon 2 project is part of a 'whole-of-government' effort to maintain and expand Canada's access to a domestic source of space-based Earth observation data. This project will provide mission continuity and deliver advanced capabilities beyond those delivered by the previous Polar Epsilon project.

The project combines the wide-area surveillance and support capability currently offered by Polar Epsilon with the Automatic Identification System data tracking system used on ships. This will allow the Canadian Armed Forces to identify and track vessels from space in near-real-time. The project will greatly enhance the Canadian Armed Forces' ability to detect, identify, and track vessels of interest in Canada's maritime areas, the Arctic region, and in support of expeditionary operations around the world. Polar Epsilon 2 will also allow identification of ships by name, in addition to their radar-detected positions, providing an integrated, near-real-time maritime situational awareness capability.

Polar Epsilon 2 will use data from the next generation of Canadian Earth-observation satellites, known as the RADARSAT Constellation Mission, set to be launched in 2018. The data obtained from the RADARSAT Constellation Mission will be used for various surveillance needs ranging from monitoring of ice flows within Canada's coastal waters, providing surveillance of Canada's ocean approaches; monitoring environmental conditions, such as floods and forest fires; and managing and mapping natural resources in Canada and around the world.

The Canadian Armed Forces rely on many systems to provide surveillance of Canada's maritime approaches, but satellite systems such as the RADARSAT Constellation Mission – combined with advanced ground segments that will be delivered by Polar Epsilon 2 project – will provide critical access to this ultimate high ground. Space-based radar data from the RADARSAT Constellation Mission will provide all-weather, day-and-night active-wide-area-surveillance in areas where other sensors are either unable to operate, or of limited capability. The Government of Canada RCM was replacing RADARSAT 2 as it reached its minimum seven year design life in 2015.

The Government of Canada has awarded MacDonald, Dettwiler and Associates Systems Ltd (MDA) a $48.5 million contract for delivery of the Polar Epsilon 2 system. The contract also includes an option to implement infrastructure to manage data, which if exercised, could increase the total contract value to approximately $63.1 million.

Polar Epsilon 2 is a key enabler in a whole of government approach that builds upon Canada's heritage in space radar to meet Canada's strategic information requirements (defense, security, environment, natural resources, disaster management, etc.). Specifically, PE2 will significantly increase the Canadian Armed Forces (CAF) near-real time situational awareness of activities in Canada's three ocean approaches and through increased surveillance persistence of Canada's Arctic and global areas of interest.

Complementing the SAR payload will be a receiver to pick up Automatic Identification System (AIS) messages. The fusion of AIS and SAR data will enable non-compliant or non-cooperative ships to be more quickly identified as targets of interest to be investigated by other Canadian Forces assets such as maritime patrol aircraft. The PE2 System will consist of two ground stations and a processing facility to exploit the Government of Canada's RADARSAT Constellation Mission (RCM). The capability will provide mission continuity and enhanced space-based surveillance and reconnaissance capabilities beyond those delivered from the PE NRTSD System.

China (CHN)

China is by geographical position, coastal area, economic and military capability one of the largest Maritime Entities. China is bordering the East China Sea, Korea Bay, Yellow Sea, and South China Sea, between North Korea and Vietnam, is the world's fourth largest country after Russia, Canada, and USA and largest country situated entirely in Asia. China strives to become the most influential power, already number three behind the combined Military Power of the European forces and the United States. Due to BREXIT China will take the second place in military forces strength with flying colors. Maritime Situational Awareness is part of the military maritime forces and closely controlled by the state.

The website Global Firepower (GFP)[43] provides since 2006 analytical display of data concerning over 135 modern military powers with the ranking based on each nation's potential war-making capability across land, sea and air fought with conventional weapons. The results incorporate values related to resources, finances, and geography with over 55 different factors ultimately making up the final rankings. In 2018 a total of 136 countries were included in the Global Firepower database ranking China third behind the USA, Russia, China and before India.

China is in international environmental agreements party to Antarctic-Environmental Protocol, Antarctic Treaty, Biodiversity, Climate Change, Climate Change-Kyoto Protocol, Desertification, Endangered Species, Environmental Modification, Hazardous Wastes, Law of the Sea, Marine Dumping, Ozone Layer Protection, Ship Pollution, Tropical Timber 83, Tropical Timber 94, Wetlands, Whaling.

The Maritime Silk Road[44] (also Maritime Silk Route) is the maritime section of the historic Silk Road that connects China to Southeast Asia, Indonesian archipelago, Indian subcontinent, Arabian peninsula, Somalia and all the way to Egypt and finally Europe, that flourished between 2nd-century BCE and 15th-century CE.

The trade route encompassed numbers of seas and ocean, including South China Sea, Strait of Malacca, Indian Ocean, Gulf of Bengal, Arabian Sea, Persian Gulf and the Red Sea. The maritime route overlaps with historic Southeast Asian maritime trade, Spice trade, Indian Ocean trade and after 8th century, the Arabian naval trade network. The network also extend eastward to the East China Sea and the Yellow Sea to connect China with the Korean Peninsula and the Japanese archipelago. On May 2017, experts from various fields have held a meeting in London to discuss the proposal to nominate "*Maritime Silk Route*" as a new UNESCO World Heritage Site.

China has been extensively building artificial islands to ensure its maritime influence in the whole Asian region and the international trade routes.

---

[43] https://www.globalfirepower.com
[44] https://en.wikipedia.org/wiki/Maritime_Silk_Road

BeiDou Navigation Satellite System

The BeiDou Navigation Satellite System (BDS)[45] is a Chinese satellite navigation system consisting of two separate satellite constellations. The first BeiDou system, officially called the BeiDou Satellite Navigation Experimental System and also known as BeiDou-1, consists of three satellites which since 2000 has offered limited coverage and navigation services, mainly for users in China and neighboring regions. Beidou-1 was decommissioned at the end of 2012.

The second generation of the system, officially called the BeiDou Navigation Satellite System (BDS) and also known as COMPASS or BeiDou-2, became operational in China in December 2011 with a partial constellation of 10 satellites in orbit. Since December 2012, it has been offering services to customers in the Asia-Pacific region.

In September 2003, China intended to join the European Galileo positioning system project and was to invest €230 million (USD296 million, GBP160 million) in Galileo over the next few years. At the time, it was believed that China's BeiDou navigation system would then only be used by its armed forces. In October 2004, China officially joined the Galileo project by signing the Agreement on the Cooperation in the Galileo Program between the "*Galileo Joint Undertaking*" (GJU) and the National Remote Sensing Centre of China (NRSCC).

Based on the Sino-European Cooperation Agreement on Galileo program, China Galileo Industries (CGI), the prime contractor of the China's involvement in Galileo programs, was founded in December 2004. By April 2006, eleven cooperation projects within the Galileo framework had been signed between China and EU. However, the Hong Kong-based South China Morning Post reported in January 2008 that China was unsatisfied with its role in the Galileo project and was to compete with Galileo in the Asian market.

In 2015, China started the build-up of the third generation BeiDou system (BeiDou-3) in the global coverage constellation. The first BDS-3 satellite was launched on 30 March 2015. As of January 2018, nine BeiDou-3 satellites have been launched. BeiDou-3 will eventually consist of 35 satellites and is expected to provide global services upon completion in 2020.

When fully completed, BeiDou will provide an alternative global navigation satellite system to the United States owned Global Positioning System (GPS), the Russian GLONASS or European Galileo systems and is expected to be more accurate than these. It was claimed in 2016 that BeiDou-3 will reach millimeter-level accuracy (with post-processing), which is ten times more accurate than the finest level of GPS.

According to China Daily, in 2015, fifteen years after the satellite system was launched, it was generating a turnover of $31.5 billion per annum for major companies such as China Aerospace Science and Industry Corp, AutoNavi Holdings Ltd, and China North Industries

---

[45] https://en.wikipedia.org/wiki/BeiDou_Navigation_Satellite_System

Group Corp.

In November 2014, Beidou became part of the World-Wide Radionavigation System (WWRNS) at the 94th meeting of The International Maritime Organization (IMO) Maritime Safety Committee, which approved the "*Navigation Safety Circular*" of the Beidou Navigation Satellite System (BDS). On 27 December, 2018, BeiDou Navigation Satellite System (Beidou-3) officially began to provide global services.

China will continue to expands its capabilities in satellite navigation for the Forces of the People's Republic of China and to ensure independence from the American and European satellite navigation systems.

Maritime Safety Administration of the People's Republic of China (CMSA)

The Maritime Safety Administration of the People's Republic of China (CMSA)[46] or Zhōnghuá Rénmín Gònghéguó Hǎishìjú is a government agency which administers all matters related to maritime and shipping safety, including the supervision of maritime traffic safety and security, prevention of pollution from ships, inspection of ships and offshore facilities, navigational safety measures (including Search and Rescue, Aids to Navigation and the GMDSS), administrative management of port operations, and law enforcement on matters of maritime safety law. It was also responsible for marine accident investigation.

CMSA was formed In October 1998 by the merger of the China Ship Inspection Bureau and the China Port Supervision Bureau into a comprehensive agency of maritime affairs, subordinate to the Ministry of Transport of the People's Republic of China. The China MSA was the only maritime administrative agency that was not merged into the new China Coast Guard (CCG) in June 2013. The CMSA retains its safety and control ("*traffic police*") remit, while the new CCG concentrates all other law enforcement and policing duties. The agency is organized into the following structure:

- Major Functions
- Safety Management of Shipping Company
- Survey of Ships
- Flag State Control
- Port State Control
- Prevention of Pollution from Ships
- Safe Carriage of Dangerous Goods
- Training, Examination and Certification of Seafarers
- Seafarers' Passports
- Aids to Navigation
- Hydrographic Survey
- Marine Traffic Control
- China Ship Reporting System (CHISREP)
- Navigational Warnings and Notices
- Vessel Traffic Service
- Maritime Search and Rescue
- Marine Accident Investigation
- Education and Training
- International Cooperation
- Law Enforcement

---

[46] https://en.wikipedia.org/wiki/China_Marine_Surveillance

CMSA is headquartered in Dongcheng District, Beijing while the MSA offices operate primarily along the PRC coastline and Yangtze River, Pearl River and Heilongjiang Rivers. The CMSA maintains 20 Regional CMSA offices, one per coastal province, under which 97 local branches have been established. Regional CMSA offices are:

- Changjiang MSA
- Fujian MSA
- Guangdong MSA
- Guangxi MSA
- Hainan MSA
- Hebei MSA
- Heilongjiang MSA
- Jiangsu MSA
- Liaoning MSA
- Shandong MSA
- Shanghai MSA
- Shenzhen MSA
- Tianjin MSA
- Zhejiang MSA

The MSA offices have about 25,000 officials, other working staff, operate a patrol force of 1,300 vessels and watercraft of various types. These include 207 patrol vessels of 20 meters and greater length, 2 are 100 meters and above, 2 are 60 meters and above, 18 are 40 meters and above, 59 are 30 meters and above and 126 are 20 meters and above.

China Marine Surveillance (CMS, Zhōngguó Hǎijiān) was a maritime surveillance agency of China. Patrol vessels from China Marine Surveillance are commonly deployed to locations in the South China Sea and East China Sea where China has territorial disputes over islands with its neighbors. The CMS has played a central role in China's increasing assertiveness in the South China Sea, encountering opposition from Japan, the Philippines and Vietnam in the disputed territories, as China tries to lock up natural resources to meet its demands as the world's largest energy consumer. The agency has been disbanded in July 2013 and has now been merged, along other similar three agencies, with the China Coast Guard (CCG)[47].

---

[47] https://en.wikipedia.org/wiki/China_Coast_Guard

Cooperation Council for the Arab States of the Gulf (GCC)

The Cooperation Council for the Arab States of the Gulf[48], originally (and still colloquially) known as the Gulf Cooperation Council (GCC)[49], is a regional intergovernmental political and economic union consisting of all Arab states of the Persian Gulf, except for Iraq. Its member states are Bahrain, Kuwait, Oman, Qatar, Saudi Arabia, and the United Arab Emirates.

All current member states are monarchies, including three constitutional monarchies (Qatar, Kuwait, and Bahrain), two absolute monarchies (Saudi Arabia and Oman), and one federal monarchy (the United Arab Emirates, composed of seven member states, each with its own emir). There have been discussions regarding the future membership of Jordan, Morocco, and Yemen.

A 2011 proposal to transform the GCC into a "*Gulf Union*" with tighter economic, political and military coordination has been advanced by Saudi Arabia, a move meant to counterbalance the Iranian influence in the region. Objections have been raised against the proposal by other countries. In 2014, Bahrain prime minister Khalifa bin Salman Al Khalifa said that current events in the region highlighted the importance of the proposal.

The original union comprised the 1,032,093-square-mile (2,673,110 km^2) of the six Persian Gulf states of Bahrain, Kuwait, Oman, Qatar, Saudi Arabia and the United Arab Emirates. The unified economic agreement between the countries of the Gulf Cooperation Council was signed on 11 November 1981 in Abu Dhabi. These countries are often referred to as "*the GCC states*".

However, Oman announced already back in December 2006 it would not be able to meet the target date. Following the announcement that the central bank for the monetary union would be located in Riyadh and not in the UAE, the UAE announced their withdrawal from the monetary union project in May 2009. The name "*Khaleeji*"[50] has been proposed as a name for this currency. If realized, the GCC monetary union would be the second largest supranational monetary union in the world, measured by GDP of the common-currency area.

The Gulf area has some of the fastest growing economies in the world, mostly due to a boom in oil and natural gas revenues coupled with a building and investment boom backed by decades of saved petroleum revenues. In an effort to build a tax base and

---

[48] https://en.wikipedia.org/wiki/Gulf_Cooperation_Council
[49] http://www.gcc-sg.org/ar-sa/Pages/default.aspx
[50] The term '*Khaleeji*' is Arabic for "*of the Gulf*", and is traditionally associated with Eastern Arabia's Gulf states. The proposed name was turned down in late 2009, and no official name was later agreed prior to the withdrawal of an agreement for a GCC common currency. Although the name "Dinar" (Arabic: دينار, from Latin, denarius) has been suggested since it is already used in the Arab world and is mentioned in the Quran due to a dinar being used during the time of the life of Mohammed.

economic foundation before the reserves run out, the UAE's investment arms, including Abu Dhabi Investment Authority, retain over $900 billion in assets. Other regional funds also have several hundreds of billions of dollars of assets under management.

The region is also an emerging hotspot for events, including the 2006 Asian Games in Doha, Qatar. Doha also submitted an unsuccessful application for the 2016 Summer Olympic Games. Qatar was later chosen to host the 2022 FIFA World Cup, but Qatar was long criticized as host of the game because of its poor human rights records.

Recovery plans have further criticized the crowding out the private sector, failing to set clear priorities for growth, failing to restore weak consumer and investor confidence, and undermining long-term stability.

A common market was launched on 1 January 2008 with plans to realize a fully integrated single market. It eased the movement of goods and services. However, implementation lagged behind after the 2009 financial crisis. The creation of a customs union began in 2003 and was completed and fully operational on 1 January 2015. In January 2015, the common market was also further integrated, allowing full equality among GCC citizens to work in the government and private sectors, social insurance and retirement coverage, real estate ownership, capital movement, access to education, health and other social services in all member states.

However, some barriers remained in the free movement of goods and services. The coordination of taxation systems, accounting standards and civil legislation is currently in progress. The interoperability of professional qualifications, insurance certificates and identity documents is also underway.

In 2014, major moves were taken to ensure the launch of a single currency. Kuwait's finance minister stated that a currency should be implemented without delay. Negotiations with the UAE and Oman to expand the monetary union were renewed. Bahrain, Kuwait, Qatar and Saudi Arabia took major steps to ensure the creation of a single currency.

Kuwait's finance minister said the four members are pushing ahead with the monetary union but said some "*technical points*" need to be cleared. "*A common market and common central bank would also position the GCC as one entity that would have great influence on the international financial system*" he added. The implementation of a single currency and the creation of a central bank is overseen by the Monetary Council.

There is currently a degree to which a nominal GCC single currency already did exist. Businesses trade using a basket of GCC currencies, just as before the euro was introduced, the European Currency Unit (ECU) was long used beforehand as a nominal medium of exchange. Plans to introduce a single currency had been drawn up as far back as 2009, however due to the financial crisis and political differences, the UAE and Oman withdrew their membership.

Companies and investors from GCC countries are active in mergers and acquisitions (M&A). Since 1999, more than 5,200 transactions with a known value of US$573 billion have been announced. They are not only active in national deals or within GCC, but also as important investors in cross-border M&A abroad. The investor group includes in particular a number of Sovereign Wealth Funds.

The Gulf Cooperation Council launched common economic projects to promote and facilitate integration. The member states have cooperated in order to connect their power grids. A water connection project was launched and plans to be partly in use by 2020. A project to create common air transport was also unveiled.

The GCC also launched major rail projects in order to connect the peninsula. The railways are expected to fuel intra-regional trade while helping reduce fuel consumption. Over $200 billion will be invested to develop about 40,000 kilometers of rail network across the GCC, according to Oman's Minister of Transport and Communications. The project, estimated to be worth $15.5 billion, is scheduled to be completed by 2018. *"It will link the six member states as a regional transport corridor, further integrating with the national railway projects, deepening economic social and political integration, and it is developed from a sustainable perspective."* stated, Ramiz Al Assar, Resident World Bank advisor for the GCC.

Saudi Arabian Railways, Etihad Rail, and national governments have poured billions into railway infrastructure to create rail networks for transporting freight, connecting cities and reducing transport times.

The supreme council is the highest authority of the organization. It is composed of the heads of the member states. It is the highest decision-making entity of the GCC. The supreme council sets the vision and the goals of the Gulf Cooperation Council. Decisions on substantive issues require unanimous approval, while issues on procedural matters require a majority. Every member state has one vote.

The Ministerial Council is composed of the Foreign Ministers of all the Member States. It convenes every 3 months. It primarily formulates policies and makes recommendations to promote cooperation and achieve coordination among the member states when implementing ongoing projects. Its decisions are submitted in the form of recommendations to the Supreme Council for its approval. The Ministerial Council is also responsible for preparations of meetings of the Supreme Council and its agenda. The voting procedure in the Ministerial Council is the same as in the Supreme Council.

The Secretariat is the executive arm of the Gulf Cooperation Council. It takes decisions within its authority and implements decisions approved by the Supreme or Ministerial Council. The Secretariat also compiles studies relating to cooperation, coordination, and planning for common action. It prepares periodical reports regarding the work done by the GCC as a whole and regarding the implementation of its own decisions.

The GCC Patent Office was approved in 1992 and established soon after in Riyadh, Saudi Arabia. Applications are filed and prosecuted in the Arabic language before the GCC Patent Office in Riyadh, Saudi Arabia, which is a separate office from the Saudi Arabian

Patent Office. GCC Standardization Organization (GSO)[51] is the standardization organization of the GCC, and Yemen also belongs to this organization.

The Gulf Organization for Industrial Consulting (GOIC) was founded in 1976 by the Gulf Cooperation Council (GCC) member states: The United Arab Emirates, Bahrain, Saudi Arabia, Oman, Qatar and Kuwait, and in 2009, Yemen joined the Organization is headquartered at Doha Qatar. The organization chart of GOIC includes the Board members and the General Secretariat. The Board is formed by member state representatives appointed by their governments.

Amidst the Bahraini uprising, Saudi Arabia and the UAE sent ground troops to Bahrain in order to protect vital infrastructure such as the airport and highway system. Kuwait and Oman refrained from sending troops. Instead, Kuwait sent a navy unit.

In September 2014 GCC members Saudi Arabia, Bahrain, UAE, Qatar plus pending member Jordan, commenced air operations against ISIL in Syria cooperation. Saudi Arabia and the UAE however are among the nations that oppose the Muslim Brotherhood in Syria, whereas Qatar has historically supported it. They also pledged other support including operating training facilities for Syrian rebels (Saudi Arabia) and allowing the use of their airbases by other countries fighting ISIL.

---

[51] https://en.wikipedia.org/wiki/GCC_Standardization_Organization

Council of the Baltic Sea States (CBSS)

The Council of the Baltic Sea States (CBSS)[52] is an overall political forum for regional cooperation consisting of 11 Member States (Denmark, Estonia, Finland, Germany, Iceland, Latvia, Lithuania, Norway, Poland, Russia & Sweden), as well as a representative of the European Union. It supports a global perspective on regional problems including politically and practically translating the UN Sustainable Development Goals, the Paris Climate Agreement, the Sendai Framework on Disaster Risk Reduction, the Palermo Protocol and the UN Convention on the Rights of the Child, into regional actions on the ground.

The Baltic Sea Region is an geographical area where all capitals are located above 52° North. The region spans from Reykjavik to Moscow, from the coastline to the mountain ranges. There is a sense and appreciation that comes from living in a place that enjoys short but light summers, tempered by long and dark winters. In this guide we make the journey from 52 degrees to 65 degrees up north. Here we have collected a number of choices that capture at this very point in time what represents the potential of the region – to present the region as a whole, in a travel guide that has not grouped these countries together before.

Throughout the year the Council of the Baltic Sea States holds a number of international and regional meetings. These culminate in two alternating high-level meetings, either a CBSS Ministerial or a Baltic Sea States Summit. The Summit is held by the Head of Government of the presiding Member State and the Ministerial is held by the Foreign Minister of the respective CBSS Member State. Both of these meetings are also attended by a representative from the European Commission. Here we provide you with an historical overview of our yearly meetings and the communiques and declarations that emanate from them.

In addition, the CBSS functions as a coordinator of a multitude of regional actors in the areas of its three long-term priorities. It is a strength of the organization, that political level agreements can be put into action through concrete projects. Set up in 1992 to ease the transition to a new international landscape, the organization today focuses on themes such as societal security, sustainability, innovation & education, as well as countering human trafficking. One of the greatest achievements has probably been the establishment of the Baltic Sea Region Border Control Cooperation (BSRBCC).

---

[52] http://www.cbss.org

Djibouti Code of Conduct

The Djibouti Code of Conduct is a Code of Conduct, first developed together with the European Union and IMO, concerning the Repression of Piracy and Armed Robbery against Ships in the Western Indian Ocean and the Gulf of Aden, also referred to as the Djibouti Code of Conduct, was adopted on 29 January 2009 by the representatives of Djibouti, Ethiopia, Kenya, Madagascar, Maldives, Seychelles, Somalia, the United Republic of Tanzania and Yemen. Comoros, Egypt, Eritrea, Jordan, Mauritius, Mozambique, Oman, Saudi Arabia, South Africa, Sudan and the United Arab Emirates have since signed bringing the total to 20 countries from the 21 eligible to sign.

On 12. November 2015, the DRTC building was officially opened in Doraleh, Djibouti. The establishment of a regional training center was originally recommended by the 2009 Djibouti Meeting, and the center is intended to play a key role in regional capacity-building initiatives under the Code. IMO and the EU will continue to help the Djibouti Regional Training Center to deliver on its objectives while investing in capacity to implement more programs in the region.

The signatory States to the Code undertook to review their national legislation with a view to ensuring that there are laws in place to criminalize piracy and armed robbery against ships and to make adequate provision for the exercise of jurisdiction, conduct of investigations and prosecution of alleged offenders.

IMO is working closely with the United Nations Office on Drugs and Crime (UNODC) as well as other international organizations and development partners to assess and assist with upgrading national legislation, focusing on empowering States' law-enforcement forces to conduct arrests and criminal investigations.

The Code provides the rules and background for sharing of piracy-related information, through its information sharing network established in 2011 and the Critical Maritime Routes Program (MARSIC)[53] was created to implement the Djibouti Code of Conduct for the suppression of piracy and armed robbery against ships in the Western Indian Ocean and the Gulf of Aden under the IMO and tin support of the European Union.

Three Maritime Information Sharing Centers were established in 2011 to support the program:

- Regional Maritime Rescue Coordination Center (RMRCC) in Mombasa, Kenya
- Maritime Rescue Coordination Center (MRCC) in Dar es Salaam, United Republic of Tanzania
- Regional Maritime Information Sharing Center (ReMISC) in Sana'a, Yemen

---

[53] https://criticalmaritimeroutes.eu/projects/marsic/

The centers are used to exchange information on piracy incidents across the region and other relevant information to help shipping and signatory States to take action to mitigate piracy threats. The Regional Maritime Information Sharing Center (ReMISC) and other centers cooperates for example with various partners like the NATO Shipping Center (NSC), the Maritime Security Center Horn of Africa (MSCHOA), the UK Maritime Trade Organization (UKMTO) and the Maritime Liaison Office (MARLO) which has played a significant role in countering piracy. IMO will continue to support the capacity of the regional network to counter piracy as well as other illicit activities at sea.

IMO is also working to develop signatory States' maritime domain awareness. Projects to increase the use of terrestrial automatic identification systems (AIS), long-range identification and tracking of ships (LRIT), coastal radar and other sensors and systems have been undertaken and continue to be implemented.

IMO has been working with partners to boost the capacity of states in the Western Indian Ocean and Gulf of Aden region to suppress piracy by supporting development of maritime infrastructure, law enforcement and implementation of the Djibouti Code of Conduct. In its endeavor to strengthen capacity in the region IMO has signed five strategic partnerships with UN agencies and the EU. These joint agreements to combat piracy reaffirm the mutual commitments to improving coordination at all levels and across all relevant programs and activities, with a view to strengthening the capacity of States in the region to deal with piracy, as well as to help develop viable and sustainable alternatives to piracy.

Since it was signed in 2009, the Code has evolved to be the major focus for facilitating transnational communication, coordination and cooperation within the region, creating a basis for technical cooperation between the signatory States, IMO and international partners that is trusted, effective and popular.

IMO continues to support Member States to implement the Djibouti Code of Conduct through its Integrated Technical Cooperation Program (ITCP) and activities funded by the Djibouti Code Trust Fund. It also maintains a presence in the region, focused on the Code, with two staff members based in Nairobi, Kenya, whose primary role is training.

Economic Community of Central African States (ECCAS)

Africa represents one of the world's fastest growing numbers in population while the industrial and economic growth is rather slow compared to other regions and continents.

The Economic Community of Central African States (ECCAS)[54] is an Economic Community of the African Union for promotion of regional economic co-operation in Central Africa. It *"aims to achieve collective autonomy, raise the standard of living of its populations and maintain economic stability through harmonious cooperation"* with its members Cameroon, Equatorial Guinea, São Tomé e Príncipe, Gabon, Congo-Brazzaville, Democratic Republic of Congo, and Angola.

The Customs and Economic Union of Central Africa was established by the Brazzaville Treaty in 1964 and formed a customs union with free trade between members and a common external tariff for imports from other countries. The treaty became effective in 1966 after it was ratified by the then five member countries, Cameroon, the Central African Republic, Chad, the Republic of Congo, and Gabon. Equatorial Guinea joined the Union on 19 December 1983. It signed a treaty for the establishment of an Economic and Monetary Community of Central Africa (CEMAC) to promote the entire process of sub-regional integration through the forming of monetary union with the Central Africa CFA franc as a common currency; it was officially superseded by CEMAC in June 1999 (through agreement from 1994). To date CEMAC has not achieved its objective of creating a customs union.

At a summit meeting in December 1981, the leaders of the UDEAC agreed in principle to form a wider economic community of Central African states. ECCAS was established on 18 October 1983 by the UDEAC members, São Tomé and Príncipe and the members of the Economic Community of the Great Lakes States (CEPGL established in 1976 by the DR Congo, Burundi and Rwanda). Angola remained an observer until 1999, when it became a full member.

ECCAS began functioning in 1985, but was inactive for several years because of financial difficulties (non-payment of membership fees by the member states) and the conflict in the Great Lakes area. The war in the DR Congo was particularly divisive, as Rwanda and Angola fought on opposing sides. ECCAS has been designated a pillar of the African Economic Community (AEC), but formal contact between the AEC and ECCAS was only established in October 1999 due to the inactivity of ECCAS since 1992 (ECCAS signed the Protocol on Relations between the AEC and the regional blocs (RECs) in October 1999). The AEC again confirmed the importance of ECCAS as the major economic community in Central Africa at the third preparatory meeting of its Economic and Social Council (ECOSOC) in June 1999.

---

[54] http://www.ceeac-eccas.org

The 10th Ordinary Session of Heads of State and Government took place in Malabo in June 2002. This Summit decided to adopt a protocol on the establishment of a Network of Parliamentarians of Central Africa (REPAC) and to adopt the standing orders of the Council for Peace and Security in Central Africa (COPAX), including the Defense and Security Commission (CDC), Multinational Force of Central Africa (FOMAC) and the Early Warning Mechanism of Central Africa (MARAC). Rwanda was also officially welcomed upon its return as a full member of ECCAS.

On 24. January 2003, the European Union (EU) concluded a financial agreement with ECCAS and CEMAC, conditional on ECCAS and CEMAC merging into one organization, with ECCAS taking responsibility for the peace and security of the sub-region through its security pact COPAX. CEMAC is not one of the pillars of the African Economic Community, but its members are associated with it through Economic Community of Central African States. The EU had multiple peacekeeping missions in the DR Congo: Operation Artemis (June to September 2003), EUPOL Kinshasa (from October 2003) and EUSEC DR Congo (from May 2005).

The 11th Ordinary Session of Heads of State and Government in Brazzaville during January 2004 welcomed the fact that the Protocol Relating to the Establishment of a Council for Peace and Security in Central Africa (COPAX) had received the required number of ratifications to enter into force. The Summit also adopted a declaration on the implementation of NEPAD in Central Africa as well as a declaration on gender equality.

On September 23, 2009, pursuant to Presidential Determination 2009-26 and as published in the Federal Register / Vol. 74, No. 183 (Presidential Documents 48363) ECCAS was made eligible under the U.S. Arms Export Control Act for the furnishing of defense articles and defense services. This makes the ECCAS organization and (theoretically) the countries under their charter eligible for U.S. Foreign Military Sales Program (i.e. government to government sales and assistance) pursuant to the Arms Export Control Act and for other such U.S. assistance as directed by a USG contract to U.S. industry for such support pursuant to the (ITAR).

In 2007, Rwanda decided to leave the organization in order to remove overlap in its membership in regional trade blocks and so that it could better focus on its membership in the EAC and COMESA. Rwanda was a founding member of the organization and had been a part of it since 18 October 1981.

ECCAS had formulated a maritime security strategy in October 2009, focusing on the creation of a Regional Coordination Center for the Maritime Security of Central Africa, the Centre Régional De Sécurité Maritime de l'Afrique Centrale (CRESMAC), tasked with pooling military and civilian skills in member countries and, promoting synergy between the Gulf of Guinea Commission and ECOWAS.

Centre Régional De Securité Maritime de l'Afrique Centrale (CRESMAC)

The Centre Régional De Securité Maritime de l'Afrique Centrale (CRESMAC)[55] was an agreement signed by the participating Chief-of-States of the Central African Countries in 2009 in Kinshasa and the Center was inaugurated in 2014, located in Pointe-Noire, Congo which was at the time the only force provider for the initiative.

The CRESMAC is a multinational and multifunctional body composed of civil servants and from military administrations of the Member States having competence in relation to the marine environment:

- the French Navy,
- the Merchant Navy and ports,
- the Fisheries Administration,
- the Police and the Gendarmerie at borders,
- Customs,
- the Administration of the environment,
- the Shippers Council.

The CRESMAC is a Community-military body for strategic and operational coordination under the supervision of the General Secretariat of the economic community of the States of Central Africa (ECCAS) and its headquarters are located in Pointe Noire, Republic of the Congo since the 20 October 2014.

The CRESMAC center is in Pointe-Noire in the Republic of the CONGO and has for mission to ensure, at the strategic level, control of maritime space of States of ECCAS of the Gulf of Guinea by:

- the protection of natural resources and areas of maritime fishing,
- the security of the maritime routes
- illegal immigration,
- drug trafficking,
- the illegal circulation of small arms,
- piracy and the taking of hostages at sea,
- marine pollution,
- the ships under standard,
- and any other task necessary for the implementation of the strategy.

The CRESMAC ensures the coordination of the control of the maritime space that extends from the border between Cameroon and NIGERIA to the North and between Angola and

---

[55] http://cresmacpointenoire.org

Namibia to South and includes the exclusive economic zones of States parties. For the performance of its tasks, the CRESMAC is at the top of an operational architecture to three levels:

- at the regional level, the CRESMAC,
- at the level of maritime zones, multinational coordination centers (CMC) responsible for planning and implementing operational and tactical,
- at the national level, Navy operational centers (COM) loads of the implementation and the tactical coordination of the national forces.

Community contingencies control distribution organic and functional of the CRESMAC by contributor whose start organization chart consists of eight offices and a central monitoring station (flow chart below). This chart should evolve 2017 in a Direction with four services (administration and finance services, operations, maritime and Logistics Administration) in accordance with the ramp-up plan.

The maritime space of central Africa was first divided into zone A, zone B, and zone D which have three Multinational Centers of Coordination (CMC) under the command of the Regional Maritime Security Center of Central Africa (CRESMAC) in Douala in Cameroon (zone D) and in ANGOLA (zone A, planned). The operational zones A and B and D are along the coasts of:

- Angola
- Burundi
- Cameroon
- Central African Republic
- Chad
- Democratic Republic of the Congo
- Equatorial Guinea
- Gabon
- Republic of the Congo
- Sao Tome and Principe

Multinational Center of Coordination (CMC) in Duoala

The Ministers of Defense of the four countries of zone D signed on the 06th May 2009 in Yaounde, a technical agreement in order to enable the navies of the zone D to control its waters, protect vital interests of the different countries in the aim to facilitate free circulation of persons and goods in this zone.

The Multinational Center of Coordination (CMC) in Douala, Cameroun, also known as the Maritime Coordination Center (MCC), is situated in the Naval Base with logistic support provided by the host nation. The base port for the patrol boat on duty is Malabo, with facilities offered by Equatorial Guinea for lodging the crew. Manning and staffing costs are covered by the member states. Douala and is linked to the Regional Staff of ECCAS in Libreville and supported in terms of logistic by the joint chief of staff of Cameroon.

The first launching of joint patrols held in Malaboon in September 2009 with the Patrol boats CABOSANJUAN of Equatorial Guinea, the AKWAYAFE of Cameroon, and the BETSENG of Gabon in the maritime space shared by Cameroon, Equatorial Guinea, Gabon, and Sao Tome and Principe.

The zone D has been firstly activated in response to the growing piracy in the area since 2006. The Multinational Center of Coordination of the zone D includes:

- Chief of the CMC (Cameroonian Navy Officer)
- An officer in charge of operations and logistics (Gabonese)
- An intelligence officer (Equatorial Guinea)
- An officer in charge of transmission (Sao Tome and Principe)

Economic Community of West African States (ECOWAS)

The Economic Community of West African States (ECOWAS)[56] is a regional group of fifteen West African countries. Founded on 28 May 1975, with the signing of the Treaty of Lagos, its mission is to promote economic integration across the region.

Considered one of the pillars of the African Economic Community, the organization was founded in order to achieve "collective self-sufficiency" for its member states by creating a single large trading bloc through an economic and trading union. It also serves as a peacekeeping force in the region. The organization operates officially in three co-equal languages French, English, and Portuguese.

The ECOWAS consists of two institutions to implement policies, the ECOWAS Commission and the ECOWAS Bank for Investment and Development, formerly known as the Fund for Cooperation until it was renamed in 2001.

A few members of the organization have come and gone over the years. In 1976 Cape Verde joined ECOWAS, and in December 2000 Mauritania withdrew, having announced its intention to do so in December 1999.

ECOWAS has similar ambitions as the Economic Community of Central African States (EC-CAS) in fostering cooperation in maritime affairs amongst its members; however it has not achieved similar goals as i.e. establishing the security centers.

ECOWAS has operational zones E (Togo, Benin, Nigeria, Niger), F (Guinea, Sierra Leone, Liberia, Côte d'Ivoire, Ghana, Burkina Faso) and G (Senegal, The Gambia, Guinea Bissau, Cape Verde, Gambia, Mali).

---

[56] https://en.wikipedia.org/wiki/Economic_Community_of_West_African_States

Regional Coordination Center for the Maritime Security of West Africa (CRESMAO)

Zone E was the initial area created in September 2012 in Lomé (Togo) including Nigeria, Benin, Togo and Niger coastal waters. A Regional Coordination <u>Center for the Maritime Security of West Africa (CRESMAO)</u>[57] will be established in Côte d'Ivoire, Abidjan, and will be served by a CMC which was opened on 13th March 2015 in Cotonou (Benin), where the INTERPOL Regional Bureau of the West African Police Information System (WAPIS) is already operational.

Here is the Center for Information and Communication (CINFOCOM) planned and the Maritime Trade Information Sharing Center - Gulf of Guinea (MTISC-GOG) has already an Initial Operating Capability established.

In accordance with the desire to bolster regional Maritime Security Architecture, the Commission of the Economic Community of West African States (ECOWAS) has taken delivery of a range of technical equipment in 2018 from the government of Germany. Consisting of 63 high tech communication, surveillance, monitoring, conferencing, visualization, logistics items and other accessories, the equipment including their components were handed over to the regional organization during a ceremony which took place at the West Africa Maritime Regional Safety Center (CRESMAO).

---

[57] https://cic-gog.org/en/glossary/cresmao/

Center for Information and Communication (CINFOCOM)

The Center for Information and Communication (CINFOCOM)[58] is a specialized division of the Maritime Organization for the West and Central Africa (MOWCA) responsible for the centralization and dissemination of comprehensive information and update on maritime policies, control measures and strategies against piracy and other illegal practices at sea.

The CINFOCOM aims to:

- Foster collaboration among MOWCA and some regional institutions in terms of safety, security and protection of the marine environment,
- Provide national and sub-regional maritime authorities detailed information necessary for the implementation of sectoral policies,
- Serve as an information portal on the actors and auxiliary maritime transport in West and Central Africa,
- Serve patent management platform and certificates of marine MOWCA space;
- Allow the establishment of a maritime public domain mapping in West and Central Africa,
- Promote the creation of a sub-regional database on small boats and the maritime public domain MOWCA space,
- Provide reliable statistics on vessel traffic in West and Central Africa,.

The CINFOCOM mission:

- The networking of all maritime and port authorities of the 25 member countries of MOWCA for the exchange of information relating to maritime safety and security.
- Permanent Monitoring and centralization of information relating to maritime and port area of West and Central Africa.
- Collection platform and continuously in real-time dissemination of information on various maritime, port activities and river-lagoon of the 25 countries of MOWCA.
- To allow the relay urgent information regarding events and flail maritime character.

The CINFOCOM is to achieve:

- The designation of focal points CINFOCOM Angola, Benin, Burkina Faso, Cameroon, Congo, Côte d'Ivoire, Guinea Conakry, Democratic Republic of Congo, Senegal, Mali, and Togo,
- The Organization of three (03) training seminars and capacity building of the focal points,

---

[58] http://cinfocom.omaoc.org

- The development of regional statistics data collection indicators,
- The creation of an integrated regional data management system available in the 25 countries of MOWCA.

Maritime Trade Information Sharing Center - Gulf of Guinea (MTISC-GOG)

The <u>Maritime Trade Information Sharing Center - Gulf of Guinea (MTISC-GOG)</u>[59] provides the regional principle point of contact for the mariner off the coast of West Africa.

MTISC-GoG operates a *"Voluntary Reporting Area"* and provides Maritime Security advice and guidance to the mariner operating in the Voluntary Reporting Area receiving reports and information on suspicious incidents from merchant shipping and shares that information with its regional and national contacts.

The MTISC-GoG information is shared via a website with details on how to use the Voluntary Reporting Area together with general maritime security guidelines for vessels operating in the region (Maritime Security Guidance - Gulf of Guinea "MSG-GoG") issued by MTISC-GoG. The website also provides a summary of maritime security incidents within the voluntary reporting area.

Further website updates will provide *"patterns of life information"* for the region which will help mariners better understand normal maritime activity within the voluntary reporting area to assist them in identifying suspicious activity. Additional, more specific guidance following post incident analysis will also be provided in further website updates.

The Voluntary Reporting Area (VRA) is administered by the MTISC-GoG with a voluntary reporting scheme for merchant vessels operating within a defined VRA. All vessels operating within the VRA are encouraged to register with MTISC-GoG. Registering with MTISC GoG establishes direct contact between the reporting vessel and the MTISC-GoG Center in Ghana. Once direct contact with MTISC-GoG had been established the Center is able to:

- Warn the vessel reporting about maritime incidents which may affect that vessel.
- Receive reports of maritime security incidents and suspicious activity from the reporting vessel.
- Share information provided by the reporting vessel to the appropriate authorities within the region.
- Advise the reporting vessel about suspect vessels in the VRA about which further information including position is sought.
- Assist with other information enquiries from the reporting vessel.

Vessels are also encouraged to report positional information daily while operating in the VRA using a series of reporting forms as detailed in the reporting section of the website.

Maritime Security Guidance - Gulf of Guinea (MSG-GoG) provides advice and guidance to improve the safety of the mariner operating in the VRA. It seeks to enhance awareness

---

[59] http://www.mtisc-gog.org/

of the maritime security situation in the VRA and offer basic guidance on how to mitigate the risks. The MSG-GoG document is located at the "MSG-GoG" section of the website. Gulf of Guinea Interregional Network (GoGIN) was the follow-on project of CRIMGO

Launched in December 2016 by the EU, the Gulf of Guinea Inter-Regional Network (GoGIN)[60] aimed to improve safety and maritime security in the Gulf of Guinea, notably by supporting the establishment of an effective and technically efficient regional information sharing network.

The project covers 19 countries across the Gulf of Guinea, although its initial actions will focus on a pilot area corresponding to zones D and E of the architecture of Yaoundé (extending from Togo to Gabon).

Building on the achievements of CRIMGO, this new four-year project will support the implementation of the Yaoundé Code of Conduct and process. To that end, GoGIN will work to improve regional capacity for dialogue and coordination in the maritime domain.

These objectives were achieved through concrete activities to support inter-sectorial coordination, as well as inter-regional maritime steering and dialogue. The EU initial investment was 7,5M Euro and the project ended on 20. October 2016.

---

[60] https://criticalmaritimeroutes.eu/projects/gogin/

Inter-Regional Coordination Center (ICC/CIC)

In June 2014 the three organizations ECCAS, ECOWAS and GGC signed the additional Protocol to establish the Inter-regional Coordination Center (ICC) on Maritime Safety and Security, which started operations in September 2014 with the Headquarter in Yaoundé, Cameroon. France is partnering nation. This step was taken to enhance the cooperation between the various Center of Coordination (CMC) with their extended areas along the African coast.

The Interregional Coordination Center (ICC)[61] or Centre Interregional de Coordination (CIC) is responsible for the implementation of regional strategy for Maritime Safety and Security in Central and West Africa and coordinates the activities of CRESMAC and CRESMAO

In accordance with the instruments adopted during the Summit of Heads of State and Government of ECCAS, ECOWAS and the GGC on safety and security in the maritime space of Central Africa and West Africa. The Centre is the body in charge of enhancing the activities geared towards cooperation, coordination, and systems interoperability as well as the implementation of the regional strategy on safety and security within the Central and West African common maritime space.

The five division of the Interregional Coordination Centre comprises five divisions:

- the Division of Political Affairs and International Cooperation
- the Division of Information Management and Communications
- the Division of Training and Practice
- the Division of Legal Affairs and Judicial Cooperation
- the Division of Administration and Finance

---

[61] https://cic-gog.org/en/about-the-centre/

European Union (EU)

Europe is by geographical position, coastal area, economic and military capability one of the largest Maritime Entities having over 32.000 km coast lines not counting the European Union (EU)[62] Overseas Countries and Territories (OCT) and Outermost Regions (OMR).

The European Union is a politico-economic union of 28 member states that are located primarily in Europe. It covers an land area of 4,324,782 km^2, with an population of over 508 million. The European Union has by far the greatest concentration of military power in the world and still today the military power of the European forces combined would rank in numbers second after the United States.

The EU operates through a system of supranational institutions and intergovernmental-negotiated decisions by the member states. The institutions are: the European Parliament, elected every five years by EU citizens, the European Council, the Council of the European Union, the European Commission, the Court of Justice of the European Union, the European Central Bank, and the Court of Auditors. Various European Agencies have been established under these institutions in order to deal with the growing tasks and challenges.

The EU has developed a single market through a standardized system of laws that apply in all member states. Within the Schengen Area, passport controls have been abolished. EU policies aim to ensure the free movement of people, goods, services, and capital, enact legislation in justice and home affairs, and maintain common policies on trade, agriculture, fisheries, and regional development. The monetary union was established in 1999 and came into full force in 2002. It is currently composed of 19 member states that use the euro as their legal tender.

The EU traces its origins from the European Coal and Steel Community (ECSC) and the European Economic Community (EEC), formed by the Inner Six countries in 1951 and 1958, respectively. In the intervening years, the community and its successors have grown in size by the accession of new member states and in power by the addition of policy areas to its remit. The Maastricht Treaty established the European Union under its current name in 1993 and introduced European citizenship. The latest major amendment to the constitutional basis of the EU, the Treaty of Lisbon, came into force in 2009.

Covering 7.3% of the world population, the EU in 2014 generated a nominal gross domestic product (GDP) of 18.495 trillion US dollars, constituting approximately 24% of global nominal GDP and 17% when measured in terms of purchasing power parity. Additionally, 26 out of 28 EU countries have a very high Human Development Index, according to the UNDP. In 2012, the EU was awarded the Nobel Peace Prize. Through the Common Foreign and Security Policy, the EU has developed a role in external relations and defense. The union maintains permanent diplomatic missions throughout the world and represents itself at the

---

[62] https://europa.eu/european-union/index_en

United Nations, the WTO, the G8, and the G-20. Because of its global influence, the European Union has been described as a current or as a potential superpower, however the different interest of the member states and the required consensus for decisions make a common decision often difficult. The result is often the smallest common nominator, most of the time not having the required impact on a subject.

Foreign policy co-operation between member states dates from the establishment of the Community in 1957, when member states negotiated as a bloc in international trade negotiations under the Common Commercial policy. Steps for a more wide ranging co-ordination in foreign relations began in 1970 with the establishment of European Political Co-operation which created an informal consultation process between member states with the aim of forming common foreign policies. It was not, however, until 1987 when European Political Cooperation was introduced on a formal basis by the Single European Act. EPC was renamed as the Common Foreign and Security Policy (CFSP) by the Maastricht Treaty.

The predecessors of the European Union were not devised as a military alliance because NATO was largely seen as appropriate and sufficient for defense purposes. 22 EU members are members of NATO while the remaining member states follow policies of neutrality. The Western European Union (WEU), a military alliance with a mutual defense clause, was disbanded in 2010 as its role had been transferred to the EU. The view of the European Union and his members however changed due to the political developments in the US and since then steps have been taken to establish a more harmonized military presence.

The Common Security and defense Policy (CSDP) of the European Union (EU), formerly known as the European Security and defense Policy (ESDP), is a major element of the Common Foreign and Security Policy and is the domain of EU policy covering defense and military aspects, as well as civilian crisis management. The ESDP was the successor of the European Security and Defense Identity under NATO, but differs in that it falls under the jurisdiction of the European Union itself, including countries with no ties to NATO.

On December 12, 2008, the EU Council approved the *"Report on the Implementation of the European Security Strategy"*. Among other decisions, the Report stressed that there is a *"… need to strengthen the (NATO-EU) strategic partnership in service of our shared security interests, with better operational cooperation, in full respect of the decision-making autonomy of each organization, and continued work on military capabilities"*.

In December 2013, the EU Council decided that *"The Common Security and Defence Policy (CSDP) will continue to develop in full complementarity with NATO in the agreed framework of the strategic partnership between the EU and NATO and in compliance with the decision-making autonomy and procedures of each"*.

The above decisions harmonize the existing military potential of two politically and economically similar but yet so different entities, the North Atlantic Treaty Organization and the European Union, enabling both to grow independent military capability while driving for best interoperability.

Since 2002, the European Union has become increasingly active abroad under the Common Security and Defense Policy (CSDP). As of February 2014, it has engaged in 30 operations, using civilian and military instruments in several countries in three continents (Europe, Africa and Asia), whereof fifteen of these operations where at the time currently ongoing, and fifteen completed.

Permanent Structured Cooperation (PESCO)

The next influential political decision in the EU took place ten years after the Lisbon Treaty entered into force, 25 EU member States established the Permanent Structured Cooperation (PESCO), arguably the biggest legal framework innovation within the Common Security and Defense Policy, even regarded the often called "*Sleeping Beauty*" of CSDP.

PESCO was officially established in December 2017 after the Council unanimously agreed. 25 EU countries decided to participate in PESCO: Austria, Belgium, Bulgaria, Czech Republic, Croatia, Cyprus, Estonia, Finland, France, Germany, Greece, Hungary, Italy, Ireland, Latvia, Lithuania, Luxembourg, the Netherlands, Poland, Portugal, Romania, Slovenia, Slovakia, Spain and Sweden. As Denmark has opted out from participation in CSDP projects, Malta and Great Britain are the only two countries that do not take part in PESCO.

PESCO is one of the most interesting CSDP instruments introduced by the 2007 reform, as it allows groups of member States to advance their defense Cooperation in the Council without a veto. However, PESCO remained a paper tiger for ten years due to political differences. It was not until the Bratislava Summit in September 2016 that the possibility of using PESCO to intensify defense Cooperation was recalled. Thanks to an unprecedented effort made by the EU High Representative, Mogherini, PESCO was officially established in December 2017 after the Council unanimously agreed.

However, the increased uncertainty and unpredictability at global level and the negative impact on EU security (especially terrorism) have heightened awareness of the importance of a more coherent EU on security and defense issues. High Representative Mogherini has therefore successfully tried to capitalize on the growing attention by launching a new security strategy based on some simple, yet crucial assumptions:

- The EU countries can no longer protect their citizens and their interests individually. However, a stronger EU at international level has the potential to resolve this issue.

- If the EU is to be a credible international actor, tor, it must pursue an independent foreign policy agenda, including the ability to use military instruments when necessary.

There are at least three driving factors within and one outside the EU for PESCO, the EU Global Strategy (EUGS), its follow-up, and Brexit and the change in the political landscape on the American Continent. Although an EU strategy was already in place (2003, revised in 2008), EU defense integration remained limited due to a lack of political will.

In summary, it can be said that deepened defense Cooperation is seen as a sine qua non in Order to protect the interests of the member States on the one hand and to strengthen the EU's international role on the other.

Since then, member States have agreed on a number of actions to be taken for EUGS' implementation. In fact, PESCO is part of a more ambitious and comprehensive defense package which consists of:

- a Coordinated Annual Review of Defense, to assess member States' defense capabilities and monitor the evolution of their national defense plans,

- a European Defense Action Plan, to sustain common investments through different economic measures: strengthen the Single Market for Defense; enhance SMEs' investments; finance innovative technologies (research window) and the joint purchase of materiel (capability window),

- reframed CSDP structures and procedures, to make civilian and military missions faster and more efficient.

Britain was one of the first sponsors of the EU security and defense policy (Saint Malo agreement, 1998). London has, however, repeatedly then blocked the further Integration of defense within the EU because GBR believes that NATO is the only Institution responsible for the defense of the EU. Britain has preferred bilateral agreements on defense over EU Cooperation. At the political level, Brexit is likely to be an opportunity as Britain can no longer veto further defense integration and finally in 2017, PESCO was founded. The view of NATO as only Institution responsible for the defense of the EU has today vanished due to the global political changes and NATO itself.

This notwithstanding, Brexit creates a factual capability gap in Europe's balance of forces. Britain out of the EU means the loss of one of the two militarily most powerful countries in Europe and one of the two nuclear powers. Therefore, if the EU considers the military element to be a key factor in its foreign policy, it would have to press ahead with Cooperation in order to balance the Brexit, a challenging ambition in times of budgets constraints, especially considering that member States have heterogeneous military capabilities. PESCO could help EU countries to improve defense efficiency, for example by reducing the number of different assets of the same type.

PESCO has a two-layered structure consisting of the Council and project levels. The Council is responsible for general policy-making and decision-making. Resolutions are adopted unanimously, with the exception of those concerning suspension of membership or new memberships (qualified majority). Each project is managed exclusively by the participating countries (the project level) in accordance with the general project management rules of the Council, which is responsible for the unity, consistency and effectiveness of PESCO. As PESCO is part of the CSDP, the institutions/agencies involved in EU defense policy will also play a role in PESCO.

The High Representative is involved in PESCO proceedings and reports annually to the Council about PESCO implementation and the participants' commitment to projects. The European External Action Service (EEAS) and the EU Military Staff provide their support as point of contact for PESCO and assess the projects' compliance with their operational needs.

The European Defense Agency (EDA) provides support to the High Representative for reporting activities and will be in charge of avoiding unnecessary duplication with other initiatives, thus facilitating capability development projects. PESCO will be closely connected to the other defense efforts, in particular, it will include capability projects identified according to CARD priorities, and eligible projects could become European Defense Fund recipients. Funds could cover up to 20% of the costs for capability projects, and up to 30% for prototypes. Participating States remain contributors to PESCO projects, but these could also receive contributions from the EU general budget in accordance with the treaties and other existing EU instruments.

The three projects dedicated to the maritime environment focus on enhancing maritime security in national and international waters. The Harbor Maritime Surveillance and Protection (HARMSPRO) proposes the Integration of maritime sensors, Software and platforms in one system to ameliorate security and safety of maritime traffic and support. The Upgrade of maritime surveillance is more focused on land, maritime and aerial Systems' Integration and a better information exchange. The Maritime (semi-)Autonomous Systems for Mine Countermeasures (MAS MCM) aim at developing a mix of (semi-)autonomous underwater, surface and aerial technologies for maritime mine countermeasures.

PESCO is still under development but has already domestic (CSPD) and international (EU-NATO) impact at the political and military level. PESCO will likely have more positive impact on CSDP at the political level. EU members have finally decided to reach a higher level of defense Cooperation through binding measures. This means that they now consider cooperation as a viable solution to face defense and security threats. If member States will maintain their commitment, PESCO will probably give the EU defense the impetus it has always demanded, but never achieved. PESCO projects could represent a potential evolution in European thinking of defense, as they are concrete steps towards integration, something that has been missing so far. Furthermore, other projects will likely be added in the months ahead. In general, the use of PESCO tools trigger further integration in the defense sector.

Conversely, PESCO will likely have a limited impact on military integration overall. As repeatedly remarked by the High Representative Mogherini, PESCO does not intend to establish an singular EU military force. Rather, it is an instrument to enhance security within and outside the EU, for example, PESCO participants commit themselves to make available formations, but this commitment "*does neither cover a readiness force, a Standing force nor a stand by force.*"

Furthermore, PESCO might be far too inclusive to have a real impact on overall military capacity. Inclusiveness has been a disputed point between Germany and France. Berlin wanted a broader PESCO in order to develop more EU defense capabilities to be used for external missions, mainly under the civilian aspect. Paris would have preferred a smaller number of participants and a deeper impact on EU military defense. The French idea was to use PESCO projects to build some EU military capabilities and gain strategy independence, especially from the US and NATO.

In addition, as member States participate on a voluntary basis, a change of government in one or more EU countries could have a negative impact on PESCO implementation. PESCO projects and their structure and goals will undoubtedly ameliorate EU capabilities in the defense domain.

As PESCO is not a first step towards an EU military force, its impact on the EU-NATO relationship will likely be limited in political terms, will not rival NATO for what concerns EU defense. Rather, being a credible military actor will help the EU to pursue its own international agenda, establish a stronger foreign policy agenda and develop its independent military capabilities accordingly (especially more ambitious and military-focused PESCO projects).

PESCO could potentially have an impact on EU-NATO relations, particularly as regards the EU's contribution to NATO. According to the PESCO factsheet, "*military capacities developed within PESCO remain in the hands of member States, which can also make them available in other contexts such as NATO or the UN.*" As the political priorities of NATO and the EU could diverge, capability programs can vary accordingly. This means that countries that are members of the EU and NATO may in the medium and long term have to choose the political agenda they want to follow and plan their economic and military contributions accordingly.

From the NATO point of view, PESCO is an opportunity to reinforce the Alliance, rather than a danger to its stability. In fact, a stronger EU defense means that EU countries could provide NATO with new capabilities by sharing burdens within the alliance. PESCO will therefore ameliorate complementarity, rather than promoting competition. Military mobility is the most compelling project in this sense, as NATO will be the primary recipient of its implementation. The Alliance faces several difficulties in assets/personnel movements within the EU, especially for the lack of adequate infrastructures. Increased mobility will enable EU countries to better serve NATO missions and improve mobility of US personnel and assets within the EU. Other projects, such as the establishment of an EU medical command and a network of logistics hubs, might also serve NATO missions.

PESCO could be a potential game changer for EU defense in the long run. However, its current goals and projects do not ameliorate EU defense capabilities. Consequently, PESCO has a high political impact, but a limited military one. From a NATO perspective, PESCO serves the Alliance's agenda, as its limited military impact keeps EU countries mainly dependent on NATO capabilities. PESCO is very inclusive and projects are aimed at strengthening the EU's Strategic independence; however, EU members are divided in terms of foreign and defense priorities.

The European maritime area is a common capital asset with regard to resources, security and ultimately prosperity of the Member States under CSDP and PESCO. There is a clear need to share maritime surveillance information. Developing the necessary means to allow for such data and information exchange should enhance the different users' situational awareness and enhance the maritime picture. Such improved situational awareness will increase the efficiency of Member States' authorities and improve cost effectiveness.

European Union Maritime Security Strategy (EUMSS)

The European Union Maritime Security Strategy[63] (EUMSS) for the global maritime domain, adopted by the European Council in June 2014, is a joint EU plan to improve the way in which the EU pre-empts and responds to the maritime challenges. It is an overarching maritime security strategy against all challenges from the global maritime domain that may affect people, activities or infrastructures in the EU, and is built upon closer collaboration within the EU, across the regional and national levels. It seeks to increase awareness and ensure higher efficiency of operations.

The EU's maritime security strategy action plan was first adopted on 16 December 2014 to help safeguard the interests of the EU and protect its member states and citizens. It addresses global maritime risks and threats, including cross-border and organized crime, threats to freedom of navigation, threats to biodiversity, illegal, unreported and unregulated fishing or environmental degradation due to illegal or accidental discharge.

A second objective is to protect EU maritime interests worldwide. The EUMSS strengthens the link between internal and external security, and couples the overall European Security Strategy with the Integrated Maritime Policy. The EUMSS is complemented with an Action Plan, a list of over 130 specific actions and a timeframe to drive the implementation of the EUMSS forward. All maritime security stakeholders in the EU – across sectors and borders – are called upon to participate directly in a cooperative setting.

The revision in 2018 adopted allows for a more focused reporting process to enhance awareness and better follow-up to the strategy. The action plan brings together both internal and external aspects of the Union's maritime security. The actions foreseen in the plan also contribute to the implementation of the EU Global Strategy, the renewed EU internal security strategy 2015-2020, the Council conclusions on global maritime security, and the joint communication on international ocean governance.

The EUMSS serves as a comprehensive framework, contributing to a stable and secure global maritime domain, in accordance with the European Security Strategy (ESS), while ensuring coherence with EU policies, in particular the Integrated Maritime Policy (IMP), and the Internal Security Strategy (ISS). The EUMSS Action Plan reflects a comprehensive and cross-sectoral approach to the maritime domain, integrating both the internal and external aspects of maritime security.

The EUMSS Action Plan is organized around five key areas of cooperation, chosen after a comprehensive and forward-looking analysis of the threats and challenges affecting maritime security: External Action; Maritime Awareness, Surveillance and Information Sharing; Capability Development; Risk Management, Protection of Critical Infrastructures and Crisis Response; Maritime Security Research and Innovation, Education and Training.

---

[63] https://ec.europa.eu/maritimeaffairs/policy/maritime-security_en

Additional benefits:

- The Baltic Sea Maritime Incident Response project (BSMIR) analyzed the level of preparedness of the eight Baltic Sea States plus Norway and Iceland vis-à-vis large-scale and multi-sectorial maritime accidents. This nine-month-long project resulted in a final report including suggestions for international cooperation.

- The European Coast Guard Functions Academy Network (ECGFA NET) strengthens international collaboration on training. It set up a network of training institutions for coastguard functions, which led to a shared qualifications framework. Run by 12 EU coastguard agencies, it has so far managed to engage 37 training institutions.

- The large amount of regional cooperation initiatives reported by EU agencies and Member States, such as MARSUR, FRONTEX operations in the Mediterranean, Baltic Sea Regional Border Control Cooperation (BSRBCC) and Black Sea Security initiatives, including promising collaboration with third nations and international forums.

Integrated Maritime Policy (IMP)

In its communication on the Integrated Maritime Policy (IMP) for the European Union, the European Commission undertook to *"take steps towards a more interoperable surveillance system to bring together existing monitoring and tracking systems used for maritime safety and security, protection of the marine environment, fisheries control, control of external borders and other law enforcement activities."*

The European Commission (EC)[64] is the executive of the European Union and promotes its general interest. Under the European Commission the Directorate-General Maritime Affairs and Fisheries (DG MARE) is responsible for EU policy on maritime affairs and fisheries. DG MARE is responsible for the development of the Integrated Maritime Policy (IMP)[65] which seeks to provide a more coherent approach to maritime issues, with increased coordination between different policy areas. It focuses on:

- Issues that do not fall under a single sector-based policy e.g. *"blue growth"* (economic growth based on different maritime sectors).

- Issues that require the coordination of different sectors and actors e.g. marine knowledge.

The EU Integrated Maritime Policy (IMP) has established itself as new approach to enhance the optimal development of all sea-related activities in a sustainable manner. It has confirmed the vision that, by joining up policies towards seas and oceans, Europe can draw much higher returns from them with a far lesser impact on the environment.

EU institutions, Member States and regions have set-up governance structures to ensure that policies related to the seas are no longer developed in isolation and take account of connections and synergies with other policy areas. Stakeholders have confirmed the considerable interest shown during the broad consultation process of 2006-07, establishing the IMP as a particularly bottom-up driven policy of the European Union. Cross-sectoral tools such as maritime spatial planning, integrated surveillance or marine knowledge have registered tangible progress and should lead to substantial improvements in the way we manage our oceans.

EU sectoral policies with a bearing on our seas and coasts, like fisheries, transport, environment, energy, industry or research policy, have all taken substantial moves in the direction of greater integration and consistency. The Commission has also made first steps to implement the IMP on a regional basis.

In short, the EU IMP is changing the way Europeans look at their seas and oceans and

---

[64] https://ec.europa.eu/commission/index_en
[65] https://ec.europa.eu/maritimeaffairs/policy_en

reaffirmed the strategic importance of the continent's seas and coastal regions. IMP covers cross-cutting policies:

- Blue growth[66]
- Marine data and knowledge[67]
- Maritime spatial planning[68]
- Integrated Maritime Surveillance[69]
- Sea basin strategies[70]

The IMP seeks to coordinate policies on specific maritime sectors and to:

- take account of the inter-connectedness of industries and human activities centered on the sea. Whether the issue is shipping and ports, wind energy, marine research, fishing or tourism, a decision in one area can affect all the others. For instance, an off-shore wind farm may disrupt shipping, which in turn will affect ports.
- save time and money by encouraging authorities to share data across policy fields and to cooperate rather than working separately on different aspects of the same problem.
- build up close cooperation between decision-makers in the different sectors at all levels of government, national maritime authorities, regional and local authorities, and international authorities, both inside and outside Europe. Many countries are recognizing this need and move towards more structured and systematic collaboration.

Another product of the DG Mare is the European Atlas of the Seas[71] is an easy and fun way for professionals, students and anyone interested to learn more about Europe's seas and coasts, their environment, related human activities and European policies. It was developed to raise awareness of Europe's oceans and seas, in the context of the EU's integrated maritime policy.

The European Atlas of the Seas is an easy to use and interactive web-based atlas on the coasts and seas within and around Europe. A new version of the Atlas has been released on the 11. June 2018, offering new features and more content. It is freely accessible on the

---

[66] https://ec.europa.eu/maritimeaffairs/policy/blue_growth
[67] https://ec.europa.eu/maritimeaffairs/policy/marine_knowledge_2020
[68] https://ec.europa.eu/maritimeaffairs/policy/maritime_spatial_planning
[69] https://ec.europa.eu/maritimeaffairs/policy/integrated_maritime_surveillance
[70] https://ec.europa.eu/maritimeaffairs/policy/sea_basins
[71] https://ec A new Maritime Spatial Planning Project added to the European Atlas of the Seas with the addition of a new project, SIMCelt, a cross-border project involving partners from the UK, Ireland and France. It aims to support cooperation between Member States on the implementation of the Maritime Spatial Planning Directive in the Celtic Seas. .europa.eu/maritimeaffairs/atlas_en

internet. Next to the English version, it will be consultable in all languages of the European Union, thanks to automatic e-Translation. The atlas offers a remarkably diverse range of information about Europe's seas, for example:

- Sea depth and underwater features
- Coastal regions geography and statistics
- Blue energies and maritime resources
- Tide amplitude and coastal erosion
- Fishing stocks, quotas and catches
- European fishing fleet
- Aquaculture
- Maritime transport and traffic
- Ports' statistics
- Maritime protected areas
- Tourism
- Maritime policies and initiatives
- Outermost regions

A common information sharing environment for the EU maritime domain constitutes at first concrete, overarching and strategic step to fulfil this mandate. It spells out guiding principles for the development of a common information sharing environment for the EU maritime domain and to launch a process towards its establishment. Two pilot projects tested in a theatre of operations how integrating maritime surveillance can work in practice; one in the Mediterranean basin and another one in the Northern European sea basins.

The Commission services responsible for initiatives relevant to Integrated Maritime Surveillance are:

- DG Energy and Transport
- DG Justice, Freedom and Security
- DG External relations
- DG Enterprise and Industry
- DG Environment
- DG Taxation and Customs Union
- Joint Research Center

The agencies and bodies involved in Integrated Maritime Surveillance are:

- European Maritime Safety Agency (EMSA)
- Community Fisheries Control Agency (CFCA)
- European Agency for the Management of Operational Cooperation at the External Borders of the Member States of the European Union (FRONTEX)
- European Defense Agency (EDA)

- European Space Agency (ESA)
- European Environment Agency (EEA)

DG MARE has in the period 2008 – 2010 invested in total over 7M EUR (BlueMassMed, MAR-SUNO, PASTAMARE) in actions, whose results represent:

1. a good knowledge basis for further development of the integration of maritime surveillance in the form of CISE,
2. some established mechanisms of integrated maritime surveillance which can serve as a basis for further development in the area and
3. understanding the capabilities of new surveillance tools that can have several cross-sectoral applications (such as the S-AIS).

Administrative Arrangement with the Joint Research Centre (JRC) to support the work on the Roadmap, in particular in the Technical Advisory Group (TAG). Contract was signed in September 2010, duration was 27 months, overall budget 600k EUR. The Commission has therefore requested 4M EUR for the three coming years to research IT requirements and possible solutions, identify legal obstacles, study impact on administrative management and related requirements, research data access rights, identify economic and financial implications, provide further information which will form part of an impact assessment (such as social and environmental implications) and, if necessary and where appropriate, support the development of existing systems. The budget was requested for two overall purposed:

- 3.3M EUR for the studies (technical, legal, administrative, economic, financial, social, environmental aspects), and
- 700.000 EUR for SafeSeaNet adaptations, which were followed by:
  – A study on the future sources of growth in the context of the Europe 2020 strategy supported by the Council in its conclusions of 12/06/2010 has been budgeted in 2010 and ran through 2011 and 2012. This study developed a number of maritime growth scenarios, based on key technologies as enablers and using foresight methodologies.
  – A study on maritime clusters (2009) assessed the structure and economic key figures of the maritime clusters in the EU by analyzing existing employment data in the maritime sectors and areas in the EU by relating this employment data to the added value created.
  – A study on maritime employment (2006) provided an analysis of employment trends, identified potentials for growth and increased employment and the type of policy action that can help realize that potential.
  – A study on the economics of climate change adaptation in EU coastal areas regions (released in May 2009) provides insights in the state-of-play and the financial dimension of the actions undertaken to prepare Europe's

coastal zones, including the Outermost regions, to the effects of climate change. The study complements the work of DG ENV- DG CLIMA, and a study on tourism facilities in ports (finalized in October 2009) aimed at analyzing the benefits for ports to invest in infrastructure and facilities for receiving tourists, notably through cruise tourism.

– A study about a database on EU-funded projects in maritime regions (finalised in December 2009) provides, in a coherent and transparent way, an overview of the EU budget spent on maritime-related projects for the previous programming period and the current programming period (projects approved before 31 December 2008). A total of almost 74.000 projects accounting for a total EU funding of almost 15 billion EUR have been identified during the 2000-2008 period.

The DG MARE Maritime Forum[72] website for stakeholder dialogue became operational in the summer of 2009 (budget 550k EUR) as an interactive tool to develop dialogue with and among maritime stakeholders. The Forum allows parties interested in the EU maritime policy to communicate on a common platform. They can publish events, documents and follow developments in their areas of interest. Information can be shared amongst a closed community or published openly.

EU maritime policy stakeholders can publish events, documents and follow developments in their areas of interest. Information can be shared amongst a closed community or published openly. Anybody can register to the forum and comment on its content.

---

[72] https://webgate.ec.europa.eu/maritimeforum

Common Information Sharing Environment (CISE)

Integrated maritime surveillance is about providing authorities interested or active in maritime surveillance with ways to exchange information and data. Sharing data will make surveillance cheaper and more effective. Currently, EU and national authorities responsible for different aspects of surveillance, e.g. border control, safety and security, fisheries control, customs, environment or defense, collect data separately and often do not share them. As a result, the same data may be collected more than once.

The Common Information Sharing Environment (CISE)[73] concept was since 2009 developed jointly by the European Commission and European Union / European Economic Area Member States including civilian and military authorities as well as the European agencies operating in the maritime field in order to enable the information sharing within the European stakeholders.

Originally named only the Common Information Sharing Environment (CISE), the prefix „Maritime" is sometimes used to distinguish between the context and stakeholders of Maritime, Land and Air. The Initiative Common Information Sharing Environment (CISE) originates from the EU Commission Communication (COM (2009) 538)) to establish Guiding principles towards the establishment by DG MARE of a:

1. Common: basic data collected only once,
2. Information: must enable user-defined situational awareness,
3. Sharing: each community receives but also provides information,
4. Environment: interconnected sectorial information systems allowing users to identify trends, threats and detect anomalies.

General aims are to link all relevant user Communities, build a technical framework for interoperability and future integration for the information exchange between civilian and military authorities and the specific legal provisions.

With the Integrated Maritime Policy (IMP) starting from 2007 with the Blue Book, the "*Guiding principles Communication*" provided principles for the development of a Common Information Sharing Environment (CISE). In a first approach the main user communities were defined in order to establish a common approach and understanding within the EU members and their agencies and organizations, as all have slightly different operational, technical and procedural specifications and needs.

CISE aims under the Integrated Maritime Policy (IMP) to support the development of the Blue Economy of the European Union, is a key innovation of the European maritime gov-

---

[73] https://www.efca.europa.eu/en/content/common-information-sharing-environment-cise

ernance, is an element of the European Digital Agenda, and, finally, is a pillar of the European Union Action Plan for the European Union Maritime Security Strategy (EUMSS).

The Common Information-sharing Environment (CISE) concept is currently being further developed jointly by the European Commission and EU/EEA members with the support of relevant agencies. It will integrate existing surveillance systems and networks and give all those authorities concerned access to the information they need for their missions at sea. The CISE will make different systems interoperable so that data and other information can be exchanged easily through the use of modern technologies. DG MARE invested in the period 2008 – 2010 a total of 6.968.558 EUR in supporting actions aiming at:

- a good knowledge basis for further development of the integration of maritime surveillance in the form of CISE,
- some established mechanisms of integrated maritime surveillance which can serve as a basis for further development in the area and
- understanding the capabilities of new surveillance tools that can have several cross-sectorial applications (such as the S-AIS).
- In June 2013 the European Commission launched a consultation on the implementation of CISE based on national and EU maritime information systems to ensure the optimal use of information already being collected and to increase the efficiency, quality, responsiveness and coordination of all maritime activities in the Europe. Contributions were submitted by using an online questionnaire.

CISE thrives on a political, organizational and legal environment to enable information sharing across the seven relevant sectors/user communities (transport, environmental protection, fisheries control, border control, general law enforcement, customs and defense) based on existing and future surveillance systems/networks infrastructure for enhanced maritime awareness for Member States and cross-sectorial cooperation.

In June 2013 the European Commission launched a consultation on the implementation of CISE based on national and EU maritime information systems to ensure the optimal use of information already being collected and to increase the efficiency, quality, responsiveness and coordination of all maritime activities in the Europe. Contributions were submitted by using an online questionnaire.

EUCISE2020 conducted several Security Research projects under the European Seventh Framework Program; it aims at achieving the pre-operational Information Sharing between the maritime authorities of the European States.

Guiding principles and recommendations to act according to them are developed in the Communication. These are briefly mentioned here together with their respective recommendations in order to emphasize that they are applied as important and basic working principles. However the pace of development and even more the integration is at a very low speed due the huge number of different stakeholders involved.

The CISE Roadmap included until 2013:

1. Member States and the Commission shall identify the participants to the information exchange. Due to the diversity in administrative organization in the EU Member States and EEA States it is necessary to focus on "functions" based on already established 'User Communities' rather than on types of national authorities;
2. Mapping of data sets and gap analysis for data exchange to ensure that there is an added value to the CISE by:
    a. drawing up a map of data exchanges already taking place at EU and national level and
    b. drawing up a gap analysis to identify the sectorial demand for data currently not matched by supply.
3. addresses the problem that sectorial User Communities classify same type of data in a different manner.

Without interfering with national data classification levels and for the purpose of enabling data exchange within the CISE, Step 3 was to identify which national classification levels correspond to each other and thus establish common grounds for data exchange under the CISE.

Step 4 was to define the supporting technical framework for the CISE, thus for setting up the interfaces between the existing and planned sectoral systems in view of enabling cross-sectorial data exchange. This should be worked out by the representatives of the various sectorial user communities based on available results of FP7 and pilot projects (e.g. MARSUNO, BluemassMed, EUROSUR pilot project on the communication network, GMES, PT MARSUR, SafeSeaNet based pilot projects).

The different components of the Common Information Sharing Environment (CISE) should be understood as follows:

- 'Common': As the information is to be shared between the different user communities, data used for this information should be collected only once.
- 'Information' must enable user-defined situational awareness. Coming from disparate user communities, information should be identifiable, accessible, understandable and usable. Processing such information with the appropriate security safeguards must be ensured.
- 'Sharing' means that each community receives but also provides information based on previously agreed standards and procedures.
- 'Environment' refers to interconnected sectorial information systems that allow users to build up their specific situational awareness pictures, which enable them to identify trends and detect anomalies and threats.

CISE participants shall be able to discover CISE services provided by other CISE partners with a service registry, which acts as a repository for managing the life cycle of different services exposed by a CISE service provider can be used to communicate its offering towards other participants. The multiple services will be based on business requirements defined on the needs of the different CISE communities. One query service is created that allows for querying the entire data model, however when offering multiple specific services, only parts of the CISE Data Model can be queried.

The Seven Maritime CISE User Communities

CISE distinguishes Seven User Communities:

1. <u>Maritime Safety</u>
    - Monitoring of compliance with regulations on the safety and prevention of pollution caused by ships (construction, equipment, crew/passengers, cargo); support of enforcement operations
    - Monitoring of compliance with regulations on the safety of navigation (vessel traffic safety); support of enforcement operations
    - Monitoring of compliance with regulations on the security of ships; support of enforcement operations
    - Supporting safe and efficient flow of vessel traffic; vessel traffic management
    - Early warning/identification of ships/persons in distress; support of response operations (search and rescue, salvage, place of refuge)
    - Early warning/identification of maritime security threats, within the scope of SOLAS Chapter XI-2; support of response operations
    - Early warning/identification of threats/acts of piracy or armed robbery; support of response operations

2. <u>Fisheries Control</u>
    - Monitoring of compliance with regulations on fisheries;
    - support of enforcement operations
    - Early warning/identification of illegal fisheries or fish landings;
    - support of response operations

3. <u>Marine Pollution</u>
    - Monitoring of compliance with regulations on the protection of the marine environment; support of enforcement operations
    - Early warning/identification of incidents/accidents that may have an environmental impact; support of pollution response operations

4. <u>Customs</u>
    - Monitoring of compliance with customs regulations on the import, export and movement of goods; support of enforcement operations
    - Early warning/identification of criminal trafficking of goods (narcotics, weapons, etc.); support of response operations

5. Border Control

    – Monitoring of compliance with regulations on immigration and border crossing;
    – support of enforcement operations
    – Early warning/identification of cases of illegal migration or trafficking in human beings; support of response operations

6. General Law Enforcement

    – Monitoring of compliance with applicable legislation in sea areas, where there is policing competence and support to enforcement and/or response operations

7. Defense

    – Monitoring in support of general Defense tasks, such as: exercising national sovereignty at sea; combating terrorism and other hostile activities outside the EU; other Common Security and Defense Policy tasks, as defined in Articles 42 and 43 TEU.

CISE Concept for a federated architecture derived from Network-Centric Approach

Cooperation Project (CoopP), CISE Handbook, CISE Incubator, and CISE Data Model

The CISE Handbook was a proposed document from the Cooperation Project (CoopP) which should include the relevant use cases, access rights matrix, data entities and services. The purpose of the CISE Handbook is the description of the process of developing services for the CISE and which data entities should be used. The structure of the services have been specified by the Cooperation Project (CoopP).

The Maritime Common Information Sharing Environment (Maritime CISE) created under the Cooperation Project (CoopP) also a list of 93 Use Cases which were defined by the Technical Advisory Group (TAG). CoopP agreed on a set of nine representative and generic Use Cases for use in the EUCISE2020 project.

The CISE Architecture Visions Document (DIGIT, DG MARE, JRC, 2013) analyzed different visions to integrate the User Communities. As a result of this study, a hybrid-vision was selected as preferred solution by the Member States.

The hybrid-vision allows Member States to decide whether to nominate a single or multiple providers of CISE services at national/sectorial level. This means that a provider of CISE services at national level may be nominated to deliver CISE services of interest for one or more User Communities. The hybrid vision applies a holistic and flexible governance model that takes into consideration both the national and the User Community perspectives.

The basic idea was that CISE participants should find CISE services provided by other CISE partners with a service registry, which acts as a repository for managing the life cycle of different services exposed by a CISE service provider can be used to communicate its offering towards other participants. Based on this assumption the participants in the CISE Incubator events concluded for the services:

- One service: One query service is created that allows for querying the entire data model.
- Multiple services based on data entities: By offering multiple more specific services, only parts of the data model can be queried.
- Multiple services based on business requirements: In this case, services are defined based on the business needs of the different CISE participants.

Based on the options above, it was recommended to define services based on business requirements. However, as this is a process which requires in-depth study and analysis of the different CISE participants, the other options could be a good starting point. By already offering services which facilitate information exchange in general, new business needs could be discovered in the future.

While it is a sign of good progress that certain sectorial systems exchange certain data for particular tasks and purposes with specific data models that already exist or are being designed, there are a number of cross-sectorial tasks and purposes, mainly at national level, identified by stakeholders (TAG, MSEsG), for which no appropriate data model exists.

In these cases a flexible 'purpose-neutral' 'translation' data model is needed.

Connecting more than 400 authorities with several data models specifically designed for particular purposes would be ineffective, not cost efficient and very time consuming. This would not allow to capture all tasks and purposes identified by stakeholders and would be difficult to impossible to implement. On the other hand a flexible 'purpose-neutral' data model can translate any type of information for any task and purpose between any system of any authority at EU and national level, which is not only feasible but also more effective and cost efficient, in particular if connecting the high number of 400 authorities.

The CISE Data Model designed in Coop Project identified seven core data entities (Agent, Object, Location, Document, Event, Risk and Period) and eleven auxiliary ones (Vessel, Cargo, Operational Asset, Person, Organization, Movement, Incident, Anomaly, Action, Unique Identifier and Metadata), which together represent the information exchanged in. The new developed CISE Data Model consisted in 2015 already of 18 data entities, 7 main and 11 complimentary, with 271 data attributes:

Seven Main Data Entities:

1. Agent
2. Object
3. Risk
4. Period
5. Location
6. Event
7. Document

Eleven Complimentary Data Entities:

1. Action
2. Movement
3. Anomaly
4. Incident
5. Vessel
6. Cargo
7. Operational Asset
8. Person
9. Organization
10. Metadata
11. Unique Identifier

The EUCISE2020 Data Model is based on the CISE Data Model version 1.0 which was defined in the Cooperation Project. The CISE Data Model comprises several special features to accommodate crosscutting concerns such as auditing, security, and data reliability and validity. It represent over 50% of the information needs identified by the Technical advisory Group (TAG) for the development of the CISE, and over 64% of its definitions are based on existing definitions from 34 standards, systems and initiatives; however it is still missing other vital format integration as i.e. from the NATO Format Message Catalog, here the ADat-P3, or the US-national Operational Specification Over-the-Horizon Targeting GOLD (OS-OTH GOLD) standard message used in NATO maritime messaging. The EUCISE2020 Consortium made the EUCISE2020 Data Model v1.0[74] document available, in order to support the participation to the H2020 Call "*SEC-19-BES-2016*" and "*SEC-20-BES-2016*".

The EUCISE2020 Data Model is licensed as Attribution-Share-A-Like (CC BY-SA). This license lets others remix, tweak, and build upon it even for commercial purposes, as long as they credit the EUISE2020 Consortium and license their new creations under the identical term.

The purpose of the later CISE Incubator (2014) was to provide a valuable input for future work on the CISE specifications (e.g. the CISE Pre-Operational Validation project or the EUCISE2020 project). The actual name was "*CISE Demonstrator*", however as there was no demonstrator to be developed at this stage but rather to investigate on the general feasibility of the concept and first approaches taken, DG DIGIT[75] proposed the term "*CISE Incubator*" instead.

The EUCISE2020 project followed the CISE Incubator as an important milestone in the roadmap for implementation of CISE, which is meant to complement existing systems, not to create a new overall EU system but rather a network.

---

[74] http://www.eucise2020.eu/media/1131/d4_3-annexb.pdf
[75] Vincent Dijkstra

European Coast Guard Functions Forum (ECGFF)

The European Coast Guard Functions Forum (ECGFF)[76] is a self-governing, non-binding, voluntary, independent and non-political forum whose membership includes the Heads of the Coast Guard or equivalents of each European Union maritime nation and associated Schengen countries, the European Commission and its Institutions and Agencies with related competencies in Coast Guard Functions. ECGFF is promoting maritime issues of importance and of common interest across borders and sectors, both civil and military.

Members include Coast Guards Authorities of 25 EU member states and Schengen Associated Countries as well the European Commission and its Institutions and Agencies with related competencies in Coast Guard Functions. The initiative of a European Coast Guard Functions Forum (ECGFF) was first launched in 2009 and is co-financed by the EU.

The overall aim of the Forum is to study, contribute to and promote understanding and development of maritime issues of importance and of common interest related to Coast Guard Functions across borders and sectors, both civil and military, and to contribute to progress in the various CGF activities. Coast Guard Functions cover a wide range of maritime issues related to maritime safety and security. Coast Guard Functions as defined by the ECGFF:

- Maritime safety, including vessel traffic management
- Maritime, ship and port security
- Maritime customs activities
- The prevention and suppression of trafficking and smuggling and connected maritime law enforcement
- Maritime border control
- Maritime monitoring and surveillance
- Maritime environmental protection and response
- Maritime search and rescue
- Ship casualty and maritime assistance service
- Maritime accident and disaster response
- Fisheries inspection and control and activities related to the above Coast Guard Functions.

A Conference of the Heads of Coast Guards Authorities of EU Member States and Schengen Associated Countries, supported by FRONTEX, was then arranged in Warsaw, just when a parallel initiative in Genoa, from the Italian Coast Guard, was giving origin to the Mediterranean Coast Guard Forum (MEDFORUM).

---

[76] https://www.ecgff.eu

In Poland, participants agreed to develop together an inter-agency cooperation, coordination and multifunctional performance for jointly approach the future challenges in the fields of maritime border security, maritime safety, search and rescue, marine environmental protection and other maritime issues. The main goal remains to make coherent the activities among Member States and relevant EU bodies in this field. Other Conferences followed: Malaga (April 2010), Malmoe (September 2011), Dublin (2012), Chios (2013) and Italy (2014) assigning to the Forum a double role of meeting place for sharing Coast Guard issues and that of possible instrument to supply effective technical counseling for the European Institutions. The VII Plenary Conference will be held in Helsinki, Finland and during the Conference UK will take over the presidency of the Forum.

The annual Plenary Conference is assisted by a Secretariat chaired by the Member State responsible for the chairmanship for the forthcoming Plenary Conference. The secretariat meets three times a year to develop the mandate assigned during the conference. The Chair will circulate national Points of Contact (POC's) for the Member States with any minutes or relevant documentation.

Coast Guard Functional activities are mainly defined for the purpose of the Forum and in no order of priority as:

- maritime safety, including vessel traffic management
- maritime, ship and port security
- maritime customs activities
- the prevention and suppression of trafficking and smuggling and connected maritime law enforcement
- maritime border control
- maritime monitoring and surveillance
- maritime environmental protection and response
- maritime search and rescue
- ship casualty and maritime assistance service
- maritime accident and disaster response
- fisheries inspection and control
- and activities related to the above Coast Guard Functions

Over the course of time the Terms of Reference have been established listing the functions related to the dialogue of the Forum and also its general objectives. Recently mandates of the Plenary Conference aimed to start the European Coast Guard Functions Academy Network project for European Sectorial Qualification's Framework for Coast Guarding and established a Group Office of the Forum in Brussels, Belgium.

The Forums annual plenary conference of the Heads of Coast Guard Functions of the EU and associated Schengen countries may result in decisions to conduct further work, projects, research or produce documents within the scope of the Forums objectives. This work may be done either by the secretariat or through special tailored working groups directed by a national lead partner supported by the secretariat.

Decisions on further work of the secretariat will be made by the chairman and decisions on creating working group(s) will be made by the annual plenary conference following the proposal of a country willing to be a national lead partner through the secretariat. Results of the works of the secretariat as well as working groups will be presented during the following Plenary Conference, or as appropriate.

With the general aim of improving the development of Coast Guard Forums across borders and sectors, the Forum has the following objectives:

- To build and maintain a network of Heads of National authorities for Coast Guard Functions and designated Officers from EU Institutions, Agencies and Directorates with related competencies in CGFs
- To agree Rules and Procedures for the Plenary Conference, Secretariat and any working groups including matters relating to funding
- To assist in the development of common operational procedures and standards in line with prescribed international norms, reinforce synergies and improved operational preparedness, cooperation and response across borders and sectors
- To consider the possibilities of promoting trust, burden sharing, asset sharing and enhanced regional cooperation
- To establish a standing forum promoting the exchange of information, expertise, technical assistance, best practice, training, exercises and education
- To provide relevant recommendations, advice and joint submissions as appropriate
- To act as a shared advice source on operational 'coast guarding' to others including the EU Institutions
- To consider, discuss and where possible develop a common understanding of operational challenges and emerging risks in the maritime domain of EU States
- To develop over time and as appropriate a coordinated and collective response through cooperation and agreement to emerging and existing risks in the EU maritime domain and
- To monitor on-going development in the maritime domain that may have an impact on Coast Guard Functions
- To establish and maintain contacts, within the frame of common objectives, with relevant institutions and organizations at the international and/or European level

European Coast Guard Functions Academy Network Projects

The general objective of the ECGFA NET projects is to enhance educational cooperation in the field of coast guard functions (CGF) and, consequently, to facilitate the interoperability and cooperation amongst different bodies carrying out coast guard functions in order to enhance the coherence and effectiveness of CGF activities.

The EU funded 14 month-long European Coast Guard Functions Academy Network Project (ECGFA NET) was completed in February of 2016 as "*ECGFA NET I Project*", taking the European Union's maritime training cooperation an important step forward. A new cooperation framework was launched: European Coast Guard Functions Training Network (ECGF Training Network). This Network can be joined by those training institutes that offer training in coast guard functions and operate in the member countries of the European Coast Guard Functions Forum. The Network will increase the visibility of the various coast guard functions related agencies and educational institutions, contributing positively to the exchange of information and best practices via networking. ECGF Training Network will create a solid foundation for maritime collaboration in the sphere of European training and security.

The ECGFA NET II Project is direct continuation for the European Coast Guard Functions Academy Network Project implemented in 01/2015-02/2016, and designed to ensure that the results of the ECGFA NET are sustainable.

The first of the specific objectives of the project is to provide support for the European Coast Guard Functions Training Network (ECGF Training Network) and the ECGF Training Portal during their start-up phase in order to increase the number of members and to enable the Network to function in an active and autonomous manner, taking ownership on the Network, its activities and future development. In order to meet this objective, the project facilitates the activities of the Governing Board (i.e. through organization of board meetings, drafting of working documents, establishment of the Secretariat), actively advocates the membership of the Network and the use of the ECGF Training Portal (i.e. through contacting of stakeholders, participation at stakeholder events and use of media) and runs the ECGF Training Portal (i.e. through technical maintenance and by functioning as the Portal Coordinator).

The second specific objective is to launch a pilot ECGF Expert Exchange Program in order to increase the sharing of knowledge and skills in the CGF sector and to assess the future needs of the Exchange Program. To meet the second objective, the project runs and facilitates a pilot exchange program for ca. 20 exchanges for the duration of five working days per exchange. ECGF Training Portal will be used as the main platform and tool to assist with the coordination. Exchange program consists of a total of eight steps, starting with the calls for the Hosts and nominations of Exchanges, as well as with the matching and selection process and preparations for the exchange. The actual exchange period is followed by a cascading step, where the Exchanges shares the newly gained knowledge and experience i.e. with ones colleagues, as well as by a feedback step, where both the

Hosts and Exchanges will report on how the objectives were met. As a last step, those Exchanges who fulfilled their obligations will be awarded with certificates. By the end of the project, the overall results and development needs for the future running of the Exchange Program will be assessed, and there will be prepared an assessment report along with recommendations.

The third specific objective is to advance the development of the Coast Guard Functions Sectorial Qualifications Framework (CGFSQF) by drafting the final structure of the CGFSQF and by completing stages 2 and 3 out of a total of 7 in the CGFSQF development process. Development of the CGFSQF is a multi-year process where the end goal of the CGFSQF is to function as a set of common standards for CGF education to assure quality of training and teaching and to be applied on a voluntary basis. In order to reach this third specific objective, the proposed project starts with the identification of a SQF working group and of different stakeholders at national and international levels, and with the defining of a working plan for the development of the CGFSQF. These are followed with activities such as the mapping of different qualifications, identification of the number of levels, drafting of the learning outcomes and as a last step ñ drafting of the CGFSQF final structure.

European Coordination Centre for Accident and Incident Reporting Systems (ECCAIRS)

The European Coordination Centre for Accident and Incident Reporting Systems (ECCAIRS) was developed by JRC to monitor transport safety in Europe and received an extension with the Pirate Incident Reporting And informaTion Exchange System (PIRATES) on the parliaments request.

ECCAIRS enables the collection and analysis of accident and incident data for three different public transport modes: aviation, maritime and railway. The ECCAIRS reporting system in the transport domain is provided free of charge to the competent transport authorities worldwide. Within the European Union, the European Commission is providing technical support, training and data exchange services to these organizations, all free of charge. Outside the European Union, these services need to be provided by third parties.

Since ECCAIRS already provides a reporting system allowing data collection and data exchange, it was suggested to study from a technical point of view if ECCAIRS could be a suitable system for addressing the above mentioned issues in the Gulf of Guinea region. The technical feasibility study is relevant for JRC because the possibility to apply ECCAIRS to another domain than transport safety was studied. Furthermore, this study was conducted to research the time and effort needed to develop a new extension within ECCAIRS.

The mission of ECCAIRS is to assist National and European transport authorities and accident investigation bodies in collecting, sharing and analyzing their safety information in order to improve public transport safety. The ECCAIRS Reporting System has been developed to support occurrence reporting in civil aviation. The system has been in use since 1998 in various different versions.

In 2008, ECCAIRS was made suitable to support aviation as well as maritime and other transport sectors. This so called ECCAIRS Common Framework (ECF) can provide the complete collection, exchange and analysis infrastructure to be used by reporting environments for each of the supported transport sectors. The transport dependent parts of the final reporting systems are concentrated in so called domain-specific extensions.

The philosophy of an ECCAIRS Common Framework is based on replaceable extensions where the user-interface components and the taxonomy have been extracted from the system and the remaining parts together form the ECF. The extension is formed by the combination of the Taxonomy contents and the User Interface where obviously the user interface part strongly depends on the taxonomy. All other tools, used for analysis, data exchange and data-integration are part of the ECF and can be applied without limitations for each Extension in the same manner.

The bottom of this diagram groups all basic functionalities of the system in a system library accessible via an internal ECCAIRS Application Programming Interface (API). Functions provided by this basic software implement access to data, security, user authentication and access to the taxonomy used to encode the stored information. ECCAIRS tools and utilities are making the functionalities available to the end-users. In the ECCAIRS architecture the majority of functions implemented are independent from the transport mode in which the system could be used. The transport dependent parts could be found in the Taxonomy (completely depending) and in both browser applications which depend only for a small part (the user-interface) on the transport mode.

The taxonomy of ECCAIRS is the catalogue of information describing what information can be stored in the ECCAIRS Repository and how this information is (possibly) encoded in the data fields. To design the taxonomy for maritime piracy, an analysis of reported piracy incidents was conducted. Due to different definitions and interpretations of what piracy and armed robbery at sea is, there were found some differences in the reporting of piracy incidents.

Whereas in the definition of the United Nations Convention on the Law of the Sea (UNCLOS) the location of incidents plays an important role, the Office of Naval Intelligence (ONI) is focusing on the type of attack and the status of the ship in their definition of maritime piracy. According to the IMB, besides the type of attack, a hostile intent has to be present and of course this is not easy to measure. Next to these differences in definitions, there are differences in categories as well. For instance, the International Maritime Bureau (IMB) uses four types of attack (attempted, boarded, fired upon and hijacked), whereas the United Kingdom Maritime Trade Organisations (UKMTO) also has a category "*suspicious vessel*". And the Regional Cooperation Agreement on Combating Piracy and Armed Robbery against Ships in Asia (ReCAAP) uses categories for the significance level of the reported incident (very significant incident, moderately significant incident, less significant incident), and has a different category for attempted attacks.

The International Maritime Bureau (IMB), part of the International Chamber of Commerce (ICC), runs the Piracy Reporting Centre (PRC) web portal which provides the most recent piracy attacks and armed robbery incidents worldwide. Each "live piracy report" consists of:

1. Attack number
2. Date/Time
3. Type of Vessel
4. Attack position on a map (Google Maps)
5. Latitude/Longitude
6. Location detail (area of attack)
7. Type of Attack
8. Narrations (description of the incident)

The PRC wants to provide a free 24/7 service to the seafarer. The main objective of the PRC is to be the first point of contact for the shipmaster to report any incident of piracy and armed robbery at sea. The incident reporting through the PRC should be done through the Piracy & Armed robbery attack report of the IMB, available at their website.

The information requested through this attack reports concerns among other things:

1. Vessel details

    - Name of ship
    - IMO number
    - Flag of ship
    - Type of Ship
    - Tonnages
    - Owner information

2. Incident details

    - date and time of incident
    - location of incident
    - nearest country
    - status of ship
    - e. weather during attack
    - f. type of attack

3. Details of raiding party

    - number of pirates
    - physical appearance
    - language spoken
    - craft used
    - violence used

4. Details of weapons used

5. Damaged caused

    - weapon type
    - details of damage

6. Other details

    - action taken by master and crew
    - anti-piracy measures employed
    - private security team embarked
    - number of crew and nationality

For the reporting of a piracy incident in PIRATES, different organizations are using different forms and reports. In ECCAIRS, it is possible to design the various forms and reports as whished for, however the structure of the underlying data has to be respected. The design of the reporting of piracy incidents in PIRATES can be seen on the next page as well. To test the piracy reporting software PIRATES, internet research was conducted to collect information on piracy incidents and to fill the PIRATES database with these data. A number of piracy data sources provide data about reported incidents related to maritime piracy.

To assist in anti-piracy measures, the International Maritime Organization (IMO) issues reports on piracy and armed robbery against ships submitted by Member Governments and international organizations. The reports, which include names and descriptions of ships attacked, position and time of attack, consequences to the crew, ship or cargo and actions taken by the crew and coastal authorities, are now circulated monthly, with quarterly and annual summaries.

Further, the IMO has opened a "piracy and armed robbery" module on the Global Integrated Shipping Information System (GISIS) in order to improve the timeliness of reporting of incidents and to enable users to generate their own search criteria and produce customizable reports. This database is now configured for public, read-only access and is searchable. Reports can be compiled in GISIS directly by Member States and registered public users. These reports can now include follow-up information, for example, dates of release of hijacked ships. Given that the new functionality in GISIS allows for user-defined piracy reports, the Committee agreed that the practice of publishing quarterly summaries was no longer warranted and would be discontinued with effect from May 2011. The total number of acts of piracy and armed robbery against ships reported to the Organization is 6,060, as of 8 August 2011.

Critical Maritime Routes Indian Ocean (CRIMARIO)

The European Union supported the implementation of a regional mechanism in the Indian Ocean under the Djibouti Code of Conduct (DCoC), agreed in 2009 by 21 littoral states of the Western Indian Ocean, and initiated the Critical Maritime Routes (CMR) program.

Under CMR the Critical Maritime Routes Indian Ocean (CRIMARIO)[77] aims to strengthen maritime safety and security in the wider IO region by supporting coastal countries in enhancing maritime situational awareness (MSA). MSA is the sharing and fusion of data from various sources to achieve a comprehensive understanding of the maritime domain, whilst an effective and sustainable MSA enables maritime stakeholders to improve security, safety and environment of this domain.

To support this, EU CRIMARIO will introduce various initiatives in the region such as a web based Information Sharing and Incident Management Network, here with the Indian Ocean Regional Information Sharing (IORIS), training & capacity building, workshops aimed at enhancing interagency and regional cooperation, and establishing a set of standards for information sharing.

FIDES does not store the actual information in a database but rather provides a reference link connecting the user with the data. It automatically acknowledges, archives and logs each information exchange (called *transaction*). Different authorized users can submit or request statistical reports on transactions from the system, but the available action and information depends on the user's access level within the system. The configuration requirement for FIDES is minimal in order to take away the added cost in time and financial resources of installing DG Fisheries software or hardware on the premises of EU Member State administrations.

---

[77] https://www.crimario.eu/en

Indian Ocean Regional Information Sharing (IORIS)

The Indian Ocean Regional Information Sharing (IORIS)[78] was initiated by the EU CRIMARIO project, as tool to enable member countries to set up a collaborative working environment to improve the understanding of the maritime domain and coordinate operations when incidents at sea occur. It expresses the cooperative approach that has been developed by the European Union to address piracy and new maritime security challenges faced by the Indian Ocean littoral states such as drugs and arms trafficking, illegal fishing, environmental damages, etc.

EU CRIMARIO developed the web based information sharing tool in collaboration with regional partners, for sharing information and managing incidents at both national and regional level as a maritime communications tool with collaboration capability for the Indian Ocean Region.

RMIFC (Regional Maritime Information Fusion Centre) based in Madagascar and RCOC (Regional Centre for Operational Coordination) based in Seychelles were created with the support of IOC by the Maritime Security (MASE) program covering the ESA-OI region (Eastern and Southern Africa-Indian Ocean).

Key functions include:

- Instant messaging (public/private): users can utilize a public or private secure chat function within the secure areas to chat with members of any group for routine exchange of information or in support of an event;
- Advanced mapping: using OpenStreet map and dedicated nautical charts, the module allows the import and export of maps and navigation markers (location of vessels, tracking, etc.), to plot incidents and polygons. It also shows the range of vessels based on the direction of travel and estimated speed to extrapolate where a vessel might be now;
- Sharing of documents: users can share any format of documents (image, text, …) with an additional option to encrypt where sensitive information requires;
- Form creation: any action can be easily documented and followed through the creation of on-line forms;
- Area management: only accessible to the designated administrator of each area, they will define the access rights and the membership of their dedicated area for centers, local organizations or incidents.

The integrated modules are designed to achieve: (a) information sharing in a secure and

---

[78] https://www.crimario.eu/en/

flexible environment; (b) real time management of incidents at sea; (c) secure communications between users (national agencies, regional centers); allowing each to control members and access rights for their designated areas.

The development of IORIS[79] was awarded to Polymorph Ltd after a global open tendering process. A second development phase of IORIS is planned which will follow a period of operational use to maximize valuable feedback from partners. The development and future phases of IORIS will be supported by EU CRIMARIO project which includes running costs (hosting, maintenance and support) until March 2020.

The IORIS platform is only accessible to authorized users designated by partners countries, maritime centers and regional organizations. The first centers to join IORIS are the two regional centers, the RMIFC based in Madagascar and the RCOC based in Seychelles. IORIS is also accessible to national maritime administrations and agencies of Kenya, Madagascar, Mauritius, Seychelles. Comoros will follow very soon.

IORIS incorporates the latest technologies for ensuring a high level of security: including over-the-wire encryption, additional encryption of documents, two-factor authentication for mobile users, IP restriction, etc. The whole system is a web-platform hosted in a secure environment to ensure the correct balance of availability vs security for regional users.

IORIS provides functionality for all national and regional maritime centers to communicate regionally, or privately 1 to 1 or with selected groups for the exchange of non-classified information in a secure environment. IORIS includes the capability for: information sharing, collaboration, incident management, training, library. The hosted solution with access to a private web based portal, will also provide a public facing interface with non-sensitive information / open source and secure login to secure area for authorized users.

IORIS manages multiple simultaneous collaboration areas that can be used for any incident as designated by the users such as oil spill, human trafficking, report suspicious activity etc. or a collaboration area for general, non – time – sensitive information. Each partner country will be able to create its own secure collaboration area for national needs and invite other registered national users to participate as required. Each collaboration area will only be accessible by the owners of that collaboration space, and the users granted, or by making the collaboration area 'public access'.

A dashboard will only show the user the collaboration areas that he either created or been granted access to, with recent updates from each collaboration area aggregated into a single page. This will enable users to see easily when each collaboration area has been updated. Users will then be able to click through to each separate, collaboration page.

The official launch for IORIS was held in Mahé on 4. September 2018. The emphasis will be to demonstrate the added value of this unique information sharing and communications tool and prepare the governance of the platform among the Indian Ocean countries and

---

[79] https://www.crimario.eu/en/information-sharing/the-ioris-network/

beyond. The IORIS collaboration area includes:

- Title and purpose for the collaboration area e.g. specific oil spill, or supporting agency;
- Contact details of the incident coordinators;
- A list of users with authorized access to the collaboration area, and the time since they were last active;
- Ability to start private chats (instant messaging) on a one-to-one basis with other users;
- Ability to import map data (from KML or CSV files);
- Ability to publicly 'chat' (instant messaging) as well as identifying certain posts as 'Alerts';
- Ability to upload and share documents incl. photographs, PDFs, Word documents etc.;
- Ability to plot individual map points and interact with the map (various filters, zoom etc.);
- The date, UTC time and local time.

CRIMARIO team will continue to train centers and agencies involved and propose a set of draft policies initially recommended and agreed with the participants. Ultimately however the decision on all system and user policies will be agreed on a regional basis as the ownership is transferred to become a truly regional platform that will enhance Maritime Situational Awareness in an efficient and cost-effective manner.

European Police Office (EUROPOL)

The European Police Office (EUROPOL)[80] is the law enforcement agency of the European Union, whose remit is to help make Europe safer by assisting law enforcement authorities in EU member states.

The Council is responsible for the main control and guidance of Europol. It appoints the Director and the Deputy Directors, and approves Europol's budget (which is part of the general budget of the EU), together with the European Parliament. It also can adopt, together with the European Parliament, regulations related to Europol's work. Each year the Council forwards a special report to the European Parliament on the work of Europol.

Europol is headed by a Director, who is Europol's legal representative and appointed by the EU Council. Europol's Management Board gives strategic guidance and oversees the implementation of Europol's tasks. It comprises one high-ranking representative from each EU country and the European Commission. Each country has a Europol National Unit, which is the liaison body between Europol and the other national agencies.

Europol employs some 100 criminal analysts who are among the best-trained in Europe. This gives it one of the largest concentrations of analytical capability in the EU. Analysts use state-of-the-art tools to support national agencies' investigations on a daily basis. To give national partners a deeper insight into the criminal problems they face, Europol produces regular long-term analyses of crime and terrorism. New dangers are also growing, such as online radicalization and trafficking in human beings. The networks behind the crimes in each of these areas are quick to seize new opportunities, and they are resilient in the face of traditional law enforcement measures.

Europol's daily business is based on its strategy. Its specific objectives are set out in the Europol annual work program. In 2010, the EU established a multi-annual policy cycle to ensure effective cooperation between national law enforcement agencies and other bodies (EU and elsewhere) on serious international and organized crime.

This cooperation is based on the Europol Serious and Organized Crime Threat Assessment (SOCTA), which identifies and assesses emerging threats; describes the structure of organized crime groups (OCGs) and the way they operate as well as the main types of crime affecting the EU. The EU Terrorism Situation and Trend Report (TE-SAT), which gives a detailed account of the state of terrorism in the EU.

Headquartered in The Hague, the Netherlands, EUROPOL assist the 28 EU Member States in their fight against serious international crime and terrorism and works with many non-EU partner states and international organizations. EUROPOL Operational Center[81] enables

---

[80] https://www.europol.europa.eu
[81] https://www.europol.europa.eu/activities-services/services-support/operational-coordination/operational-centre

work 24/7 in a high-security environment. Benefiting from its central position in the European security architecture, Europol offers a unique range of services:

- support for law enforcement operations on the ground
- a hub for information on criminal activities
- a center of law enforcement expertise.

Large-scale criminal and terrorist networks pose a significant threat to the internal security of the EU. Terrorism, cybercrime and people smuggling, to name just a few, pose a severe threat to the safety and livelihood of its people. The biggest security threats according to EUROPOL come from:

- Terrorism;
- International Drug Trafficking and Money Laundering;
- Organized Fraud;
- Counterfeiting of Euros;
- People Smuggling.

The information exchange benefits:

- Law enforcement agencies, who get 24/7 operational support.
- Government departments and private companies working in partnership with Europol.
- EU Member States, supported in their investigations, operational activities and projects to tackle criminal threats.

International crime and terrorist groups operate worldwide and make use of the latest technology. To ensure an effective and coordinated response, EUROPOL needs to be equally flexible and innovative, and to make sure its methods and tools are up to date. That is the reason for a number of specialized bodies and bespoke systems.

- EUROPOL Operational Center, the 24/7 hub for data exchange among Europol, EU Member States and third parties;
- the European Cybercrime Centre (EC3), which aims to strengthen the law enforcement response to cybercrime in the EU and thus help protect European citizens, businesses and governments from online crime;
- the Joint Cybercrime Action Taskforce (J-CAT), which drives intelligence-led, coordinated action against key cybercrime threats and top targets by stimulating and facilitating the joint identification, prioritization, preparation and initiation of investigations;
- the European Counter Terrorism Centre (ECTC), an operations center and hub of expertise that is a central part of the EU's efforts to enhance its response to terror;

- the European Migrant Smuggling Centre (EMSC), which supports EU Member States in targeting and dismantling the complex and sophisticated criminal networks involved in migrant smuggling;
- the Intellectual Property Crime Coordinated Coalition (IPC3), which is central to the EU's efforts to stem the tide of intellectual property crime within and outside the EU.

The specialized systems, which offers fast and secure capabilities for storing, searching, visualizing and linking information, comprises a sophisticated crime-fighting toolbox, which include:

- Financial Intelligence Units (FIU.net), a decentralized and sophisticated computer network supporting the financial intelligence units (FIUs) in the EU in their fight against money laundering and the financing of terrorism;
- The Secure Information Exchange Network Application (SIENA), a state-of-the-art platform that meets the communication needs of EU law enforcement;
- the Europol Platform for Experts (EPE), a secure, collaborative web platform for specialists working in a variety of law enforcement areas;
- the European Information System, the reference system for offences, the individuals involved in them, and other related data.

Since Europol can respond flexibly, the focus is on different areas of criminal and terrorist activity vary from one year to the next, depending on requirements. Main priorities, however, tend to remain relatively stable, reflecting those of international criminal and terrorist groups. Over the years EUROPOL have built up substantial experience in fighting drug trafficking, illicit immigration networks and trafficking in human beings, illicit vehicle trafficking, cybercrime, money laundering and currency forgery.

In the world of counterfeit money, the most common culprits are the 20s, be they pounds, dollars, euros or pesos. EUROPOL is the central European office for combatting euro counterfeiting. Euro banknotes, the common currency of the 19 euro area countries, are produced with sophisticated printing technology. They also have a number of prominent security features like the simple *"feel, look and tilt"* making them easy to distinguish from counterfeit notes without the use of special equipment, and thus deter counterfeiters.

European Union Patrol Network (EPN)

Special European forces of rapidly deployable border guards were created by EU interior ministers in April 2007 to assist in border control, particularly on Europe's southern coastlines. The European Patrol Network (EPN) began work under Frontex in the Canary Islands in May 2007. Patrols carried out by Frontex are contributed to not only by EU members, but by other Schengen area countries such as Iceland, which sent patrols to the Mediterranean in 2010. The European Patrol Network (EPN) is a permanent regional security on the coordination of individual national interests, efforts and measures in the context of the EU.

The European Patrol Network (EPN) was established by Frontex and unites the members of the agency's operational branch of the *"Joint Maritime Operations"*. EPN is implemented in two phases. First, there are only based on the existing management activities on the shores of the Mediterranean Sea and Atlantic Ocean, coordinated through POC's. In the second phase EPN will be created as an organizational structure, the National Coordination Centers and the EPN is part of a European Surveillance System.

FRONTEX and the eight Member States Portugal, Spain, France, Italy, Slovenia, Malta, Greece, Cyprus, Bulgaria and Romania located at the southern and south-eastern maritime external borders cooperate in the framework of the European Patrol Network (EPN). The objective of the EPN is to establish a permanent regional border security concept at these borders, enabling the synchronization of national measures of the Member States and their integration into joint European activities. Patrolling activities of Member States covering defined coastal areas of the Mediterranean Sea and the Atlantic Ocean are planned, coordinated and implemented through a system of a national contact points in the Member States together with FRONTEX.

EPN is being further developed by establishing National Coordination Centers as envisaged in FRONTEX and EUROSUR.

European Union SECRET (EU SECRET) Network

There are actually several EU SECRET Networks, one being the backbone of the EU OPS WAN system which can be used for transmission up to EU SECRET. The SECRET EU Network is regularly inspected to maintain its classification level. Each inspection identifies updates to follow the evolutions of risks, standards and regulations. Modifications were conducted in 2009 with the corresponding first inspection in 2010. On this basis, the interconnection between EUSC operational network and the EU OPS WAN system as well as many other EU agencies and EU member states organization is foreseen.

Aside from this EU SECRET Network exist other enclaves, in agencies and institutions, which are sometimes not connected to the main network. The European Union has a list of approved cryptographic products for SECRET UE/EU SECRET[82].

---

[82] https://www.consilium.europa.eu/en/general-secretariat/corporate-policies/classified-information/information-assurance/eu-secret/

European Union Naval Force SOMALIA (EU NAVFOR SOMALIA)

Forces of Operation ATALANTA, the European Union Naval Force SOMALIA (EUNAVFOR SOMALIA)[83] operate under the EUNAVFOR Operation ATALANTA, a European Union military operation in the region of Somalia. Piracy impacts on international trade and maritime security and on the economic activities and security of countries in the region of Somalia. As a result, and as part of its Comprehensive Approach to Somalia, the EU launched the Operation ATALANTA in December 2008 within the framework of the European Common Security and Defense Policy (CSDP) and in accordance with relevant UN Security Council Resolutions (UNSCR) and International Law.

EU NAVFOR SOMALIA operates in an Area of Operations covering the Southern Red Sea, the Gulf of Aden and a large part of the Indian Ocean, including the Seychelles, Mauritius and Comoros. The Area of Operations also includes the Somali coastal territory, as well as its territorial and internal waters.

Due to the BREXIT the OHQ EU NAVFOR SOMALIA, Operation ATALANTA, moved to a new OHQ in ROTA (ESP).

---

[83] http://eunavfor.eu

European Union Naval Force Mediterranean (EUNAVFOR MED)

As a consequence of the April 2015 Libya migrant shipwrecks, the EU launched the military operation European Union Naval Force Mediterranean (EUNAVFOR MED)[84] under the Operation SOPHIA, with the aim of neutralizing established refugee smuggling routes in the Mediterranean. The operational headquarters is located in Rome.

The European Council stressed that the Union will mobilize all efforts to prevent further loss of life at sea, tackle the root causes of the human emergency in the Mediterranean - in cooperation with the countries of origin and transit - and fight human smugglers and traffickers.

The Council approved the Crisis Management Concept for a military Common Security and Defense Policy (CSDP) operation to disrupt the business model of human smuggling and trafficking networks in the Southern Central Mediterranean (Council Decision 2015/778 dated 18. May 2015). As a result, and as part of the European Union's Comprehensive Approach, the EU launched on 22. June 2015 a European Union military operation in the Southern Central Mediterranean (EUNAVFOR MED). The aim of this military operation is to undertake systematic efforts to identify capture and dispose of vessels as well as enabling assets used or suspected of being used by migrant smugglers or traffickers.

The missions of the European Union Naval Force SOMALIA (EU NAVFOR SOMALIA) and the European Union Naval Force Mediterranean (EUNAVFOR MED) have similar constrains as all EU missions compared to NATO: they lack a thoroughly dedicated and pre-established EU Mission Network linking the operational headquarters down to the unit level.

While EUNAVFOR MED setup a restricted ATALANTA Mission Network (AMN), EMSA provided SafeSeaNet (SSN) to the OHQ SOPHIA of EUNAVFOR MED and the FHQ Afloat aside from the existing NATO RMP information of the Italian Navy. The SafeSeaNet data-stream is used in a deviant of the Service-oriented infrastructure for MARitime Traffic tracking (SMART) of the Italian Navy in order to provide information in the OHQ and down to the unit level.

In 2017 the Maritime Surveillance Networking (MARSUR) demonstrated the capability of linking the units equipped with a Stand-Alone Workstation (MARSUR User Interface) with the OHQ via the MARSUR Exchange System (MEXS) in a demonstration.

---

[84] http://eur-lex.europa.eu/legal-content/EN/TXT/PDF/?uri=CELEX:32015D0778&qid=1435825940768&from=EN

European Union Maritime Safety Agency (EMSA)

The European Maritime Safety Agency (EMSA)[85] is a is a European Union agency tasked with reducing the risk of maritime accidents, marine pollution from ships and the loss of human lives at sea by helping to enforce the pertinent EU legislation. It is headquartered in Lisbon, there since June 2009 in the new premises near Cais do Sodré in central Lisbon.

EMSA was founded in 2002, after the EU adopted substantial packages of legislation relating to maritime security in the wake of major shipping disasters in European waters, such as those involving the ferry ESTONIA and the oil tankers ERIKA and PRESTIGE. These incidents resulted in huge environmental and economic damage in the Baltics and to the coastlines of Spain and France. They also acted as a reminder to decision-makers that Europe needed to invest in better preparation for a large-scale oil spill, i.e. above-and-beyond the resources available at individual Member State level.

The European Maritime Safety Agency provides technical assistance and support to the European Commission and Member States in the development and implementation of EU legislation on maritime safety, pollution by ships and maritime security. It has also been given operational tasks in the field of oil pollution response, vessel monitoring and in long range identification and tracking of vessels. For these tasks EMSA closely cooperates with the Member States' maritime services, has a staff of just under 200 and operates a small network (at the end of 2009, 16 vessels) of stand-by oil recovery vessels contracted from the commercial sector, designed to provide top-up capacity to Member States' own response resources.

EMSA has the following mission:

- assist the Commission in preparing EU legislation in the field of maritime safety and prevention of pollution by ships,
- assist the Commission in the effective implementation of EU legislation on maritime safety and maritime security, in particular by monitoring the overall functioning of the EU port State control regime,
- organize training activities, develop technical solutions and provide technical assistance related to the implementation of EU legislation,
- help develop a common methodology for investigating maritime accidents,
- provide data on maritime safety and on pollution by ships and help improve the identification and pursuit of ships making unlawful discharges.

The 2008 budget for EMSA was just over EUR 50 M€ of which over a third, EUR 18 M€ is specifically used for at sea pollution response tasks. The Executive Director is supported by

---

[85] http://www.emsa.europa.eu

three departments and an Executive Office unit:

Currently EMSA has 11 units, 10 of which come under these 3 departments:

- Department A: Corporate Services (Human Resources & Internal Support; Legal, Financial & Facilities Supports; Operations Support)
- Department B: Safety and Standards (Visits & Inspections; Ship Safety; Environment & Capacity Building)
- Department C: Operations (Pollution Response Services; Vessel & Port Reporting; Maritime Surveillance; Digitalization & Application Development)

The Administrative Board supervises the work undertaken by the Agency and the Executive Director. In particular, the Administrative Board adopts the Agency's Single Programming Document covering a 3-year period, and within its competence in the framework of the budgetary procedure, the budget and the establishment plan. The Administrative Board also adopts and assesses the Consolidated Annual Activity Report of the Agency, as requested by the Financial Regulation on achievement of objectives and performance output relating to the principles of cost-effectiveness, efficiency and sound financial management.

Within the Administrative Board there are representatives of all EU Member States, Iceland and Norway (EFTA countries) and four representatives from the Commission, plus four non-voting representatives from different sectors of the maritime industry.

The concept of a European Maritime Safety Agency (EMSA) as a regulatory agency originated in the late 1990s along with a number of other major European maritime safety initiatives. EMSA was established by Regulation (EC) No 1406/2002 as a major source of support to the Commission and the Member States in the field of maritime safety and prevention of pollution from ships, and subsequent amendments have refined and enlarged its mandate.

The cooperation with other Agencies and bodies is one of the tools for the Agency to avoid duplication of work and foster synergies in its relevant fields of activities. The cooperation developed by the Agency with different bodies at technical level confirms that EMSA is considered a reliable partner.

In most of the cases the final objective of these cooperation arrangements is to improve the quality of services offered by the Agency to the Member States and the Commission, within the limits of its mandate.

EMSA has installed a procedures for requesting EMSA data[86] from maritime applications.

---

[86] http://www.emsa.europa.eu/emsa-documents/data-request-procedure.html

EMSA manages external requests for data from the EMSA applications including Long Range Identification and Tracking (LRIT), SafeSeaNet, CleanSeaNet and Satellite AIS data, where member states receive access to already submitted national data from contributing member states plus information provided by other network participants. The requests are made by European Union (EU) Member State's national government authorities and EU institutions and bodies, and by projects or programs established by these parties and working on issues of public interest.

For each specific data type, a different procedure is applied. The applicable procedures are described in this document, which provides the following information:

- What data is made available,
- How to apply,
- Conditions for obtaining access to or receiving the data,
- What happens once a request has been approved.

Different agreements signed by EMSA relate also to exchange of information and data, relevant in the field of maritime safety, prevention of pollution from ships, pollution preparedness, detection and response. These international and European agreements are the core foundation of the EMSA operational work and enables a wide network for information exchange.

The list of third party agreements[87], which are either with an open end or to be renewed, includes i.e.:

- Interspill Agreement between the European oil spill industry trade associations, IPIECA and EMSA to hold Interspill conferences and exhibitions
- Ministère de l'Ecologie - Direction des affaires maritimes: EQUASIS Supervisory Committee - Management of EQUASIS system
- Ministère de l'Ecologie - Direction des affaires maritimes: EQUASIS - IT SERVICES
- Paris MoU: Agreement on updating SSN with information on Banned Vessels/Agreement on EMSA technical database management for THE-TIS
- EUNAVFOR-Athena Atalanta: Delivery of an integrated maritime monitoring service
- EUNAVFOR-MED: Data Access Agreement defining the condition for the use of SAT-AIS data provided by EMSA for the purpose of the EUNAVFOR MED operation
- CEDRE-CEFIC: Establishing the MARICE service (Network of chemical experts for HNS marine pollution)

---

[87] http://www.emsa.europa.eu/partnerships/operational-agreements.html

- Norwegian Coastal Administration: Service Level Agreement between the Norwegian Coastal Administration and the EMSA for the hosting, maintenance and operation of the North Atlantic AIS Regional Server and its connection with SafeSeaNet
- Italian Coast Guard: Service Level Agreement between the Italian Coast Guard and the EM-SA for the hosting, maintenance and operation of the Mediterranean AIS Regional Server and its connection with SafeSeaNet
- Danish Maritime Authority: Service Level Agreement between the Danish Maritime Authority and the EMSA for the hosting, maintenance and operation of the HELCOM and the North Sea AIS Regional Servers and their connection with SafeSeaNet (former)
- European Fisheries Control Agency (EFCA): MARSURV-3 Monitoring Services
- European Space
- Agency (ESA): Cooperation for the use of space-based systems and data in support of maritime activities
- Frontex: Provision of services for implementation of Concept of Operations with-in EUROSUR
- IMSO: LRIT Services Agreement IDE
- IMSO: LRIT Services Agreement EU CDC
- SMHI: Development and implementation of an operational capability be-tween oil spill models and CNS DC
- Joint Research Centre (JRC): Provision of services for the EMCIP platform
- RBINS: Cooperation Agreement between the Royal Belgian Institute of Natural Sciences, Operational Directorate Natural Environment and EMSA
- European Telecommunications Standards Institute (ETSI)   Memorandum of Understanding on Marine equipment
- MARETEC-IST: Cooperation Agreement between MARETEC-IST and EMSA regarding oil spill modelling
- MAOC-N: Agreement with Maritime Analysis and Operations Centre - Narcotics
- DG ECHO: Working arrangement on cooperation in the framework of maritime emergencies, including marine pollution preparedness, monitoring and response
- DG ENV: Cooperation Agreement between the European Commission, DG Environment and EMSA for developments in support of the implementation of the Sulphur Directive (2012/33/EU) (THETIS-S) and relevant technical assistance
- DG ENV: Cooperation Agreement between the European Commission, DG Environment and EMSA for the development of inventories of shipping emissions
- DG NEAR: Preparatory measures for the participation of Enlargement countries
- DG NEAR: Grant Contract 2012/308-813 for the implementation of SAFEMED III

- DG NEAR: Grant Contract ENPI/2013/334-385 for the implementation of *"TRACECA Maritime Safety and Security II"*
- DG NEAR: Cooperation Agreement between the European Commission, Directorate-General for Internal Market, Industry, Entrepreneurship and SMEs (DG GROW) and EMSA on the implementation of the maritime surveillance component of the Copernicus Security Service

CleanSeaNet (CSN)

CleanSeaNet (CSN)[88] is a European satellite-based oil spill and vessel detection service, set up and operated by the EMSA since April 2007, which offers assistance to participating States for the following activities:

- Identifying and tracing oil pollution on the sea surface;
- Monitoring accidental pollution during emergencies;
- Contributing to the identification of polluters.

The CleanSeaNet (CSN) service is based on the regular ordering of Synthetic Aperture Radar (SAR) satellite images, providing night and day worldwide coverage of maritime areas independent of fog and cloud cover. Data from these satellites is processed into images and analyzed for oil spill, vessel detection and meteorological variables. The information retrieved includes among others: spill location, spill area and length, confidence level of the detection and supporting information on the potential source of the spill (i.e. detection of vessels and oil and gas installations). Optical satellite images can also be acquired upon request, depending on the situation and user's needs.

CSN therefore uses different from SSN and STIRES satellite imagery to detect marine pollution and track the perpetrators. Here the EU Thetis report supports the implementation of the new Port State Control (PSC) inspection regime laid down in a number of EU Directives that support the Paris Memorandum of Understanding on PSC (Paris MOU) which includes Canada, Iceland, Norway and the Russian Federation.

To facilitate the planning of inspections, CleanSeaNet (CSN) is linked to SafeSeaNet (SSN) which provides information on ships in, or expected at, all ports of the Member States. In co-operation with users, EMSA's Earth Orbit (EO) team plans and orders satellite imagery to meet their service coverage requirements. After image acquisition trained operators assess the images, together with supporting information (meteorological, oceanographic and ancillary information such as AIS and vessel detection) to identify possible pollutions, to determine the likelihood of the presence of oil on the sea surface and to assist in identifying the source of the pollution.

When a possible oil spill is detected in European waters, an alert message is sent to coastal States. Analyzed images are available to national contact points in near-real time and are sent to the national authorities who then follow up on the alert report.

A distributed network of CleanSeaNet services providers process and analyze images on a routine basis from ENVISAT, RADARSAT 1 and 2 and in future from the SENTINEL-1 satellite. Service timeliness and quality control is an incentive to guarantee best service. CleanSeaNet has the capacity to acquire image segments from 200 km long up to 1400

---

[88] http://www.emsa.europa.eu/csn-menu.html

km with a nominal "*Near Real Time*" performance of 30 minutes for a 400 km long acquisition.

Based on rules decided by the Member States, CleanSeaNet disseminates specific alerts to the national users; the alert level is defined for each spill by combining likelihood information, culprit information and impact information. When a recent or ongoing spill is detected with a potential source connected to it or in its vicinity, the relevant authorities in are informed immediately. CleanSeaNet combines the oil spill detections with vessel information from SafeSeaNet and backtracking models from national and regional centers to help identify the source of pollution.

CSN near-real time service capabilities are therefore crucial to a rapid response by coastal states as well as to increase the likelihood of catching the polluter red-handed. In case of oil spill related accidents or emergencies the affected coastal State can request additional satellite images to monitor the spill area over an extended period of time, capturing the evolution of the spill and supporting response and recovery operations.

The CleanSeaNet Data Centre (CSN-DC) is the core system to receive, manage, distribute and visualize the data products of the CleanSeaNet service: oil spill and vessel detection data, new satellite missions on an ad-hoc basis, Electronic Nautical Charts (ENC) and additional information layers - also provided by the user community.

The CSN-DC provides a single user interface (GIS viewer):

- for web based image planning and allocation by the Member States,
- for accessing images and the analysis results,
- for accessing the CleanSeaNet data base
- and for communication between users allowing the exchange of knowledge.

The CSN-DC allows flexible navigation between the products: it is possible to search not only for satellite images but also for potential oil spills and vessel detection results within a certain time interval and area of interest. Results of these queries are displayed both on the map and in tables with a direct interaction between both.

The CleanSeaNet service is available to all participating States including EU Member States and their overseas territories, candidate countries and the European Free Trade Association (EFTA) Member States.

Each coastal State has access to the CleanSeaNet service through the SafeSeaNet Ecosystem Graphical User Interface (GUI) from STIRES, which enables them to view ordered images. Users can also access a wide range of supplementary information through the interface, such as oil drift modelling (forecasting and backtracking), optical images, and oceanographic and meteorological information.

Electronic Quality Ship Information System (EQUASIS)

The aim of the Electronic Quality Ship Information System (EQUASIS)[89] of the European Maritime Safety Agency (EMSA) is to improve the quality and security of international maritime transport through the exchange of information in the maritime industry and public bodies. The voluntary data is available on the Internet.

The role of the industry in promoting quality and safety in marine transport was at the heart of the Quality Shipping Campaign, launched by the European Commission and the UK Government in November 1997. The Campaign's aim was to bring together all players involved in the various fields of marine business in an effort to improve maritime safety. It was based upon dialogue between all the marine industry and public authorities and its tools were, primarily, voluntary measures. As the Quality Shipping Campaign demonstrated, one of the greatest impediments to a genuine quality culture in shipping is the lack of transparency in the information relating to the quality of ships and their operators.

While much relevant information is collected and available, it is scattered and often difficult to access. One of the main conclusions of the Quality Shipping Conference in Lisbon in June 1998, was an unanimous call from the participants representing the whole range of industry professionals (including ship-owners, cargo owners, insurers, brokers, classification societies, agents, ports and terminals), to make such information more accessible.

In response to this call, the European Commission and the French Maritime Administration decided to cooperate in developing an information system which collates existing safety-related information on ships from both public and private sources, and makes it available on the Internet. The main principles associated with the set-up of the EQUASIS information system were as follows:

- EQUASIS should be a tool aimed at reducing substandard shipping, and it should be limited to safety-related information on ships;
- EQUASIS addresses a public concern and should act accordingly;
- EQUASIS should be an international database covering the whole world fleet;
- Active co-operation with all players involved in the maritime industry is needed;
- EQUASIS will be a tool used for better selection of ships, but it will be used on a voluntary basis and there will be no legal pressure for industry to use it.

The set-up and effective operation of EQUASIS will help promote the exchange of unbiased information and transparency in maritime transport and thus allow persons involved in maritime transport to be better informed about the performance of ships and maritime organizations with which they are dealing. The EQUASIS website went online on 17. May 2000.

---

[89] http://www.equasis.org

Integrated Maritime Data Environment (IMDatE)

The European Maritime Safety Agency (EMSA) seems to strive for information superiority in the European Union in respect to Maritime Traffic Data. For this reason EMSA operates and manages a wide suite of systems which receive, process, and distribute information on vessel traffic reports (LRIT, SafeSeaNet), satellite monitoring (CleanSeaNet), and Port State Control (Thetis). They will be integrated within a technical framework, the Integrated Maritime Data Environment (IMDatE)[90].

The Integrated Maritime Data Environment (IMDatE) is a technical framework currently under development. In future, it will combine and process data from EMSA's maritime applications SafeSeaNet (SSN), CleanSeaNet (CSN), Thetis and other external sources as well as the IMO established Long Range Identification and Tracking (LRIT) to provide a more complete maritime picture to users.

IMDatE combines different data sources to provide a complete near real-time maritime picture, including AIS, LRIT, Satellite AIS, coastal radars, VMS, and Earth Observation data. IMDatE provides data fusion functionality in order to provide enhanced information from the combination of data sources. IMDatE provides access to different data sets and services according to individual user access rights. These may be delivered via a user friendly web interface via SafeSeaNet (SSN) or distributed automatically to authorized external systems.

- Integrated Ship Profile Service. This service provides a combined view of all information related to a ship or fleet based on information available in the different systems which are connected to the IMDatE.
- Area Centric Service. This service provides a complete maritime and oceanographic picture of a selected area, built-up from different layers of information, such as ship traffic data (full range of available ship position reports), satellite SAR picture of the defined area, optical image of the area, weather forecast associated to the area, oceanographic data (currents, waves, sea temperature, algae, etc.).
- Maritime Surveillance Service. This service allows users to analyses all available ship traffic information in order to identify activities of interest for the purposes of Maritime Surveillance activities.
- EU Common Maritime Space Monitoring Service. This specific service supports the implementation of EU Common Maritime Space (CMS) applications. In particular, the service will monitor ships engaged in EU (coastal) trade and ships (ferries and coasters) engaged in scheduled and/or regular services between EU ports.

---

[90] http://www.emsa.europa.eu/lrit-main/lrit-home/item/489-integrated-maritime-data-environment-imdate.html

The project started early 2011 and the first year was mainly focused on detailing and clarifying the technical specifications, evaluating available technology and designing the system architecture. Second year (2012) was the year of technical development and upgrading whilst the 2013 is dedicated to operational testing and delivery of sustainable services. The guiding principle of the project was to move from the existing operational status of stand-alone systems delivering individual services towards a flexible service-oriented approach, where users can access a portfolio of services without the burden of being involved in the technical details of the system(s).

Using the Service Oriented Architecture (SOA) design, OGC (Open Geographical Consortium) standards and standard web-services (XML, SOAP, WMS, WFS) we have reconverted the existing systems into modules of a horizontal platform where services can be designed and data integrated through various combinations of existing resources. This was a complex process involving Concurrent Engineering (CE) and agile software development approach for Integrated Product Development (IPD) (David Rainey, 2008).

It was an iterative process of collecting user requirements, propagating early conceptual designs, running computational models, creating prototypes, developing and delivering products, testing and analyzing needs and performance, incremental development and restarting the full cycle considering new governance, funding, work force capability and priorities from one year to another.

SafeSeaNet (SSN)

SafeSeaNet (SSN)[91], sometimes also referred to as Safety@SeaNet, is a European platform for Maritime Data Exchange between Member States' maritime authorities. Since 1993, the Commission has initiated over 15 proposed Directives or Regulations concerning passenger vessels' safety, prevention of pollution, port state control, requirements for seafarers, etc. Their implementation includes the collection and dissemination of maritime data send from 27 nations which SafeSeaNet supports in order to enhance:

- Maritime safety
- Port and maritime security
- Marine environment protection
- Efficiency of maritime traffic and maritime transport

SSN has been set up as a network for maritime data exchange, linking together maritime authorities from across Europe. It enables European Union Member States, Norway, and Iceland, to provide and receive information on ships, ship movements, and hazardous cargoes.

SSN's main objective is to aid the collection, dissemination and harmonized exchange of maritime data. The network assists communication between authorities at local/regional level and central authorities thus contributing to prevent accidents at sea and, by extension, marine pollution, and that the implementation of EU maritime safety legislation will be made more efficient. The SSN network involves many maritime authorities across Europe, each with their own IT infrastructure and objectives. This invariably leads to varying data formats distributed across different systems throughout Europe.

EMSA has through SSN developed a community vessel traffic monitoring and information system according to Directive 2002/59/EC. In addition, it incorporates data exchange requirements from other EU Directives such as those relating to:

- Port reception facilities for ship waste and
- Port state control inspections in ports of the European Union.

SafeSeaNet (SSN) started services in 2004 with a major upgrade 2009, AIS, port- and Hazmat notifications, cargo lists are based on the terrestrial system with at that time 727 AIS coastal stations with daily 13.000 ships tracked in European waters and an average of over 100.000.000 AIS positions per month in track history. SSN has 2.326 data providers and 556 authorized users, VTS's, MRCC's, Port Authorities, Coast Guards, Pollution survey center, national administrations of 22 EU coastal states, Norway and Iceland. In June 2009 STIRES was delivered to the EU and the EU LRIT Data Center started sending LRIT reports in 2010.

---

[91] http://www.emsa.europa.eu/ssn-main.html

SafeSeaNet (SSN) and the related SafeSeaNet Traffic Information Relay and Exchange System (STIRES) are the first EU systems for an EU maritime picture.

On the technical side, the core of the SafeSeaNet architecture is the EIS (European Index Server). This acts as a secure and reliable index system within a "*hub and spoke*" network (including authentication, validation, data transformation and logging) which sends requests to, and receives notifications and responses from, approved users. SSN has further implemented EIS to a Central Index System that stores only references to the data locations and not the actual data itself. It functions as a central hub for all communication between data requesters and data providers - somewhat like a telephone switchboard.

The Central Index needs to know what information each data provided holds. Data providers connected within the SSN network send information by means of a notification mechanism. The data provider, upon receiving queries from the data requester routed through the Central Index, retrieves the data from their local database. In this way the Central Index acts as the sole point of contact. The main information elements that are contained in the system and made available to users are as follows:

- Automatic Identification System (AIS) based near-real-time ship positions,
- Archived historical ship positions (over several years),
- Additional information from AIS-based ship reports (e.g. identification name/numbers, flag, dimensions, course, speed, dimensions, destination and ship type),
- Estimated/actual times of arrival/departure,
- Details of hazardous goods carried on board,
- Information on safety-related incidents affecting ships,
- Information on pollution-related incidents affecting ships,
- Details of waste carried on board/to be offloaded (from June 2015),
- Ship security-related information (from June 2015),
- Information on the location of remaining single hulled tankers,
- Information on the location of ships that have been banned from EU ports,
- Digital map layers (containing information on depths, navigation aids, traffic separation schemes, anchorages, AIS station locations, etc.).

SSN ensures that access to the Central Index is restricted and secure yet available 24/7 on TESTA (TESTA II Network, Trans-European Services for Telematics between Administrations)). Confidentiality is guaranteed by the use of PKI. The heart of the SSN is the XML messaging system with a Yellow Pages Index System (Address Book), with authentication, validation, data information, logs and auditing. Data exchange is via four main messages, such as ship based on MRS Notification, Alert Notifications, Notifications HAZMAT, Port Notifications. Ship Notifications based on AIS are to be delivered via a feed from STIR. Position update is currently so every 2 hours.

The EIS provides two different interfaces to enable users to exchange messages, first the

XML message-based interface, which enables the applications of Member States to communicate programmatically with the SafeSeaNet system (i.e. information is automatically exchanged between systems), and second the browser-based web interface, which provides a pan-European vessel traffic image and enables users to visualize information stored in the EIS.

The EIS is able to locate and retrieve information on vessels related to one Member State in response to a query or request made by another. The information exchanged is extensive, but the main notification reports submitted to SafeSeaNet are as follows:

- Ship Notifications: These are used to provide SafeSeaNet with voyage and cargo information. Notifications are based on two types of message. Automatic Identification System (AIS) messages are sent automatically by the ships through very high frequency (VHF) radio signals, and received by coastal stations within range. Mandatory ship reporting systems (MRS) can be established by governments, with approval from the International Maritime Organization, for certain types of vessel transiting through defined areas. MRS messages are sent by ship masters to coastal stations. Information includes ship identification, course, speed, and cargo.
- Port Notifications: These are used to notify SafeSeaNet when vessels arrive and depart from ports. The estimated time of arrival, actual time of arrival, actual time of departure and the number of persons on board are included in the message.
- Hazmat Notifications: These are used to notify SafeSeaNet that vessels are carrying hazardous materials (dangerous or polluting goods) on board, and that the data provider has detailed information on these goods.
- Incident Reports: These are used to provide SafeSeaNet with information on incidents involving ships. These might be related to ship safety and seaworthiness (e.g. collisions, groundings, equipment failures), the environment (e.g. pollution incidents) or other pre-defined categories (e.g. banned ships, ships not reporting according to rules).
- Waste Notifications: These are provided to users via SafeSeaNet in compliance with the EU Reporting Formalities Directive, and they allow interested parties to find out the different types of waste on board, and when and where they are to be discharged.
- Security Notifications: These are also provided to users via SafeSeaNet in compliance with the EU Reporting Formalities Directive, and they provide information on security issues that relate, in particular, to avoiding the ship being used as a weapon.

One challenge in SafeSeaNet are the Hazardous Material Reports (Hazmat) from the port authorities, which are partly supplied in PDF-format, which cannot be automatically read by the system. A new VTS version is based on the SafeSeaNet Traffic Information Relay and Exchange System (STIRES). SSN has considerably improved data exchange with better standardization and a profusion of transfer and contributes to the efficient implementation of the EU maritime safety legislation.

When accessing SafeSeaNet via the web interface, a map-based graphical interface (GIS), users have the capability to zoom in and out to display the images from EU-level to individual quays in ports. Historical vessel positions and selected information on ships can be obtained in a number of different ways. The information is presented on high-quality nautical charts containing a range of useful maritime information and users can provide and/or request data in SSN.

The SafeSeaNet Ecosystem Graphical User Interface (SEG) is the common web interface providing access to EMSA's maritime applications and data sets including SafeSeaNet, Integrated Maritime Services, Long Range Identification and Tracking and CleanSeaNet.

The SEG replaces the SSN Graphical Interface (GI), the IMS Web User Portal (WUP), the LRIT User Web Interface (UWI) and the CSN GIS Viewer. The SEG is designed for both mobile and desktop/laptop devices. The existing graphical interfaces associated with the individual systems will be phased out after a transitional period of one year.

SafetySeaNet Tracking Information Relay & Exchange System (STIRES)

SafetySeaNet Tracking Information Relay & Exchange System (STIRES) is a merchant ships tracking system that allows maritime administrations of member states of the European Union, Norway and Iceland to locate and control their fleets. It provides information on cargoes, ship safety records and port destinations, thus enabling the tracking of ship movements along the entire European Union coastline. STIRES facilitates exchange of information regarding merchant ships by interfacing with several existing and planned systems for supporting safety at sea, and for protecting the maritime environment as well as economic efficiency.

The STIRES system, one of the applications included currently in the SSN v.2 system, is conceived as an enhancement to the SafetySeaNet (SSN) system for facilitating relaying and exchanging information between the EU Member States, Norway and Iceland.

STIRES is based on article 9 of the Directive 2002/59 requires MS to build up all necessary equipment and shore-based installations and ensure that the appropriate equipment for relaying the information to and exchanging it between, the national systems of Member States shall be operational by the end of 2008. In addition the IMO has adopted amendments to the SOLAS convention in relation to LRIT.

The Transport, Telecommunications and Energy Council Meeting in Luxemburg agreed in October 2007, in addition to agreeing the setting-up of a European LRIT Data Centre, to a number of actions related to AIS data and AIS system development including:

1. encouraging integration of AIS data into the LRIT system; and
2. progressing integration of LRIT and AIS information in the context of an EU AIS Master Plan.

Three Regional Systems (RS) already existed in the European Union before the STIRES was designed:

- the HELCOM covering the Baltic Sea,
- the HELCOM covering the North Sea,
- and the Mediterranean Sea (now renamed MARES).

All three RS have established the exchange of AIS data between the MSs for the whole of their corresponding areas. It was therefore decided (during the initial design phase of STIRES in 2008), to take advantage of this and interface STIRES to the three RS in order for them to collect the AIS data provided by the MS instead of establishing new interfaces directly to each MS. A software application, currently called National Proxy (NPR) was developed to facilitate the connection between RS and the STIRES server. The proxies must be installed at the Regional premises (typically in a redundant configuration) and manages the secure connection which allows the RS and the STIRES to exchange AIS data.

The standard IEC 61162 protocol (R1) was simply adapted for the exchange of AIS data between the RS and STIRES because of the already wide utilization (with all 3 RS in common, already complying with the IEC 61162). However, the IEC 61162 standard, which was primarily designed for on-board AIS equipment, has some serious drawbacks and limitations when used in complex environments like SSN. For this reason, the STIRES core module (COR) and the proxy are both supporting the Comment Block (CB) extension as defined in the IEC 62320-1 standard [R2], which provides for the AIS Base Stations interface. In particular, CBs are already used in SSN to extend the content of the standard IEC 61162 AIS messages with complete timestamp information (only partially complete when transmitted by ships).

The experience from the use of NPRs within the STIRES architecture during the period that followed its launch, revealed that comment blocks could be used to expand further the usage of the NPR as an interface with Regional servers and Member States to improve the monitoring of the reception/ distribution streams; this would facilitate the exchange of position messages originated from systems other than terrestrial AIS (e.g. Satellite AIS, VMS, etc).

Furthermore, recent developments in the AIS standardization field (altering the use of some fields in messages 1, 2 and 3) as well as an assessment on the consequences of recent decisions of IMO (reference is made to SN.1/ Circ.289-2/6/2010 providing guidance on the use of AIS application-specific messages) calls for an evolution to the proxy application towards a more open and future-proof design approach.

Contrary to SafeSeaNet the STIRES is designed from the start for raw AIS data exchange and contains a full Graphic Information System (GIS) with standardized electronic navigation maps. In June 2009 STIRES was delivered to the EU and the EU LRIT Data Center started sending LRIT reports in 2010. SafeSeaNet (SSN) and STIRES are the first EU systems for an EU maritime picture.

As mentioned before, the SafeSeaNet Ecosystem Graphical User Interface (GUI) is the common web interface providing access to EMSA's maritime applications and data sets including SafeSeaNet, Integrated Maritime Services, Long Range Identification and Tracking and CleanSeaNet. The SafeSeaNet Ecosystem GUI (SEG) replaces the SSN Graphical Interface (GI), the IMS Web User Portal (WUP), the LRIT User Web Interface (UWI) and the CSN GIS Viewer. The SEG is designed for both mobile and desktop/laptop devices. The existing graphical interfaces associated with the individual systems will be phased out after a transitional period of one year.

Thetis

Thetis is a report system of ships earmarked for inspection and records the results. These reports are made available via Thetis to all Port State control authorities in the Community and the Paris MOU.

Thetis also interfaces with a number of other maritime safety-related databases including those of the EU-recognized classification societies, Community and national information systems and other Port State control regimes so as to exchange data and provide a full picture for the inspector. Inspection results are also available through a public website.

Overall, EMSA's suite of maritime surveillance systems is enabling EU Member States as well as the participation nations of Norway, Iceland, and including Canada and Russia under the Paris MoU and other European and international agreements, to provide and receive information on ships, ship movements and hazardous cargoes.

The main information elements that are contained in the system and made available to users are focused on safety. Positional data relies on Automatic Identification System (AIS) and LRIT, which are cooperative systems, which renders the system insufficient for security purposes.

In the end of the day, Thetis, like many other singular existing systems should be an integral part of the Integrated Maritime Data Environment (IMDatE) provided by EMSA or within the Common Information Sharing Environment (CISE) for the EU Member States in order to prevent a further fragmentation of systems, which is counter-productive in Maritime Situational Awareness.

EMSA has developed a new module in THETIS, namely THETIS-MRV[92], enabling companies responsible for the operation of large ships using EU ports to report their CO_2 emissions under the Regulation (EU) 2015/757 on Monitoring, Reporting and Verification of CO_2 from marine transport.

Through this web-based application all relevant parties foreseen by the Regulation can fulfil their monitoring and reporting obligations in a centralized and harmonized way.

THETIS-MRV includes a mandatory and a voluntary module: through the mandatory module, companies will generate Emission Reports which will then be assessed by Verifiers who will issue a Document of Compliance in system; through the voluntary module, companies may draft their monitoring plans and the system will make them available for verifier' assessment. The system is available since 7. August 2017.

---

[92] https://mrv.emsa.europa.eu

Fisheries Areas Network (FARNET)

Fisheries Areas Network (FARNET)[93] is the community of people implementing Community-Led Local Development (CLLD) under the European Maritime and Fisheries Fund (EMFF). This network brings together Fisheries Local Action Groups (FLAGs), managing authorities, citizens and experts from across the EU.

This CLLD funding is delegated to local partnerships that bring together the private sector, local authorities and civil society organizations. Known as Fisheries Local Action Groups (FLAGs), these partnerships fund local projects within the framework of a strategy, developed in response to specific needs and opportunities identified locally.

Continuing the area-based approach initiated by Axis 4 of the European Fisheries Fund (EFF), CLLD under the EMFF brings new opportunities for local communities to tackle the multiple challenges faced by coastal and fisheries areas across the EU. Building on the 10.000+ projects supported in the 2007 – 2014 period, local communities will now also have the possibility to combine CLLD funds from the EMFF with those from other European Structural and Investment Funds: the European Social Fund (ESF), European Regional Development Fund (ERDF) and European Agricultural Fund for Rural Development (EARDF).

The FARNET Support Unit has been set up by the European Commission to assist in the implementation of CLLD under the EMFF. Working closely with the Directorate-General for Maritime Affairs and Fisheries (DG MARE), the member state administrations, the national fisheries areas networks, the FLAGs and others involved in the implementation of CLLD, the FARNET Support Unit strives to build a *"learning network"* that connects the growing knowledge and experience in CLLD from fisheries areas across Europe.

The European Fisheries Areas Network (FARNET) includes all fisheries areas supported by Priority Axis 4[94] of the European Fisheries Fund (EFF) in a community networking. Through information exchange and a dedicated support unit, this network aims to assist the different stakeholders involved in the sustainable development of fisheries areas at local, regional, national and European level.

The core of the network is made up of over 300 Fisheries Local Action Groups (FLAGs). These public-private partnerships, set up at local level, work towards the sustainable development of their areas. Based in 21 Member States, these FLAGs each manage a budget to support a range of projects proposed and carried out by a wide variety of local stakeholders.

Fisheries areas across the EU are confronting significant challenges. A decline in income and employment in the fishing sector has underlined the urgent need for new, sustainable

---

[93] https://ec.europa.eu/fisheries/cfp/eff/farnet_en
[94] There are five priority areas (axes) for EFF funding: https://ec.europa.eu/fisheries/cfp/eff_en

and inclusive solutions.

Europe's oceans, seas and inland waters, and the environmental and cultural assets linked to them, represent a vast and rich resource which can provide new opportunities for economic and social renewal. A key challenge, therefore, is to re-evaluate the potential of these assets in order to identify and develop new, more diversified and more sustainable economic activities and employment.

Priority Axis 4 of the European Fisheries Fund (EFF) provides support for the sustainable development of fisheries areas. This involves adding value and creating jobs at all stages of the fisheries and aquaculture supply chain, as well as supporting diversification inside and outside fisheries. Furthermore, Axis 4 aims to provide local people in general, and fishing communities in particular, with a tool for participating in and benefiting from the potential of blue growth and coastal development.

FARNET connects local people from all sectors – public, private and civil society - into Fisheries Local Action Groups (FLAGs) that design and implement integrated local development strategies that meet their needs. The aim of the FARNET Support Unit is to support and facilitate this process.

The FARNET Support Unit is the technical assistance team established by the European Commission to assist in the implementation of Axis 4. It consists of a permanent team of 10 people, based in Brussels, assisted by 21 regional experts.

Fishery Data Exchange System (FIDES)

Fishery Data Exchange System (FIDES)[95] provides the communication network between the European Commission and EU Member States, as they put this policy into practice, safeguarding fish stocks and the fishery sector. FIDES is a one-stop shop which automates the management of fishery data using Internet technologies, accessible by national administrations in the EU Member States and the European Commission. It offers a wide range of alternatives such as web, e-mail and file transfer.

With worldwide fish stocks declining, the European Commission's DG Fisheries and relevant administrations in the European Union face the challenge of striking a sustainable balance between available marine resources and their exploitation through the Common Fishery Policy. FIDES provides the communication network between the European Commission and EU Member States, as they put this policy into practice, safeguarding fish stocks and the fishery sector.

FIDES is a one-stop shop which automates the management of fishery data using Internet technologies, accessible by national administrations in the EU Member States and the European Commission. It offers a wide range of alternatives such as web, e-mail and file transfer.

Overall, FIDES aims to improve the operation of the Community's Common Fishery Policy through a technology enhanced communications infrastructure linking DG Fisheries and corresponding administrations in the EU Member States. There is also the possibility of expanding the system to other countries in the future. Member States can send and retrieve data to and from the FIDES application through several communication channels (e-mail, HTTP etc.). The FIDES system acts as an enabler, providing the link between Member State business processes and DG Fisheries.

Member States can send and retrieve data to and from the FIDES application through several communication channels (e-mail, HTTP etc.). The FIDES system acts as an enabler, providing the link between Member State business processes and DG Fisheries. It does not store the actual information in a database but rather provides a reference link connecting the user with the data. It automatically acknowledges, archives and logs each information exchange (called "transaction"). Different authorized users can submit or request statistical reports on transactions from the system, but the available action and information depends on the user's access level within the system.

The configuration requirement for FIDES is minimal in order to take away the added cost in time and financial resources of installing DG Fisheries software or hardware on the premises of EU Member State administrations.

---

[95] http://ec.europa.eu/idabc/en/document/2254/5926.html

FRONTEX

FRONTEX[96] (abbreviation from French: Frontières extérieures for English "External Borders", is the European Agency for the Management of Operational Cooperation at the External Borders of the Member States of the European Union) and was developed on the basis of the Schengen Agreement to support the aim of a Europe without border controls. FRONTEX promotes, coordinates and develops European border management in line with the EU fundamental rights charter applying the concept of Integrated Border Management.

During the 1980s five Member States (Belgium, France, Germany, Luxembourg and the Netherlands) decided to create a territory without internal borders. They signed the first agreements in a small town in Luxembourg called Schengen, hence the "*Schengen area*", a territory in which the free movement of persons is guaranteed. The original agreement was complemented in 1990 by a convention.

When this convention entered into force in 1995 it abolished checks at the internal borders and created a single external border. Whatever their location, officers working at the external border perform border checks in accordance with identical procedures. The rules governing visas and the right to asylum are also common for all Schengen countries.

In order to keep a balance between freedom and security the participating member states agreed to introduce so-called "*compensatory measures*". These are focused on cooperation and coordination of the work of the police and judicial authorities. Organized crime networks do not respect borders; therefore this cooperation is a key to safeguard internal security.

In 1999, with the signing of the Treaty of Amsterdam, this intergovernmental cooperation was incorporated into the EU framework. Since 1999 the European Council on Justice and Home Affairs has taken several steps towards strengthen cooperation in the area of migration, asylum and security.

FRONTEX helps border authorities from different EU countries work together and started operations in 2004 to reinforce and streamline cooperation between national border authorities. In pursuit of this goal, FRONTEX has several operational areas which are defined in the founding FRONTEX Regulation and a subsequent amendment.

While fulfilling its mandate, FRONTEX liaises closely with other EU partners involved in the development of the area of Freedom, Security and Justice such as Europol, EASO, Eurojust, FRA or CEPOL, as well as with customs authorities in order to promote overall cohesion. In the border management field this led to the creation of the External Border Practitioners Common Unit - a group composed of members of the Strategic Committee on Immigration, Frontiers and Asylum (SCIFA) and heads of national border control services.

---

[96] https://frontex.europa.eu

FRONTEX Missions and tasks:

- Joint Operations — FRONTEX plans, coordinates, implements and evaluates joint operations conducted using Member States' staff and equipment at the external borders (sea, land and air).
- Training — FRONTEX is responsible for developing common training standards and specialist tools. These include the Common Core Curriculum, which provides a common entry-level training rationale for border guards across the Union, and mid- and high-level training for more senior officers.
- Risk Analysis — FRONTEX collates and analyses intelligence on the ongoing situation at the external borders. These data are compiled from border crossing points and other operational information as well as from the Member States and open sources including mass media and academic research.
- Research — FRONTEX serves as a platform to bring together Europe's border-control personnel and the world of research and industry to bridge the gap between technological advancement and the needs of border control authorities.
- Providing a rapid response capability — FRONTEX has created a pooled resource in the form of European Border Guard Teams (EBGT) and an extensive database of available equipment which brings together specialist human and technical resources from across the EU. These teams are kept in full readiness in case of a crisis situation at the external border.
- Assisting Member States in joint return operations — When Member States make the decision to return foreign nationals staying illegally, who have failed to leave voluntarily, FRONTEX assists those Member States in coordinating their efforts to maximize efficiency and cost-effectiveness while also ensuring that respect for fundamental rights and the human dignity of returnees is maintained at every stage.
- Information systems and information sharing environment — Information regarding emerging risks and the current state of affairs at the external borders form the basis of risk analysis and situational awareness for border control authorities in the EU. FRONTEX develops and operates information systems enabling the exchange of such information, including the Information and Coordination Network established by Decision 2005/267/EC as European Border Surveillance System (EUROSUR).

The Common Unit coordinated national projects of the first Ad-Hoc Centers for Border Control. Their task was to oversee EU-wide pilot projects and common operations related to border management.

Two years after the establishment of the first Ad-Hoc Centers for Border Control the European Council decided to go a step further. With the objective of improving procedures and working methods of the Common Unit, on the 26 October 2004 the European Agency for the Management of Operational Cooperation at the External Borders of the Member

States of the European Union (FRONTEX) was established and is today an integral part of the European Security.

The original six ad-hoc centers are:

- Risk Analysis Center (Helsinki, Finland)
- Center for Land Borders (Berlin, Germany)
- Air Borders Center (Rome, Italy)
- Western Sea Borders Center (Madrid, Spain)
- Ad-hoc Training Center for Training (Traiskirchen, Austria)
- Center of Excellence (Dover, United Kingdom)
- Eastern Sea Borders Center (Piraeus, Greece)

Currently, as of 2014, the European Union extends along approximately 12,000 km of land borders and 45,000 km of maritime borders. In contrast to many other policy areas, the common policies for the EU's external borders do not adhere to the aims originally set at their creation in the 1950s. On the contrary, from the very beginning, the EU's founding fathers rooted four fundamental freedoms in the treaties. Frontex is not a border policing body, but an agency of the European Union, contributing to the management of these EU's external borders.

Frontex continues to expand its capacity to support EU Member States in returns after launching the European Centre for Returns in 2018. In the first half of 2018, Frontex has coordinated and co-financed 165 return operations by charter flights, returning nearly 6400 non-EU nationals to non-EU countries.

The Frontex European Center for Returns provides operational and technical support to EU Member States and Schengen associated countries in carrying out return operations. It also provides assistance with pre-return activities, such as acquisition of the necessary travel documents.

While Frontex supports return operations, the decision about who should be returned is always taken by the judicial or administrative authorities of the individual Member States. Although most return operations coordinated by Frontex are done on planes chartered by Member States, the agency has also been conducting a pilot project to organize returns by scheduled commercial flights. So far, 11 EU Member States have benefitted from the pilot project, which has been extended to 30. June 2019. Frontex is also about to begin to charter planes by itself for return operations, giving more flexibility to the agency and Member States.

The original mandate of Frontex was limited in purpose, focusing on the coordination of border control operations, preparing risk assessments and assisting member states with training and return missions. Fairly quickly, the EU border management agency saw an expansion of competences with the creation of Rapid Border Intervention Teams (RABIT) in 2007 to provide rapid operational assistance to member states under exceptional migration pressures. However, Frontex's most significant boost in competences came in the aftermath of the 2015 migration crisis. After only about a year of negotiations, EU member states turned Frontex into a European Border and Coast Guard Agency (EBCGA). Launched in October 2016, the EBCGA was dubbed "Frontex+" because of its additional resources and competences such as the power to initiate EU return flights for irregular migrants and more responsibilities countering organized crime and smuggling of human beings.

In the first 11 months of 2018, the number of irregular border crossings into the EU fell by 30% from a year ago to about 138 000, mainly because of lower migratory pressure in the Central Mediterranean. At the same time the UN statistics show 2262 drowned while trying to migrate. Until 2027 an additional 10,000 border guards are under consideration in the EU.

Bulgaria, Cyprus, Ireland, Romania and the UK and are not members in FRONTEX, while the non-EU members Norway, Iceland and Switzerland do participate.

European Border Surveillance System (EUROSUR)

According to the Frontex study on concepts and systems of the authorities for border control are overlapping and inefficient. Consequently, the establishment of a European Border Surveillance System was adopted and the a "European Border Surveillance System[97]" (EUROSUR) became the first project as a system-of-system and regional approach for becoming a component of the Integrated Maritime Policy of the EU. Due to road of the development taken, there is however no longer any system-of-systems approach left, rather it has to be seen as server farms and hubs connecting the national systems.

The EU Commission outlines a three-phase common technical framework for setting up EUROSUR to support the Member States in their efforts to reduce the number of illegal immigrants entering the European Union by improving their situational awareness at their external borders and increasing the reaction capability of their information and border control authorities. EUROSUR was implemented in three phases:

1. PHASE: Interconnect and rationalize border surveillance systems at national level.
2. PHASE: Improve the performance of surveillance tools at EU level.
3. PHASE: Creation of a common monitoring and information-sharing environment for the EU maritime domain.

EUROSUR was gradually establishing a mechanism whereby Member States' authorities carrying out border surveillance can cooperate and share operational information with one other and with FRONTEX. The political priority given to EUROSUR was confirmed by the Stockholm Program and the Action Plan for its implementation.

European Border Surveillance System (EUROSUR) is today the information-exchange framework designed to improve the management of Europe's external borders, a system used by Frontex, the European Agency for the Management of Operational Cooperation at the External Borders of the Member States of the European Union. It aims to support Member States by increasing their situational awareness and reaction capability in combating cross-border crime, tackling irregular migration and preventing loss of migrant lives at sea.

The backbone of EUROSUR is the network of National Coordination Centers (NCCs). Each member state establishes an NCC, which groups the authorities responsible for border control in a given member state. The main role of the NCC is to coordinate the border surveillance activities on national level and serve as a hub for the exchange of information. The EUROSUR Fusion Services include automated vessel tracking and detection capabilities, software functionalities allowing complex calculations for detecting anomalies and predicting vessel positions, as well as precise weather and oceanographic forecasts.

---

[97] Search EUROSUR: https://eur-lex.europa.eu/browse/summaries.html

The practical application of the necessary technical and legal requirements of EUROSUR, the creation of central databases with the involvement of the Data Protection Authorities took place in 1995. The agreement governs the compensatory measures that ensure a single area for security and justice. FRONTEX is unifying via EUROSUR the rules for entry and short stays by foreigners in the Schengen Area (Schengen Visa).

- Asylum (determining MS's responsibilities for asylum)
- Measures to prevent cross-border drug trafficking
- Police cooperation (hot pursuit), and
- Cooperation between the Schengen States in the judiciary system

When the Schengen Agreement entered into force in 1993, a singular system for the Agency was in the planning process, however it was not foreseen that it would require the integration of Maritime Surveillance/MSA as a major data source into EUROSUR and align the respective developments.

Therefore the European Commission has published an evaluation of the European Border Surveillance System (EUROSUR), which recommends that the system be expanded for the *"systematic inclusion"* of all border crossing points; the monitoring of *"secondary movements"* of migrants within the EU; and to develop new services and better cooperate with *"third parties"*, for example through *"big data analysis"* of EU databases such as the Schengen Information System, the Visa Information System and Europol's computer systems.

Galileo

Galileo[98] is the global navigation satellite system (GNSS) that is being created by the European Union (EU) through the European GNSS Agency (GSA), headquartered in Prague in the Czech Republic, with two ground operations centers; Oberpfaffenhofen near Munich in Germany and Fucino in Italy. The project is named after the Italian astronomer Galileo Galilei with an initial budget of 10B €.

One of the aims of Galileo is to provide an independent high-precision positioning system so European nations do not have to rely on the U.S. GPS, or the Russian GLONASS systems, which could be disabled or degraded by their operators at any time. The use of basic (lower-precision) Galileo services will be free and open to everyone. The higher-precision capabilities will be available for paying commercial users. Galileo is intended to provide horizontal and vertical position measurements within 1-meter precision, and better positioning services at higher latitudes than other positioning systems. Galileo is also to provide a new global search and rescue (SAR) function as part of the MEOSAR system.

The first Galileo test satellite, the GIOVE-A, was launched 28 December 2005, while the first satellite to be part of the operational system was launched on 21 October 2011. As of July 2018, 26 of the planned 30 active satellites are in orbit. Galileo started offering Early Operational Capability (EOC) on 15 December 2016, providing initial services with a weak signal, and is expected to reach Full Operational Capability (FOC) in 2019. The complete 30-satellite Galileo system (24 operational and 6 active spares) is expected by 2020.

In 1999, the different concepts of the three main contributors of ESA (Germany, France and Italy) for Galileo were compared and reduced to one by a joint team of engineers from all three countries. The first stage of the Galileo program was agreed upon officially on 26 May 2003 by the European Union and the European Space Agency. The system is intended primarily for civilian use, unlike the more military-oriented systems of the United States (GPS), Russia (GLONASS), and China (BeiDou-1/2). The European system will only be subject to shut down for military purposes in extreme circumstances (like armed conflict). It will be available at its full precision to both civil and military users. The countries that contribute most to the Galileo Project are Germany and Italy.

Galileo is intended to be an EU civilian GNSS that allows all users access to it. Initially GPS reserved the highest quality signal for military use, and the signal available for civilian use was intentionally degraded (Selective Availability). This changed with President Bill Clinton signing a policy directive in 1996 to turn off Selective Availability. Since May 2000 the same precision signal has been provided to both civilians and the military.

Since Galileo was designed to provide the highest possible precision (greater than GPS) to anyone, the US was concerned that an enemy could use Galileo signals in military strikes

---

[98] https://en.wikipedia.org/wiki/Galileo_(satellite_navigation)

against the US and its allies (some weapons like missiles use GNSSs for guidance). The frequency initially chosen for Galileo would have made it impossible for the US to block the Galileo signals without also interfering with its own GPS signals. The US did not want to lose their GNSS capability with GPS while denying enemies the use of GNSS. Some US officials became especially concerned when Chinese interest in Galileo was reported.

An anonymous EU official claimed that the US officials implied that they might consider shooting down Galileo satellites in the event of a major conflict in which Galileo was used in attacks against American forces. The EU's stance is that Galileo is a neutral technology, available to all countries and everyone. At first, EU officials did not want to change their original plans for Galileo, but have since reached the compromise that Galileo is to use a different frequency. This allowed the blocking or jamming of either GNSS without affecting the other.

One of the reasons given for developing Galileo as an independent system was that position information from GPS can be made significantly inaccurate by the deliberate application of universal Selective Availability (SA) by the US military. GPS is widely used worldwide for civilian applications; Galileo's proponents argued that civil infrastructure, including airplane navigation and landing, should not rely solely upon a system with this vulnerability.

On 2 May 2000, SA was disabled by the President of the United States, Bill Clinton; in late 2001 the entity managing the GPS confirmed that they did not intend to enable selective availability ever again. Though Selective Availability capability still exists, on 19 September 2007 the US Department of Defense announced that newer GPS satellites would not be capable of implementing Selective Availability; the wave of Block IIF satellites launched in 2009, and all subsequent GPS satellites, are stated not to support SA. As old satellites are replaced in the GPS Block IIIA program, SA will cease to be an option. The modernization program also contains standardized features that allow GPS III and Galileo systems to interoperate, allowing receivers to be developed to utilize GPS and Galileo together to create an even more accurate GNSS.

In June 2004, in a signed agreement with the United States, the European Union agreed to switch to a modulation known as BOC 1.1 (Binary Offset Carrier 1.1) allowing the coexistence of both GPS and Galileo, and the future combined use of both systems. The European Union also agreed to address the "*mutual concerns related to the protection of allied and US national security capabilities.*"

Galileo is to provide a new global search and rescue (SAR) function as part of the MEOSAR system. Satellites will be equipped with a transponder which will relay distress signals from emergency beacons to the Rescue coordination center, which will then initiate a rescue operation. At the same time, the system is projected to provide a signal, the Return Link Message (RLM), to the emergency beacon, informing them that their situation has been detected and help is on the way. This latter feature is new and is considered a major upgrade compared to the existing COSPAS-SARSAT system, which does not provide feedback to the user. Tests in February 2014 found that for Galileo's search and rescue function, operating as part of the existing International COSPAS-SARSAT Program, 77% of simulated distress locations can be pinpointed within 2 km, and 95% within 5 km.

In 2014, ESA and its industry partners begun studies on Galileo Second Generation satellites, presented to the EC for the late 2020s launch period. One idea is to employ electric propulsion, which would eliminate the need for an upper stage during launch and allow satellites from a single batch to be inserted into more than one orbital plane.

Each Galileo satellite has two master passive hydrogen maser atomic clocks and two secondary rubidium atomic clocks which are independent of one other. As precise and stable space-qualified atomic clocks are critical components to any satellite-navigation system, the employed quadruple redundancy keeps Galileo functioning when onboard atomic clocks fail in space. The onboard passive hydrogen maser clocks' precision is four times better than the onboard rubidium atomic clocks and estimated at 1 second per 3 million years (a timing error of a nanosecond or 1 billionth of a second (10^{-9} or $1/1,000,000,000$ s) translates into a 30 cm (11.8 in) positional error on Earth's surface), and will provide an accurate timing signal to allow a receiver to calculate the time that it takes the signal to reach it.

The Galileo satellites are configured to run one hydrogen maser clock in primary mode and a rubidium clock as hot backup. Under normal conditions, the operating hydrogen maser clock produces the reference frequency from which the navigation signal is generated. Should the hydrogen maser encounter any problem, an instantaneous switchover to the rubidium clock would be performed. In case of a failure of the primary hydrogen maser the secondary hydrogen maser could be activated by the ground segment to take over within a period of days as part of the redundant system. A clock monitoring and control unit provides the interface between the four clocks and the navigation signal generator unit (NSU). It passes the signal from the active hydrogen master clock to the NSU and also ensures that the frequencies produced by the master clock and the active spare are in phase, so that the spare can take over instantly should the master clock fail. The NSU information is used to calculate the position of the receiver in trilateration the difference of the received signals from multiple satellites.

The onboard passive hydrogen maser and rubidium clocks are very stable over a few hours. If they were left to run indefinitely, though, their timekeeping would drift, so they need to be synchronized regularly with a network of even more stable ground-based reference clocks. These include active hydrogen maser clocks and clocks based on the cesium frequency standard, which show a far better medium and long-term stability than rubidium or passive hydrogen maser clocks.

These clocks on the ground are gathered together within the parallel functioning Precise Timing Facilities in the Galileo Control Centers in Fucino and Oberpfaffhofen. The ground based clocks also generate a worldwide time reference called Galileo System Time (GST), the standard for the Galileo system and are routinely compared to the local realizations of UTC, the UTC(k) of the European frequency and time laboratories.

Maritime Security (MASE) Program /Projects

The European Union launched the program to promote regional Maritime Security (MASE)[99], and continues to play a key role throughout its implementation. Start-up MASE, a precursor to the full MASE Program, was completed in June 2013, leading the way for the commencement of the MASE Program. The project, funded by the European Union Development Fund, has a budget of 37 million EUR and will be implemented by the Indian Ocean Commission (IOC) and other regional organizations (COMESA, EAC, and IGAD) over a five year period.

The contribution of the European Union is 37M EUR and the duration of the project will take five years, it started from 2012 and will end in 2018. The overall objective of the program is to enhance maritime security in the ESA-IO region hence contribute to global security and create a favorable environment for the economic development of the ESA-IO region and beyond. The Specific objective of the program is to strengthen the capacity of the ESA-IO region in the implementation of the Regional Strategy and Action Plan against Piracy and for Maritime Security. The program has five results to be achieved in four different locations:

1. Alternative livelihoods through vocational development initiatives and advocacy against piracy are supported; maritime coordination mechanisms are reinforced in Somalia (IGAD).
2. National and regional capacities in legal matters, legislation and infrastructure for the arrest, transfer, detention and trial of pirates are developed and / or strengthened (EAC).
3. The regional capacity to break the financial networks of pirates and their sponsors and to decrease structural and sustainable economic impact of piracy are reinforced (COMESA).
4. National and regional capacities for sea action are improved (IOC).
5. Regional coordination and exchange of information are improved (IOC).

The targets of the project are the criminal justice institutions of Kenya, Seychelles, Mauritius and Tanzania which are involved in prosecuting piracy and maritime crime, namely the police/coastguard, prosecution, courts and prisons. In addition to supporting criminal justice institutions, some support will also be targeted directly at individuals being prosecuted for piracy and maritime crime in those states. MASE is in support of the UNODC Global Maritime Crime Program (GMCP) which assists states to strengthen their capacity to combat maritime crime. The MCP was formed in 2009 as the UNODC "Counter Piracy Program" (CPP), which was established in response to Security Council resolutions calling for a concerted international response to the scourge of piracy off the Horn of Africa.

---

[99] https://eeas.europa.eu/headquarters/headquarters-homepage/8407/program-promote-regional-maritime-security-mase_en

Piracy, Maritime Awareness & Risks (PMAR)

Piracy, Maritime Awareness & Risks (PMAR)[100] was an EU project trial and implementation (PMAR-MASE[101]) for the Maritime awareness for Authorities responsible for piracy in the regions of Africa, here the Indian Ocean. It covers parallel and/or alternative to deployment of out-of-region forces and investigates the technical possibilities.

The PMAR approach based on the fusion of AIS and LRIT data from several sources, with satellite AIS being the most valuable data type, supplemented by a limited number of satellite SAR images.

In October of 2012 Indian Ocean Commission (IOC) organized the first regional workshop against money-laundering and piracy in the Seychelles. The UN Office on Drugs and Crime (UNODC) is helping IOC implement the terms of the EU funding agreement. The objectives are to strengthen de-valued capabilities, such as: investigations, indictment and imprisonment in regional countries.

The IOC also works in partnership with the EU on the Smartfish program for the Implementation of a Regional Fisheries Strategy in the Eastern and Southern Africa and Indian Ocean (ESA-IO) region, which aims to contribute to an increased level of social, economic and environmental development and deeper regional integration in the ESA-IO through the sustainable exploitation of marine and lake fisheries resources.

In addition, monthly ship density maps have been produced, and a number of satellite images have been analyzed to assess the presence of non-reporting ships. The purpose of the project was familiarization of maritime authorities in the Eastern-Southern Africa/Indian Ocean region with region-wide maritime monitoring, providing hands-on experience, and developing an understanding of what kind of information level is attainable and how to use the information.

Aim was to familiarize operators and decision makers in the Regional Maritime Rescue Coordination Centre (RMRCC) in Mombasa (DCoC ISC) and in the IOC Anti-Piracy Unit (APU) in the Seychelles with region-wide maritime surveillance capabilities, build capacities and culture with operators, analysts and IT staff, formulate requirement, get user feedback, and enable scoping of future operational systems.

PMAR first ended in October 2015 and demonstrated the current level of MSA (Maritime Situational Awareness) and the Maritime Security achievable at regional scales for counter-piracy, bridging the current gap of foreign naval forces, and building up a maritime surveillance culture.

---

[100] https://ec.europa.eu/jrc/en/publication/eur-scientific-and-technical-research-reports/pmar-piracy-maritime-awareness-risks-trial-implementation-under-mase
[101] https://ec.europa.eu/jrc/sites/default/files/Status Update.pdf

During one year, from September 2014 to September 2015, the PMAR-MASE project produced the maritime traffic picture of the reporting ships that use the AIS or LRIT automatic position reporting system over the entire Western Indian Ocean, and delivered it via a web viewer to two authorities in Africa with a regional maritime security responsibility:

- the Anti-Piracy Unit of the Indian Ocean Commission in the Seychelles,
- and the Regional Maritime Rescue Coordination Centre of the Kenya Maritime Authority in Mombasa.

First PMAR-MASE project implementation ended Oct 2015, it had laid the foundation for the Maritime Security (MASE) Program and projects. PMAR envisioned in the MASE program managed by Indian Ocean Commission (IOC):

- PMAR-1 at the Horn of Africa from 2010-2012
- PMAR-2 at the Gulf of Guinea from 2012-2013
- PMAR-MASE in East Africa from 2014-2015

PMAR used SeaVision as trial system for the AIS-feed in Maritime Situational Awareness (MSA) to provide the maritime situation in the entire Western Indian Ocean basin to:

- Regional Maritime Rescue Coordination Centre (RMRCC) in Mombasa
- IOC Anti-Piracy Unit (APU) in Seychelles

A main goal of the MASE Project is to implement the Somalia Inland Action Plan, which is meant to address the root causes of piracy. The objective of the strategy is to develop organic solutions to prevent and repress piracy off the coast of Somalia, to support Somali regional and federal governments as well as local communities to address piracy in a holistic manner and to further develop capacity of Somali institutions.

Eastern and Southern Africa and Western Indian Ocean region (ESA-IO) is increasingly subject to security challenges linked to piracy, maritime insecurity and organized crime. The International Maritime Bureau (IMB) reported in 2009[1] that the ESA-IO region had become the most pirate-infested waters in the world. Somali pirates have been attacking vessels in the marine areas of Kenya, Tanzania, Seychelles, and Madagascar and further off into the Indian Ocean. An insecure marine area affects trade, food security, fisheries, other marine resources and tourism (tourism revenues in the Seychelles alone have declined by 10% and fisheries revenues by 30% over the past few years), not only for the region itself but also for the broader international community.

Maritime security and law enforcement in the Indian Ocean region are of international importance due to the high level of trade routed via sea and the threat that piracy and armed robbery at sea represents for crews and passengers, and for the security of navigation. Piracy and lack of maritime security continue to increase transport costs, trade costs and insurance. This ultimately affects the regional integration process by undermining development efforts which negatively impact on the inclusion of the region in the global economy (90% of ESA-IO regional trade by volume is transmitted by maritime transport in 2008). Furthermore, piracy increases the risks of terrorism and smuggling of weapons and drugs.

The ESA-IO coastline and marine area are too extensive to be permanently patrolled individually by countries. The region itself does not have an effective Coast Guard capacity or a sustained deep-water maritime surveillance and relies heavily on outside forces to protect maritime commerce and shipping.

As a result, much of this area is in effect, ungoverned and vulnerable to piracy, smuggling, terrorism as well as unlicensed and illegal fishing. Even if the region is capable to make arrests, or, as in the case of countering piracy, foreign navies make arrests, the criminal justice framework and law enforcement capacity at the regional and national level are not in all cases able to cope with the necessary detention, prosecution, trials and imprisonment. In addition, there is limited capacity to cope with maritime disasters or search and rescue operations and basic safety navigation and marine pollution prevention.

The increasing concern in the region from a political, social and economic perspective, especially in the economically vulnerable small island states prompted the region to attribute to all the Regional Organizations a specific mandate for the fight against piracy. The 13th COMESA (Common Market for Eastern and Southern Africa) Summit held in June 2009 in Zimbabwe condemned acts of piracy, adopted an Action Plan and called upon the International Community to adopt a coordinated approach in a holistic manner through the UN.

The Eastern Africa Community (EAC) Summit as well as Council of Ministers meetings held in November and December 2010 took note of the Strategy and Action plan. Meanwhile, the Intergovernmental Authority on Development (IGAD) has adopted an Inland Somali Action Plan and the African Union (AU) is developing its continental African Integrated Maritime Strategy (AIMS) 2050 addressing Africa's maritime challenges, threats and opportunities and also providing a continent-wide framework for the protection of African Maritime Domain including responding to threats such as piracy off the continent's east and west coast, illegal fishing, pollution and human trafficking.

A High Level Regional Ministerial Meeting on Piracy was held on 7. October 2010 in Mauritius and adopted the ESA-IO Regional Strategy and Action Plan against Piracy and for Promoting Maritime Security. This Regional Action Plan is a 'rolling' process which would be updated annually by the ESA-IO Ministerial meeting on the basis of results achieved and evolution of the situation. In line with the Action Plan, a Program for the Implementation of a Regional Maritime Security Strategy (MASE) for the ESA-IO region is proposed in accordance with the provisions of the ESA-IO 10th EDF Regional Strategy Paper/Regional Indicative Program.

Since the adoption of the Djibouti Code of Conduct concerning the prevention and repression of piracy and armed robbery against ships ("*Djibouti Code of Conduct*") under the auspices of International Maritime Organization (IMO) in 2009, mainly through efforts of regional countries such as Kenya and Seychelles and implementing organizations as the United Nations Office on Drug and Crime (UNODC), fight against piracy have become important areas of joint and multilateral cooperation in the region and at the international level.

- Maritime security and law enforcement require involvement of a wide variety of stakeholders in countries across the region and the support of the UN and African Union. Quick wins are important to show the difference that a maritime strategy and activities in this field can have for the countries key players.
- Piracy and armed robbery at sea today is an international problem which requires a comprehensive, multilateral solution with an integrated short, medium and long-term strategy including the establishment of a functioning Somali government. Only a holistic approach can address these maritime issues.
- This holistic approach should also addresses the need for a functional government in Somalia while strengthening other regional governments to take a leading role in controlling pirate activity in the Indian Ocean.
- EU's partnerships, such as with IMO on counter piracy, INTERPOL on sharing information and UNODC on addressing crime and criminal justice, need to be used to ensure all stakeholders work closely together. Close coordination needs to be paid to the Contact Group on Piracy off the Cost of Somalia (CGPCS) and its working groups, as they can ensure addressing piracy in the region, both on land and at sea.

The MASE Program build on the EUR 2.0 million MASE Start-up project started in January 2012 as well as on efforts already undertaken in the region and work in close cooperation with the implementing organizations/authorities in the region.

The MASE program is part of the EU Strategic Framework for the Horn of Africa which includes the EU Action Plan against Piracy recently approved and the appointment of EU Special Representative for the Horn of Africa.

The Program will ensure coordination with the EU funded capacity building projects financed through the Instrument for Stability (IfS) short and long-term components. Actually, the MASE Program will secure the continuity of the short term actions already financed under the IfS at the national level. It will also ensure coordination with the on-going long term IfS "*Critical Maritime Routes*" program in the region. Namely, activities related to maritime information-sharing system and networking will be coordinated with the MARSIC project (Enhancing maritime security and safety through information sharing and capacity building) and CRIMARIO (information sharing between western Indian Ocean region and South East Asia).

Activities related to financial flows will have to be coordinated with the "*Law enforcement in East Africa*" (CRIMLEA) project focusing on effective pro-active investigation on piracy organizers and financers. Activities related to Somalia Inland Action Plan will be complementary with EU Somalia Unit's Economic Development, Governance and Rule of Law program and those related to maritime capacity building will be in line with the ESA-IO Fishery program implemented by IOC and FAO fisheries training activities.

Close collaboration is established with the civilian EU Common Security and Defence Policy (CSDP) mission for regional maritime capability building, EUCAP NESTOR. The action will ensure complementarities with the on-going UNODC Counter Piracy Program. MASE is also expected to coordinate with other counter-piracy program at national level supported by other donors.

MASE, in particular IOC, will also take into account the results of the European Parliament requested Pilot Project on Piracy, Maritime Awareness and Risks (PMAR) which explores technical tools helping to improve maritime awareness in the Horn of Africa region, primarily for counter-piracy needs, and will support its follow on. As the MASE program will only cover a part of the overall region affected by piracy in the Indian Ocean, there will be need to explore coordination with activities in the Western Indian Ocean countries.

Full coordination was ensured with stakeholders in the region dealing with maritime capacity building through the Capacity Building Coordination Group (CBCG) of Working Group 1 of the Contact Group on Piracy off the Coast of Somalia (CGPCS). The CBCG's function is to facilitate the coordination of regional and international maritime capacity building projects in the West Indian Ocean region including through its Web portal (an internal data base) where all capacity building projects (including MASE activities) are to be uploaded The IMO sponsored Djibouti Code of Conduct will also be taken into account to foster information sharing, capacity building and provide a framework for regional cooperation.

Collaboration will be established with Pan African and International partners in particular with the EU Common Security and Defence (CSDP) Naval Force (EUNAVFOR) Operation ATALANTA, as well as with the EU CAP NESTOR mission for regional maritime capability building.

Close collaboration will be developed with the Ports Management Authority of Eastern and Southern Africa (PMAESA) and the Djibouti Code of Conduct, the Eastern African Stand-by Force (EASF) maritime cell. Steps will be taken to ensure conformity and adherence to the implementation of the AU African Maritime Transport Charter and Plan of Action, the AU Durban Declaration on Maritime Safety and Security. Activities will be coordinated with other bilateral partners (Norway, US, Japan, UK, Germany ...).

An annual conference of all the key stakeholders will be organized by the MASE program in order to ensure closer and improved coordination of activities. The overall objective of the program is still to enhance maritime security in the ESA-IO region hence contribute to global security and create a favorable environment for the economic development of the ESA-IO region and beyond. The Specific objective of the program is to strengthen the capacity of the ESA-IO region in the implementation of the Regional Strategy and Action

Plan against Piracy and for Maritime Security.

The project was to achieve five results in line with the Regional Strategy and Action Plan as adopted by the ESA-IO Ministerial Meeting in Mauritius in 2010, as follows:

- Result 1: Somalia Inland Action Plan is implemented;
- Result 2: National/regional legal, legislative and infrastructural capability for Arrest, Transfer, Detention and Prosecution of Pirates is developed and/or strengthened;
- Result 3: Regional capacity to disrupt the financial networks of pirate leaders and their financiers while also addressing the structural vulnerability factors and minimize the economic impact of piracy is strengthened;
- Result 4: National and regional capacity for maritime tasks and support functions are enhanced;
- Result 5: Regional coordination and information exchange is improved.

The Somalia Inland Action Plan (SIAP) is implemented to address the root causes of piracy. The Inland Strategy and Action Plan to counter piracy is based on field research, commissioned by the IGAD Capacity Building Program Against Terrorism (ICPAT) in Puntland which involved Puntland authorities, imprisoned pirates, officials of regional organizations and civil society figures; focal group discussions with local communities, clan and religious figures, civic society representatives; observations at Puntland prisons and former pirate centers, as well as document analysis.

The objective of the strategy is to initiate home-grown solutions to prevent and repress piracy in Somalia, to support Somali administrations and communities in addressing piracy in a holistic manner and to reinforce the capacity of the institutions of Somali administrations. Activities of the first result will also focus on Central-South Somalia to complement EDF and thematic programs actions that support return to the rule of law in Somalia.

The SIAP will build upon and will be complementary to the on-going and planned cooperation activities of the EU Delegation to Kenya, Somalia Unit, which is highly involved in providing support to Somalia in this field. It will also coordinate with the UN agencies present in the field, as well as with other key actors. It will build on successful models of grassroots' based counter piracy measures already undertaken by Puntland administration of Somalia. It will borrow from IGAD's experiences in early warning, information sharing and response at the community, national and regional levels with pastoralism, with the aim of applying them to the counter piracy operations.

These activities will build on IGAD's experiences in working within traditional value systems for conflict prevention management and resolution. The Inland Action Plan within the MASE program, to be implemented by IGAD as a distinct program, will include: support to the Kampala Framework as a successful platform for dialogue among Somalis; promotion of the home-grown solutions; collection, analysis, and dissemination of community level data on piracy; counter piracy initiatives based on Somali traditional values and customs; identification and support to home-grown alternative livelihoods.

The activities will be conducted in coordination and cooperation and complementary with the work of the EUCAP NESTOR mission on regional maritime capacity building, as well as with the EU Delegation to Kenya, Somalia's Unit support for Somalia in collaboration with EU Djibouti Delegation. Activities will also be in coordination, cooperation and complementary to the activities conducted by UNODC, UNDP Somalia, UN Political Office for Somalia (UNPOS) and INTERPOL.

The second resulting point is the national/regional legal, legislative and infrastructural capability for Arrest, Transfer, Detention and Prosecution of Pirates, which should be developed and strengthened (Lead partner: EAC) enhancing the capacity needs of the respective law enforcement agencies based on regional common standards, lessons learned and existing best practices relevant to the ESA-IO region, which is a challenging process.

This result will focus both on developing and implementing joint legislative frameworks and on strengthening relevant institutions that relate to the arrest, transfer, detention of pirates, consistent with on-going UN initiatives and the Djibouti Code of Conduct. The activities for this result area will include:

- Continuation of existing support to trials of suspected pirates in countries of the region;
- Agreement on a harmonized legal framework for maritime law enforcement with the possibility to allow for national adjustment of laws and regulations;
- Develop training and mentoring program as well as provision for equipment for police, prosecutors, and courts staff on a variety of maritime law enforcement angles, including human resource development;
- Develop training and mentoring for prison staff (including attention for human rights) and develop staff living conditions and human resources;
- Support for the implementation Prisons Reform Programs. UNODC (through Contribution Agreement) and INTERPOL (contract with EAC) will be the key partners in the implementation of the activities. MASE will coordinate its activities with EUCAP NESTOR to ensure complementarity.

Third the strengthening of the regional capacity to disrupt the financial networks of pirate leaders and their financiers while addressing the structural vulnerability factors and minimizing the economic impact of piracy (Lead Partner: COMESA) is addressed. This seeks to sustainably strengthen the region's capacity to detect and prevent the financial sources and vehicles that support, facilitate or propagate piracy, particularly money laundering.

One of the key objectives is to provide the regional states with the tools and training to disrupt the financial networks of pirate leaders and their financiers. Therefore, tracking and disrupting of illicit financial flows as well as confiscation of assets and proceeds of criminal activity are important tools to deter acts of piracy and to apprehend and prosecute those who finance piracy. Thought, the regional states will be able to identify how the proceeds from ransom payments are laundered, identify vulnerabilities in current regional Asset Forfeiture and Money Laundering regimes and develop the capacity to address these vulnerabilities, including the development of Financial Intelligence Units.

It will also address structural factors including strengthening of legal and policy frameworks to discourage the flow of proceeds from piracy into the region and break business model of piracy. The implementation will continuously draw from outputs of the 5th Working Group of the ICG on Piracy, which is closely linked to international efforts to track financial flows and implement in line with UNSC resolutions 1976 and 2020; ensuring to build on ongoing work by Critical Maritime Routes (CRIMLEA) project implemented by INTERPOL and other relevant initiatives. Implementation will be done in close collaboration with INTERPOL, the Eastern and Southern Africa Anti-Money Laundering Group (ESAAMLG), World Bank, UNODC Global Money Laundering Group and other actors to ensure to draw on synergies.

Focus areas:

- Strengthen financial analytical capacity in the region by setting up and strengthening of national Financial Intelligence Units, and the financial institutions in the region to analyze, detect and track illicit financial flows linked to piracy. Developing common, coordinated and inter agency framework on anti-money laundering and piracy funding, including supporting linkages with national Financial Intelligence Units;
- Drafting or amending national money laundering laws and regulations and adopt regional policy frameworks to counter piracy and to address economic impact of piracy;
- Assess and strengthen regional capacity to investigate and prosecute financial crimes at national and international levels. The COMESA Governors of Central Bank, the COMESA Monetary Institution, the COMESA Clearing House, the Africa Trade and Insurance and the COMESA Reinsurance company and the COMESA Court of Justice will be valuable supportive institutions.

The MASE Program will promote regional maritime security focusing on maritime capacity building in its broader sense ensuring a comprehensive and integrated multi-sectoral approach for the ESA-IO region and maximum complementarity with existing and planned EU and IMO projects that helps the region to counter piracy and to mitigate the phenomenon in a sustainable way, including fishing fleets and rescue capabilities, which are not covered part in the Djibouti Code of conduct (DCoC).

This approach will contribute to develop secure maritime routes, pollution response, and sustainable marine resources and fisheries, search and rescue. In particular, the MASE program will facilitate the strengthening of the coast guarding functions of all the coastal States in the region to secure their maritime zones and ESA-IO waters for all seafarers, and improved control and surveillance of the Exclusive Economic Zones, including improvement of search and rescue capacity.

Close cooperation and alignment to related projects and activities such as those of UN, IMO, will avoid duplication of efforts. Activities may benefit from the exchange of information with the European Defence Agency (EDA) in light of their expertise and developed technological solutions in the field of maritime surveillance (MARSUR). MASE may build on the successful regional operational patrol system against illegal fishing, implemented by IOC with EU funding, and other existing maritime security operations, such as Operation ATALANTA. In addition, coordination with international partners will be ensured through an ad-hoc coordination process comprising of IOC, IMO and the EU. Focus will be on:

- Improved operational ability of the maritime law enforcement agencies in the region;
- Improved maritime domain awareness;
- An assessment of existing means and an exchange of lessons learnt in the domain of maritime safety, search and rescue;
- The development of a joint operational patrol initiative.

Close coordination will be undertaken with the IOC led ESA-IO Monitoring Control and Surveillance mechanism undertaken under the Fisheries Program as well as the World Bank funded Maritime Route Program implemented by IOC and South African Maritime Safety Authority (SAMSA) and the Critical Maritime Routes CRIMLEA project. The EU, PMAESA and IMO will be the key partners in the implementation of the activities.

Furthermore, IOC will work in close cooperation with the EUCAP NESTOR mission on regional maritime capacity building. MASE and EUCAP NESTOR should agree on complementarity activities to avoid duplication and reinforce cooperation. MASE can contribute to the sustainability of the EUCAP NESTOR. The IOC will also take into account the work of the Commission in developing a maritime situational picture for the Western Indian Ocean under the PMAR Project and support its follow on.

Information exchange, including exchange of operational information on incidents of piracy and armed robbery coordination and networking are prerequisites for the successful implementation of the ESA-IO Regional Strategy against Piracy and for Maritime Security. MASE will build on other initiatives such as those of the Djibouti Code of Conduct that focuses on information sharing and capacity building and aims at:

- Reinforcing a strategic framework and system for exchange of information and coordination for the ESA-IO region; and
- Utilizing the training activities of the Djibouti Regional Training Centre as well as other related training opportunities for the whole ESA-IO region.

Existing information networks (such as MARSIC, GRIMARIO) will be fully used and complemented where necessary. Provision will be made in each Result Area for specialized information exchange.

The activities of the project will be implemented according to the following principles:

gender neutrality in its employment and outsourcing actions, and gender balance in all of its committees, workshops and training interventions; participatory approach on direct involvement of civil society and the private sector; promotion of good governance issues; visibility ensured notably by publicity campaigns and publications valorizing program results.

Risks and assumptions are as follows:

- All ESA-IO countries perceive the urgency and importance of addressing piracy and armed robbery at sea. (Not all the ESA-IO countries are affected to the same extent. Most countries have a high degree of common interest, but do exhibit individualistic tendencies);
- Firm commitment from all partners in the region to collaborate on a regional basis and political will to cooperate remains;
- Political instability in the region, conflicts or other effects do not harm the benefits generated. It is assumed that other partner countries will operate in a stable security and government environment;
- The International community will honor their commitments and pledges for support;
- National authorities will continue to invest in their capabilities;
- Sustained support for the Kampala Framework for dialogue among Somalis and initiate home grown solutions;
- Other support to help Somali security forces and AMISOM is also maintained;
- The new Federal Government is provided support for restoration of functioning State, good governance and rule of law and the relief of poverty;
- Increased coordination between the new Federal Government and International Community;
- Increased Inter-States and Inter-Agency cooperation and coordination on money laundering activities related to piracy and other criminal acts;
- Strong political will to address impunity related to trans-national crimes;
- Somalia is provided support for the delineation of its EEZ and its control and surveillance.

Modalities will be explored to involve and work closely with SADC. There will be need for extensive coordination with EU led actions (including its Common Defense and Security Policy missions, in particular EUCAP NESTOR) and key stakeholders such as AU, SADC, UN, IMO, UN Security Council mandated Contact Group against Piracy off the Coast of Somalia (CGPCS), the Ports Management Authority of Eastern and Southern Africa (PMAESA), INTERPOL and a representative of the Interregional Coordination Committee (IRCC) Secretariat. The MASE will make provision for a coordination mechanism to be set-up.

All Member States of the region should contribute to the burden sharing regardless of whether they share a coastline or not because maritime insecurity is a regional problem and the program also covers security on inland water systems.

The MASE will be implemented jointly by IGAD, COMESA, EAC and IOC and the ESA-IO Member States based on the principle of complementarity and subsidiarity. IGAD will be the RAO. IGAD has a large experience in managing 10th EDF funds (about EUR 100 million in partially decentralized cooperation). However, each of the ROs and their member States will report on progress on the implementation of the MASE within its own national Policy Organs. The ESA-IO Regional Ministerial Meeting on Piracy will meet every year to provide overall political guidance.

The ESA-IO Member States will have identified focal points and ensure full involvement in all the result areas as active participants. The regional Network of focal points on Piracy (NFPP) of the ROs and ESA-IO countries will form a Technical Steering Committee (TSC) for MASE, to oversee and validate the overall direction and orientation and for the timely and effective implementation. The TSC will be under the chairmanship of the RAO. EU representatives (such as Delegations of Djibouti, Mauritius, Somalia Unit of Delegation in Kenya, Tanzania, Zambia, Commission, EEAS) will be attending as observer the meetings of the TSC. The TSC shall be supported by a Project Implementation Committee (PIC) composed of the ROs, Implementing Partners and representatives of EU.

The NFPP/TSC will analyses the MASE strategic priorities, review progress on implementation of the project to ensure flexibility and sustainability and will be a platform for coordination and harmonization between donors, other Technical Assistance providers, recipient countries and regional organizations.

MASE will have specific components led by the four beneficiary Regional Organizations (ROs). Each RO will be responsible for the specific component they have the lead; the relevant RO and relevant EU Delegation will be responsible for the validation of the reports, as follows: component 1 on Inland Action Plan for Somalia (IGAD; EU Delegation to Kenya, Somalia's Unit in collaboration with EU Djibouti Delegation; component 2 for detention, transfer, prosecution (EAC; EU Delegation in Tanzania); component 3 for addressing economic impact of piracy (COMESA; EU Delegation in Zambia); and components 4 for maritime security and 5 for capacity building and information exchange (IOC; EU Delegation in Mauritius.

The program will also make provision for administration and support and office support personnel, technical support and backstopping needed for ensuring quality and consistency in their advice and activities and a pool of short-term experts for specific assignments at the RAO and each RO based on identified needs. The ESA-IO Member States will be fully involved in the implementation of the program through their focal points which are part of the TSC/NFPP.

The budget available to implement the action is detailed in the Appendix. The total costs of the MASE Program was EUR 37.500.000 financed from the 10th EDF RIP for the ESA-IO region in the framework of the revised ACP-EU Partnership Agreement. The operational duration of the project was 60 months from the signature of the Financing Agreement. The project's budget make due allowance for Information, Communications, and Visibility activities.

These will be aligned with the requirements for visibility of EU aid while demonstrating the regional ownership of this project outputs. The IRCC's and ESA-IO Regional Organizations Internet websites including the dedicated website for program shall be the main tool to ensure communication, visibility and dissemination of the project's achievements and of its current and forthcoming events.

The project will implement its own communication strategy and develop specific awareness-raising, information and dissemination activities (also giving due consideration to Non State Actors) in order to inform partner countries, regional and national research centers and other potential stakeholders of the opportunities that it provides. Specific attention will be given to the linguistic necessities in the region.

Piracy Incident Reporting and Information Exchange System (PIRATES)

The <u>Pirate Incident Reporting And informaTion Exchange System (PIRATES)</u>[102] is an extension within European Coordination Centre for Accident and Incident Reporting Systems (ECCAIRS), developed by JRC to monitor transport safety in Europe, and both are situated in Brussels, Belgium.

Maritime piracy presents a significant threat to the global shipping industry. In the first six months of 2012, International Maritime Bureau (IMB) Piracy Reporting Centre received 177 incident reports worldwide. In this period, 20 vessels were actually hijacked with 334 crew members taken hostage, and 80 vessels were boarded, 25 vessels were fired upon and 52 vessels reported attempted pirate attacks. Five African countries along the African continent's sea coast have been listed by the IMB as risky areas for the movement of vessels and ships due to attacks by pirates.

The countries participating in PIRATES are Somalia, Nigeria, Benin, Guinea and Cameroon, which are, besides Somalia, all in the Gulf of Guinea of West Africa. Ships in this region, including oil and chemical tankers, are increasingly being attacked with automatic weapons and rocket propelled grenade launchers. The Gulf of Guinea region is seen as one piracy hotspot after Somalia.

Among different initiatives to counter piracy in the Gulf of Guinea region, the European Parliament has in 2010 requested additional attention for the fight against piracy. By addressing improved maritime governance capacities and improved safety and security at sea in the Gulf of Guinea region, the Pilot Project on Piracy, Maritime Awareness and Risks – Gulf of Guinea (PMAR-GOG) aims to implement the Parliament's request.

The general objective of PMAR-GOG is to identify and assess technical means and practices that can be used in the near and medium future by the stakeholders of the Gulf of Guinea region to contribute to their maritime awareness, chiefly by supporting the gathering and sharing of information. Specifically, in the Gulf of Guinea region, one of the main problems is the lack of good quality information on piracy attacks and there is no formal system to gather information on piracy attacks. Furthermore, no formal system of information exchange exists between States or organizations in the Gulf of Guinea region.

Therefore, Joint Research Center (JRC) of the EU developed a software system to collect and store information on piracy incidents with a functionality to share this information with other States or organizations. This prototype called Pirate Incident Reporting And informaTion Exchange System (PIRATES) is an extension within European Coordination Centre for Accident and Incident Reporting Systems (ECCAIRS) which was developed by JRC to monitor transport safety in Europe, both situated in Brussels, Belgium.

---

[102] https://ec.europa.eu/jrc/en/publication/eur-scientific-and-technical-research-reports/maritime-piracy-incident-reporting-technical-feasibility-study

Smartfish Program

SmartFish[103] is one of the largest regional programs for fisheries in Africa covering 20 beneficiary countries in the Eastern, Southern Africa and the Indian Ocean (ESA-IO) region. Funded by the European Union and implemented by the Indian Ocean Commission (IOC) jointly with the Food and Agricultural Organization of the United Nations (FAO).

The first assignment of the SmartFish Program was the Trade Readiness Document (TRD), which forms part of the overall Work plan for the Regional Fisheries Strategy for the Eastern-Southern Africa and India Ocean Region (IRFS) launched in February 2011. The overall program with its four main components is commonly referred to as 'SmartFish'. The SmartFish Program is financed by the European Union under the 10th European Development Fund (EDF 10) within a total financial contribution of Euro 21 million.

The SmartFish Program encompasses four Regional Economic Communities (RECs) and is executed by the Indian Ocean Commission (IOC) as the Regional Authorizing Office in collaboration with the Common Market for East and Southern Africa (COMESA), the East Africa Community (EAC) and the Inter-Governmental Authority on Development (IGAD). Other regional institutions that may be consulted include the Southern African Development Community (SADC) and regional fisheries management organizations, such as the Indian Ocean Tuna Commission (IOTC), the Southwest Indian Ocean Fisheries Commission (SWIOFC), the Lake Victoria Fisheries Organization (LVFO), and the Lake Tanganyika Authority (LTA).

The first phase of the Program was planned to be implemented over a period of 31 months (March 2011 - September 2013). The region defined as the SmartFish beneficiary comprises 19 countries: Burundi, Comoros, DR Congo, Djibouti, Eritrea, Ethiopia, Kenya, Madagascar, Malawi, Mauritius, Rwanda, Seychelles, Somalia, Sudan, South Sudan, Swaziland, Tanzania, Uganda, Zambia and Zimbabwe. All countries are members of COMESA (except Tanzania and Somalia) and the Tripartite Group negotiation process for the establishment of one large Free Trade Areas (FTA) for 26 countries.

It is noted that the individual RECs are at different stages of trade liberalization and integration, e.g. EAC is a Customs Union (CU) by name, COMESA (including also most IOC countries) and SADC are established FTAs, whereas IGAD has not taken steps in this direction and is a structure with no particular intra-country trade agreement. The actual implementation of the CU and FTAs, however, indicates an ongoing process where regulations and procedures are not fully in place.

Thus areas such as, common external tariffs, Rules of Origin (RoO) for certain products and lists of sensitive products, are not yet harmonized. Important aspects such as border crossing facilities and procedures also differ, and trade in highly perishable products such as

---

[103] http://commissionoceanindien.org/activites/smartfish/

fish still suffer from obstacles. In short the trade enabling environment has shortcomings that need concrete actions for enhancement.

Main activities included discussions with public sector authorities that constitute the trade enabling environment, and interviews with the private sector operators active in the fish product value chain. Region Economic Communities (RECs) such as COMESA and IOC have been consulted regarding strategic visions, to ensure that the outcome of the TRD study is in tandem.

Competitiveness of the involved economies was assessed based a set of commonly accepted principles. The level of competitiveness and the nationally provided trade enabling environment constitutes the operating framework for the private sector. A brief review of the most important factors of competitiveness as the macro-economic environment, institutional set-up, training, access to finance and market issues revealed that 3 of 4 countries belong to the group of factor-driven economies.

Such countries compete based on their factor endowments—primarily unskilled labor and natural resources. Companies compete on the basis of price and sell basic products or commodities, with their low productivity reflected in low wages. The fourth country has moved up ladder and is becoming efficiency-driven with more competitive and efficient production processes. Increases in product quality become necessary because wages have risen without the option to increase prices as the country is not yet price setting.

Products traded regionally mostly involve simple processing done to local tastes so that imported replacements are virtually non-existent. The exception is Tilapia, where price relationships are determinants for the level of competitiveness against the imported frozen product. Despite higher prices for local products regional consumer tastes now seems to favor the local product over the imported.

In factor driven economies and the efficiency driven ones are lacunae and gaps in parts of the public sector service delivery system that constitute the enabling environment. Note that the variations in the seriousness of such between the pilot TRD countries, the more advanced the economy the less serious the problem. The most important problems are institutional:

- A Competent Authority is in place, but there is a lack of capacity regarding fish relevant SPS issues in border areas, and outreach services are limited.
- With one exception there is no systematic education in trade documentation, customs clearance and border procedures offered to SMEs.
- Targeted training for SMEs in management, technical aspects and quality assurance is low or missing as is Trade Promotion related support.
- There is virtually no support to product development.
- Access to funding and credit for SMEs is difficult if not prohibitive.

Initiatives from COMESA and other RECs are now underway to improve trade facilitation and assist the removal of Non-Tariff Barriers to trade (NTB). Tariff levels for fish products still exist, but are now handled under simplified trade regimes making trade easier. The COMESA template for removing remaining tariffs and the NTBs is generally accepted by the member countries.

Regarding the private sector operators and trade readiness differences were found for the two main economic groups:

- Large enterprises often based on foreign capital, and
- SMEs locally owned and operated.

For both groups a set of categories of business activities was applied to determine the level of vertical integration. Large companies display vertical integration and thereby indicated a good understanding of the importance and practice of trade readiness. Several large companies are now operating in the regional markets as product volumes and price levels are found still more attractive. Large companies know how to apply trade readiness in their strategic approach to market development, whereas SMEs displayed limited understanding and virtually no practice. The round of company interviews led to a modification of the trade readiness questionnaire template, which can be used in a full analysis of trade readiness in the Eastern-Southern Africa and Indian Ocean region (ESA-IO region).

The role of functioning associations was pivotal for distributing the Trade Readiness Questionnaire, disseminating results and acting as vehicles for training and extension. They also play a vital role in making the fish value chain operators heard in the trade policy debate. SMEs do suffer from a situation where the owner(s) is also the daily management, marketing manager and financial manager. This mix of roles centralizes decision making and virtually halts development ideas. For aquaculture the lack of guidance by the government in planning and production techniques has led to loss of investment. SMEs should be allowed to learn from 'star' large companies via centers of excellence. There was found a distinct lack of understanding of SPS requirements, trade and border crossing procedures, and trade readiness. The interviews therefore helped the process of adapting the TRQ. For the SME segment, associations are geared for public debate.

The report lists the important issues with common style pointers to the four countries. The five pointers will need to be detailed to the country specific situation. For the private sector operators the Trade Readiness Questionnaire was adapted to the SME condition, and can be disseminated to all 20 countries.'

The TRD is conceived as a pilot intervention that aims to identify the key requirements and obstacles to trade that have to be removed for the timely implementation of government policy that support a trade enabling environment. Improved trade readiness for the private sector operators is one tool to be enhanced, in particular for Small and Medium-sized Enterprises (SME). It involves missions to four countries to discuss with relevant authorities and the resulting guideline should be replicable to the needs of the remaining 16 beneficiary countries of SmartFish. Towards this end a group of four counties was selected for visits by the TRD team.

The key findings with recommendations of SmartFish were:

- Informal trade and border procedures
- In adequate capacity for SPS control in border areas
- High cost of trade due to poor trade facilitation
- Poor market intelligence trade promotion
- Market adaptation and product development

The activities and the output was to be coordinated with the sector strategic work already carried out by COMESA and IOC, so these RECs were also visited. The mission program was originally planned to be fairly continuous in time, but due to other urgent SmartFish activities requesting also the participation of the TRD Team members the elapsed time spanned from end Nov. 2011 to May 2012.

Informal trade and border procedures concerns the only very limited intra-regional trade registered due to the practice of informal trade, thus much more intra-regional trade than what is officially recorded exists. Informal trade means that the border crossing of fish products take place in small non-dutiable quantities and cannot be recorded in statistics under the existing reporting systems. Informal trade practices provide a somewhat diffuse and even confusing picture to the national authorities when trying to analyze statistics for planning and policy development purposes.

Informal trade lead to insufficient information and statistics on intra-regional fish trade in volume and value and by different product types. The value-chain and actual extent of the informal trade should be resolved by committing the national authorities to practice and promote:

Linking up with the COMESA ICBTA program to collect systematic data and aim to formalize the trade and promote the establishment of ICBTAs

- firm statistical data collection and reporting practices by training in collection methods and analysis of statistical data would be needed for policy and planning purposes;
- A regional fish trade information and database system should be established at the level of COMESA members but with a view to be expanded to the Tripartite region, the data collection points should be clearly identified and mandated and with transfer of operating knowledge via information, communication and outreach programs;

The inadequate capacity for SPS control in border areas result from insufficient institutional capacity as human resources with the right training, and limited funding is an important issue. This was identified as a key problem by both large scale and SMEs; in particular the outreach to SME operators is missing or weak. The result is that not all countries implement the SPS sub-agreement which also applies to intra-regional and domestic trade. It follows that fish control and inspection is inadequate and the fish products for human consumption may not be safe to eat. In fact the local trader and operator also violate the SPS related commitments as there clearly is no documentation for traceability, which in on element of the Pre Requisite Program (PRP).

It is important that guidelines, protocols, PRP manuals, HACCP plans and their means of verification are standardized, harmonized and designated/gazette. This may lead towards compliance with the SPS agreement for fish products that are traded mostly intra-regionally (smoked, sundried, salted products, chilled and even some frozen products):

- Standards equal to the SPS and Codex Alimentarius or otherwise relevant ISO standards for capture fisheries and aquaculture should be promoted to operators.
- The associations should be informed about these requirements including traceability in an easy to understand form and be able to disseminate these to the members;
- Build capacity with the CA in particular in the border areas, so that inspection and control is done by adequately trained staff.

High cost of trade due to poor trade facilitation is an issue regarding implementation of the commitments under existing FTAs signed by REC members, in particular for COMESA countries where a generally template for a Customs Union exist. As a result, transaction costs for fish products are very high in continental Africa, partially due to the poor quality of physical infrastructure, but also as a result of border crossing procedures and issues. The latter may involve up to four days of waiting for a truck to pass. However, part of this issue is addressed by the large scale program under the Tripartite region's and COMESA's leadership.

Political will and good governance, exercised by the participating members of the Tripartite negotiation process for a region-wide FTA, is a key element to its success. It is therefore important that the fishery sector is not marginalized in the deliberations at national, regional or international level and that Trade Readiness is enhanced. To hear the voice of particularly SME in fisheries as regards Trade Readiness the following should be done:

- Train association members in STR requirements and procedures.
- Establish or equip existing professional associations with mandates to disseminate and promote participation in fish trade policy debate as regards trade readiness;
- Mobilization of SME operator associations in the fish products value chain to form discussion fora locally as a 'cluster' and help define the key actions required to enhance trade readiness; Capacity building of regional traders and processors associations/organizations in areas such as STR based border procedures, self-policing, and how to access financing.

Poor market intelligence trade promotion deals with the need for access to up-dated market intelligence on regional markets. This encompasses understanding dynamics in consumer preferences for product types, presentation and taste, and regarding changes in price levels for the different fish products. Also information on changes in supply and prices of input to aquaculture as fingerlings and feed are missing. Thus the industry is deprived of key information upon which to base or adapt their market strategy. In some countries the gathering of market intelligence is mandated to public sector agencies, but there are actually no such activities going on.

Market adaptation and product development refers to intra-regional trade found to be low, thus depriving the consumers of the use of white protein from fish as a supplement to food security. Products such as sundried fish suffer high postharvest losses, in particular during the rainy season due the simple processing methods.

Part of this issue is also related to inadequate infrastructure for intra-regional fish trade in terms of lacking or poorly equipped collection centers, distribution facilities, receiving centers, and retail markets. Such facilities are not regularly inspected by the CA, which is also the case at some border crossings posts, making the products from these facilities non-compliant to the SPS requirements.

Missing market intelligence can lead to arbitrary and wrong market strategies at company level. In a historic situation, where the demand for fish products is growing regionally, combined with a slow increase in purchasing power, this in not only missing an opportunity it can be devastating to the industry and hence cause a misuse of scarce national resource wealth. If the products made are not known outside the national border regional trade in such products will be slowed down.

Vessel Detection System (VDS) in the European Union

Vessel Detection Systems (VDS)[104] started in 2006 on a voluntary basis as an additional tool which Member States of the European Union could use to supplement their surveillance picture if this proved to be cost-efficient. However some claimed that the system penalizes those who obey the rules since it cannot monitor those vessels with the system being switched off or malfunctioning and also it is not able to identify vessels from non-EU countries that do not have the VMS.

For this reason the EU Fisheries Council of December 2002 asked Member States to carry out pilot projects to assess the use of remote sensing for control. Today all European Union vessels above 15 meters in length are to be fitted with a Vessel Monitoring System (VMS) and similar rules apply all over the world in national, international or organizational cooperation.

The European Commission Regulation 1461/2003 of 18 August 2003 states that the aim of the proposed Vessel Detection System (VDS) is to (a) determine the number of fishing vessels and their position in a given area; (b) cross-check the positions of the fishing vessels detected by VDS with position reports from VMS, and (c) signal the possible presence of fishing vessels from which no position reports have been received through VMS. European Council Regulation 1966/2006 of 21 December 2006 foresees operational use of satellites in contexts where cost-effectiveness can be proven, starting in January 2009. First VDS was developed by the Joint Research Center (JRC).

More as variants and additions are different defined system in the EU that collect information on vessels or goods, which can also be found under regulations of national, international or organizational cooperation.

---

[104] http://ec.europa.eu/research/press/2007/maritime-briefing/pdf/43-vessel-detection-system-fisheries_en.pdf

Vessel Monitoring System (VMS)

Vessel Monitoring System (VMS)[105] for fisheries control started in the early 1990s on a pilot project basis in the EU and, following successive applications, it is presently deployed in fishing vessels above 15m in length and it is envisaged to be extended it to all fishing vessels larger than 10m. Vessel monitoring systems (VMS) are used in commercial fishing to allow environmental and fisheries regulatory organizations to monitor, minimally, the position, time at a position, and course and speed of fishing vessels according to EC Regulation 2244/2003 and the Council Regulation 1224/2009 establishing a Community control system (VMS) for ensuring compliance with the rules of the common fisheries policy.

VMS relates here specifically to fisheries management systems while the term VMS is sometimes also used as an informal synonym for other such things as the Automatic Identification System (AIS) in general or a Vessel Traffic Service (VTS).

AIS and VTS however are quite different from VMS, although they may be complementary may apply to marine oversight and sensing programs that deal with the safety of navigation, hazardous material spills, and environmental threats such as algal blooms. VMS uses different radio technologies, is long-range, and handles commercially sensitive information. All European Union vessels above 15 meters in length are fitted with a Vessel Monitoring System (VMS).

Similar systems are operational or being brought into operation in other fishing areas and by other fishing nations. The system relies on satellite navigation and communication technologies. A box installed on board the vessel transmits the GPS-derived vessel position by satellite to the Fisheries Monitoring Centre (FMC) in the flag state which then communicates the information to the state or regional fisheries body in whose waters the vessel is fishing.

The period between transmissions varies but is normally between one and two hours. The vessel information can also be pulled by the FMC, here a VMS message is sent on request via satellite communication to the vessel's Flag State authorities. Often INMARSAT-C is used, sometimes EUTELSAT or ARGOS (the last one did in the past not allow the pull request). The Flag State will forward the VMS message to the Coastal State in which waters (EEZ) the ship is located. The operational authority that handles the VMS is the Fisheries Management Centre (FMC) in each MS.

In this way, the national FMC is continuously aware of all its national VMS carrying fishing vessels wherever they are on the globe, and of all VMS-carrying fishing vessels in the waters under its jurisdiction (i.e. in most cases its own EEZ which extends out to 200 nm). This allows the authorities to determine the position of all vessels fitted with VMS within a certain area at a certain time. The system enables the fishermen to demonstrate their compliance with

---

[105] https://en.wikipedia.org/wiki/Vessel_monitoring_system

regulations on days at sea, closed areas or closed seasons.

The EU regulation requires that VMS data are stored for a period of 3 years (although it does not indicate whether the complete data or a subset needs to be stored). There seems to be a small loss of accuracy in the data so transferred (fewer digits in the geographic position). These transmissions occur between FMCs via X.25 link, currently migrating to https, and are routine and automatic. VMS data are also forwarded to Regional Fisheries Management Organizations (RFMOs) by Flag States whose vessels are active in the waters controlled by the RFMO. This typically happens at longer intervals, e.g. 6-hourly.

Vessel Traffic Monitoring (VTM)

Vessel Traffic Monitoring or Vessel Traffic Managing (VTM) refers to any Vessel Traffic Monitoring System (VTMS) and application for specific Monitoring Solutions to facilitate traffic management and planning, vessel monitoring, environmental protection, security provision, coastal surveillance, collision avoidance, shore-based pilots, SAR Coordination and Vessel Traffic Monitoring (VTM) Vessel Traffic Service (VTS), Vessel Detection System (VDS) and their operator training. EU Directive 2002/59 of the European parliament and oft the Council of 27 June 2002 was establishing a Community vessel traffic monitoring and information system and repealing European Council Directive 93/75/EEC.

Vessel Traffic Services (VTS) are in general all Vessel Monitoring or Detection Systems intended to establish maritime safety in particular areas of dense shipping. IMO and IALA are instrumental in their global standardization. The official IMO definition is: "*a service implemented by a competent authority, designed to improve the safety and efficiency of vessel traffic and to protect the environment*". They are primarily operated in ports, and in coastal regions where there is an increased risk, the latter often in association with Traffic Separation Schemes. The VTS infrastructure typically consists of a station onshore where the staff maintains a picture of the local maritime traffic. To that end, it uses primarily radar and communications links with the passing ships by VHF radio, fax or phone.

In general also visual observations are made, sometimes aided by optical or infrared cameras. Auxiliary sensors may be available, such as Radio Direction Finder (RDF) to locate the bearing of a radio transmission. In most cases the control center and the sensors (radar, cameras) are co-located, but sometimes, for the larger VTS systems, the sensors can be located away from the control center to enlarge the coverage.

Vessel Traffic Monitoring System (VTMS)

Vessel Traffic Monitoring and Information System (VTMIS)

Vessel Traffic Monitoring System (VTMS), or Traffic Monitoring and Information System (VTMIS), are a generic terms for systems providing information on maritime traffic and are key part of monitoring control and surveillance (MCS) programs at the national and international levels. VMS may be used to monitor vessels in the territorial waters of a country or a subdivision of a country, or in the Exclusive Economic Zones (EEZ) that extend 200 nautical miles (370.4 km) from the coasts of many countries.

Detail of VMS approved equipment and operational use will vary with the requirements of the nation of the vessel's registry, and the regional or national water in which the vessel is operating, while non-EU-Vessels are covered by the Vessel Detection System (VDS).

Since VMS components are expensive, most countries subsidize the purchase of equipment, although the vessel owner may have to pay for installation, maintenance, and continuing communications costs. The communications service used can optionally support other functions such as voice, electronic mail, and other applications over which the communications cost can be amortized. VMS cost is a significant obstacle to its use in developing countries, but there are a number of subsidy programs to encourage VMS use.

VMS from marine electronics suppliers may also offer a variety of other optional functions that can help in navigation, economic analysis, safety, and finding fish to harvest. Vessel Traffic Monitoring System (VTMS) are systems and/or applications for specific Vessel Traffic Monitoring or Vessel Traffic Managing (VTM) Solutions according to EU Directive 2002/59 to facilitate traffic management and planning, vessel monitoring, environmental protection, security provision, coastal surveillance, collision avoidance, shore-based pilots, SAR Coordination and Vessel Traffic Monitoring (VTM) Vessel Traffic Service VTS, Vessel Detection System (VDS) and their operator training.

VTMS and Monitoring Solutions provide extensive situational awareness and navigation information, facilitate vessel and other object identification and tracking, and enable comprehensive traffic planning in coastal and regional areas. Land-based the term refers to or Vehicle Tracking and Monitoring System (VTMS). In the EU community the vessel traffic monitoring and information systems under the EU Directive 2002/59 are the response on the call for integration.

Vessel Traffic Service (VTS)

A <u>Vessel Traffic Service (VTS)</u>[106] is a marine traffic monitoring system established by harbour or port authorities, similar to air traffic control for aircraft. Typical VTS systems use radar, closed-circuit television (CCTV), VHF radiotelephony and automatic identification system to keep track of vessel movements and provide navigational safety in a limited geographical area. A service implemented by a competent authority, VTS is designed to improve the safety and efficiency of navigation, safety of life at sea and the protection of the marine environment. VTS is governed by SOLAS Chapter V Regulation 12 together with the Guidelines for Vessel Traffic Services (IMO Resolution A.857(20)] adopted by the International Maritime Organization on 27 November 1997.

The VTS traffic image is compiled and collected by means of advanced sensors such as radar, AIS, direction finding, <u>CCTV</u> and VHF or other co-operative systems and services. A modern VTS integrates all of the information in to a single operator working environment for ease of use and in order to allow for effective traffic organization and communication. Operators working in HELCOM from Helsinki, Finland, VTS are covering i.e. the area from Emäsalo to Inkoo.

A VTS should always have a comprehensive traffic image, which means that all factors influencing the traffic as well as information about all participating vessels and their intentions should be readily available. By means of the traffic image, situations that are developing can be evaluated and responded upon. The data evaluation depends to a great extent on the quality of the data that is collected and the ability of the operator to combine this with an actual or developing situation. The data dissemination process exists of conveying the conclusions of the operator.

---

[106] http://en.wikipedia.org/wiki/Vessel_Traffic_Service

Calais-Dover Reporting System (CALDOVEREP)

The Dover Strait/Pas de Calais and its approaches are one of the busiest waterways in the world, and it poses severe safety problems to ships because of the density of traffic and the proximity of navigational hazards. In 1977 the traffic separation scheme, in the Dover Strait and adjacent waters, became compulsory.

When entering the area covered by the system, all ships over 300GT report to Dover Coastguard, which deals with south-west bound traffic, or to Gris Nez Traffic (in France), which handles north-east bound traffic. The reporting system is mandatory, and the short title for the system is CALDOVEREP. The following description is from the IMO's publication Ship's Routing:

- The CNIS processing and display system receives inputs from the radar and VHF DF equipment, processes the information and presents it on any or all of six displays. Each display shows processed images (tracks) from any of the three radar inputs overlaid on a synthetic map of a selected area. New targets entering radar range are automatically tagged with a unique track number. The position course and speed information of up to 300 racks is automatically updated and recorded, for each of the three radars, throughout the vessel's passage through the CNIS area, giving the CNIS a 900-track capability.
- DOVER COASTGUARD maintain a continuous watch on traffic in the Dover Strait/Pas de Calais. Operators can add vessel information to the IPRS (information processing and retrieval system) database (such as name and cargo) and can display that supporting information on a separate screen. CNIS is capable of providing an automatic alarm to identify any track, which strays into an unauthorised area. VHF DF vectors appear when a VHF radio transmits on the frequency selected on the VHF DF equipment. Recording equipment automatically stores information from all tracks which can either be replayed on the system or specific track movements can be plotted onto an A0-size sheet of paper.

CNIS was introduced in 1972. It provides a 24-hour radio service for all shipping in the Dover Strait and is operated from the MRCC at Langdon Battery near Dover.

CNIS broadcasts on VHF radio channel 11, every 60 minutes (every 30 minutes in poor visibility), and gives warnings of navigational difficulties and unfavorable conditions likely to be encountered in the Dover Strait.

Container Tracking System (CTS)

The Container Tracking System (CTS) is the equivalent to the Vessel Tracking System (VTS) for tracking and monitoring the movement of goods by container, however mainly using GPS instead of AIS. CTS are not mandatory in the European Union but exist in different configurations all over in the world, especially in the ports.

Container Tracking Suite (CTS) conglomerate shipping and logistic companies to track all full and empty containers in and out transactions at their freight stations. They use web browser based applications designed to integrate with clients system to pull and store basic shipping details like shipping date, delivery date, shipping port, destination freight station etc. of a container.

At any of the freight stations gate respective user will get a notification of planned incoming or outgoing container, based on this alert user can create requisite in/out gate pass. At any point of time CTS provides its users with container inventory details like full containers, empty containers, damaged containers, delivery expired containers as well as missing containers. As this application deals with price sensitive data a mechanism to automatically trigger medium and critical alerts via E-Mails and SMS built-in.

Container numbers have the format XXXU1234567 and can be tracked via several online services, as i.e the service track-trace[107] for 122 companies.

---

[107] https://www.track-trace.com/container

Five Power Defense Arrangements (FPDA)

The Five Power Defense Arrangements (FPDA)[108] was set up following the termination of the United Kingdom's defense guarantees of Malaysia and Singapore under the Anglo-Malayan Defense Agreement, as a result of the UK's decision in 1967 to withdraw its armed forces east of Suez. The FPDA is a low-profile regional security institution between Australia, Malaysia, New Zealand, Singapore and the United Kingdom.

The FPDA are a series of defense relationships established by a series of multi-lateral agreements between the Commonwealth members of the United Kingdom, Australia, New Zealand, Malaysia and Singapore and was signed in 1971, whereas the they are to consult each other immediately in the event or threat of an armed attack on any of the five countries for the purpose of deciding what measures should be taken jointly or separately in response. There is no specific commitment to intervene militarily.

The Five Powers Defense Arrangements do not refer to exclusive economic zones (EEZ) and the enforcement of a state's EEZ rights is a matter for that state; a state may request the assistance of other states in so doing. Exercise BERSAMA is an annual FPDA Training conducted in Singapore, Malaysia or the South China Sea.

The FPDA provides defense co-operation between the countries, establishing an Integrated Air Defense System (IADS) for Peninsular Malaysia and Singapore based at RMAF Butterworth under the command of an Australian Air Vice-Marshal. RMAF Butterworth, was under the control of the Royal Australian Air Force until 1988, and is now run by the Royal Malaysian Air Force but hosts rotating detachments of aircraft and personnel from all five countries.

In 1981, the five powers organized the first annual land and naval exercises. Since 1997, the naval and air exercises have been combined. In 2001, HQ IADS was predesignated Headquarters Integrated Area Defense System. It now has personnel from all three branches of the armed services, and co-ordinates the annual five-power naval and air exercises, while moving towards the fuller integration of land elements. An annual FPDA Defense Chiefs' Conference (FDCC) is hosted by either Malaysia or Singapore, and is the highest military professional forum of the FPDA and serves as an important platform for dialogue and exchange of views among the Defense Chiefs. There is also a Five Powers Defense Arrangements Ministerial Meeting (FDMM).

The FPDA has continued to fulfil vital security roles to the benefit of not only its members but also the wider security and stability of South-east Asia. But an important question for its members on how to develop the FPDA in the future.

One important motivation for setting up the arrangements in the first place was hedging

---

[108] https://en.wikipedia.org/wiki/Five_Power_Defence_Arrangements

against the resurgence of an unstable and threatening Indonesia which might again endanger the security of Malaysia and Singapore, and even the wider South-east Asian balance of power. A second implicit role was to provide channels of communication on defense matters between Malaysia and Singapore, and to build strategic confidence between them. And in the early 1970s, there was much anxiety in South-east Asia over the impending US military withdrawal from Vietnam and prospect of communist victories throughout Indochina.

However, this regional strategic environment has been transformed and the original reasons for the FPDA have become much less relevant or, indeed, redundant. While there may sometimes be irritants in relations with Jakarta, it would be fanciful to suggest that Indonesia would again seek to destabilize its smaller neighbors. Meanwhile, relations between Malaysia and Singapore are probably the best they have ever been. And while Cambodia, Laos and South Vietnam fell to communism and the subsequent Third Indochina War posed serious security threats, since the 1990s the three Indochinese countries have been integrated into ASEAN and Vietnam is now a good partner for Malaysia and Singapore in many areas.

There is much concern over international terrorism, and the potential for extremist groups to occupy weakly governed spaces such as the Sulu zone in the southern Philippines. Cyberthreats to critical national infrastructure are ever more worrying. Natural disasters, sometimes exacerbated by climate change, also pose major threats. Even more seriously, the regional order which has underpinned Asia-Pacific prosperity as well as security since the 1960s is under severe challenge. Potential regional flash points in Korea and the South China Sea threaten regional stability more than ever. ASEAN has struggled to find a convincingly strong common position on regional security, particularly in relation to the South China Sea. Amid this uncertainty, most states in the region are seeking to increase their military capabilities.

The deteriorating regional security environment requires boosting the interoperability of forces, particularly in the maritime and air spheres. Malaysia is building new frigates for its navy, while Singapore is commissioning new Littoral Mission Vessels. Australia's air warfare destroyers and P-8A maritime patrol aircraft are entering service and it has ordered F-35 Joint Strike Fighters. New Zealand is on the brink of ordering new patrol aircraft. The first of the UK's new aircraft carriers (which will carry F-35s) is expected to deploy operationally in 2021. There is much potential for the FPDA nations to bring these new ships and aircraft into their joint exercises to synergize their capabilities. More ambitiously, the FPDA might play a more active role in maritime security and counter-terrorism, benefiting other Southeast Asian countries.

France (FRA)

France as always been an European maritime power with today the French Navy being the largest Western European Navy after the British Royal Navy. Both nations also represent the only European Nuclear Military Powers.

France is bordering the Bay of Biscay and English Channel, between Belgium and Spain, southeast of the UK; bordering the Mediterranean Sea, between Italy and Spain and includes:

- French Guiana: Northern South America, bordering the North Atlantic Ocean, between Brazil and Suriname
- Guadeloupe: Caribbean, islands between the Caribbean Sea and the North Atlantic Ocean, southeast of Puerto Rico
- Martinique: Caribbean, island between the Caribbean Sea and North Atlantic Ocean, north of Trinidad and Tobago
- Mayotte: Southern Indian Ocean, island in the Mozambique Channel, about halfway between northern Madagascar and northern Mozambique
- Reunion: Southern Africa, island in the Indian Ocean, east of Madagascar

France is largest West European nation; most major French rivers – the Meuse, Seine, Loire, Charente, Dordogne, and Garonne – flow northward or westward into the Atlantic Ocean, only the Rhone flows southward into the Mediterranean Sea.

France is in international environmental agreements party to Air Pollution, Air Pollution-Nitrogen Oxides, Air Pollution-Persistent Organic Pollutants, Air Pollution-Sulfur 85, Air Pollution-Sulfur 94, Air Pollution-Volatile Organic Compounds, Antarctic-Environmental Protocol, Antarctic-Marine Living Resources, Antarctic Seals, Antarctic Treaty, Biodiversity, Climate Change, Climate Change-Kyoto Protocol, Desertification, Endangered Species, Hazardous Wastes, Law of the Sea, Marine Dumping, Marine Life Conservation, Ozone Layer Protection, Ship Pollution, Tropical Timber 83, Tropical Timber 94, Wetlands, Whaling

Système d'Information et de Commandement 21 (SIC 21)

In February 2004, the French armaments procurement agency (DGA) awarded Thales a $141 million (€113 million) contract for SIC 21 development and operational support for a ten years period. The SIC 21 deployment started in 2006 with 125 systems to be integrated on a wide range of platforms, i.e. surface ships, ground stations and aircrafts, ranging from single notebooks to the full-size systems integration in the operations room on the Charles de Gaulle aircraft carrier. Eventually the SIC21 is to replace the Army and Navy's legacy command information systems with a single, harmonized system to meet the growing need for force interoperability.

SIC 21 deployments include more than 90 sites, i.e. land-based and shipborne, 150 user locations and 2000 different users. The company is also the lead system integrator for the Australian Defence Forces' SEA 1442 maritime intranet system. Thales experience grounds on and also provides the Fully Integrated Communications Systems (FICS) for the Royal Navy's second batch of Type 45 anti-air destroyers in the United Kingdom. The company's communication suites provide the full range of internal and external communications capabilities, including radio and satellite links, internet, videoconferencing and tactical datalinks, and have also been selected for the FREMM multi-mission frigate program.

There are several French command information systems:

- SICF: (Système d'Information pour le Commandement des Forces) is the Army's command information system. It has been in service with land forces at tactical levels 1 to 3 (headquarters to brigade) for more than 10 years. SICF is used on over 3,000 workstations and is deployed in remote theatres.
- SIC 21: (Système d'Information et de Commandement) is the Navy's command information system. It is currently installed on most French Navy vessels. SIC 21 is an operational information system for the command and control of naval operations in a joint forces and international context.
- SIA: Evolving command structures and cost reduction targets have prompted the Ministry of Defence to streamline Army, Air Force and Naval information systems within a single system. The purpose of the SIA program is to rapidly deliver a fully tri-service operational information system. SIA will deliver improvements on three fronts: operational efficiency, MoD organization, and technical maturity of the system.

The Système d'information des armées Commandement et Contrôle (SIA C2, armed forces' information system – command and control) will facilitate information exchange at all levels by delivering a set of common tools for joint forces operations, while meeting all the specific requirements of land forces command. Legacy systems will continue to be maintained and supported until the SIA C2 system is rolled out across the two forces.

The Système d'Information et de Commandement 21 (SIC 21) is a Command and Control System for Naval Forces from the Thales Group designed for battle management / command aides and developed to meet the requirements of the French Navy. The system is to ensure secure liaison and communications between the French national authorities; French expeditionary forces deployed overseas and allied forces.

The SIC 21 modular architecture allows integration on the full spectrum of fighting and flag ships owned by the French Navy. Its primary functions comprise messages delivery, theater situational awareness, mission planning, operations course assessment, and logistics management.

The SIC 21 system is fully compliant with French and NATO security standards allowing seamlessly integration of French forces within coalition forces. The technical/COTS standards are i.e. XML, SQL, Microsoft® Office©, Outlook®, RepliWeb, Iris forms™, Java™, COTS GIS and the military standards include APP6-A, STANAG 4420, Adat-P3 V11, OTHT-Gold, DISAC, FORMAREN, IP ZONEX.

SIC 21 features

- Operational command & control in joint and international coalitions
- Maritime and land-based deployments
- Worldwide networked users with other CCIS interoperability
- A common technical core to federate maritime information applications

Collaborative working information services

- Geospatial information
- Multimedia communication with messaging, video conferencing, chat
- Office web-based portal, document engineering, shared database access

Naval mission-oriented applications

- Command & control assets management
- Formatted military message handling
- Recognized maritime picture and shared situational awareness
- Maritime space control : ZONEX, deployment areas
- Planning & control of air operations
- Intelligence access & exploitation

SIC 21 rests on naval communications and network infrastructure

- Dedicated network for RIFAN tactical internet, RLM, IP-SOCRATE
- Satellite (SYRACUSE, INMARSAT, ARISTOTE, ATHREIS), HF and UHF communications

SIC 21 exchanges with legacy information systems, notably:

- Naval information systems like SENIT 8, SLPM, CENTAC 2, TRELO, MARIE, MELODIE, SPATIONAV
- Joint (SICA), Land (SICF), Air (SCCOA) Command & Control Information Systems
- Eased access to NATO, joint and naval relevant information NATO MMHS and C2IS: CRONOS, MCCIS, ICC.

Germany (DEU)

Germany is from its geographical setting and coastline length no foremost maritime powers, however the economical dependency within Europe and world trade influence and imprint to a certain degree also a global maritime role. Germany is bordering the Baltic Sea and the North Sea, between the Netherlands and Poland, south of Denmark. With these nations it shares a strategic location on the North European Plain and along the entrance to the Baltic Sea.

The importance of maritime dependencies as a business location is an important factor for a high-technology and export-dependent nation like Germany. The maritime industry in the international environment and the role of the maritime defense industry in an international environment and international cooperation, safe maritime trade routes, industrial strategic interest in a comprehensive security system and the strategic importance of the Baltic Sea and the North Sea are key factors in Germany's role as the largest economic and one of the largest military nations on the Baltic Sea and in Europe. Therefore there are various players in German MSA of which not all can be mentioned here.

Germany is in international environmental agreements party to Air Pollution, Air Pollution-Nitrogen Oxides, Air Pollution-Persistent Organic Pollutants, Air Pollution-Sulfur 85, Air Pollution-Sulfur 94, Air Pollution-Volatile Organic Compounds, Antarctic-Environmental Protocol, Antarctic-Marine Living Resources, Antarctic Seals, Antarctic Treaty, Biodiversity, Climate Change, Climate Change-Kyoto Protocol, Desertification, Endangered Species, Environmental Modification, Hazardous Wastes, Law of the Sea, Marine Dumping, Ozone Layer Protection, Ship Pollution, Tropical Timber 83, Tropical Timber 94, Wetlands, Whaling.

Kiel International Sea Power Symposium (KISS)

The <u>Kiel International Sea Power Symposium</u>[109] (KISS) is a one-day international conference held during Kiel Week first conducted in 2017, Northern Europe's largest annual maritime festival and a traditional naval event. The symposium's objective is to discuss current matters pertaining to sea power, which is broadly understood as a broad set of maritime issues and those naval, commercial, and policy implications stemming from it.

KISS aims to contribute to Germany's increasingly important role in Europe and the NATO Alliance. The symposium is an opportunity to meet thought leaders as well as to articulate and critically discuss novel naval / maritime ideas. The event builds on a string of similar conferences. It draws on Kiel's rich maritime and naval heritage: as a naval base; a renowned port city; home to leading submarine technology, maritime strategic research, and oceanographic expertise; and as the namesake to the busiest artificial waterway in the world, the Kiel Canal.

The acronym KISS is a deliberate reference to the US Navy's principle established in the 1960s by the late Rear Admiral (USN) Paul D. Stroop, *"Keep it short and simple"* (sometimes also as *"Keep simple and straightforward"*). The symposium is guided by the idea that in Europe, there are too many conferences where commonplaces are the norm; the Kiel International Sea Power Symposium offers a forum for straight talk founded in sound academic research and hard operational experience.

KISS brings together hand-picked experts from the following fields: allied and international military (in particular naval, army, and air force ranks OF-4 and up), academia (in particular maritime security experts from the fields of political science, history, and law; from civilian universities and service academies), policy-makers (in particular from Berlin, Brussels, multinational commands, and allied capitals), and selected industry representatives from the maritime defense sector.

In particular, KISS strengthens the civilian-military dialogue among experts and reinforces ties between academia and navies. It also aims to include selected young professionals to sea power by providing hand-picked next-generation decision-makers with an opportunity to interact with senior operators, scientists, and policy-makers. Participation at this event is by invitation only and deliberately limited to allow for proper dialogue.

The Kiel International Sea Power Symposium is conducted under the "Chatham House Rule". The sea power strategic enterprise is the key objective and topics, panelists, and discussants should be selected based on this purpose. The organizers are committed to provide participants and speakers with an opportunity to learn something new and to stir up lively discussions.

---

[109] https://www.kielseapowerseries.com/en/

National Maritime Conference (NMK)

The National Maritime Conference (German "*Nationale Maritime Konferenz*, NMK) is a series of events organized by the Federal Ministry for Economic Affairs and Energy (BMWi) on maritime issues and German interests. The National Maritime Conference is under the patronage of the Federal Chancellor. It is organized, organized and chaired by the Maritime Coordinator of the Federal Government and focuses on topics such as shipbuilding, port management, maritime shipping and technology, offshore topics and climate and environmental protection.

The first National Maritime Conference in Emden was opened on 13 June 2000, further conference venues were Warnemünde (2001), Lübeck (2003), Bremen (2005), Hamburg (2006), Rostock (2009), Wilhelmshaven (2011), Kiel (2013), Bremerhaven (2015) and Hamburg (2017). The National Maritime Conferences are dialogue forums, idea exchanges and showrooms and includes the subject of Maritime Security (Maritime Sicherheit).

The 11. National Maritime Conference was for the first time held in a landlocked country to underline the importance of the conference for the whole economy of the entire Federal Republic and the dependency of the all the countries.

German Maritime Forces Staff (DEUMAR FOR)

The German Maritime Forces Staff (DEU MARFOR)[110] is the future operations staff located in Rostock, Germany. Though actually a national command, it will include a number of staff positions earmarked for personnel from maritime Allies. The aim is to generate some strong multinational cooperation for the command. The command will initially have a core staff of around 100, and up to 25 of those posts are to be filled by personnel from other countries.

Starting in 2020, a further enlarged DEU MARFOR will then transform the organization into the Baltic Maritime Component Command (BMCC). After an establishment, training and certification phase, the new command is to be operational from 2025. It will have a capacity of up to 170 staff.

The new Maritime Operation Center (MOC) is currently under construction and will combine for the first time DEU MARFOR as a national staff with international representatives and organizational structures which will then be able to plan and conduct maritime operations on the northern flank of NATO. This national task force is intended to grow into a much larger international leadership and then become the Baltic Maritime Component Command (BMCC). The construct can be offered to NATO as a maritime command for operations on the Alliance's northern flank, but also in other regions, for the purposes of national and alliance defense.

The Bundeswehr is reflecting to a renewed focus on national and Alliance defense strategies and will be able to offer NATO and the EU the use of the BMCC as a tactical command for multinational operations. Based in Rostock, the command staff will also provide the German Navy with effective command and control for operations in the Baltic and beyond. This will mean additional protection for the Alliance, especially on its maritime northern flank.

This development represents a new level of dovetailing between the German Navy and its European partners. The Navy's various integration measures now range from the upcoming German-Norwegian cooperation on submarines, to the German-Dutch collaboration on amphibious operations, to the Slovak boarding teams on German frigates.

In practical terms, DEU MARFOR will be formed by amalgamating the Navy's existing three smaller commands. Based in Kiel, Wilhelmshaven and Rostock, they each currently have staffs around one third as large as the future command.

---

[110] https://www.marine.de

Community-Of-Interest specific Maritime Situational Awareness (COI Specific MSA)

German Armed Forces are undertaking a program called HAFIS (Harmonisierung Führungs- und Informationssysteme) aiming in providing all required services for the forces on one platform. HAFIS requires Network Interconnection Areas (NIA) between domains with different security levels with the two subareas Shared Services Area (SSA) and Domain Transfer Area (DTA). The SSA includes a Layer 3 firewall and a dedicated service gateways for each service provided (shared service). The DTA provides among other services the prevention mechanism for non-authorized data flows (Data Loss Prevention / Information Protection Services).

The Community-Of-Interest specific Maritime Situational Awareness (COI Specific MSA) is the HAFIS aligned White Shipping system of the German Navy. Besides the Maritime Safety and Security System (MSSIS) the Water- and Shipping Directorate (WSD, Wasser- und Schifffahrtsverwaltung) provides two additional data streams (German Coastal and HELCOM) in NMEA-Format via the Integrated Sea Surveillance System. The WSD has further two possibilities for streaming Radar data in ASTERIX (All Purpose Structured Eurocontrol Surveillance Information Exchange) to the COI Specific MSA. The ASTERIX library has a number of Parts; each one describes a specific application. Part 1 contains the basic principles and rules to be followed when implementing ASTERIX. Parts 2 and higher describe how to encode data for a specific application. These parts are commonly referred to as "*ASTERIX Categories*"; they are available for applications in these areas:

- 000-127: Standard Civil and Military Applications
- 128-240: Special Civil and Military Applications
- 241-255: Civil and Military Non-Standard Applications

The allocation of category numbers, ranging from 000 to 127, is the responsibility of the ASTERIX Maintenance Group (AMG). Categories in the range from 128 to 240 can be allocated by national authorities. In order to allow us to maintain a full overview over existing ASTERIX categories, please notify the ASTERIX team about any category number allocated locally. Data Categories between 241 and 255 are designed for specific users and not do not form part of the ASTERIX Standard Document.

The WSD provides radar data in ASTERIX Category 10 and Category 62:

- Category 10 is used to transmit the single data stream of each the station with roughly calculated 1MB/s x station number. The WSV uses "Asterix-Recorder" to limit the overall bandwidth requirement of the 48(?) Radar stations for their user.
- Category 62 is used to send/receive the fused radar data without the possibility to identify a single source (radar station)

The basic use request is an overall fused/correlated stream in the Category 62, however, in case an operator would need to investigate a case, it is necessary to select a dedicated radar stream, hence in the Category 10.

Neither one of the ASTERIX categories originating from EUROCONTROL is suited for providing information within the maritime domain. In order to be able to distribute a Recognized Maritime Picture (RMP) the format/stream i.e. needs to include name, callsign, MMSI, IMO-number, etc. of the tracks in order for the operator to associate the track in the RMP, which ASTERIX does not provide.

Implementation of ASTERIX Category 10 and 62 in the Maritime Awareness Tactical Information System (MATIS) as core system in the Community-Of-Interest specific Maritime Situational Awareness (COI Specific MSA) supports raw sensor data collection for the operators to make use the WSD radars almost as they would be sensors of the Navy. Additionally MATIS also supports a proprietary ASTERIX message that has been implemented for coastal surveillance purposes.

MATIS is a comprehensive solution that covers a myriad of bases with operational flexibility in mind. MATIS is a derivate of the Finnish system for Coast Guard, Customs and the Navy, which is using it in the scaled version SEALION, a Joint & Common C4 Information Systems developmental for the Finnish Defence Forces.

Maritime Safety and Security Center (MSSC)

The <u>Maritime Safety and Security Center (MSSC)</u>[111] (German: Maritimes Sicherheitszentrum des Bundes und der Länder in Cuxhaven (MSZ Cuxhaven)) is a communication and cooperation network for the operational forces of the Federal Government and the German coastal states, which ensures maritime safety and security. The different specialist and capabilities are united in the Joint Emergency Reporting and Assessment Centre Sea (GLZ-See).

In the interest of national and international shipping as well as the safety and security of the German coast, relevant information is exchanged and operations are coordinated on the basis of common situation assessments.

Special incidents in the field of security (in terms of hazard prevention) and safety (in terms of damage control) have been motivating a continuous development in the course of the past decades leading to the establishment of the Maritime Safety and Security Centre. The respective framework was provided by political and legal decisions.

In April 2013 the Federal Control Center for Maritime Security starts their active operations. Since its commencement of work in 2007, the Joint Emergency Reporting and Assessment Centre Sea has been housed in the office building of the Waterways and Shipping Administration in Cuxhaven.

The new building of the Maritime Safety and Security Center was constructed under the overall control of the Federal Ministry of Transport and Digital Infrastructure. The new Maritime Safety and Security Centre is located on the premises of the Cuxhaven Waterways and Shipping Office, in the immediate vicinity of the existing office building.

In addition to the Joint Emergency Reporting and Assessment Center, the new building was planned to provide office spaces for the Central Command for Maritime Emergencies as well as for the heads of the coordination centers of the Federal Police Department of Maritime Security, the Waterways and Shipping Administration, the Fishery Protection Agency, the Federal Customs Administration, the German Navy and the Waterways Police. The groundbreaking ceremony took place in February 2017, operations started in February 2018.

The control over operations conducted as part of everyday routine or in special situations will be exercised by the authority being locally and technically responsible in each case. The partners assure their support within the framework of the pertinent legal provisions. In particularly serious operational situations, special organizational procedures will be activated and policing tactics will be applied. Complex emergency situations will be managed by the operational staff of the Central Command for Maritime Emergencies.

---

[111] https://www.msz-cuxhaven.de/EN

The Federal Waterways and Shipping Administration ensures communication with its competent agencies, particularly as regards traffic-related measures.

The network partners ensure that their agencies and institutions in the MSSC exchange all information required for the fulfilment of tasks without delay and that, in special situations, they pass on all relevant information to the operational commander and provide further information on request. Information is transferred using the MSSC data platform.

The following agencies work joint in the MSSC Joint Emergency Reporting and Assessment Centre Sea on a 24/7/365 basis:

- the Maritime Emergencies Reporting and Assessment Center of the Central Command for Maritime Emergencies
- the German Waterways Police Reporting and Coordination Centre of the German coastal states
- the Federal Police coordination center of Maritime Security
- the Federal Customs Administration coordination center
- the Coordination Centre of the Fishery Protection Agency of the Federal Office for Agriculture and Food
- the Federal Waterways and Shipping Administration including the international Point of Contact
- a liaison element of the German Navy

The Joint Emergency Reporting and Assessment Centre Sea has above all the following tasks:

- providing the network partners with relevant data and information
- compilation of common situation pictures
- joint initial assessment of the overall situation
- mutual consultation
- mutual assistance in handling special operational situations
- logistic support of law enforcement and operational forces
- cooperation with situation centers of other authorities and institutions
- participation in the preparation, conduct and evaluation of joint exercises

The Area of Responsibility of the Joint Emergency Reporting and Assessment Center Sea as a joint institution of the Federal Government and the Germany coastal states comprises the Section 3 (1) of the administrative agreement on a Maritime Safety and Security Centre the cooperation of partners:

1. within the exclusive economic zone of the Federal Republic of Germany;
2. on maritime waterways as defined in Section 1 (2) sentence 1 of the Federal Waterways Act;

3. on the navigable waterways of the River Elbe (excluding the Hamburg area), the Kiel Canal, the Rivers Trave, Warnow and Weser (under the German Traffic Regulations for Navigable Maritime Waterways) and, on the River Ems, (under Section 1 of the regulation on the introduction of the shipping regulation concerning the estuary of the River Ems).

In the Maritime Safety and Security Center (MSSC), partners from federal and coastal state institutions work together in the form of a network in a 24-hour shift system, 365 days a year, here the:

- Federal Police Department of Maritime Security
- Federal Customs Administration
- Federal Office for Agriculture and Food
- Federal Waterways and Shipping Administration
- German Navy
- Waterways Police forces of the German federal and coastal states
- Central Command for Maritime Emergencies

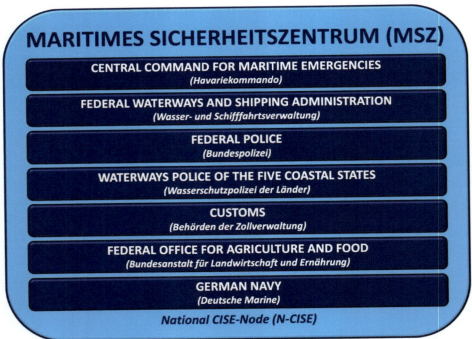

The German Navy participates in the MSSC with its White Shipping System, the Community-of-Interest Specific Maritime Situational Awareness (COI Specific MSA) tool, which is also used in a derivate for the EUCISE2020 project to connect to as a National CISE-Node, while the Federal Waterways and Shipping Administration relies on its Maritime Traffic System (System Maritime Verkehrstechnik (SMV)) connecting the coastal radars and AIS-Base-Stations and linking them into the MSSC. The Maritime Traffic System is to be replaced by the Integrated Sea Surveillance System.

Most other partners rely on the maritime information systems of these two partners and connect primarily with commercial phone, fax and e-mail.

System Maritime Verkehrstechnik (SMV)

System Maritime Verkehrstechnik (SMV) is the System for Maritime Traffic and Transportation of the German Water and Shipping Directorate (Wasser- und Schifffahrtsverwaltung (WSV) Nord und Nord-West) integrated the AIS base stations in the German territorial waters.

The German Water- and Shipping Directorate has VTS data centers in the port of Hamburg, in the Travemunde Control Center and is currently developing a new system for connecting the German WSV-AIS-Base stations and Coastal Radar stations with new functionalities.

Additional sensor information is e.g. retrieved from radar, video, DGPS, UHF). In 2008 about eleven AIS base stations were fielded (in ~16 in 2009) and the data stream started to HELCOM and to the German SAR Organization (Deutsche Gesellschaft zur Rettung Schiffbrüchiger (DGzRS)). In 2009 SMV received a GIS and processing functionality and provides AIS data links and sensor data (radar, video) and interfaces for external clients. The hard- and software (Dell, Debian Linux, NAVICON) were also used on the platforms in windpark BARD Offshore 1.

Integrated Sea Surveillance System

The new "*Integrated Sea Surveillance System*" of the Water and Shipping Directorate under development is to provide information on shipping, airborne surveillance, ship traffic messages (for example by pilots) and notices issued by the traffic centers as central point of contact for navigation including radar, AIS, radio communication.

The various pieces of information are combined in a situational picture for the entire German coast. In the MSSC in Cuxhaven the ship traffic information, radar, AIS, hydrometeorological data and event data are merged, edited and saved. Thus, the law enforcement forces will always have the history available required in the case of accident and later prosecution.

The core components may include the Innovative Navigation products like:

- Radar Video Processing
- AIS Network
- Data Processing
- High Level Data Integration
- Simulation
- People Tracking
- inVTS Systems
- Navigation

Gulf of Guinea Commission (GGC)

The Gulf of Guinea Commission (GGC)[112] was established by the Treaty signed in Libreville, Gabon, on 3 July 2001 by Angola, Congo, Gabon, Nigeria and Sao Tome and Principe. This constituted in a permanent Institutional framework for cooperation amongst the countries bordering the Gulf of Guinea in order to defend their common interest and promote peace and socio-economic development based on the bases of dialogue, consensus, and ties of friendship, solidarity and fraternity.

Cameroun and Democratic Republic of Congo joined the Gulf of Guinea Commission in 2008. Membership of Gulf of Guinea Commission is open to other states in the Gulf of Guinea for purposes of transforming the sub region into a Zone of Peace and Security. With an area of 5,629,471 km² and a population of about 260 million people, the sub-region covered by the Gulf of Guinea Commission is one of the largest and most populated Geopolitical spaces.

ECOWAS, ECCAS and GGC signed in 2013 the Yaoundé Declaration to support the Maritime Strategy 2050 and to promote cooperation, coordination and pooling of resources between the members. The Headquarter is located in Luanda, Angola and the members of the Gulf of Guinea Commission are Angola, Cameroon, Democratic Republic of the Congo, Equatorial Guinea, Gabon, Nigeria, Republic of the Congo and Sao Tome and Principe.

The Summit is the supreme organ of the Commission. It is composed of Heads of State and Government or their duly authorized representatives. It meets once a year, in ordinary session, and at any time, in extraordinary session, under to the agreement of two thirds of the Commission's Member States.

The Summit has the following duties:

- Set the general policy and the Commission's broad guidelines;
- Control the operation and the Commission;
- Analyze the Council's reports and produce relevant decisions;
- Decide ultimately on all issues for which the Council could reach a decision;
- Create any body or Specialized Committees of the Commission,
- Determine the budget of the Commission;
- Appoint and dismiss the Executive Secretary
- Provide the Commission Headquarter

The Summit takes decisions by consensus or, failing that, by a majority of 2/3 of the States

---

[112] http://cggrps.org/en/

present. To meet and legitimately deliberate it is required quorum of 2/3 of the Commission's Member States.

The Council of Ministers is composed of the Ministers of Foreign Affairs or any other Minister or authority designated by the Member States. The Ministers of Economy, Hydrocarbons, Fishery Resources, Mines, Environment or any other Minister may also meet in case of need.

The Council meets once a year in ordinary session and at any time, in special session, by the request of any Member State and subject to the agreement of the majority of 2/3 of Member States. The Council is assisted in carrying out its tasks by specialized committees of this Treaty. The Council takes decisions by consensus or failing that, by a majority of two thirds of the States present. To meet and deliberate legitimately, the required quorum is 2/3 of the Commission's Member States.

The Commission's Member States Council tasks are:

- Prepare the sessions of the Summit;
- Promoting any actions aimed at achieving the objectives set out in Article 2 of this Treaty, within the framework of the general policy set by the Summit;
- To this end, develop and propose appropriate general policy measures;
- Know all the issues that the Summit designates;
- Implement cooperation policy in accordance with the general policy defined by the Summit;
- Create the Committees and establish the competences;

The Secretariat is headed by an Executive Secretary appointed by the Summit for a term of three years, renewable once. The secretariat functions are:

1. Ensure the proper functioning of the Commission,
2. implement the decisions of the Summit and Council,
3. Prepare reports, decision projects and agreements, for consideration by the Summit and the Council
4. Formulate recommendations likely to contribute to the functioning and effective and harmonious development of the Commission;
5. Ensure the technical services for meetings of the Summit and the Council, as well as of the specialized committees;
6. Assume the role of custodian of the documents and the Commission's assets;
7. Prepare the budget of the Commission;

Perform such other functions as the Conference or the Council entrust to the Secretariat. The Summit establishes Specialized Committees to address, by the request of the Summit or the Council, the specific issues pertaining to the achievement of the objectives of this Treaty.

The Summit may, if it deems necessary, to restructure the existing Committees or create new ones according to the needs of the Commission. Each Committee can, if necessary, create sub-committees to assist in fulfilling its mandate. The Committee Determines the composition of the sub-committees.

An *ad hoc* Arbitral Mechanism is created within the Commission. The rules of procedure and other issues concerning the *ad hoc* Arbitral mechanism will be established by the Council and adopted by the Conference.

Helsinki Commission (HELCOM)

The Baltic Marine Environment Protection Commission (HELCOM)[113], also known as Helsinki Commission), is an intergovernmental organization governing the Convention on the Protection of the Marine Environment of the Baltic Sea Area[114] (Helsinki Convention). HELCOM is encompassing various measures for the prevention and elimination of pollution of the Baltic Sea.

HELCOM (Baltic Marine Environment Protection Commission - Helsinki Commission) is the governing body of the Convention on the Protection of the Marine Environment of the Baltic Sea Area, known as the Helsinki Convention. The contracting Parties are Denmark, Estonia, the European Union, Finland, Germany, Latvia, Lithuania, Poland, Russia and Sweden.

At a later stage Norway and Czech Republic joined, however mostly Time-Late-AIS-Data-Feeds from nations under the HELCOM Treaty. The first Convention on the Protection of the Marine Environment of the Baltic Sea Area was signed by Denmark, Finland, West Germany, East Germany, Poland, USSR and Sweden in 1974 and entered into force on 3 May 1980. A new convention was signed in 1992 by Czechoslovakia, Denmark, Estonia, the European Community, Finland, Germany, Latvia, Lithuania, Poland, Russia and Sweden. The Convention on the Protection of the Marine Environment of the Baltic Sea Area, 1992, entered into force on 17 January 2000.

The States-Parties to the Convention agreed individually or jointly to take all appropriate legislative, administrative or other relevant measures to prevent and eliminate pollution in order to promote the ecological restoration of the Baltic Sea Area and the preservation of its ecological balance.

The States-Parties to the Convention agreed individually or jointly to take all appropriate legislative, administrative or other relevant measures to prevent and eliminate pollution in order to promote the ecological restoration of the Baltic Sea Area and the preservation of its ecological balance.

HELCOM's vision for the future is a healthy Baltic Sea environment with diverse biological components functioning in balance, resulting in a good ecological status and supporting a wide range of sustainable economic and social activities.

The Helsinki Commission meets annually, with the Heads of Delegation (HOD) representing the Contracting Parties. The Commission adopts Recommendations for the protection of the marine environment, decides on the budget and makes other key decisions. Decisions

---

[113] http://www.helcom.fi
[114] https://en.wikipedia.org/wiki/Helsinki_Convention_on_the_Protection_of_the_Marine_Environment_of_the_Baltic_Sea_Area

are made by consensus. In addition, Ministerial level meetings are held every few years.

The chairmanship of the Commission rotates between the Contracting Parties every two years, according to their alphabetical order in English. The working structure of HELCOM consists of the meetings of the Helsinki Commission, the Heads of Delegation, and the eight main groups. The Secretariat coordinates HELCOM work, in which the parties agreed to apply:

- Precautionary principle, that is, to take preventive measures when there is reason to assume that substances or energy introduced, directly or indirectly, into the marine environment may create hazards to human health, harm living resources and marine ecosystems, damage amenities or interfere with other legitimate uses of the sea;
- Best Environmental Practice and Best Available Technology;
- Polluter pays principle, that is, make the party responsible for producing pollution responsible for paying for the damage done to the environment.

The aim of the States-Parties to the Convention is to prevent and eliminate pollution of the marine environment of the Baltic Sea Area caused by harmful substances from all sources, including:

- from land-based sources (measures set out in Annex III);
- from ships (measures outlined in Annex IV);
- from incineration and dumping (exemptions from dumping provisions set out in Annex V);
- from exploration and exploitation on the seabed (measures related to the offshore exploration and exploration activities set out in Annex VI).

The States-Parties to the Convention are obligated to notify and enter into consultations with each other when an environmental impact assessment of a proposed activity is likely to cause a significant adverse impact on the marine environment of the Baltic Sea Area. Similarly, they are to notify and consult each other whenever a pollution incident in their territory is likely to cause pollution to the marine environment of the Baltic Sea Area outside its territory and adjacent maritime area.

The Convention sets up a Baltic Marine Environment Protection Commission (HELCOM) whose responsibilities are to implement the Convention, make recommendations to the Parties, define pollution control criteria and objectives and promote additional measures in co-operation with respective governmental bodies of the Parties.

The Parties also undertake to implement measures to maintain adequate ability and to respond to pollution incidents in order to eliminate or minimize the consequences of these incidents and regularly report to the HELCOM commission on and inform the general public of the measures taken in accordance with the Convention.

Helsinki Commission AIS System (HELCOM AIS System)

The Helsinki Commission AIS System (HELCOM AIS System) is the regional VTS network for the maritime safety and environmental monitoring in the Baltic region and located in Helsinki/FIN. In contrast to the SafeSeaNet the HELCOM AIS System enables the exchange of AIS ship information (NEMA 0183) in near real time. The data is stored for analysis and statistics and each and every nation decides on the information it sends; however the set time late of data is about 16 minutes due to the send interval of national systems to the HELCOM AIS System.

The Helsinki Vessel Tracking Service (VTC Helsinki) in Finland is administrated by the Finnish Maritime Administration (FMA) and is the Finnish feed for the HELCOM AIS System and also further processed and fed into the SUCFIS. Finnish Navy has track data based on radar observations and additional information sources as well as from center on the West Coast, in the Archipelago, in the Bay of Bothnia (Finnish: Perämeri, Swedish: Bottenviken), in Helsinki and Saimaa. It works with AIS track data (NMEA 0183). The Finnish AIS coverage alone has 22 Base Stations (coastal) and seven inland in the lake districts.

Contracting parties of HELCOM, here Denmark, Estonia, Finland, Germany, Latvia, Lithuania, Poland, and Sweden, who are exchanging their AIS data within the HELKOM network.

India (IN)

India is buy geographical location and economic power one of the biggest Maritime Entities and has initiated and launched the Indian Ocean Naval Symposium (IONS)[115] in February 2008. India is bordering the Arabian Sea and the Bay of Bengal, between Burma and Pakistan, dominates the South Asian subcontinent, and lays in a strategic position near important Indian Ocean trade routes.

The website Global Firepower (GFP)[116] provides since 2006 analytical display of data concerning over 135 modern military powers with the ranking based on each nation's potential war-making capability across land, sea and air fought with conventional weapons. The results incorporate values related to resources, finances, and geography with over 55 different factors ultimately making up the final rankings. In 2018 a total of 136 countries were included in the Global Firepower database ranking India number forth behind the USA, Russia, and China.

India is in international environmental agreements party to Antarctic-Environmental Protocol, Antarctic-Marine Living Resources, Antarctic Treaty, Biodiversity, Climate Change, Climate Change-Kyoto Protocol, Desertification, Endangered Species, Environmental Modification, Hazardous Wastes, Law of the Sea, Ozone Layer Protection, Ship Pollution, Tropical Timber 83, Tropical Timber 94, Wetlands, Whaling.

---

[115] https://en.wikipedia.org/wiki/Indian_Ocean_Naval_Symposium
[116] https://www.globalfirepower.com

Indian Ocean Naval Symposium (IONS)

The <u>Indian Ocean Naval Symposium (IONS)</u>[117] are a series of biennial meetings among the littoral states of the Indian Ocean region. Indian Ocean Naval Symposium (IONS) provides a forum to increase maritime security cooperation, providing a forum for discussion of regional maritime issues and promote friendly relationships among the member nations.

The states are represented by their Navy chiefs, the forum was initiated and launched by India in February 2008. It is a similar approach as by the Italian Navy with the Trans-Regional Maritime Network (T-RMN) with the Virtual Maritime Traffic Center(V-RMTC).

The opening ceremony of the Indian Ocean Naval Symposium 14. February 2008 in New Delhi was held with the presence of navies from 27 countries and the chairmanship of the first IONS was given to the Indian Navy. The subject of this international symposium was contemporary seafaring and national challenges.

IONS is a security construct for the Indian Ocean region which is similar to the Western Pacific Naval Symposium (WPNS). It is a voluntary initiative among the navies and maritime security agencies of the member nations. In addition to the symposiums, numerous other activities like workshops, essay competitions and lectures are also held under the umbrella of the organization.

There are two other major information sharing cooperation's in the region of the IONS, the Information Fusion Center (IFC) and the Regional Cooperation Agreement on Combating Piracy and Armed Robbery against Ships in Asia (ReCAAP). There are 32 nations represented in IONS with 23 members in sub-regions or littorals and nine observers:

- South Asian Littorals: Bangladesh, India, Maldives, Pakistan, Seychelles, and Sri Lanka, and the United Kingdom of Great Britain and Northern Ireland
- West Asian Littorals: Iran, Oman, Saudi Arabia, UAE, (and possible new members Bahrain, Iraq, Kuwait, Qatar, Yemen)
- East African Littorals: France, Kenya, Mauritius, Mozambique, South Africa, Tanzania, (and possible new members Djibouti, Egypt, Eritrea, Comoros, Madagascar, Somalia and Sudan
- South East Asian and Australian Littorals: Australia, Indonesia, Myanmar, Singapore, Thailand and Timor Leste, (Malaysia)

The nine states with observer status are: China, Germany, Italy, Japan, Madagascar, Malaysia, Netherlands, Russia and Spain.

Collaboration to enhancing maritime security in the Indian Ocean has a long background. List of countries contributing in the IONS, reflects the importance of this symposium. IONS

---

[117] https://en.wikipedia.org/wiki/Indian_Ocean_Naval_Symposium

can play a critical role as a security and collaborative mechanism in the region.

There are important and strong economies in the Indian Ocean region, however these countries are not willing to play the primarily role to improving security in the region. It is obvious that lack of dominant or responsible power in the region leads to invite intra-regional powers. Nowadays countries in the Indian Ocean region have to avoid power competition. IONS can be an effective and comprehensive security mechanism and reduce intra-regional powers effects.

IONS Working Groups (IWG):

- Counter Piracy IWG
- Humanitarian Assistance and Disaster Relief IWG
- Information Sharing and Interoperability IWG
- IONS Preparatory Workshop

IONS is growing fast and attracted not only Indian Ocean region countries, but also Pacific Ocean powers (China and Japan), Europe, South America, and Mediterranean countries showed their interest as well. It is also worth to mention that countries without coastal boarders are interested in this symposium. The 6. IONS was held in 2018 in Tehran, Iran, and proves the importance of the forum and events.

The main objectives of IONS are:

- Promoting a common understanding of Indian Ocean issues
- Developing a common series of designed strategies to increase regional maritime security
- Promoting Oceanic navy forces to face with existing challenges
- Predicting prospects of maritime security and stability
- Cooperation in order to decrease maritime security anxieties in the Indian Ocean
- Developing collaborative capabilities in education, methods, organizational systems, logistics and Operational processes
- Developing the capacity of region's rapid response navy
- Collaborating in humanitarian and disaster relief operations throughout the Indian Ocean region

Information Management and Analysis Center (IMAC)

In the neighborhood of the national capital, the Information Management and Analysis Center (IMAC)[118] was inaugurated in Gurgaon, where the Indian Navy and Coast Guard share responsibilities and coordinate actions.

The IMAC is the single point center interlinking the newly formed coastal radar chain and is manned by the IND Navy to function under the National Security Adviser (NSA) where the National Command Control Communication and Intelligence System (NC3I) will eventually become the backbone of National Maritime Domain Awareness (NDMA).

The NC3I will link 20 naval and 31 Coast Guard monitoring stations to generate a seamless real-time picture of the nearly 7,500-km long coastline. The system currently comprises 46 radars and 30 additional radars are planned to fill all the gaps in coastline security.

The hubs are linked by high speed optical fiber networks and satellite links serve as a back-up in case of emergency. Apart from coastal radars and optical sensors, it also draws information from automatic identification systems fitted on merchant ships and has a comprehensive shipping database of world registers of shipping for analysis of traffic.

While the IMAC will be the technical fusion hub of all maritime intelligence, all activities are coordinated under the National Committee for Strengthening Maritime and Coastal Security (NCSMDS), which was created in 2009 with representatives from the Ministry of Interior, Merchant Navy, fishing fleet, Navy, Coastguard, Directorate General of Lighthouses and Chief Secretaries of the coastal states and under the presidency of a Cabinet Secretary who reports to the Indian Prime Minister.

---

[118] http://currentaffairs.gktoday.in/defence-minister-inaugurates-information-management-analysis-centre-imac-gurgaon-11201415852.html

National Command Control Communication and Intelligence System (NC3I)

The National Command Control Communication and Intelligence System (NC3I) of the Information Management and Analysis Center (IMAC) will link 20 naval and 31 Coast Guard monitoring stations with IMAC to generate a real-time picture of the nearly 7,500-km long Indian coastline.

The system at first listed 46 radars and 30 additional radars are connected sequentially to fill all the gaps in coastline security. In this system the remote hubs are linked with the center by high speed optical fiber networks and satellite links- for backup in case of emergency.

Apart from coastal radars and optical sensors, it also draws information from automatic identification systems fitted on merchant ships and has a comprehensive shipping data-base of world registers of shipping for analysis of traffic. The system receives information and intelligence from external sources, such as Lloyds, LRIT, AIS, ships and aircraft. The Navy and Coast Guard, with the help of the respective state governments intend to install transponders in all Indian private fishing boats to locate them while at sea. These transponders will be eventually connected through the NC3I system.

The network was built by Bharat Electronics Limited (BEL) which has sourced customized software from the U.S. Company Raytheon. The software has added filters to identify threats from the vast number of ocean-going vessels by correlation and data fusion. Approved by the Defence Acquisition Council in 2012, it has become operational in 15 months.

The software on which the coastal surveillance will be carried out incorporates hi-tech features like data fusion, correlation and decision support features thus facilitating better decision making. Thus, Information Management and Analysis Center (IMAC) project would go a long way in beefing up the maritime surveillance, thereby, enhancing the Indian National Maritime Domain Awareness Project.

Further upgrading of the system and agreement for data sharing with 24 countries in the Indian Ocean region was awaiting clearance from the Prime Minister-led Cabinet Committee on Security (CCS).

Indian Regional Navigation Satellite System (IRNSS)

The Indian Regional Navigation Satellite System (IRNSS)[119], also named NAVIC ("*sailor*" or "*navigator*" in Sanskrit, Hindi and many other Indian languages), is an autonomous regional satellite navigation system that provides accurate real-time positioning and timing services. It covers India and a region extending 1,500 km (930 mi) around it, with plans for further extension. An Extended Service Area lies between the primary service area and a rectangle area enclosed by the 30th parallel south to the 50th parallel north and the 30th meridian east to the 130th meridian east, 1,500–6,000 km beyond borders. The system at present consists of a constellation of seven satellites, with two additional satellites on ground as stand-by.

The constellation is in orbit as of 2018, and the system was expected to be operational from early 2018 after a system check. NAVIC will provide two levels of service, the "standard positioning service", which will be open for civilian use, and a "restricted service" (an encrypted one) for authorized users (including military). Due to the failures of one of the satellites and its replacement, no new date for operational status has been set.

The system was developed partly because access to foreign government-controlled global navigation satellite systems is not guaranteed in hostile situations, as happened to the Indian military in 1999 when it was dependent on the American Global Positioning System (GPS) during the Kargil War. The Indian government approved the project in May 2006.

The system is intended to provide an absolute position accuracy of better than 10 meters throughout Indian landmass and better than 20 meters in the Indian Ocean as well as a region extending approximately 1,500 km (930 mi) around India. The Space Applications Centre in 2017 said NAVIC will provide standard positioning service to all users with a position accuracy up to 5 m. The GPS, for comparison, had a position accuracy of 20–30 m. Unlike GPS which is dependent only on L-band, NAVIC has dual frequency (S and L bands). When low frequency signal travels through atmosphere, its velocity changes due to atmospheric disturbances. US banks on atmospheric model to assess frequency error and it has to update this model from time to time to assess the exact error. In India's case, the actual delay is assessed by measuring the difference in delay of dual frequency (S and L bands). Therefore, NavIC is not dependent on any model to find the frequency error and is more accurate than GPS.

The Space segment consists of 7 satellites, three of the seven satellites are located in geostationary orbit (GEO) at 32.5° East, 83° East, and 131.5° East longitude, approximately 36,000 km (22,000 mi) above earth surface. Remaining four satellites are in inclined geosynchronous orbit (GSO). Two of them cross equator at 55° East and two at 111.75° East.

---

[119] https://en.wikipedia.org/wiki/Indian_Regional_Navigation_Satellite_System

The four GSO satellites will appear to be moving in the form of an "8".

Ground Segment is responsible for the maintenance and operation of the IRNSS constellation. The Ground segment comprises:

- IRNSS Spacecraft Control Facility (IRSCF)
- ISRO Navigation Centre (INC)
- IRNSS Range and Integrity Monitoring Stations (IRIMS)
- IRNSS Network Timing Centre (IRNWT)
- IRNSS CDMA Ranging Stations (IRCDR)
- Laser Ranging Stations
- IRNSS Data Communication Network(IRDCN)

The INC established at Byalalu performs remote operations and data collection with all the ground stations. 14 IRIMS are currently operational and are supporting IRNSS operations. CDMA ranging is being carried out by the four IRCDR stations on regular basis for all the IRNSS satellites. The IRNWT has been established and is providing IRNSS system time with an accuracy of 2 ns (2.0×10^{-9} s) (2 sigma) w.r.t UTC. Laser ranging is being carried out with the support of ILRS stations around the world. Navigation Software is operational at INC since 1 August 2013. All the navigation parameters viz. satellite ephemeris, clock corrections, integrity parameters and secondary parameters viz. iono-delay corrections, time offsets w.r.t UTC and other GNSS, almanac, text message and earth orientation parameters are generated and uplinked to the spacecrafts automatically. The IRDCN has established terrestrial and VSAT links between the ground stations. Seven 7.2 m FCA and two 11 m FMA of IRSCF are currently operational for LEOP and on-orbit phases of IRNSS satellites.

A messaging interface is embedded in the NavIC system. This feature allows the command center to send warnings to a specific geographic area. For example, fishermen using the system can be warned about a cyclone.

India's Department of Space increased the number of satellites in the constellation from 7 to 11 for extending coverage. Development of space qualified atomic clocks was initiated, along with study & development initiative for All Optical Atomic Clock (ultra stable for IRNSS and Deep Space Communication).

Study and analysis for Global Indian Navigational System (GINS) was initiated as part of the technology and policy initiatives in the 12th FYP (2012–17). The system is supposed to have a constellation of 24 satellites, positioned 24,000 km (14,913 mi) above Earth. As of 2013, the statutory filing for frequency spectrum of GINS satellite orbits in international space, has been completed.

Indian Ocean Commission (IOC)

The Indian Ocean Commission (IOC)[120] is an intergovernmental organization that was created in 1982 at Port Louis, Mauritius and institutionalized in 1984 by the Victoria Agreement in the Seychelles. The IOC is composed of five African Indian Ocean nations: Comoros, Réunion (a department of France), Madagascar, Mauritius and Seychelles.

Notwithstanding their different characteristics (Reunion as a French department; Mauritius and Seychelles as Middle-Income Countries whereas Comoros and Madagascar are amongst the Least-Developed Countries), the five islands share geographic proximity, historical and demographic relationships, natural resources and common development issues.

IOC's principal mission is to strengthen the ties of friendship between the countries and to be a platform of solidarity for the entire population of the African Indian Ocean region. IOC's mission also includes development, through projects related to sustainability for the region, aimed at protecting the region, improving the living conditions of the populations and preserving the various natural resources that the countries depend on. Being an organization regrouping only island states, the IOC has usually championed the cause of small island states in regional and international fora.

The IOC works on four pillars which have been adopted in 2005 by the Summit of Heads of States:

- Political and diplomatic cooperation,
- Economic and commercial cooperation
- Sustainable development in a globalization context, cooperation in the field of agriculture, maritime fishing, and the conservation of resources and ecosystems
- Strengthening of the regional cultural identity, cooperation in cultural, scientific, technical, educational and judicial fields.

EU partnership with the IOC has been effective for as long as 25 years where EAS-IO provides the strategic framework for cooperation with four regional organizations – COMESA, EAC, IOC and IGAD.

The original ideas were to encourage trade and tourism. Recently, cooperation has focused on marine conservation and fisheries management. The COI has funded a number of regional and national conservation and alternative livelihoods projects through the Regional Program for the Sustainable Management (ReCoMAP) of the Coastal Zones of the Countries of the Indian Ocean (PROGECO in French). This project ended in 2011. An example of these projects is project to catalyze the development of sea cucumber and

---

[120] https://en.wikipedia.org/wiki/Indian_Ocean_Commission

seaweed aquaculture in South West Madagascar with the NGOs, Transmad, Blue Ventures, and Madagascar Holothuria.

The Commission has a Secretariat which is located in Mauritius and headed by a Secretary General. Political and strategic orientations of the organization are under the responsibility of the Council of Ministers which meets annually. The organization also has a system of rotating presidency of each Member State. The Presidency is currently ensured by Comoros. The highest level of the organization's structure is the Summit of Heads of States whose last meeting was held in Madagascar in 2005.

EU partnership with the IOC has been effective for as long as 25 years. The successive EDFs have financed programs implemented by the IOC to the tune of EUR 100 million, mainly in the area of environment and natural resources. The EU, which is the main development partner of the IOC, accounting for app. 70% of total financial support to IOC, has scaled up its assistance to IOC during the last few years. The EU has three main on-going program with the IOC for a total amount of EUR 48 million:

- The SMARTFISH Program for the Implementation of a Regional Fisheries Strategy in the Eastern-Southern Africa and Indian Ocean region (ESA-IO region). Its overall objective is to contribute to an increased level of social, economic and environmental development and deeper regional integration in the ESA-IO through the sustainable exploitation of marine and lake fisheries resources;
- The Islands program for the Implementation of the Mauritius Strategy for Small Islands Developing States (SIDS) of the ESA-IO region, which will contribute to an increased level of social, economic and environmental development and deeper regional integration through the sustainable development of SIDS in the ESA-IO region. This program will assist beneficiary countries towards this year's United Nations Conference on Sustainable Development (Rio +20). A grant agreement of EUR 470.000 was recently signed between the EU and the United Nations Department of Economic and Social Affairs for the design and implementation of a Monitoring and Evaluation System to track countries progress in implementing the Mauritius Strategy.
- A program to support the IMF Regional Technical Assistance Centers (AFRITACs) of the ESA-IO, namely AFRITAC East, based in Tanzania, and AFRITAC South, based in Mauritius. The program's objective is to contribute to the regional economic integration process, and higher growth and poverty reduction in the ESA-IO region. More specifically, the program aims at improving the design, implementation, and monitoring of sound macroeconomic policies, and enhanced regional harmonization and integration, in ESA-IO Member States and Regional Organizations. Two grant agreements amounting to EUR 14.675.000 were signed between the EU and the IMF on 17. October 2011 on the occasion of the inauguration of AFRITAC South in Mauritius.

A Start up Project to promote regional maritime security is the MASE Project, a Financing Agreement, which amounts to EUR 2 million, was signed on 13 December 2011 by the Indian Ocean Commission and Mauritius, the Seychelles and the Union of Comoros. The objective of the Startup MASE Project is to support the implementation of short term actions of the Regional Strategy and Action Plan against Piracy and for Maritime Security as adopted in the Regional Ministerial Conference on Maritime Piracy Meeting in 2010 in Mauritius, co-chaired by the EU High Representative and Vice President of the European Commission, and with the Inter-Regional Coordinating Committee (IRCC) to coordinate program and strategies within EAS-IO to be implemented.

The IOC also implements part of two 10th EDF programs which are under the responsibility of COMESA, namely: The Inter-Regional Coordination Committee (IRCC) Project to support the Indian Ocean Commission in ensuring its tasks within the IRCC configuration. Activities which have been undertaken by the IOC since 2011 include:

- support to the preparation of new projects;
- support to the strengthening and restructuring of the IOC Secretariat;
- setting up of the IOC Project Monitoring and Evaluation System;
- facilitation for the development of an economic strategic plan for IOC;
- and drawing up of IOC Strategic Development Plan for 2012-2016.

Regional Integration Support Program (RISP)

The Regional Integration Support Program (RISP) which aims at advancing the regional economic integration in the ESA-IO region. The program is jointly implemented by the COMESA, the EAC, the IGAD and the IOC for a period of three years from July 2010 to June 2013. Under the IOC annual work plans, the focus is on activities to facilitate Member States participation in COMESA regional integration agenda, and in the overall regional integration agenda process, while promoting the specificity of small island states, based on the principle of subsidiarity.

France/Reunion is not a beneficiary of EDF but may participate in program via its own funds, and contribute with its expertise in several areas. The collaboration is of primary importance to facilitate a good economic integration of countries and regions pertaining to the same geographical areas. It is pursued vigorously and positively by France/Reunion.

One should think that the Indian Ocean Commission (IOC), initiated by France, would be closely linked to the Indian Ocean Naval Symposium (IONS), initiated by India, however the reality shows IOC closer related to European affairs and efforts, while IONS routes back to the requirements and efforts by a local actor and stakeholder. The website of the Indian Ocean Commission (IOC) with background information and the activities is therefore kept purely in French with no translations. Positive effects can be observed. Although in disagreement on the issue of the island of Mayotte, over which the Union of the Comoros claims sovereignty, France and Comoros remain key partners. Human ties, as shown by the many citizens with dual nationality, and neighbor relations (migratory pressure on Mayotte) are key points.

On 21. June 2013, the Presidents of France and the Comoros signed the *"Paris declaration on friendship and cooperation between France and Comoros"* to revive the bilateral relationship. A strengthened political dialogue was opened with the creation of a *"High Joint Council"* (HCP) which has so far met five times.

At the regional level, there are however many positive examples of cooperation and ad hoc regional organizations in the Indian Ocean region created to manage a distinct maritime issue, such as:

- the Gulf organization on port security and control (2004 Riyadh Memorandum of Understanding, headquartered in Oman),
- an anti-pollution institution of eight Persian Gulf states based in Kuwait (1981 Regional Organization for the Protection of the Marine Environment - ROPME),
- a fisheries management initiative based in Kenya (the 2004 South West Indian Ocean Fisheries Commission),
- or an anti-piracy coalition to guarantee free movement in the Strait of Malacca (2004 MALSINDO - the Malaysia, Singapore, Indonesia organization).

These efforts have narrow mandates and one can readily identify the stakeholders and the official counterparts in each participating state.

The hope is on support for such initiatives, but also that these initiatives can be linked with other national and regional maritime players to enhance the information sharing and co-ordination. Often well-intended actions in one sphere can unintentionally have adverse effects on another, for example the destruction of mangroves, coastal tourism and other activities to promote development, for example, exacerbated the effects of the tsunami in Sri Lanka and in other Indian Ocean islands.

Overall it is one of many efforts in the region of the Indian Ocean that will need coordination with other events.

Indian Ocean Rim Association (IORA)

The Indian Ocean Rim Association (IORA)[121] is a dynamic inter-governmental organization aimed at strengthening regional cooperation and sustainable development within the Indian Ocean region through its 21 Member States and 7 Dialogue Partners.

The vision for IORA originated during a visit by late President Nelson Mandela of South Africa to India in 1995, where he said: *"The natural urge of the facts of history and geography should broaden itself to include the concept of an Indian Ocean rim for socioeconomic cooperation."*

This sentiment and rationale underpinned the Indian Ocean Rim Initiative in March 1995, and the creation of the Indian Ocean Rim Association (then known as the Indian Ocean Rim Association for Regional Co-operation) two years later, in March 1997.

As the third largest ocean woven together by trade routes, commands control of major sea-lanes carrying half of the world's container ships, one third of the world's bulk cargo traffic and two thirds of the world's oil shipments, the Indian Ocean remains an important lifeline to international trade and transport.

Home to nearly 2.7 billion people in Member States with coastlines at the ocean that are rich in cultural diversity and richness in languages, religions, traditions, arts and cuisines, they vary considerably in terms of their areas, populations and levels of economic development. The area can also be divided into a number of sub-regions (Australasia, Southeast Asia, South Asia, West Asia and Eastern & Southern Africa), each with their own regional groupings (such as ASEAN, SAARC, GCC and SADC, to name a few). Despite such diversity and differences, these countries are bound together by the Indian Ocean.

IORA's apex body is the Council of Foreign Ministers (COM) which meets annually. The Republic of South Africa will assume the role for 2017-2019, followed by the United Arab Emirates. A committee of Senior Officials (CSO) meets twice a year to progress IORA's agenda and consider recommendations by Working Groups and forums of officials, business and academics to implement policies and projects to improve the lives of people within the Indian Ocean Member States.

The year 2017 was a landmark for IORA, celebrating the 20th Anniversary of the Association as a proactive inter-governmental organization with an ever growing importance within the Indian Ocean region. Strengthening the ties that bind Member States whose shores are washed by the Indian Ocean waters, IORA remains committed to build and expand understanding and mutually beneficial cooperation through a consensus based evolutionary and non-intrusive approach in the rapid changing environment faced by the region.

---

[121] http://www.iora.net/en

Today, IORA is a dynamic organization of 21 Member States and 7 Dialogue Partners, with an ever-growing momentum for mutually beneficial regional cooperation through a consensus-based, evolutionary and non-intrusive approach. The Association will facilitate and promote economic co-operation, bringing together inter-alia representatives of Member States' governments, businesses and academia. In a spirit of multilateralism, the Association seeks to build and expand understanding and mutually beneficial co-operation through a consensus-based, evolutionary and non-intrusive approach.

The Secretariat of the Indian Ocean Rim Association (IORA) is hosted by the Government of the Republic of Mauritius which is based in Cyber City, Ebène, Mauritius. It manages, coordinates, services and monitors the implementation of policy decisions, work programs and projects adopted by the Council of Ministers. The Secretariat is responsible for the servicing of all IORA meetings, the representation and promotion of the Association, the collation and dissemination of information, the maintenance of an archive, depository and registry for IORA documentation and research material and the mobilization of resources.

The Secretariat is headed by a Secretary-General, who is assisted by four Directors and Experts, on voluntary seconded experts from Member States. A Chair in Indian Ocean Studies has been established to support the work of the Association through rigorous, in-depth research on the six priority areas of IORA. The Secretariat is also supported by locally recruited staff focusing on various areas of development.

Priority and focus are in flag ship areas:

- The Indian Ocean Dialogue
- The Somalia and Yemen Development Program (SYDP)
- The IORA Sustainable Development Program (ISDP)

These priority areas were re-invigorated at the 11th COM meeting, in Bengaluru, India on 15 November 2011. To promote sustained growth and balanced development in the Indian Ocean Region, IORA focused on Maritime Safety & Security, Trade & Investment Facilitation, Fisheries Management, Disaster Risk Management, Tourism & Cultural Exchange, Academic, Science & Technology, Blue Economy, Women's Economic Empowerment.

The IORA has been addressing especially Maritime Safety & Security in the Indian Ocean through a broad range of activities to enhance international cooperation in security and governance to successfully tackle the challenges faced by the region in MSS. The IORA Leaders' Summit held in March 2017, in Jakarta, Indonesia, highlighted the prioritization of these concerns through its theme, *"Strengthening Maritime Cooperation for a Peaceful, Stable, and Prosperous Indian Ocean."*

Intergovernmental Authority on Development (IGAD)

The Intergovernmental Authority on Development (IGAD) in Eastern Africa was created in 1996 to supersede the Intergovernmental Authority on Drought and Development (IGADD) which was founded in 1986.

The recurring and severe droughts and other natural disasters between 1974 and 1984 caused widespread famine, ecological degradation and economic hardship in the Eastern Africa region. Although individual countries made substantial efforts to cope with the situation and received generous support from the international community, the magnitude and extent of the problem argued strongly for a regional approach to supplement national efforts.

In 1983 and 1984, six countries in the Horn of Africa - Djibouti, Ethiopia, Kenya, Somalia, Sudan and Uganda - took action through the United Nations to establish an intergovernmental body for development and drought control in their region. The Assembly of Heads of State and Government met in Djibouti in January 1986 to sign the Agreement which officially launched IGADD with Headquarters in Djibouti. The State of Eritrea became the seventh member after attaining independence in 1993.

In April 1995 in Addis Ababa, the Assembly of Heads of State and Government made a Declaration to revitalize IGADD and expand cooperation among member states. On 21 March 1996 in Nairobi the Assembly of Heads of State and Government signed the Letter of Instrument to Amend the IGADD Charter / Agreement establishing the revitalized IGAD with a new name "The Intergovernmental Authority on Development". The Revitalized IGAD, with expanded areas of regional cooperation and a new organizational structure, was launched by the IGAD Assembly of Heads of State and Government on 25 November 1996 in Djibouti, the Republic of Djibouti.

The IGAD mission is to assist and complement the efforts of the Member States to achieve, through increased cooperation:

- Food Security and environmental protection
- Promotion and maintenance of peace and security and humanitarian affairs, and,
- Economic cooperation and integration.

IGAD will be the premier regional organization for achieving peace, prosperity and regional integration in the IGAD region. The objectives of IGAD are to:

- Promote joint development strategies and gradually harmonize macro-economic policies and programs in the social, technological and scientific fields;
- Harmonize policies with regard to trade, customs, transport, communications, agriculture, and natural resources, and promote free movement of goods, services, and people within the region.

- Create an enabling environment for foreign, cross-border and domestic trade and investment;
- Achieve regional food security and encourage and assist efforts of Member States to collectively combat drought and other natural and man-made disasters and their natural consequences;
- Initiate and promote programs and projects to achieve regional food security and sustainable development of natural resources and environment protection, and encourage and assist efforts of Member States to collectively combat drought and other natural and man-made disasters and their consequences;
- Develop and improve a coordinated and complementary infrastructure, in the areas of transport, telecommunications and energy in the region;
- Promote peace and stability in the region and create mechanisms within the region for the prevention, management and resolution of inter-State and intra-State conflicts through dialogue;
- Mobilize resources for the implementation of emergency, short-term, medium-term and long-term programs within the framework of regional cooperation;
- Promote and realize the objectives of the Common Market for Eastern and Southern Africa (COMESA) and the African Economic Community;
- Facilitate, promote and strengthen cooperation in research development and application in science and technology.

The IGAD Strategy was elaborated in 2003 and adopted by the 10th Summit of Heads of State and Government.

The Strategy provides a coherent framework aimed at guiding IGAD priority development programs in pursuit of its mandate, and in moving it forward as the premier regional economic organization for achieving peace, prosperity and regional integration in the IGAD region. It draws its spirit and substance from the Member States' desire to attain a viable regional economic cooperation in the IGAD region. It incorporates and benefits from:

- experiences gained in the past;
- current framework conditions of development cooperation; and;
- regional as well as global challenges and emerging issues facing the Region.

In preparing the Strategy, IGAD has embraced and employed the principles ownership, participation and partnership. With this in mind, IGAD has involved to the maximum extent possible, its staff, Member States and IGAD Partners in preparing the Strategy.

The Strategy document is composed of five main sections. Section I introduces an overview of the profile of the Region highlighting the economic and social situation. A glimpse of IGAD's history and a summary of its mandate provide the reason for establishing this regional organization. The Section cites some of the strengths of IGAD as a regional development vehicle. It also, mentions the challenges and constraints that were encountered and the valuable lessons learned which will be utilized in focusing and sharpening the Strategy.

Section II of the Strategy presents IGAD's vision and mission statements. It also defines some other important aspects of the Strategy among which are a set of principles and values that IGAD will abide by as it pursues its mandate; IGAD's strategic approach and partnerships and comparative advantages of IGAD. Also, the Section presents characteristics of the Strategy with regard to its flexibility and dynamism, and framework conditions.

Details of the core outputs of the Strategy are provided in Section III. The Section identifies the need to address a number of key strategic issues pertaining to the complex and ever changing nature of regional cooperation. These include policy matters; development information sharing; capacity building; establishing partnerships and alliances; and facilitating research and technology development.

Section IV is the main thrust of the Strategy. It presents the programs under the three priority sectors of IGAD: agriculture and environment, political and humanitarian affairs and economic cooperation. Fairly detailed descriptions of the programs and other crosscutting themes like gender mainstreaming are presented in the Section.

Section V underscores the important issue of implementing the strategy. It identifies that successful implementation of the strategy can only be realized through the concerted collaborative effort of the IGAD member states, the Secretariat and IGAD Partners Forum, IPF.

Finally it is recognized that the strategy was first and foremost a tool to guide the IGAD Secretariat in implementing its mandate but also owned by the Member States and supported by the IPF who finance most of the programs. At various stages of the process of formulating the Strategy, it became clear that a more focused Strategy did not mean reduction of regional program priorities but rather the level of intervention and outputs commensurate with Secretariat's capacity. It should be underlined in this juncture that the capacity of IGAD encompasses both the technical and institutional capacity in the Member States that are to the IGAD disposal.

The Intergovernmental Authority on Development is comprised of four hierarchical POLICY organs:

- THE ASSEMBLY OF HEADS OF STATE AND GOVERNMENT is the supreme policy making organ of the Authority. It determines the objectives, guidelines and programs for IGAD and meets once a year. A Chairman is elected from among the member states in rotation.
- THE COUNCIL OF MINISTERS is composed of the Ministers of Foreign Affairs and one other Focal Minister designated by each member state. The Council formulates policy, approves the work program and annual budget of the Secretariat during its biannual sessions.

- THE COMMITTEE OF AMBASSADORS is comprised of IGAD member states' Ambassadors or Plenipotentiaries accredited to the country of IGAD Headquarters. It convenes as often as the need arises to advise and guide the Executive Secretary.
- THE SECRETARIAT is headed by an Executive Secretary appointed by the Assembly of Heads of State and Government for a term of four years renewable once. The Secretariat assists member states in formulating regional projects in the priority areas, facilitates the coordination and harmonization of development policies, mobilizes resources to implement regional projects and programs approved by the Council and reinforces national infrastructures necessary for implementing regional projects and policies.
- The Executive Secretary is assisted by four Directors heading Divisions of Economic Cooperation & Social Development; Agriculture and Environment; Peace and Security; and Administration and Finance plus twenty two regional professional staff and various short-term project and Technical Assistance Staff.

The Intergovernmental Authority on Development (IGAD) superseded the Intergovernmental Authority on Drought and Development (IGADD) established in 1986 by the then drought afflicted six Eastern African countries of Djibouti, Ethiopia, Kenya, Somalia, Sudan and Uganda. The State of Eritrea was admitted as the seventh member of the Authority at the 4th Summit of Heads of State and Government in Addis Ababa, September 1993.

Although IGADD was originally conceived to coordinate the efforts of member states to combat drought and desertification, it became increasingly apparent that the Authority provided a regular forum where leaders of the Eastern African countries were able to tackle other political and socioeconomic issues in a regional context. Realizing this the Heads of State and Government of Djibouti, Eritrea, Ethiopia, Kenya, Sudan and Uganda, at an extra- ordinary Summit on 18 April 1995, resolved to expand the mandate of IGADD and made a declaration to revitalize IGADD and expand cooperation among member states. The revitalized IGADD was renamed the Intergovernmental Authority on Development (IGAD).

IGAD Civil Society forum was established in the 2003.However; very little activity was carried out since then. The IGAD program officer responsible for this sector was also recruited only in October 2007.

In order to be able to resuscitate the interface between IGAD Secretariat and the NGOs-CSOs in the region the division has decided to have consultancy service. The objective of the consultancy is to involve in a concrete manner NGOs-CSOs of the region in IGAD thematic areas namely Peace and Security, Agriculture and Environment and economic Co-operation and Integration.

Accordingly, the PSD prepared a draft term of reference and as the issue of CSOs/NGOs is a crosscutting one proposed to the Executive Secretary the setting up of an Inter-divisional committee consisting of representatives from the three divisions of the secretariat i.e. Peace and Security, Economic Cooperation and Social Development & Agriculture and Environment.

With a compared small number of members as Ethiopia, Uganda, South Sudan, Kenya, Somalia, Eritrea, Sudan, the IGAD represents a rather massive organizational construct with a small output on Maritime Security issues. However, with the support of the EU and neighboring countries the political ties are established.

International Association of Lighthouse Authorities (IALA)

The <u>International Association of Lighthouse Authorities (IALA)</u>[122] is a nonprofit, international technical association. Established in 1957, it gathers together marine aids to navigation authorities, manufacturers, consultants, and, scientific and training institutes from all parts of the world and offers them the opportunity to exchange and compare their experiences and achievements.

IALA encourages its members to work together in a common effort to harmonize aids to navigation worldwide and to ensure that the movements of vessels are safe, expeditious and cost effective while protecting the environment.

Taking into account the needs of mariners, developments in technology and the requirements and constraints of aids to navigation authorities, a number of technical committees have been established bringing together experts from around the world.

The work of the committees is aimed at developing common best practice standards through publication of IALA Recommendations and Guidelines. This work ensures that mariners have aids to navigation which will meet their needs both now and in the future. Thus IALA contributes to a reduction of marine accidents, increased safety of life and property at sea, as well as the protection of the marine environment.

IALA also encourages cooperation between nations to assist developing nations in establishing aids to navigation networks in accordance with the degree of risk for the waterway concerned. The purpose of IALA is to ensure that seafarers are provided with effective and harmonized marine Aids to Navigation services worldwide to assist in safe navigation of shipping and protection of the environment.

To achieve this aim IALA:

- constantly reviews the services provided for shipping by Aids to Navigation and Vessel Traffic Service Authorities; the international regulations pertaining to the operating practices and equipment carriage requirements of shipping; and new and developing technologies that may improve the effectiveness of aids to navigation services or reduce the cost of providing the service without degrading its effectiveness;
- provides detailed information to Authorities in Handbooks, such as the NAVGUIDE and VTS Manual, which are regularly reviewed and updated every four years.

IALA develops and publishes:

---

[122] http://www.iala-aism.org

- recommendations on technologies and practices that are intended to improve the services being provided, such as Training of VTS personnel, the provision of AIS facilities by shore authorities, the implementation of DGNSS stations and the Photometry of Marine Aids to Navigation Signal Lights;
- model courses to supplement the Recommendations on the training of VTS personnel; and
- Guidelines that either supplement recommendations or provide advice and guidance on how new and developing technologies can be used to improve the operational or cost effectiveness of services.

IALA also studies the methods of introducing new and improved risk management techniques to assist in determining the optimum arrangement of aids to navigation in waterways.

IALA recognized that the maritime industry is going through a stage of rapid technological development and change, including developments in the design and operations of ships. IALA members face the challenge of providing appropriate aids to navigation to cater for high speed vessels, both currently in operation and in the future, and may use IALA risk assessment techniques in their planning.

The increasing complexity and amount of information available to the navigator emphasizes the need to take into account the possibility of information overload and confusion during the design of new and innovative aids to navigation.

Protection of the marine environment and security needs may lead to regional VTS systems being merged into a global network of systems. IALA itself provides for the International Association of Lighthouse Authorities Network (IALA-Net) and is involved with regional test beds for evaluating possible technology to meet the needs of marine transport in the future.

International Association of Lighthouse Authorities Network (IALA-Net)

The International Association of Lighthouse Authorities Network (IALA-Net) is a near real time maritime data exchange service, provided through the Internet provided by the International Association of Lighthouse Authorities (IALA). It has AIS data storage and statistics capability.

IALA-Net is a worldwide service available only to national competent authorities who provide maritime data from their area of responsibility in exchange for data from other participants. The IALA-net statistics engine has the following capabilities:

- Ship report (vessel movements within specified period).
- Passage line statistics (count of vessel passages), histogram (where the vessels pass and in which direction) and data (a table of vessel data for each passage).
- Density plot (traffic density and routes in a given area).

The network is intended to assist participating authorities in fulfilling their duties in relation to maritime safety, security, protection of the marine environment, and the efficiency of navigation. Participation in IALA-NET is free of charge. The data provided by IALA-NET and associated services is not verified. IALA-NET is simply a data exchange mechanism similar to MSSIS with limited data storage up to a few months.

IALA national members or national aids to navigation authority, and even participants in other regional or global data exchange of information systems can join the IALA-NET and the IALA Members in Europe, here Bulgaria, Denmark, Estonia, Finland, France, Ireland, Latvia, Montenegro, Norway, Poland, the Faroe Islands participate in the network; data from Australia, China, Oman, the USA, Iraq, Greenland, Ukraine and Chile is sourced.

At the core of IALA-NET are a number of servers through which the data is exchanged. Initially there was one server (Denmark) but at least three servers were envisaged with automatic failover to ensure high system reliability. The servers are planned to be hosted in Denmark, China and USA. However, the Danish Maritime Authorities are no longer hosting the IALA-NET servers and the homepage has been shut down.

International Association of Ports and Harbors (IAPH)

The <u>International Association of Ports and Harbors (IAPH)</u>[123] is the global trade association for seaports worldwide. Formed in 1955, it is now recognized as the NGO representing ports worldwide. With over 200 ports in membership, as well as numerous national port representative bodies, it now has consultative status with 5 UN agencies, including the United Nations Conference on Trade and Development (<u>UNCTAD</u>[124]) and the IMO.

The IAPH membership is divided into 3 regions, African/European, American and Asian/Oceanian. The Secretariat is based in Tokyo, Japan and is headed by a Secretary General who is appointed by the Board of Directors.

---

[123] https://www.iaphworldports.org
[124] https://en.wikipedia.org/wiki/UNCTAD

International Chamber of Shipping (ICS)

The <u>International Chamber of Shipping (ICS)</u>[125] is the principal international trade association for merchant shipowners and operators, representing all sectors and trades and over 80% of the world merchant fleet. ICS membership comprises national shipowners' associations in Asia, Europe and the Americas whose member shipping companies operate over 80% of the world's merchant tonnage.

Established in 1921, ICS is concerned with all technical, legal, employment affairs and policy issues that may affect international shipping. ICS represents shipowners with the various intergovernmental regulatory bodies that impact on shipping, including the International Maritime Organization. ICS also develops best practices and guidance, including a wide range of publications and free resources that are used by ship operators globally.

The Asian Shipowners' Association (ASA) and the European Community Shipowners' Associations (ECSA) are Regional Partners with memberships that overlap with ICS and which also comprise national shipowners' associations. The aim of ICS is to promote the interests of shipowners and operators in all matters of shipping policy and ship operations and ICS will:

- Encourage high standards of operation and the provision of high quality and efficient shipping services.
- Strive for a regulatory environment which supports safe shipping operations, protection of the environment and adherence to internationally adopted standards and procedures.
- Promote properly considered international regulation of shipping and oppose unilateral and regional action by governments.
- Press for recognition of the commercial realities of shipping and the need for quality to be rewarded by a proper commercial return.
- Remain committed to the promotion of industry guidance on best operating practices.
- Cooperate with other organizations, both intergovernmental and nongovernmental, in the pursuit of these objectives.
- Anticipate whenever possible and respond whenever appropriate to policies and actions which conflict with the above.

---

[125] http://www.ics-shipping.org/

International Cargo Handling Coordination Association (ICHCA)

The International Cargo Handling Coordination Association (ICHCA)[126], founded in 1952, is an independent, not-for-profit organization dedicated to improving the safety, security, sustainability, productivity and efficiency of cargo handling and goods movement by all modes and through all phases of national and international supply chains.

ICHCA International's privileged non-government organization (NGO) status enables it to represent its members, and the cargo handling industry at large, in front of national and international agencies and regulatory bodies. Its Expert Panel provides technical advice and publications on a wide range of practical cargo handling issues.

ICHCA International operates through a series of autonomous national and regional chapters – including ICHCA Australia, ICHCA Japan and ICHCA Canarias/Africa (CARC) – plus Correspondence and Working Groups to provide a focal point for informing, educating, networking, shaping and sharing industry views to improve knowledge and best practice across the global cargo chain.

ICHCA International is dedicated to improving the safety, security, sustainability, productivity and efficiency of cargo handling and transport by all modes.

- Promote efficient cargo handling in appreciation of its vital function to world trade, national economies, business and industry, and the livelihood and wellbeing of many hundreds of thousands of workers right around the globe
- Assist co-operation between many different parties as a critical factor to the safety and overall performance of today's increasingly complex cargo chains, which often reach across multiple continents, cultures, jurisdictions, transport modes and logistics nodes
- Provide a platform, both internationally, and through its various chapters, to co-ordinate the dialogue and build relations between the many private and public sector stakeholders, to foster greater cross-party understanding, and to shape and share good practice for the benefit of all

---

[126] https://ichca.com/

International Fishery Organizations (IFO)

There are various <u>International Fishery Organization (IFO)</u>[127] that promote international co-operation to achieve effective, responsible marine stewardship and ensure sustainable fisheries management. International Fishery Organizations (IFO) are established under the Food and Agriculture Organization of the United Nations (FAO).

Atlantic Ocean Regional Fisheries Management Organizations

- International Commission for the Conservation of Atlantic Tunas (ICCAT) is an intergovernmental fishery organization responsible for the conservation of tunas and tuna-like species in the Atlantic Ocean and its adjacent seas. The Commission is responsible for providing internationally coordinated research on the condition of these species and developing regulations for their sustainable management.
- North Atlantic Salmon Conservation Organization (NASCO) is an international organization whose objective is to conserve, restore, enhance, and rationally manage Atlantic salmon through international cooperation, taking into account the best available scientific information.
- Northwest Atlantic Fisheries Organization (NAFO) is an intergovernmental fisheries science and management body responsible for the management of most resources in the region, salmon, tunas/marlins, whales, and sedentary species. NAFO's objective is to ensure the long term conservation and sustainable use of the fishery resources in the Convention Area and to safeguard the marine ecosystems in which these resources are found.
- Western Central Atlantic Fisheries Commission (WECAFC) promotes the effective conservation, management, and development of the living marine resources of the area of competence of the Commission, in accordance with the United Nations Food and Agriculture Organization Code of Conduct for Responsible Fisheries, and address common problems of fisheries management and development faced by members of the Commission.

Indian Ocean Intergovernmental Organizations

- Indian Ocean Tuna Commission (IOTC) is a regional fisheries management organization mandated to manage tuna and tuna-like species in the Indian Ocean and adjacent seas. Its objective is to promote cooperation among its members by ensuring, through appropriate management, the conservation and optimum utilization of stocks and encouraging sustainable development of fisheries based on

---

[127] http://www.fao.org/docrep/012/i1493e/i1493e00.htm

such stocks. IOTC has authority over tuna and tuna-like species, with a main focus on albacore, bigeye, and yellowfin tunas.
- Indian Ocean South East Asian Marine Turtle Memorandum of Understanding (IOSEA) is an intergovernmental agreement that aims to protect, conserve, replenish, and recover marine turtles and their habitats of the Indian Ocean and South-East Asian region, working in partnership with other relevant nations and organizations.

Pacific Ocean Regional Fisheries Management Organizations

- Agreement on the International Dolphin Conservation Program (AIDCP) is a legally binding agreement for dolphin conservation and ecosystem management in the Eastern Tropical Pacific Ocean. The objectives of the Agreement are to reduce incidental dolphin mortalities in the tuna purse-seine fishery through the setting of annual limits, to seek alternative means of capturing large yellowfin tunas not in association with dolphins, and to ensure the long-term sustainability of tuna stocks and marine resources in the Eastern Tropical Pacific Ocean.
- Inter-American Tropical Tuna Commission (IATTC) is an international commission responsible for the conservation and management of tuna and other marine resources in the eastern Pacific Ocean. The overall objective of this commission is to ensure the long-term conservation and sustainable use of the fish stocks covered by the Convention, in accordance with the relevant rules of international law.
- North Pacific Anadromous Fish Commission (NPAFC) was established under the Convention of Anadromous Stocks in the North Pacific Ocean. The NPAFC serves as a forum for promoting the conservation of anadromous stocks and ecologically related species—including marine mammals, sea birds, and non-anadromous fish in the high seas area of the North Pacific Ocean.
- Convention on the Conservation and Management of Pollock Resources in the Central Bering Sea. The Pollock Convention is an intergovernmental agreement aimed at establishing an international regime for conservation, management, and optimum utilization of pollock resources in the Bering Sea.
- Pacific Salmon Commission (PSC) was created under a treaty between the United States and Canada that aims to establish and implement fishery management regulations for the international conservation and harvest of North Pacific salmon stocks. The Commission provides regulatory advice and recommendations for sustainable management of Pacific salmon stocks.
- Western and Central Pacific Fisheries Commission (WCPFC) provides a forum for management, long-term conservation and sustainable use of highly migratory fish stocks, such as tunas, billfish, and marlin, in the western and central Pacific Ocean.

- International Pacific Halibut Commission (IPHC) is an international fisheries organization responsible for the management of Pacific halibut stocks within the Pacific waters of its member states (the United States and Canada). It was founded in 1923 by an international treaty to conserve, manage, and rebuild the halibut stocks in the Convention Area to levels that would achieve and maintain the maximum sustainable yield from the fishery.

Southern Ocean Intergovernmental Organizations

- Commission for the Conservation of Antarctic Marine Living Resources (CCAMLR) was established by international convention in 1982 with the objective of conserving Antarctic marine life. This was in response to increasing commercial interest in Antarctic krill resources, a keystone component of the Antarctic ecosystem.
- Convention for the Conservation of Antarctic Seals (CCAS) aims to promote and achieve the protection, scientific study, and rational use of Antarctic seals, and to maintain a satisfactory balance within the ecological system of Antarctica.

The organization under the Food and Agriculture Organization of the United Nations (FAO) are frequently exchanging their information in the FISHINFONetwork (FIN).

FISHINFONetwork (FIN)

The FISHINFONetwork (FIN)[128] consists of seven independent intergovernmental and governmental organizations plus the Food and Agriculture Organization of the UN (FAO) based GLOBEFISH unit. GLOBEFISH is an integral part of FIN and plays a critical role in the network's activities.

In 1986, in its first session, the FAO Sub-Committee on Fish Trade recognized the work of the four existing FIN offices (INFOPESCA created in 1977, INFOFISH in 1981, INFOPECHE and GLOBEFISH in 1984), marking the organizations as the start of a worldwide network for fish marketing development support.

Today, FIN has more than 70 full-time staff members and works with over 100 additional international consultants in all fields of fisheries. 50 national governments have signed international agreements with the different FIN services and are using the expertise of these services to develop the fishery sector worldwide.

Created to develop the fisheries and aquaculture sector particularly in developing countries and countries in transition, the network provides services to private industry and to governments. The execution of multilateral and bilateral projects is one of the main activities of the network. It is also widely known for its range of publications and periodicals as well as for the organization of international conferences, events, workshops and training seminars on seafood commodities and developing trends, offering experts the opportunity to share their views in person. In particular, FIN promotes trade in fish products by:

- providing up-to-date information on markets and prices;
- bringing buyers and sellers together in international conferences;
- training industry and government on quality requirements of the main markets.

FIN offers information in several languages on international fish trade including market studies, fish price reports and statistical information. The experiences of GLOBEFISH and FIN have been included in a publication highlighting FAO success stories, demonstrating how the network has played a vital role in collecting and sharing fish trade information worldwide.

Organizations in Cooperation:

- GLOBEFISH: GLOBEFISH is the unit within the FAO Fisheries and Aquaculture Department responsible for information and analysis of international fish trade and markets.

---

[128] http://www.fao.org/in-action/globefish/background/fishinfonetwork/en/

- INFOPESCA (South and Central America): INFOPESCA is the oldest organization in FIN and was established in 1977. It is based in Uruguay and concentrates on the fisheries in Latin America.
- INFOFISH (Asia and Pacific region): INFOFISH is the leading source of marketing support for fish producers and exporters in the Asia (Pacific region), which includes some of the largest fishing nations in the world. It was established in 1981 with the headquarters located in Malaysia.
- INFOPECHE (Africa): INFOPECHE is the African Fisheries Information service and is based in in Cote d'Ivoire. It was established in 1984 as a project implemented by FAO.
- INFOSA (Southern Africa): INFOSA is the Marketing Information and Technical Advisory Services for the Fisheries Industry in Southern Africa, and was created in 2001 as a sub-regional office of INFOPECHE. Offices are located in Namibia and its services cover the Southern African region.
- INFOSAMAK (Arab countries): INFOSAMAK was established in 1986, and is based in Morocco. INFOSAMAK provides the Arab speaking countries with information and assistance on trade in fishery and aquaculture products and encourages the promotion of investments in the sector.
- EUROFISH (Eastern and Central Europe): EASTFISH, a FAO project, was created in 1996. In 2002, EASTFISH then became EUROFISH, an international organization to serve the needs of the fisheries sector in Central and Eastern Europe.
- INFOYU (China): In 1997, GLOBEFISH assisted China in setting up the special information unit INFOYU, which then became a part of FIN. INFOYU is the Chinese member organization cooperating with the network based in Beijing.

International Hydrographic Organization (IHO)

The <u>International Hydrographic Organization (IHO)</u>[129] is an intergovernmental consultative and technical organization that was established in 1921 to support safety of navigation and the protection of the marine environment.

International cooperation in the field of hydrography began with the first International Maritime Conference held in Washington in 1889, followed by two others in Saint Petersburg, in 1908 and 1912. In 1919, twenty-four nations met in London for a Hydrographic Conference, during which it was decided that a permanent body should be created. The resulting International Hydrographic Bureau began its activity in 1921 with eighteen Member States, including the British Empire then composed of the United Kingdom and Australia. At the invitation of H.S.H. Prince Albert I. of Monaco, a noted marine scientist, the Bureau was provided with headquarters in the Principality of Monaco. The Organization has remained in Monaco ever since, thanks to the continuing and very generous support of the Prince's successors.

The official representative of each Member Government within the IHO is normally the national Hydrographer, or Director of Hydrography, who, together with their technical staff, meet at 5-yearly intervals in Monaco for an International Hydrographic Conference. The Conference reviews the progress achieved by the Organization through its committees, sub committees and working groups, and adopts the programs to be pursued during the ensuing 5-year period. A Directing Committee of three senior hydrographers is elected to administer the work of the Organization during that time.

The object of the Organization is:

- The coordination of the activities of national hydrographic offices
- The greatest possible uniformity in nautical charts and documents
- The adoption of reliable and efficient methods of carrying out and exploiting hydrographic surveys
- The development of the sciences in the field of hydrography and the techniques employed in descriptive oceanography

The Directing Committee, together with a small international staff of technical experts in hydrography and nautical cartography, makes up the International Hydrographic Bureau in Monaco. The IHB is the secretariat of the IHO, coordinating and promoting the IHO's programs and providing advice and assistance to Member States and others.

International cooperation in the field of hydrography began with a Conference held in Washington in 1899, followed by two others in Saint Petersburg, in 1908 and 1912. In 1919, twenty-four nations met in London for a Hydrographic Conference, during which it was

---

[129] https://www.iho.int/

decided that a permanent body should be created. The resulting International Hydrographic Bureau began its activity in 1921 with nineteen Member States. At the invitation of H.S.H. Prince Albert I of Monaco, a noted marine scientist, the Bureau was provided with headquarters in the Principality of Monaco. The Organization has remained in Monaco ever since, thanks to the continuing and very generous support of the Prince's successors.

In 1970, an intergovernmental Convention entered into force which changed the Organization's name and legal status, creating the International Hydrographic Organization (IHO), with its headquarters (the IHB) permanently established in Monaco. The Organization currently has a membership of eighty-five maritime States, with several others in the process of becoming Members.

In 2016, several amendments to the Convention entered into force. The principal changes to the administrative arrangements of the IHO were:

- The term International Hydrographic Bureau (IHB) used to describe the headquarters and the secretariat of the IHO ceased to be used and was replaced by the term IHO Secretariat;
- The Directing Committee, comprising a President and two Directors ceased to lead the IHB (Secretariat of the IHO). Instead, the Secretariat of the IHO is now led by a Secretary-General assisted by two subordinate Directors;
- The term International Hydrographic Conference used to designate the principal organ of the Organization, composed of all Member States, was replaced by the term Assembly. The ordinary sessions of the Assembly are held every three years instead of every five years for the Conference.
- The subsidiary organs will report to the Council that will then refer their proposals to the Assembly or to the Member States, for adoption, through correspondence;
- The planning cycle for the IHO work program and budget will change from a five-year to a three-year cycle.
- For States wishing to join the IHO that are already Member States of the United Nations there is no longer a requirement to seek the approval of existing Member States of the IHO;
- The strict eligibility requirements for candidates seeking election as the Secretary-General or a Director have been relaxed; and
- Where voting by correspondence is required, through the Council, decisions will be taken based on a majority of the Member States that cast a vote, rather than the previous arrangements where a majority of all the Member States entitled to vote was required. However a minimum number of at least one-third of all Member States eligible to vote must vote positively for the vote to stand.

A principal aim of the IHO is to ensure that all the world's seas, oceans and navigable waters are surveyed and charted. The Mission of the IHO is to create a global environment in which States provide adequate and timely hydrographic data, products and services and ensure their widest possible use.

The Vision of the IHO is to be the authoritative worldwide hydrographic body which actively engages all coastal and interested States to advance maritime safety and efficiency and which supports the protection and sustainable use of the marine environment.

The IHO sets the standards for the Electronic Chart Display and Information System (ECDIS) and the performance of the chart data in an ECDIS as well as traditional paper charts. The IHO standards have been revised a number of times since ECDIS was introduced to meet the revised ECDIS Performance Standards adopted by the International Maritime Organization (IMO) and also to improve how the chart data is displayed in ECDIS.

In January 2010, a document (IHO S-100) was published to define the 21st Century framework data structure for hydrographic and related data, based on the ISO 19,100 series of the various worldwide geographic standards. It provides the data framework for the development of the next generation of Electronic Navigational Chart (ENC) and other related digital products required by the hydrographic, maritime and Geographical Information Service (GIS) communities. This will be of direct benefit to the implementation of e-navigation.

The verified data provided by the IHO is intended to help mariners identify if their particular ECDIS is able to display all the latest IMO-approved features required on charts. It is also designed to highlight if particular display problems are present in the ECDIS being tested. It is a collaborative process. The checks are not exhaustive and mariners should report any anomalous operation of their ECDIS to their Flag State authority as requested in IMO MSC.1/ Circ. 1391.

The IHO Catalogue of publications (links to IHO Publications for the IHO Publications Catalogue) contains a list and brief description of all IHO publications. These publications have been arranged under classification criteria agreed by Member States in 2009 and reported in IHO Circular Letter 13 / 2009. This classification grouping is as follows:

- B - Bathymetric Publications (Mainly related to GEBCO)
- C - Capacity Building Publications
- M - Miscellaneous (Base Regulatory Publications)
- P - Periodic Publications
- S - Standards and Specifications

International Maritime Bureau-Piracy Reporting Center (IMB-PRC)

The International Maritime Bureau-Piracy Reporting Center (IMB-PRC)[130] was established as independent and non-governmental agency in Kuala Lumpur in 1992 and follows the definition of Piracy as laid down in Article 101 of the 1982 United Nations Convention on the Law of the Sea (UNCLOS) and Armed Robbery as laid down in Resolution A.1025 (26) adopted on 2 December 2009 at the 26th Assembly Session of the International Maritime Organization (IMO).

Before 1992, shipmasters and ship operators had nowhere to turn to when their ships were attacked, robbed or hijacked either in port or out at sea. Local law enforcement either turned a deaf ear, or chose to ignore that there was a serious problem in their waters.

IMB PRC offers a 365/24 hour free service for shipmasters to report any piracy, armed robbery or stowaway incidents. Aim of a report is to raise awareness within the shipping industry of high risk areas with pirate attacks and specific ports/anchorages where armed robberies on board ships have occurred.

PRC acts as a single point of contact for shipmasters anywhere in the world whose vessels have been attacked or robbed by pirates. All information received is immediately relayed to the local law enforcement agencies requesting assistance. Information is also immediately broadcast to all vessels in the Ocean region, providing vital intelligence and increasing awareness.

The information is shared with the IMO, governmental, inter-governmental and law enforcement agencies including industry bodies to understand the nature of piracy and reduce its effects on crew, vessel and cargo.

As a trusted point of contact, PRC is able to immediately identify any trends and shift in patterns and alert all concerned parties. The services are free and funded purely by donations. Donations are acknowledged on the website and in the quarterly and annual piracy and armed robbery reports.

---

[130] https://icc-ccs.org/piracy-reporting-centre

International Mobile Satellite Organization (IMSO)

The establishment of the International Mobile Satellite Organization (IMSO)[131] was based on two international public law instruments developed under the auspices of the International Maritime Organization (IMO). These are:

- Convention on the International Maritime Satellite Organization (INMARSAT) between States Parties to the Convention; and
- Operating Agreement between telecommunications entities public or private (one per Party) called "*Signatories*" designated by a State.

Both instruments entered into force on 16 July 1979. The purpose of INMARSAT was to make provision for the space segment necessary for improved maritime communications and, in particular, for improved safety of life at sea communications and the Global Maritime Distress and Safety System (GMDSS). This purpose was later extended through amendments to the Convention and Operating Agreement to provide the space segment for land mobile and aeronautical communications, and the name of the organization was changed to the International Mobile Satellite Organization to reflect the amended purpose.

INMARSAT was structured with three principal organs:

- the Assembly of Parties (one State, one vote), which dealt with general policy matters and the long term objectives of the Organization;
- the Council, composed of 22 Signatories, or groups of Signatories. It decided on all financial, operational, technical and administrative matters, and made provision for the space segment for carrying out the purposes of the Organization. Signatories' voting rights were linked to their utilization of the system via investment shares;
- the Directorate which was the executive body of the Organization headed by a Director General who was the Chief Executive Officer and legal representative of the Organization.

After twenty years of successful operation, Member States and Signatories of the intergovernmental organization INMARSAT decided to challenge rapidly growing competition from private providers of satellite communications services and pioneered the first ever privatization of all assets and business carried on by the intergovernmental organization while adhering to the continuous provision of the public service obligations and governmental oversight as a pre-requisite of the privatization.

At its Twelfth Session in April 1998, the INMARSAT Assembly adopted amendments to the INMARSAT Convention and Operating Agreement which were intended to transform the

---

[131] http://www.imso.org

Organization's business into a privatized corporate structure, while retaining intergovernmental oversight of certain public service obligations and, in particular, the Global Maritime Distress and Safety System (GMDSS).

The restructuring of INMARSAT involved the incorporation of holding and operating companies, located in England and registered under British law on 15. April 1999, as planned. On the same day, the Headquarters Agreement between the UK Government and the IMSO was signed.

The Assembly and Council of INMARSAT subsequently decided to implement the amendments as from 15. April 1999, pending their formal entry into force. In doing so, it was recognized that early implementation of the new structure was needed to maintain the commercial viability of the system in a rapidly changing satellite communications environment, and thereby ensure continuity of GMDSS services and other public service obligations, namely: peaceful uses of the system, non-discrimination, service to all geographical regions and fair competition.

The restructuring amendments entered into force on 31. July 2001 and became binding upon all Parties, including those which had not accepted them, and the Operating Agreement terminated on the same date.

A Public Services Agreement between IMSO and the privatized INMARSAT was also executed with immediate effect. The Operating Agreement was terminated and the Signatories received ordinary shares in the privatized INMARSAT in exchange for their investment shares. Capital requirements are met from existing shareholders, strategic investors and public investment through a listing of the shares on a stock exchange (IPO).

The INMARSAT satellites and all other assets of the former intergovernmental organization have been transferred to the privatized operating Company which continues to manage the global mobile satellite communications system, including maritime distress and safety services for GMDSS at either no cost or at a special rate.

IMSO as of today is the residual intergovernmental organization IMSO continues with 100 Parties, operating through:

- the Assembly of Parties, which meets every two years;
- the Advisory Committee (comprising around one third of the Member States appointed by the Assembly every two years) which meets on a regular basis, at least twice a year; and
- a small Directorate, headed by the Director General who is the Chief Executive Officer and legal representative of the Organization.

Under the relevant provisions of the Convention, the Public Services Agreement and the Articles of Association of INMARSAT, IMSO is charged with overseeing, and under some circumstances may enforce fulfilment of INMARSAT's public service obligations and, in particular, GMDSS services.

In performing this role, IMSO acts as the natural ally of IMO and watchdog of proper provisions and implementation of IMO's requirements in respect of GMDSS by INMARSAT. To facilitate these functions, an Agreement of Cooperation has been concluded between IMSO and IMO. Under a similar Agreement with the International Civil Aviation Organization (ICAO), IMSO ensures that INMARSAT takes into account the applicable ICAO Standards and Recommended Practices in line with the Public Services Agreement and regularly informs ICAO accordingly.

The IMSO Convention has been amended with effect from 6 October 2008 to extend IMSO's oversight to any provider which may be recognized by IMO to provide mobile satellite services for use in the GMDSS. The revised Convention also defines IMSO's role as LRIT Coordinator, appointed by IMO to audit and review the performance of the LRIT System to ensure its worldwide operation system.

Administrative Arrangements have also been signed between the Secretary-General of the International Telecommunication Union (ITU) and the IMSO Director General. These provide the Organization with direct access to the relevant bodies of the ITU, enabling IMSO to play an active role in the development of international telecommunication policies.

The horizons of mobile satellite communications have been expanding with ever-increasing speed, and there are several different options for the design and capability of new services. The adoption by the IMO Assembly of Resolution A.1001(25) - Criteria for the Provision of Mobile Satellite Communication Systems in the Global Maritime Distress and Safety System (GMDSS), provided a clear indication of IMO's intention to consider opening up provision of GMDSS services to any satellite operator whose system fits these Criteria.

Expansion of the market is most likely to happen in the context of a revision of Chapter IV (Radio communications) of the Safety of Live at Sea (SOLAS) Convention and will provide the opportunity for specifying more effective services in a way that permits the use of evolutionary capabilities and non-geostationary satellite constellations.

At present, INMARSAT, with the satellite communications system which it operates, is the sole global provider of these services, although, after the restructuring of INMARSAT, the process of liberalization and privatization of global and regional satellite communications services is fast developing. IMSO is currently conducting the assessment of the application in relation to the provision of GMDSS by Iridium.

International Police (INTERPOL)

International Police (INTERPOL)[132] is the world's only and largest international police organization, with 190 member countries. Its role is to enable police around the world to work together to make the world a safer place. The mission – what INTERPOL does to achieve its vision *"Preventing and fighting crime through enhanced cooperation and innovation on police and security matters"* is, to:

- facilitate the widest possible mutual assistance between all criminal law enforcement authorities.
- ensure that police services can communicate securely with each other around the world. - enable global access to police data and information.
- provide operational support on specific priority crime areas.
- foster continuous improvement in the capacity of police to prevent and fight crime and the development of knowledge and skills necessary for effective international policing. - strive for innovation at all times, in the areas of police and security matters.

The idea of INTERPOL was born in 1914 at the first International Criminal Police Congress, held in Monaco. Officially created in 1923 as the International Criminal Police Commission, the Organization became known as INTERPOL in 1956.

The first International Criminal Police Congress was held 1914 in Monaco. Police officers, lawyers and magistrates from 24 countries meet to discuss arrest procedures, identification techniques, centralized international criminal records and extradition proceedings.

1923 followed the creation of the International Criminal Police Commission (ICPC) with headquarters in Vienna, Austria, on the initiative of Dr. Johannes Schober, president of the Vienna Police. In 1926 the General Assembly, held in Berlin, proposes that each country establish a central point of contact within its police structure: the forerunner of the National Central Bureau (NCB).

In 1932, following the death of Dr. Schober, new statutes put in place creating the post of Secretary General. The first was Austrian Police Commissioner Oskar Dressler. The Organization's international radio network was launched in 1935, providing an independent telecommunications system solely for the use of the criminal police authorities at national level.

The ICPC falls in 1942 completely under German control and is relocated to Berlin. After the war Belgium leads in 1946 the rebuilding of the organization after the end of World War II. A new headquarters set up in Paris, and 'INTERPOL' chosen as the organization's tele-

---

[132] http://www.interpol.int

graphic address. Democratic process to elect the President and Executive Committee instituted. In 1949 the United Nations grants INTERPOL consultative status as a non-governmental organization.

Following the adoption of a modernized constitution, the ICPC becomes in 1956 the International Criminal Police Organization-INTERPOL, abbreviated to ICPO–INTERPOL or just INTERPOL. The Organization becomes autonomous by collecting dues from member countries and relying on investments as the main source of funding. In 1989 INTERPOL moves its General Secretariat to Lyon, France.

INTERPOL vision today is that of a world where each and every law enforcement professional will be able through INTERPOL to securely communicate, share and access vital police information whenever and wherever needed, ensuring the safety of the world's citizens. We constantly provide and promote innovative and cutting-edge solutions to global challenges in policing and security.

INTERPOL facilitates the widest possible mutual assistance between all criminal law enforcement authorities. It ensures that police services can communicate securely with each other around the world. It enables global access to police data and information and provides operational support on specific priority crime areas.

INTERPOL fosters continuous improvement in the capacity of police to prevent and fight crime and the development of knowledge and skills necessary for effective international policing. It strives for innovation at all times, in the areas of police and security matters.

An automated search facility for remote searches of INTERPOL databases introduced in 1992, in 1998 INTERPOL Criminal Information System (ICIS) database was created. The I-24/7 web-based communication system was launched in 2002, significantly improving NCBs' access to INTERPOL's databases and services.

For the exchange of information within the organization as well as the member states INTERPOL provides the Information Exchange System for Interpol 24/7 (INTERPOL-I24/7) which has been extended over the past years in order to fight international terrorism and crime.

Canada is the first country to connect to the system. Official inauguration of the Command and Coordination Centre at the General Secretariat, enabling the organization to operate 24 hours a day, seven days a week came in 2003, supported by a technology known as MIND/FIND, which allows frontline officers to connect directly to INTERPOL's systems.

The Strategic Framework 2017-2020 serves as a roadmap for INTERPOL's global activities. Containing the Organization's vision, mission, values, goals and objectives, it is the structure through which all the activities are defined, executed and evaluated.

Set for a period of four years, the Strategic Framework is approved by INTERPOL's supreme governing body, the General Assembly with the strategic goals of:

1. Serve as the worldwide information hub for law enforcement cooperation. The exchange of police information lies at the core of INTERPOL's mandate. INTERPOL manages the secure communication channels that connect National Central Bureaus in all member countries, along with other authorized law enforcement agencies and partners, and which give access to a range of criminal databases.
2. Deliver state-of-the-art policing capabilities that support member countries to fight and prevent transnational crimes. A large number of policing capabilities – such as forensics and training – underpin the three Global Programs to fight crime (Counter-terrorism, Cybercrime, and Organized and emerging crime). INTERPOL works directly with National Central Bureaus and specialized agencies, but also through multidisciplinary groups, to become a recognized hub for providing expertise, facilitating the exchange of best practices, and hosting training facilities.
3. Lead globally innovative approaches to policing. INTERPOL is committed to enhancing the tools and services provided, and to act as an incubator for the research and development into solutions and standards for international policing. INTERPOL strives to reinforce the role as the global police think tank and a forum for exchange at expert level, with an emphasis on future trends and strategic foresight.
4. Maximize INTERPOL's role within the Global Security Architecture to bridge information gaps in the Global Security Architecture, strengthening cooperation between relevant sectors and entities, and raising political awareness and support for INTERPOL's Program.
5. Consolidate resources and governance structures for enhanced operational performance in order to keep up with the evolving law enforcement landscape. INTERPOL will focus its efforts on implementing a sustainable funding model for the Organization, addressing risks at the organizational level, fostering change management, promoting a sustainable human resources strategy, and re-assessing the governance framework.

Information Exchange System for Interpol 24/7 (INTERPOL-I24/7)

Information Exchange System for Interpol (INTERPOL-I24/7)[133] has core functions is to enable the world's police to exchange information securely and rapidly. The organization's I-24/7 global police communications system connects law enforcement officials in all 188 member countries and provides them with the means to share crucial information on criminals and criminal activities.

As criminals and criminal organizations are typically involved in multiple activities, I-24/7 can fundamentally change the way law enforcement authorities around the world work together. Pieces of seemingly unrelated information can help create a picture and solve a trans-national criminal investigation. Using I-24/7, National Central Bureaus (NCBs) can search and cross-check data in a matter of seconds, with direct access to databases containing information on suspected terrorists, wanted persons, fingerprints, DNA profiles, lost or stolen travel documents, stolen motor vehicles, stolen works of art, etc.

These multiple resources provide police with instant access to potentially important information, thereby facilitating criminal investigations. The I-24/7 system also enables member countries to access each other's' national databases using a business-to-business (B2B) connection. Member countries manage and maintain their own national criminal data. They also have the option to make it accessible to the international law enforcement community through I-24/7.

Although I-24/7 is initially installed in NCBs, INTERPOL is encouraging member countries to extend their connections to national law enforcement entities such as border police, customs and immigration, etc. NCBs control the level of access other authorized users have to INTERPOL services and can request to be informed of enquiries made to their national databases by other countries.

West African Police Information System (WAPIS)

The West African Police Information System (WAPIS) is the INTERPOL Regional bureau in Côte d'Ivoire, Abidjan.

---

[133] http://www.interpol.int/Public/NCB/I247/default.asp

International Telecommunication Union (ITU)

International Telecommunication Union (ITU)[134] is the United Nations specialized agency for the Information and Communications Technology (ICT). ITU allocates the global radio spectrum and satellite orbits, develops the technical standards that ensure networks and technologies seamlessly interconnect, and strive to improve access to ICTs to underserved communities worldwide. ITU protects and supports everyone's fundamental right to communicate.

ITU allocates global radio spectrum and satellite orbits, develop the technical standards that ensure networks and technologies seamlessly interconnect, and strive to improve access to ICTs to underserved communities worldwide.

An organization based on public-private partnership since its inception, ITU currently has a membership of 193 countries and almost 800 private-sector entities and academic institutions. ITU is headquartered in Geneva, Switzerland, and has twelve regional and area offices around the world. ITU membership represents a cross-section of the global ICT sector, from the world's largest manufacturers and telecoms carriers to small, innovative players working with new and emerging technologies, along with leading R&D institutions and academia.

ITU was founded in Paris in 1865 as the International Telegraph Union. It took its present name in 1934, and in 1947 became a specialized agency of the United Nations. Founded on the principle of international cooperation between governments (Member States) and the private sector (Sector Members, Associates and Academia), ITU is the premier global forum through which parties work towards consensus on a wide range of issues affecting the future direction of the ICT industry.

As the UN specialized agency for ICTs, ITU is the official source for global ICT statistics. Find out more about how we produce and disseminate data, our main events and products. The ICT Data and Statistics (IDS) Division is part of ITU's Projects and Knowledge Management Department within the Telecommunication Development Bureau (BDT). One of the core activities of the Division is the collection, verification and harmonization of telecommunication/ICT statistics for about 200 economies worldwide.

ITU set for example a list of standards and specifications for the Automated Identification Systems (AIS Standards) as of: IALA Technical Clarification of ITU-R M.1371-1 (Ed.1.3); IEC 60945 ; IEC 61108-1; IEC 61162-1; IEC 61162-2; IEC 61162-3; IEC 61162-400; IEC 61993-2; IEC 62287; IEC 62320; IMO Resolution MSC.74(69), Annex 3; ITU-R M.1084-3; ITU-R M.1371; ITU-R M.493; ITU-R M.823-2; ITU-R M.825-3.

---

[134] https://www.itu.int/en

Maritime Mobile Access and Retrieval System (MARS)

The main objective of the ITU Maritime Mobile Access and Retrieval System (MARS)[135] webpage is to provide users with the means to access and retrieve operational information registered in the ITU maritime database. This information has been notified, to the Radio Communication Bureau, by Administrations of the Member States of the ITU. The ITU maritime database currently contains information concerning:

- Ship stations (including those that participate in the Global Maritime Distress and Safety System (GMDSS));
- Coast stations
- Addresses of Accounting Authorities;
- Addresses of Administrations which notify information;
- MMSI assigned to Search and Rescue (SAR) aircraft; and
- MMSI assigned to AIS Aids to Navigation (AtoN).

MARS also links to other maritime related sites within the Radio Communication Bureau and makes available the various notification forms used for the submission of data. In addition, this system provides administrations with the possibility to download a file containing all their ship stations that are currently registered with ITU.

---

[135] https://www.itu.int/en/ITU-R/terrestrial/mars/Pages/default.aspx

Italy (ITA)

Italy is by geographical location and economical power one of the larger Maritime Entities, however the position in the Mediterranean favored an approach of the challenges within international frameworks. Here, Italy and the Italian Navy have achieved remarkable results. Italy has a strategic position in Southern Europe, as a peninsula extending into the central Mediterranean Sea, northeast of Tunisia, and the territory includes the islands Sardinia and Sicily. The strategic location is dominating central Mediterranean as well as southern sea and air approaches to Western Europe.

From the earliest settlements of the over the Roman Empires and the maritime republics (Italian: repubbliche marinare) of the Mediterranean Basin, the thalassocratic city-states during the Middle Ages, among best known are the maritime republics of Venice, Genoa, Pisa, and Amalfi, the lesser known being Ragusa, Gaeta, Ancona, and Noli, to todays Italian Republic the peninsula and it's islands had a strategic position in the Mediterranean, are one important entrance for the trade routes from and to Africa and the Middle East. From the 10th to the 13th centuries fleets of ships for own protection and to support extensive trade networks across the Mediterranean were build, providing an essential role in the Crusades, the sea battles in the 1. and 2. WW fill the history books.

Today Italy faces mass immigration, economic difficulties while still providing an essential power to Europe. Italy also faces air pollution from industrial emissions such as sulfur dioxide; coastal and inland rivers polluted from industrial and agricultural effluents, acid rain damaging lakes, inadequate industrial waste treatment and disposal facilities.

Italy is in international environmental agreements party to Air Pollution, Air Pollution-Nitrogen Oxides, Air Pollution-Persistent Organic Pollutants, Air Pollution-Sulfur 85, Air Pollution-Sulfur 94, Air Pollution-Volatile Organic Compounds, Antarctic-Environmental Protocol, Antarctic-Marine Living Resources, Antarctic Seals, Antarctic Treaty, Biodiversity, Climate Change, Climate Change-Kyoto Protocol, Desertification, Endangered Species, Environmental Modification, Hazardous Wastes, Law of the Sea, Marine Dumping, Ozone Layer Protection, Ship Pollution, Tropical Timber 83, Tropical Timber 94, Wetlands, Whaling.

Although the international cooperation within the Virtual-Regional Maritime Traffic Center (V-RMTC) communities and the later Trans-Regional Maritime Network (T-RMN) are definitely aimed as multinational approach, the Virtual Maritime Traffic Center(V-RMTC) is situated in the Italian Navy Headquarter and has, in addition to the international cooperation benefits, an immense national value due to the geographical situation of Italy.

Trans-Regional Maritime Network (T-RMN)

The Trans-Regional Maritime Network (T-RMN) is the worldwide extension of the Italian Virtual Maritime Traffic Center (V-RMTC) as a global network with 28 participating nations, originated and hosted by the Italian Navy.

V-RMTC and T-RMN base on similar idea and approach like the Indian Ocean Naval Symposium (IONS) by India and is situated under the International Sea Power Symposium (ISS) meetings and has established a network for exchange of maritime information for the participation members.

Signature nations of the Operational Arrangement (OA) concerning the establishment of T-RMN for the global maritime information sharing were Albania, Belgium, Bulgaria, Chile, Croatia, Cyprus, France, Georgia, Germany, Greece, Jordan, Israel, Italy, Malta, Montenegro, The Netherlands, Peru, Portugal, Romania, Senegal, Slovenia, Spain, Turkey, United Kingdom of Great Britain and Northern Ireland, the United States of America, Brazil, India, and Singapore.

On the 18th International Sea Power Symposium (ISS) in Newport/USA the V-RMTC Wider Mediterranean Community (WMC) was in this context extended with the participation of Brazil, Singapore and India. The importance lays more on the political rather than operation level, as you find European, African, North- and South American as well as Asian countries and they different religious backgrounds combined in one Organization. Peru joined T-RMN on the ISS 2014, Chile on 2017, India in 2018, Pakistan has applied to join.

Virtual-Regional Maritime Traffic Center (V-RMTC)

The Virtual-Regional Maritime Traffic Center (V-RMTC)[136] started as a project of the Italian Navy in order to provide White Shipping information, mainly AIS-Information, to national units and the Headquarter in Rome. Later the Italian Navy added Information from Radar for the national network and linked other national institutions like the Coast Guard and Guardia di Finanzia (GdF) with its Radar and AIS sources. Furthermore the Department of Information Engineering and Mathematics (DIISM) added its expertise on the evaluation of the collected information.

The Italian Navy then established the Mediterranean Network for the Maritime Security in the Mediterranean and Black Sea under participation of Bulgaria, Croatia, Cyprus, France, Greece, Italy, Malta, Portugal, Romania, Slovenia, Spain and Turkey. In 2005 V-RMTC pro-

---

[136] http://www.marina.difesa.it

vided connections between the Naval Operational Centers (NOC's) for the data exchange at unclassified level regarding the merchant vessel traffic.

The Virtual-Regional Maritime Traffic Center (V-RMTC) established through the signatures of additional countries the group of the Virtual Regional Traffic Centre Wider Mediterranean Community (V-RMTC WMC), connecting the operational headquarters of the participating navies. This regional approach serves the Mediterranean areas interest, however the MSA required an wider approach due to the international trade and traffic. Therefore the Italian Navy Mediterranean established the Trans-Regional Maritime Network (T-RMN) aiming at an global information exchange with possibly different requirements and political background.

Virtual Regional Traffic Centre Wider Mediterranean Community (V-RMTC WMC)

The Virtual-Regional Maritime Traffic Center (V-RMTC) established through the signature of additional countries the group of the Virtual Regional Traffic Centre Wider Mediterranean Community (V-RMTC WMC), enlarging the area of interest and connecting the operational headquarters of the participating navies to cover the requirements and political background of the more European and African partners.

The Virtual-Regional Maritime Traffic Center (V-RMTC) started in September 2006 with the data exchange between the participating countries of Portugal, Spain, France, Slovenia, Croatia, Montenegro, Albania, Greece, Turkey, Cyprus, Malta, Jordan, Israel, Romania, Britain and the USA. A demonstration took place in 2008 and the system was considered operational in November 2008 and was later enlarged to the V-RMTC WMC.

However, today most of the partners in the Virtual Maritime Traffic Center 5+5 (V-RMTC 5+5), Virtual Maritime Traffic Center 8+6 (V-RMTC 8+6), Virtual Maritime Traffic Center Italy-Lebanon (V-RMTC Italy-Lebanon) and the V-RMTC Wider Mediterranean Community (V-RMTC WMC) just refer to it as the V-RMTC. Having left the Mediterranean geographical area already with the V- RMTC WMC, Italy changed the global initiative of his network extension into the Trans-Reginal Maritime Network (T-RMN). Since the data from the various communities is all hosted on the Italian system and requires separate data, the Italian Navy has proposed to combine the two networks of the Virtual-Regional Maritime Traffic Center (V-RMTC) and Trans-Reginal Maritime Network (T-RMN), possibly also the V-RMTC 5+5.

The annual estimated operational costs were in 2009 an estimated EUR 300,000 without further development and improvements, which are all solely covered by Italy. The Parent Fusion Centers of the "Wider Community" are identical with the V-RMTC and all other communities created by an Operational Arrangement (OA). Information sharing under the OA was established between Albania, Belgium, Bulgaria, Croatia, Cyprus, France, Georgia, Germany, Greece, Jordan, Israel, Italy, Malta, Montenegro, The Netherlands, Portugal, Romania, Senegal, Slovenia, Spain, Turkey, United Kingdom of Great Britain and Northern Ireland, and the United States of America. The 23 nations connected later to Brazil, India and

Singapore via the Trans-Regional Maritime Network (T-RMN) initiative. Pakistan has applied to join T-RMN/V-RMTC.

The Virtual-Regional Maritime Traffic Center (V-RMTC) is primarily a data collection and database center of the Italian Navy with Internet Web- or client-based accessible database for shipping tracks, here mainly the ships name, IMO number, call sign, positions, flag, arrival, departure and possible other voyage data, send by the partner nations. The system relies to an big extend on Automatic Identification System (AIS), and the military OS-OTH GOLD Format.

V-RMTC actually host several communities within the Italian cooperation frameworks, with the Data Fusion Center as the central hub located in the Italian Navy Fleet Command (CINCNAV) in Santa Rosa, Rome. All V-RMTC initiatives and the T-RMN are based on Service-oriented infrastructure for MARitime Traffic tracking (SMART) which collects and fuses the information originating from various sources, organizations and nations.

Based on SMART and the new developed SMART FENIX Italy is developing a system for Inter-agency Integrated Maritime Security (DIIMS) with additional sensor data from satellite, radar, Vessel Traffic Systems and other sources, making data and information exchange available to the Italian authorities (Military Police (Carabinieri), Coast Guard, State Police, Custom Police and Customs).

Virtual Maritime Traffic Center 5+5 (V-RMTC 5+5)

The Italian Navy tried to extend the V-RMTC towards the North African coast. Due to political reasons the new associated partners would not agree to the data exchange with the whole Virtual Regional Traffic Centre Wider Mediterranean Community (V-RMTC WMC). This resulted in two parallel-developed networks, the separated Virtual Maritime Traffic Center 5+5 (V-RMTC 5+5) and the enlargement into the V-RMTC Wider Mediterranean Community (V-RMTC WMC), followed by the bilateral Virtual Maritime Traffic Center Italy-Lebanon (V-RMTC Italy-Lebanon) in UNIFIL. In 2007 the V-RMTC 5+5 enabled the data exchange between (A1) Algeria, (B1) France, (B2) Italy, (A2) Libya, (B3) Malta, (A3) Mauritania, (A4) Morocco, (A5) Tunisia, (B4) Portugal and (B5) Spain.

Virtual Maritime Traffic Center 8+6 (V-RMTC 8+6)

The Italian Navy tried to extend the V-RMTC network towards the Gulf Region. Due to political reasons the new associated partners would not agree to the data exchange with the whole Virtual Regional Traffic Centre Wider Mediterranean Community (V-RMTC WMC). This resulted in yet another initiative for the Virtual Maritime Traffic Center 8+6 (V-RMTC 8+6) with Bahrain, France, Germany, Greece, Italy, Kuwait, Netherlands, Oman, Portugal, Qatar, Saudi Arabia, Spain, United Arab Emirates and United Kingdom. The V-RMTC

8+6 political talks and discussions never resulted in any active information exchange.

Virtual Maritime Traffic Center Lebanon (V-RMTC Lebanon)

The Italian Navy established the bilateral V-RMTC Lebanon network with the mission under UNIFIL. Due to political reasons the Italian Navy never intended to be enlarged into a wider cooperation. The information exchange is continued until today.

Service-oriented infrastructure for MARitime Traffic tracking (SMART)

Service-oriented infrastructure for MARitime Traffic tracking (SMART) is taking over where the V-RMTC (Virtual-Regional Maritime Traffic Centre) has technically left off providing the next spiral evolutionary step for the Italian Navy (ITN) Maritime Situational Awareness (MSA). SMART is built upon the standards, strategies, & capabilities of V-RMTC to build & deploy Maritime Domain Awareness (MDA) capabilities. The Service-oriented infrastructure for MARitime Traffic tracking (SMART) is not to be mistaking with the NATO SMART under Tidepedia.

SMART provides a centralized Common Operational Picture (COP) which is updated as inputs from various sources are received and correlated/fused. SMART includes a toolsets for collaboration and coordination for interoperability and Maritime Domain Awareness (MDA) to support MSA systems (i.e. Documents, Forum, Chat, etc.), view and export the in COP. The system takes external data sources (either SOA and traditional) and fuses them under a mediated meta-data layer. This allows for data mining, manual and automated analysis, and various visualization capabilities in a single application. SMART is a near real-time system, the COP is updated as inputs from various sources are received.

The SMART system connects Naval Operational Centers (NOCs) through virtual networks in order to allow the exchange of MSA information through a Data Fusion Hub (DFH). SMART allowed multiple partners to share information on Maritime Situational Awareness (MSA) technology systems within an unclassified environment in near real-time. SMART used external data sources and fused them under a mediated metadata layer that allowed for data mining, manual and automated analysis, and various visualization capabilities within a single application.

SMART provided geospatial situational awareness capacities and included toolsets for collaboration and coordination for interoperability and MSA features. SMART has secure encrypted communications channels and encrypted systems passwords. SMART used secure Hypertext Transfer Protocol Secure (https) for Internet access and data transfers.

SMART increment 3 is built upon the standards, strategies and capabilities of V-RMTC and SMART increment 1 and 2 systems, to build and deploy a modern and effective ITN Maritime Domain Awareness (MDA) capability. SMART will collaborate and interoperate with multiple Navy partners, working within both classified and unclassified environments, for obtaining, sharing and processing data related to Maritime Situational Awareness (MSA) in a single portal.

The MERchant vessels SITuation (MERSIT) Client-Application can manage and store data of each NOC, on a local PC and deliver national reports to CINCNAV (Data Fusion Hub (DFH), while the Converter (AIS-2-MERSIT) transforms AIS-data-feeds (NMEA 0183) into the MERSIT-Format for streaming it to the Data Fusion Hub in ITA. Other protocols used or formatted reports (i.e. XML, XCTC, CSV, OS-OTH GOLD, Locator, SISTRAM, Oasis, etc., pp.) can be imported depending on national systems output by dedicated PLUG-IN after being developed.

SMART is used in V-RMTC/T-RMN together with AIS (Automatic Identification System), the NATO MSSIS (Maritime Safety and Security Information System), can make use of formatted messages as i.e. the NATO/USA OS OTH-GOLD, Locator, while the MERSIT-Client (MERchant vessels SITuation) is used for automatic or manual transformation of the message information on the arrival and departure for feed into the Fusion Hub in the COMMCEN in Santa Rosa, Italy. The Merchant Vessel Situation (MERSIT) data format is used for information exchange in V-RMTC:

- MERSIT ARR: to notify a single arrival, at a country port, of one merchant vessel;
- MERSIT DEP: to notify a single departure, from a country port, of one merchant vessel;
- MERSIT NAV: to notify the contact of one merchant vessel located at sea.

Each report is built by one or plus additional MERSIT, and sent via a single Summary to the assigned mailbox of the Data Fusion Hub (DFH). Other protocols used by countries involved in the communities (XML, XCTC, CSV, SISTRAM). Based on the Operational Arrangement (OA) each member of the community is required to send the track data to the Data Fusion Hub (The Italian NOC), which collects the data for use via a web-browser for the community. The authorities in ITA get the information combined with national sensors information.

Based on the Operational Arrangements (OA) in V-RMTC/T-RMN each member of the community is required to send the track data to the Data Fusion Hub (The Italian NOC), which collects the data for use via a web-browser for the community. The authorities in ITA use the information from the various partner sources and combine it with national sensors information.

The Service-oriented infrastructure for MARitime Traffic tracking (SMART) used in the COMMCEN in Santa Rosa, connected to various national authorities like, Guardia Costiera, Guardia Civil, Guardia Finanza, a.o., provides for interoperability on ITA inter-agency level, but is also used in EUNAVFOR MED, Operation SOPHIA.

Units from the multinational Task Force provide a Public IP Address in order to connect via the Client-Software provided by ITA to the centralized SMART-RMP and collaboration tools at the OHQ.

- The MERSIT Client (MERchant vessels SITuation) provides for a manual submission for the arrival and departure, position, and vessel name, in formatted messages of USA/NATO (OS OTH GOLD, Locator).
- The AIS-Converter-Tool can be installed on ship systems for an automated feed into SMART from the unit-level.
- The SMART-Web access provides basic tools for viewing the OHQ RMP, MAP, FORUM, CHAT AND ENCRYPTED/SIGNED EMAIL down to the unit level.

SMART FENIX

The SMART FENIX upgrade in 2019 brought new developments in the software and had higher technical requirements on the hardware in which the Italian Navy shows its determination to continue its efforts in MSA.

MARSUR via ITA MEXS-Node and SMART in support to EU NAVFOR MED

The continued testing and improvement of connectivity in Maritime Surveillance Networking (MARSUR Networking) is in support for the track streaming service requested by ITA for the ITA MEXS-Node in order to forward UNCLASS information via the Italian SMART system to the Mission Classified Network of the OHQ EU NAVFOR MED Sophia and the Operation EU NAVFOR MED SOPHIA.

Information flow from MARSUR Member States national MEXS-Nodes is delivered to ITA MEXS-Node, transferred to the ITA SMART in COMCEN, Santa Rosa, and received on a Client in OHQ SOPHIA for forwarding the information to FHQ Afloat and Units in EU NAVFOR MED:

a) The OHQ for EU NAVFOR MED Sophia uses the Italian SMART System as CIS architecture in a Mission Classified Network. The mission environment and organizational command structure established information exchange between national Maritime Operational Centers (MOC) and ITA MEXS-Node.

b) The ITA-MEXS-Node is used to provide the information from the federated architecture of the Maritime Operation Centers of the MARSUR Member States to the hierarchical distribution within the Italian SMART System in the national COMMCEN in Santa Rosa.

c) The ITA MEXS-Node and MARSUR Networking are not part of EU NAVFOR MED/CSDP Mission Network architecture and divided by a security diode. Information from MARSUR Networking is one source of unclassified information stove-piped through the diode to the OHQ EU NAVFOR MED SOPHIA and the Mission Classified Network.

For the JOC EU NAVFOR MED SOPHIA the ITA Navy has setup an additional e-mail-account and a common chat room with a remote Client (UNCLASS) in the OHQ SOPHIA. MARSUR Member States can receive and send UNCLASS messages to the JOC using the secure MARSUR Networking VPN and the MARSUR Networking Core Services. The JOC EU NAVFOR MED SOPHIA received first connectivity to MARSUR Networking in via E-mail and Chat and the services are being continuously improved.

Japan (JPN)

Japan is by geographical position and economical power one of the most influential entities in its region. Japan lays as island chain between the North Pacific Ocean and the Sea of Japan, east of the Korean Peninsula, and includes Bonin Islands (Ogasawara-gunto), Daito-shoto, Minami-jima, Okino-tori-shima, Ryukyu Islands (Nansei-shoto), and Volcano Islands (Kazan-retto).

Japan has a strategic location in northeast Asia; composed of four main islands - from north Hokkaido, Honshu, Shikoku, and Kyushu (the "Home Islands") - and 6,848 smaller islands and islets. With virtually no natural energy resources, Japan is the world's largest importer of coal and liquefied natural gas, as well as the second largest importer of oil.

Japan is in international environmental agreements party to Antarctic-Environmental Protocol, Antarctic-Marine Living Resources, Antarctic Seals, Antarctic Treaty, Biodiversity, Climate Change, Climate Change-Kyoto Protocol, Desertification, Endangered Species, Environmental Modification, Hazardous Wastes, Law of the Sea, Marine Dumping, Ozone Layer Protection, Ship Pollution, Tropical Timber 83, Tropical Timber 94, Wetlands, Whaling.

Since the late 1990s, the Japan Coast Guard (JCG) has countered a myriad of 'outlaw' threats at sea including piracy, terrorism, the proliferation of Weapons of Mass Destruction (WMD) and the threat posed by 'rogue states'. Japan's innovative strategy has transformed maritime security governance in Southeast Asia and beyond.

To counter the new maritime threats Japan approved a new ocean policy that highlights maritime security, amid perceived growing threats from North Korea and China, in a reversal from the previous version which focused largely on sea resource development. The program cited threats from North Korea's launching of ballistic missiles, and operations by Chinese vessels around the Japan-controlled and China-claimed Senkaku Islands in the East China Sea.

The contents of the third Basic Plan on Ocean Policy are expected to be reflected in the government's defense buildup guidelines that are set to be revised in December. Since its first adoption in 2008, the ocean policy has been reviewed every five years. The policy pointed out that the maritime security situation facing the nation is *"highly likely to deteriorate, if no measure is taken."*

The government also plans to make use of coastal radar equipment, aircraft and vessels from the Self-Defense Forces and the Japan Coast Guard, as well as high-tech optical satellites of the Japan Aerospace Exploration Agency, to strengthen the nation's intelligence gathering abilities.

The policy underscores the need for cooperation between the coast guard and the Fisheries Agency to enhance responses to illegal operations by North Korea and fishing vessels from other countries, amid a surge in the number of such cases in the waters surrounding

Japan.

To ensure sea lane safety, it also stipulates the government's promotion of the "free and open Indo-Pacific" strategy for maintaining and strengthening a free and open order in the region based on the rule of law.

Japan has made a radical shift in its oceans policy to concentrate on national security instead of economic development as maritime tensions with China rise. Cyber-attacks are increasingly being seen as a threat to maritime situational awareness in the Asia Pacific region, and highlights how the Japanese and Taiwanese governments have been developing both offensive and defensive cyber operations to ameliorate vulnerabilities in their naval fleets vis-a-vis their respective relationships with China. China, the US, Japan and Taiwan are developing new capabilities in this area and new military doctrine to avoid disruptions to maritime operations.

Japan committed more than $500 million in 2013 to expand a satellite system to enhance GPS signals over its territory, is eyeing a new constellation of spacecraft for regional maritime surveillance, according to a government official.

Disaster monitoring has emerged as a Japanese priority following passage of a 2008 law that overhauled Japan's space policy. In addition to permitting Japan to deploy military space systems, the law, which was updated in January, redirects Japanese space development away from technology-focused projects to those with practical applications and economic benefits.

Outlined priority investment area for Japan is a $3 billion-per-year space program for positioning, navigation and timing, remote sensing, including military reconnaissance and weather; communications; and launch vehicles. Military space also is a priority given recent developments involving North Korea.

Quasi-Zenith Satellite System (QZSS)

The Quasi-Zenith Satellite System (QZSS)[137] (準天頂衛星システム, Juntencho eisei shisutemu) is a project of the Japanese government for the development of a four-satellite regional time transfer system and a satellite-based augmentation system for the United States operated Global Positioning System (GPS) to be receivable in the Asia-Oceania regions, with a focus on Japan. The goal of QZSS is to provide highly precise and stable positioning services in the Asia-Oceania region, compatible with GPS. Four-satellite QZSS services (QZS-4) are available on a trial basis as of 12. January 2018, and are scheduled to commence as a production service on 1. November 2018.

In 2002, the Japanese government authorized the development of QZSS, as a three-satellite regional time transfer system and a satellite-based augmentation system for the United States operated Global Positioning System (GPS) to be receivable within Japan. A contract was awarded to Advanced Space Business Corporation (ASBC), that began concept development work, and Mitsubishi Electric, Hitachi, and GNSS Technologies Inc. However, ASBC collapsed in 2007, and the work was taken over by the Satellite Positioning Research and Application Center (SPAC), which is owned by four Japanese government departments: the Ministry of Education, Culture, Sports, Science and Technology, the Ministry of Internal Affairs and Communications, the Ministry of Economy, Trade and Industry, and the Ministry of Land, Infrastructure, Transport and Tourism.

The first satellite "*Michibiki*" was launched on 11 September 2010. Full operational status was expected by 2013. In March 2013, Japan's Cabinet Office announced the expansion of QZSS from three satellites to four. The $526 million contract with Mitsubishi Electric for the construction of three satellites was scheduled for launch before the end of 2017. The third satellite was launched into orbit on 19 August 2017, and the fourth was launched on 10 October 2017. The basic four-satellite system is planned to be operational in 2018.

QZSS uses three satellites, in highly inclined, slightly elliptical, geosynchronous orbits. Each orbit is 120° apart from the other two. Because of this inclination, they are not geostationary; they do not remain in the same place in the sky. Instead, their ground traces are asymmetrical figure-8 patterns (analemmas), designed to ensure that one is almost directly overhead (elevation 60° or more) over Japan at all times.

The primary purpose of QZSS is to increase the availability of GPS in Japan's numerous urban canyons, where only satellites at very high elevation can be seen. A secondary function is performance enhancement, increasing the accuracy and reliability of GPS derived navigation solutions.

The Quasi-Zenith Satellites transmit signals compatible with the GPS L1C/A signal, as well as the modernized GPS L1C, L2C signal and L5 signals. This minimizes changes to existing

---

[137] https://en.wikipedia.org/wiki/Quasi-Zenith_Satellite_System

GPS receivers. Compared to standalone GPS, the combined system GPS plus QZSS delivers improved positioning performance via ranging correction data provided through the transmission of submeter-class performance enhancement signals L1-SAIF and LEX from QZSS. It also improves reliability by means of failure monitoring and system health data notifications. QZSS also provides other support data to users to improve GPS satellite acquisition.

According to its original plan, QZSS was to carry two types of space-borne atomic clocks; a hydrogen maser and a rubidium (Rb) atomic clock. The development of a passive hydrogen maser for QZSS was abandoned in 2006. The positioning signal will be generated by a Rb clock and an architecture similar to the GPS timekeeping system will be employed. QZSS will also be able to use a Two-Way Satellite Time and Frequency Transfer (TWSTFT) scheme, to gain fundamental knowledge of satellite behavior and other purposes.

COSPAS-SARSAT

The Kosmitscheskaja Sistema Poiska Awarinych Sudow i Samaljotow - search and rescue satellite-aided tracking (COSPAS-SARSAT[138]) program was established by Canada, France, the United States, and Russia and provides accurate, timely, and reliable distress alert and location data to help search and rescue authorities assist persons in distress. COSPAS (КОСПАС) is an acronym for the Russian words "*Cosmicheskaya Sistema Poiska Avariynyh Sudov*" (Космическая Система Поиска Аварийных Судов), which translates to "Space System for the Search of Vessels in Distress", while SARSAT is the acronym for Search And Rescue Satellite-Aided Tracking.

The objective of the COSPAS-SARSAT system is to reduce, as far as possible, delays in the provision of distress alerts to SAR services, and the time required to locate a distress and provide assistance, which have a direct impact on the probability of survival of the person in distress at sea or on land.

To achieve this objective, COSPAS-SARSAT Participants implement, maintain, co-ordinate and operate a satellite system capable of detecting distress alert transmissions from radio-beacons that comply with COSPAS-SARSAT specifications and performance standards, and of determining their position anywhere on the globe. The distress alert and location data is provided by COSPAS-SARSAT Participants to the responsible SAR services.

COSPAS-SARSAT co-operates with the International Civil Aviation Organization, the International Maritime Organization, the International Telecommunication Union and other international organizations to ensure the compatibility of the COSPAS-SARSAT distress alerting services with the needs, the standards and the applicable recommendations of the international community.

The four original countries jointly helped develop the 406 MHz Emergency Position-Indicating Radio Beacon (EPIRB), an element of the GMDSS designed to operate with COSPAS-SARSAT system. The first system satellite 'COSPAS-1' (Kosmos 1383) was launched from Plesetsk Cosmodrome on June 29, 1982. COSPAS-SARSAT began tracking the two original types of distress radio beacons in September, 1982, specifically:

- EPIRBs (Emergency Position-Indicating Radio Beacons), which signal maritime distress; and
- ELTs (Emergency Locator Transmitters), which signal aircraft distress

More recently, a new type of distress radio beacon became available (in 2003 in the USA) and these PLBs (Personal Locator Beacons), are for personal use and are intended to indicate a person in distress who is away from normal emergency services (i.e. Phone-Based

---

[138] http://www.cospas-sarsat.int/en/

Services, such as 1-1-2 or 9-1-1).

The four founding countries led development of the 406 MHz marine EPIRB for detection by the system. The EPIRB was seen as a key advancement in SAR technology in the perilous maritime environment. Prior to the founding of COSPAS-SARSAT, the civilian aviation community had already been using the 121.5 MHz frequency for distress, while the military aviation community utilized 243.0 MHz as the primary distress frequency with the 121.5 MHz frequency as the alternate. ELTs for general aviation were constructed to transmit on 121.5 MHz, a frequency monitored by airliners and other aircraft. Military aircraft beacons were manufactured to transmit at 243.0 MHz, in the band commonly used by military aviation.

Early in its history, the COSPAS-SARSAT system was engineered to detect beacon-alerts transmitted at 406 MHz, 121.5 MHz and 243.0 MHz. More recently, the COSPAS-SARSAT system has been designed to detect only alerts transmitted at 406 MHz (see below). This allows the system to be optimized for the increasingly sophisticated 406 MHz beacons, and avoids problems (including false alerts) from the less-sophisticated legacy 121.5 MHz and 243.0 MHz beacons. Many ELTs include both a 406 MHz transmitter, for satellite detection, and a 121.5 MHz transmitter that can be received by local search crews using direction-finding equipment.

The design of distress radio beacons as a whole has evolved significantly since 1982; the newest 406 MHz beacons incorporate GPS receivers; such beacons transmit highly accurate positions of distress almost instantly to SAR agencies via the GEOSAR satellites. The advent of such beacons has created the current motto of SAR agencies "Taking the 'Search' out of Search and Rescue.".

Automatic-activating EPIRBs are now required on vessels subject to requirements of the International Convention for the Safety of Life at Sea (so-called SOLAS-class vessels), commercial fishing vessels, and all passenger ships. Beacons can have vessel identification information pre-programmed into the distress transmission. Or, if the beacon has been properly registered with authorities in advance, Rescue Coordination Centers will be able to retrieve crucial vessel identification and contact information from a beacon registration database.

The automatic-activating EPIRBs, now required on SOLAS ships, commercial fishing vessels, and all passenger ships, are designed to transmit to alert rescue coordination centers via the satellite system from anywhere in the world. The original COSPAS/SARSAT system used polar orbiting satellites but in recent years the system has been expanded to also include 4 geostationary satellites. Newest designs incorporate GPS receivers to transmit highly accurate positions (within about 20 meters) of the distress position.

The original COSPAS/SARSAT satellites could calculate EPIRB position to within about 3 nautical miles (5.6 km) by using Doppler techniques. By the end of 2010 EPIRB manufacturers may be offering AIS (Automatic Identification System) enabled beacons. The serviceability of these items is checked monthly and annually and has limited battery shelf life between

2 to 5 years using mostly Lithium type batteries. 406 MHz EPIRB's transmit a registration number which is linked to a database of information about the vessel.

The system consists of a ground segment and a space segment:

- Distress radio beacons to be activated in a life-threatening emergency
- SAR signal repeaters (SARR) and SAR signal processors (SARP) aboard satellites
- Satellite downlink receiving and signal processing stations called LUTs (local user terminals)
- Mission Control Centers that distribute to Rescue Coordination Centers distress alert data (particularly beacon location data) generated by the LUTs
- Rescue Coordination Centers that facilitate coordination of the SAR agency and personnel response to a distress situation.

The space segment of the COSPAS-SARSAT system currently consists of SARR instruments aboard seven geosynchronous satellites called GEOSARs, and SARR and SARP instruments aboard five low-earth polar orbit satellites called LEOSARs.

Lloyds of London

Lloyd's of London[139], generally known simply as Lloyd's, is an insurance and reinsurance market located in London, United Kingdom. Unlike most of its competitors in the industry, it is not an insurance company; rather, Lloyd's is a corporate body governed by the Lloyd's Act 1871 and subsequent Acts of Parliament and operates as a partially-mutualized marketplace within which multiple financial backers, grouped in syndicates, come together to pool and spread risk. These underwriters, or "*members*", are a collection of both corporations and private individuals, the latter being traditionally known as "*Names*".

The business underwritten at Lloyd's is predominantly general insurance and reinsurance, although a small number of syndicates write term life assurance. The market has its roots in marine insurance and was founded by Edward Lloyd at his coffee house on Tower Street in c. 1686. Today, it has a dedicated building on Lime Street within which business is transacted at each syndicate's "*box*" in the underwriting "*Room*", with the insurance policy documentation being known traditionally as a "*slip*".

The market's motto is Fidentia, Latin for "*confidence*", and it is closely associated with the Latin phrase "*uberrima fides*", or "*utmost good faith*", representing the relationship between underwriters and brokers.

Having survived multiple scandals and significant challenges through the second half of the 20th century, most notably the asbestosis affair, Lloyd's today promotes its strong financial "chain of security" available to pay valid claims. At the end of 2017, this chain consisted of £51.1 billion of syndicate-level assets; £25.6bn of members' funds at Lloyd's (FAL); and over £3bn in a third link which includes the Central Fund.

In 2017 there were 85 syndicates managed by 56 managing agencies that collectively wrote £33.6bn of gross premiums on risks placed by 287 approved brokers. Fifty per cent of premiums emanated from North America and 29 per cent from Europe. Direct insurance represented 68 per cent of the premiums, mainly covering property and liability ("casualty"), while the remaining 32 per cent was reinsurance. The market reported a pre-tax loss of £2bn and a combined ratio result of 114 per cent for 2017.

There are two classes of people and firms active at Lloyd's. The first are members, or providers of capital. The second are agents, brokers, and other professionals who support the members, underwrite the risks and represent outside customers (for example, individuals and companies seeking insurance, or insurance companies seeking reinsurance).

---

[139] http://lloyds.com

Lloyd's Register Group Limited (LR)

Lloyd's Register Group Limited (LR) is a technical and business services organization and a maritime classification society, wholly owned by the Lloyd's Register Foundation, a UK charity dedicated to research and education in science and engineering. The organization dates to 1760. Its stated aims are to enhance the safety of life, property, and the environment, by helping its clients (including by validation, certification, and accreditation) to ensure the quality construction and operation of critical infrastructure.

Historically, as Lloyd's Register of Shipping, it was a specifically maritime organization. During the late 20th century, it diversified into other industries including oil and gas, process industries, nuclear, and rail. Through its 100% subsidiary Lloyd's Register Quality Assurance Ltd (LRQA), it is also a major vendor of independent assessment services, including management systems certification for quality certification to ISO9001, ISO14001 and OSHAS18001. Lloyd's Register is unaffiliated with Lloyd's of London.

In July 2012, the organization converted from an industrial and provident society to a company limited by shares, named Lloyd's Register Group Limited, with the new Lloyd's Register Foundation as the sole shareholder. At the same time the organization gifted to the Foundation a substantial bond and equity portfolio to assist it with its charitable purposes. It will benefit from continued funding from the group's operating arm, Lloyd's Register Group Limited.

Lloyd's Register Fairplay

Lloyd's Register Fairplay is the pre-eminent brand name in the maritime information industry and the only organization that provides comprehensive details of the current world merchant fleet (tankers, cargo, carrier and passenger ships) and a complete range of products and services to assist the world's maritime community. Lloyd's Register-Fairplay uniquely meets international demand for information requirements on ships, companies, ports, real-time vessel movements and related news, and also provides research and consultancy services critical to the shipping industry. Lloyd's Register owns as a separate entity 50 percent share of Lloyd's Register Fairplay while IHS acquired with Prime Publications Limited another 50 percent share. HIS integrated Lloyd's Register Fairplay into the new developed tool MDA WatchKeeper.

MDA WatchKeeper

MDA WatchKeeper is a cost effective, web based, maritime domain awareness solution for individuals that identify, analyze and mitigate maritime threats. It provides a timely and accurate view across your area of responsibility. It's a visualization tool for articulating the risk profile of a ship, to help defense, law and regulatory enforcement agencies identify safety, security, health and welfare risks at sea.

MDA WatchKeeper is based upon a foundation of comprehensive maritime data and can be used as visualization tool for articulating the risk profile of a ship, to help military security co-operations identify safety, security, health and welfare risks at sea.

IHS Fairplay is the only organization to provide complete details of the current world merchant fleet of 100 Gross Tons and above. The maritime data included in the system can be grouped into six categories:

1. Ships: All propelled seagoing merchant vessels of 100 Gross Tons (GT) and above are available in the database. This includes ships on order and under construction, the current trading fleet, as well as details of ships in casualty, lost or broken up.
2. Owners and Operators: Seven levels of ownership ranging from group beneficial owner to registered owners with full address and communication details.
3. Movements: Daily coverage includes historic movement information from 10,000 ports and terminals and positional reports from over 66,000 ships, which represents the majority of all commercial vessels that are actively trading. Each record includes longitude, latitude, time, date and time zone of the movement.
4. Automatic Identification System (AIS): Detailed positional data provided by AIS transponders located on ships exceeding 100GT. These provide near real-time knowledge of a ship's position and include such useful data elements as longitude, latitude, time, course, speed and ETA to destination and general cargo information.
5. Port/Facility: Detailed information covering over 10,000 ports world-wide, encompassing all ports with significant traffic. Data available includes location, physical layout (berth maps), services available, capabilities, owner and operator contact information.
6. IMO: IHS Fairplay is the originating source for the IMO Ship Number and is the sole authority with responsibility for assigning and validating these numbers.

MDA WatchKeeper runs at the unclassified level and has a graphical user interface that addresses the multiple challenges that organizations face in the deployment and use of technology to support MDA. One can define up to 60 criteria (i.e. vessel age, flag of convenience) to provide a risk profile on each ship and send E-mail alerts configured to notify multiple parties.

Maritime Analysis Operations Center – Narcotics (MAOC – N)

The <u>Maritime Analysis and Operations Center – Narcotics (MAOC-N)</u>[140] was established in 2007 to create a dynamic operational platform with the co-location of experienced investigators and military attachés from the seven EU Members Portugal, Spain, United Kingdom, Ireland, France, the Netherlands, Italy and the United States of America. It supports the information/intelligence exchange from naval and military assets of the partners to intercept trans-Atlantic vessels trafficking bulk shipments and cocaine smuggling, by air and by sea, from Latin American and Caribbean (LAC) countries across the Atlantic Ocean.

MAOC-N supports, plans for and tasks operations to interdict illegal drugs being moved by not canalized maritime and air conveyances. The goals of such operations are to prevent drugs from reaching European markets, arrest perpetrators and to deny traffickers their revenue from the delivery of those drugs, and in general to provide the long-term deterrence of illicit drug smuggling.

MAOC-N works with all agencies, countries and third parties to find practical and pragmatic solutions that identify and counter the threats from organized crime in order to break the drug supply route when the drugs are in bulk form and at their most valuable. The interceptions of these bulk quantities of drugs (before the product has been adulterated on the streets of Europe) is when it has the most impact on the higher echelons of organized crime, in terms of lost revenue/proceeds of crime.

MAOC-N is located in Lisbon, Portugal and protects the interests of all EU Member States, takes an active role in the current EU Planning Cycle 2014-17 and works closely with Europol, Eurojust and Interpol. It works closely with the European Monitoring Center for Drugs and Drug Addiction (EMCDDA) and the European Maritime Safety Agency (EMSA) which are also based in Lisbon. The has been active since April 1, 2006. The Centre aims to use intelligence on drug smuggling operations, by air or sea, to apply the most suitable military and/or law enforcement teams to respond to situations that arise. Its area of operations is the eastern Atlantic, from the Cape of Good Hope in Southern Africa to the Norwegian Sea.

The US is represented, and there are observers from Morocco, Greece, Germany, Canada and Cape Verde. The United Kingdom has allocated a team of three from its Serious Organized Crime Agency (SOCA), a SOCA officer serves as the Centre's first Director until October 2009. For the British, this was expected to build on the existing strong relationship between SOCA and the Royal Navy. Ireland's representation is a combined one from the Irish Drugs Joint Task Force of An Garda Síochána (Irish police), the Irish Customs Service and the Irish Naval Service. Personnel from Europol (on behalf of the European Union) and the US Joint Interagency Task Force (JIATF) are also present at the Center to offer liaison

---

[140] http://www.maoc.eu/

and technical expertise.

The seven participating nations are France, Italy, Ireland, Netherlands, Portugal, Spain and UK. The partners will:

- Fuse and respond to actionable intelligence and multinational law enforcement direction.
- Coordinate counter drug interdiction operations in order to support military and law enforcement units from participating countries within and adjacent to the areas of responsibility.
- Support surveillance as required and coordinate the interdiction and arrest phase under the Tactical Control of the nation providing the asset whilst receiving technical advice from the JOCC (Joint Operation and Coordination Center). This will ensure that resources are employed in a manner consistent with the parent organization's policies, directives, rules of engagement and legal authorities.

The costs of the MAOC are shared between the participating states (aggregate 30% of total) and an EC Action Grant which covers the remaining 70% (EURO 661,000). In 2007-8, each state contributed an initial EURO 35,000 followed a year later by a further EURO 5,800 to cover a shortfall. About twenty operations were executed in MAOC's first six months, resulting in 10 seizures in total of more than 10 tons of cocaine. In an operation in June, 2007, 840 kilograms of cocaine were seized aboard a Brazilian flagged vessel which was intercepted by the French Navy, acting on UK intelligence. The vessel, drugs and prisoners were returned to Brazil for prosecution. The Irish Joint Task Force on Drug Interdiction co-operated with MAOC and SOCA in Operation Sea Bight which intercepted a large consignment of cocaine off the Irish coast. MAOC-N uses the Web Enabled Temporal Analysis System (WebTAS).

Centre de Coordination pour la Lutte Anti-Drogue en Mediterranée (CeCLAD-M)

The Centre de Coordination pour la Lutte Anti-Drogue en Mediterranée (CeCLAD-M) is another intelligence-led anti-narcotics law enforcement supported military platform. CeCLAD is a French initiative, Toulon based, aimed at intercepting drug trafficking, by sea and by air, from North and West Africa in the Western Mediterranean Sea. CeCLAD-M has been launched by an International Conference organized by the French Presidency of the EU in Toulon, on the 24-26 September 2008, and co-financed by the EC.

In that forum the countries belonging to of the Inter-ministerial Conference of the Western Mediterranean (CIMO) have been requested to join CeCLAD, which should start interdiction operations before the end of 2008. In a second phase all interested littoral EM MS will have the possibility to provide their operational contribution into CeCLAD. CeCLAD is committed at strengthening intelligence exchange as well as dismantling criminal organization.

Web Enabled Temporal Analysis System (WebTAS)

Web Enabled Temporal Analysis System (WebTAS) is an analysis tool used in the Maritime Analysis Operations Center – Narcotics (MAOC-N) operation center for processing maritime and open source information.

Intelligent Software Solutions (ISS) software is known as Web-Enabled Temporal Analysis System (WebTAS). WebTAS gathers and analyzes intelligence information to determine and improve mission situational awareness during intelligence, surveillance and reconnaissance (ISR) mission planning and execution. The technology assess how ISR efforts are supporting United States Central Command (USCENTCOM) operations and other U.S. government customers, like the Department of Defense, Department of Homeland Security, National Intelligence Organizations, i.e. under the National Maritime Intelligence Information Office (NMIO), as well as NATO and foreign government customer for improving situational awareness.

The national US WebTAS-TK contract encompasses over 100 projects for 70 different user communities. Associated projects range from efforts supporting advanced research and development of machine learning and complex event processing to the development and deployment of state-of-the-art command and control applications, as well as intelligence analysis tools. In addition to providing a vehicle for many software development efforts, this program enables ISS to maintain its ongoing support to combat operations in Afghanistan and Iraq, as well as on-site support at many locations around the United States and internationally. The ISS team currently has more than 120 personnel in Iraq and Afghanistan supporting the bundled tools under this contract.

Maritime Multi-National IP Interoperability (M2I2)

The Maritime Multi-National IP Interoperability (M2I2) Steering Group is a coalition working level group that was formed in 2004, consisting of allied and coalition partner representatives from nations that take part in maritime multi-national operations and exercises, mainly with the Combined Enterprise Regional Information Exchange System (CENTRIXS). The M2I2 organizational structure consists of individuals representing allied and coalition nations supporting operational maritime forces, providing technical, information assurance, training requirements, and planning associated with IP network interoperability.

The M2I2 has been established to develop and supervise the implementation of mutually desired operational and technical solutions for the efficient and secure operation of all CENTRIXS enclaves, including communities of interest (COI). While the M2I2 focus revolves around CENTRIXS-M networks, the group is not limited to CENTRIXS only. Allied/Coalition IP networks in general is the focus of this steering group. The M2I2 is the maritime subset of the larger DISA Multi-National Information Sharing (MNIS) Working Group.

Activities that are developed and coordinated at the M2I2 are typically provided to the MNIS working groups for consideration and planning as there are CENTRIXS Maritime Multi-National IP Interoperability (M2I2) Operations and Plans (OPS & PLANS) and CENTRIXS Maritime Multi-National IP Interoperability (M2I2) Technical Engineering.

The CENTRIXS Maritime Multi-National IP Interoperability (M2I2) Steering Group is a coalition working level group that was formed in 2004, consisting of allied and coalition partner representatives from nations that take part in maritime multi-national operations and exercises. The M2I2 organizational structure consists of individuals representing allied and coalition nations supporting operational maritime forces, providing technical, information assurance, training requirements, and planning associated with IP network interoperability.

The M2I2 has been established to develop and supervise the implementation of mutually desired operational and technical solutions for the efficient and secure operation of all CENTRIXS enclaves, including communities of interest (COI). While the M2I2 focus revolves around CENTRIXS-M networks, the group is not limited to CENTRIXS only. Allied/Coalition IP networks in general is the focus of this steering group. The M2I2 is the maritime subset of the larger DISA Multi-National Information Sharing (MNIS) Working Group. Activities that are developed and coordinated at the M2I2 are typically provided to the MNIS working groups for consideration and planning.

- CENTRIXS Maritime Multi-National IP Interoperability (M2I2) Operations and Plans (OPS & PLANS)
- CENTRIXS Maritime Multi-National IP Interoperability (M2I2) Technical Engineering

Global Counter-Terrorism Force (GCTF)

The Global Counter-Terrorism Force (GCTF) is a United States (US) led multinational naval partnership using CENTRIXS and consisting of more than 80 nations, which some are Albania, Armenia, Austria, Australia, Azerbaijan, Bahrain, Bangladesh, Belgium, Bosnia-Herzegovina, Canada, Croatia, Czech Republic, Denmark, Djibouti, Dominican Republic, Egypt, El Salvador, Eritrea, Estonia, Ethiopia, Fiji, Finland, France, Georgia, Germany, Greece, Honduras, Hungary, Iceland, India, Ireland, Italy, Japan, Jordan, Kazakhstan, Kenya, Kuwait, Kyrgyzstan, Latvia, Lithuania, Luxembourg, Macedonia, Mauritius, Moldova, Mongolia, Morocco, Netherlands, Nepal, New Zealand, Nicaragua, Norway, Oman, Pakistan, Philippines, Poland, Portugal, Qatar, Romania, Russian Federation, Saudi Arabia, Serbia-Montenegro, Seychelles, Singapore, Slovakia, Slovenia, South Korea, Spain, Sweden, Switzerland, Tajikistan, Thailand, Tonga, Turkey, Turkmenistan, Ukraine, United Arab Emirates, United Kingdom, United States, Uzbekistan, and Yemen.

Combined Maritime Forces (CMF)

Combined Maritime Forces (CMF) using CENTRIXS based in Bahrain is a United States (US) led multinational naval partnership consisting of 30 nations, which exists to promote security, stability and prosperity across approximately 3. 2 million miles of international waters in the Arabian Gulf, Gulf of Oman, Gulf of Aden, Red Sea and Western Indian Ocean, an area which encompass some of the world's most important shipping lanes.

The countries represented within CMF include Middle Eastern regional nations and Indian Ocean nations as well as countries that span the globe providing a truly global partnership. CMF is still an evolving and growing organization, and on any given day CMF has task forces and their assigned ships and aircraft conducting maritime security operations in the Indian Ocean. CMF has its roots in the US response to the 9/11 terrorist attacks 2001, in support of United Nations Security Council Resolution (UNSCR) 1377 (Threats to international peace and security caused by terrorist acts).

Initially formed as part of the Global Counter Terrorism Force, more specifically as the tactical maritime flank of Operation ENDURING FREEDOM, CMF has grown and evolved beyond that operations scope to encompass and address commonly perceived maritime security threats to member states, including measures to suppress Somali piracy and general maritime security operations in the Arabian Gulf and Indian Ocean.

Since its inception membership has grown steadily, with Yemen being the last nation to join in 2013. CMF is open to all nations with a common interest. It is unique in that it is a coalition of the willing, focused on non-state threats and is committed to making the region safer by working together with the aims to defeat terrorism prevent piracy deter illegal trafficking and promote the maritime environment as a safe place for all mariners with legitimate business.

CMF works with regional and other partners to improve overall security and stability helps strengthen regional nations 'maritime capabilities and, when requested, responds to environmental and humanitarian crises. CMF has no political or military mandate or overarching policy framework as there is no legal mandate from the UN Security Council for conducting maritime security operations in the region. Some UNSC resolutions such as 2244 (Somali Arms and Charcoal embargoes) and 1846 (Threat of Somali Piracy) provide maritime enforcement powers but the majority of CMF maritime security operations are conducted using customary international law.

This is both a strength, as member navies only participate in activities that meet their national priorities, and a weakness as the lack of a strong legal basis for ongoing operations creates unique operational challenges by inhibiting the optimization of forces that can be applied to a specific maritime security challenge. With no formal CMF rules of engagement, member nations operate within their national mandates and permissions, which are respected at all times, doing what they want, where they want, when they want.

The early partner nations were Australia, Bahrain, Belgium, Canada, Denmark, France, Germany, Greece, Italy, Japan, Jordan, Republic of Korea, Kuwait, The Netherlands, New Zealand, Malaysia, Pakistan, Portugal, Saudi Arabia, Singapore, Spain, Thailand, Turkey, United Arab Emirates, United Kingdom, and the United States of America. In 2015 CMF extended as an US led multinational naval partnership to naval operations with 30 nations, with the last one, Yemen, joining in 2013.

The Combined Maritime Forces (CMF) conducts operations through three international task forces with a yearly command rotation between the partners:

- Combined Task Force 150 for maritime security and counter-terrorism
- Combined Task Force 151 for counter-piracy
- Combined Task Force 152 for the Arabian Gulf security and cooperation

Combined Task Force 150

Combined Task Force 150 (CTF-150) for maritime security and against counter-terrorism is a multinational coalition naval task force working under the 25 nation coalition of Combined Maritime Forces (CMF) and is based in Bahrain established to monitor, inspect, board, and stop suspect shipping to pursue the "Global War on Terrorism" and in the Horn of Africa region (HOA) (includes operations in the North Arabia Sea to support operations in the Indian Ocean. These activities are referred to as Maritime Security Operations (MSO).

Countries presently contributing to CTF-150 include Canada, Denmark, France, Japan, Germany, United Kingdom and Northern Ireland, and the United States. Other nations who have participated include Australia, Italy, India, Malaysia, Netherlands, New Zealand, Portugal, Singapore, Spain, Thailand and Turkey. The command of the task force rotates among the different participating navies, with commands usually lasting between four and six months. The task force usually comprises 14 or 15 ships.[2] CTF-150 is coordinated by the Combined Maritime Forces (CMF), a 25 nation coalition operating from the US Navy base in Manama, Bahrain.

Command tours are typically 4 months, although there is also the possibility to retain command for one year, since the regional nations have much knowledge of local operating conditions so it is of great benefit to have an CTF 150.

There are a number of terrorist and violent extremist organizations active in the CMF area of operations. They include the Taliban, Al Qaeda (and its affiliates), ISIS, and Al Shabaab. The majority of these groups have expressed intent to attack global interests and some have either already demonstrated a maritime attack capability or are attempting to develop such a capability. Therefore, CMF remains vigilant for direct maritime terrorist attacks but the majority of daily CTF 150 operations, apart from providing a constant deterrent, target other illicit activities in the region that either directly or indirectly supports terrorist and violent extremist organizations, principally as mechanisms to fund their activities.

The link between the Afghan narcotics trade and the financing of terrorist and violent extremist organizations has been highlighted in a number of United Nations reports that demonstrate the dependence that the Taliban has on Afghan narcotics for financing their activities, with over a third of their funding sourced from the trade in 2011. Given the recent increase in heroin cultivation and larger heroin harvests, narcotics funding is likely to now be proportionally higher. Narcotic routes out of Afghanistan are well established and targeted by multiple agencies with varying degrees of success.

Increasingly, drug trafficking organizations are seeking the most economical routes to supply an increasingly sophisticated global distribution network. One such route is referred to as the "Southern Route" by the United Nations Office on Drugs and Crime (UNODC) in their inaugural report on this growing distribution network in June 2015. This route emanates from Afghanistan, transiting via Pakistan and Iran, to global markets, of which a sizable proportion occurs via dhows embarking drugs from the Makran Coast states of Iran and Pakistan, thence across the Indian Ocean to East Africa. The UNODC reports that despite increasing production in Afghanistan, less is being detected through the established Balkan and Northern Routes, likely due to the security situation in the transit countries of Iraq and Syria. This implies that heroin trafficking is likely to continue, if not expand, across the Indian Ocean. The expanding narcotics trade also presents significant longer term governance, security and health challenges for east Africa, and the Indian Ocean region in general.

The present primary CTF150 focus is on the most lucrative of all the illicit activities that occur in the area of operations-the narcotics trafficking from Afghanistan, It travels by sea from the Makran Coast of Iran and Pakistan to Yemen, for hashish, or to East Africa for heroin, primarily destined for Tanzania, although CMF believes that approximately one fifth is likely bound for other destinations including Mozambique and Madagascan. In September 2015 CMF also achieved a significant weapon seizure of anti-tank missiles and associated guidance and launcher systems from a dhow that, according to the master, sailed from Iran and was intercepted on route to Somalia.

Counter-narcotics operations are traditionally a constabulary role performed by law enforcement agencies, however the lack of maritime law enforcement capacity in the region has resulted in CMF being one of the only credible forces capable of exerting any type of pressure on the Indian Ocean narcotics trade. CMF conducts these operations due to the nexus between the narcotics trade and funding for terrorist and violent extremist organizations, but as the primary focus is in support of counter terrorism operations, CMF does not attempt any law enforcement outcome-if narcotics are found, they are seized and destroyed, and the dhow and its crew released. Since 2012 CMF has had increasing success in targeting the trade. CMF assesses that it intercepts approximately 15 per cent of the overall trade entering east Africa.

CMF operations at the ship and headquarters level have become more sophisticated over time and CMF now also works with the UNODC, primarily through the Global Maritime Crime program initiative - the Indian Ocean Forum on Maritime Crime Heroin Working Group, and law enforcement agencies working at the Regional Narcotics Interagency Fusion Centre, also located at Bahrain-to ensure the maximum value is gained from CMF seizures. Increasingly it is the number of ships available, and the capability of these ships to conduct this task, that is the limiting factor to CMF success. Generally, it is larger sea going vessels with good sensors, endurance, organic helicopters and/or unmanned aerial vehicles, good command and control and well trained boarding parties with specialist equipment; that consistently achieve success.

Combined Task Force 151

Combined Task Force 151 (CTF-151) or Combined Task Force One Five One is a multinational naval task force for counter-piracy as a response to piracy attacks in the Gulf of Aden and off the eastern coast of Somalia. Its mission is to disrupt piracy and armed robbery at sea and to engage with regional and other partners to build capacity and improve relevant capabilities in order to protect global maritime commerce and secure freedom of navigation. It operates in conjunction with the EU's Operation ATALANTA and NATO's Operation Ocean Shield, hence CTF 150 and 151 have overlapping areas.

CTF 151 came into being to directly counter the threat of Somali pirates, established in January 2009 with a specific piracy mission-based mandate under the authority of UNSCR 1846. The task force works alongside the European Union (EU) and NATOS counter-piracy task forces, along with ships from nations such as Iran, China, Russia and India. Collectively there are usually between 5 and 8 warships assigned to the three task forces that are on patrol daily in the Indian Ocean region.

The last successfully pirated ship was the MV Smyri in May 2012 and in 2013 two vessels were attacked but none successfully pirated. The rapid decline in piracy attacks since the peak of attacks in 2010, when 45 vessels were pirated, has been a result of a combination of activities. These include the deployment of armed security teams on merchant vessels travelling through the region, the implementation of best management practices and the coordination of military naval patrols conducted by the task forces, as well as the independently deployed units.

Despite these measures, CMF continues to observe suspicious activity and has witnessed a surge in the number of false alarms reported by merchant vessels since 2014. CTF 151 therefore regularly conducts focused operations aimed at reducing false alarm rates by educating the fishing and shipping communities on each other's standard operating procedures.

Whilst the threat from piracy has been contained, the current situation still requires vigilance. The Somali pirates retain the capability and intent to conduct pirate attacks. What they have lacked has been the opportunity. With EU and NATO counter-piracy mandates ending in December 2016, the continued suppression of piracy in the region is a strategic concern.

As of September 2013 the task force consists of six ships from Australia, Pakistan, South Korea, Turkey, the UK, and the USA under the command of Commodore Jeremy Blunden RN based on RFA Fort Victoria. Pirate attacks in the area have declined from "over 170" in 2010 to a quite smaller number in 2013.

Another more recent developing theme faced by CTF151 is that of mass migration as a result of the deteriorating situation in Yemen and ongoing instability in Somalia. Whilst humanitarian assistance, in isolation, is not a CMF mission, CMF does monitor and report all migrant contacts to the United Nations High Commissioner for Refugees and is ready to react to any mass maritime humanitarian situation needed.

CMF conducts maritime security operations in a complex and dynamic operational environment to tackle threats in the maritime domain. CTF 150 and CTF 151 maritime security operations are focused on maritime security threats in the western Indian Ocean. These threats are also relevant to the member nations of the Indian Ocean Naval Symposium (IONS). Moving forward, there is potential to develop an Indian Ocean maritime strategy with all stakeholders working together to develop a global approach to the problem of transnational maritime threats operating around the Indian Ocean rim. CMF is here for the long haul and willing to engage with like-minded navies, as collectively we can achieve more by working together, than working in isolation.

Combined Task Force 152

Combined Task Force 152 (CTF-152) is a multinational coalition naval task force, established in March 2004, for the Arabian Gulf security and cooperation. CTF 152 coordinates Theatre Security Cooperation (TSC) activities with regional partners, conducts Maritime Security Operations (MSO), and remains prepared to respond to any crisis that may develop.

The Arabian Gulf is a 989 kilometer-long inland sea that separates Iran from the Arabian Peninsula. Countries with a coastline on the Arabian Gulf, called the Gulf States, are Iran, United Arab Emirates, Saudi Arabia, Qatar on a peninsula off the Saudi coast, Bahrain on an island, and Kuwait and Iraq in the northwest.

Today the Gulf is one of the most strategic waterways in the world due to its importance in world oil transportation. It contains in the region of 700 billion barrels of proven oil reserves, representing over half of the world's oil reserves, and over 2,000 trillion cubic feet of natural gas reserves (45% of the world total). Arabian Gulf countries maintain about one-third of the world's productive oil capacity. The majority of the oil exported from the Arabian Gulf is transported by sea.

CTF-152 has included participation from Kuwait, Bahrain, United Arab Emirates, Saudi Arabia, Qatar, Italy, Australia, the United Kingdom and the United States. Participation is purely voluntary. No nation is asked to carry out any duty that it is unwilling to conduct and the command of CTF 152 is rotated between participatory nations on a three to six month basis.

Combined Enterprise Regional Information Exchange System (CENTRIXS)

The US-initiated <u>Combined Enterprise Regional Information Exchange System (CENTRIXS)</u>[141], formerly the Coalition Enterprise Regional Information Exchange System (CENTRIXS), supports intelligence and classified operations information exchange and sharing up to SECRET Releasable (REL). CENTRIXS is federated among global and command enterprise environments. The future Operational Concept in CENTRIXS envisages a single global coalition network via VPN connecting multiple communities of interest on a single work station. CENTRIXS started under the term CAS - Collaboration at Sea.

CENTRIXS is based on the LotusNotes-Suite which is also know in the military environment as Collaboration at Sea (CAS). The global environment is managed by DISA to serve and interconnect command enterprise elements. The command enterprises consist of servers, applications, and encryption systems that form essentially autonomous service environments interconnecting command enclaves through existing regional communications networks for bilateral or multilateral access among cooperating nations and international organizations.

Joint warfighting operations demand responsive information exchange across combined forces and unified commands for planning, unity of effort, decision superiority, and decisive global operations. In a concerted endeavor to provide this support to the warfighting commands, the Combined ENTerprise Regional Information eXchange System (CENTRIXS), has been fielded and designed to meet the U.S. needs for multinational information sharing networkslike the i.e. the Global Counter-Terrorism Force (GCTF) or Combined Naval Forces Central Command (CNFC).

CENTRIXS networks are a combination of global, multilateral and bilateral, virtually separate networks supporting multinational efforts including OPERATION ENDURING FREEDOM (OEF), OPERATION IRAQI FREEDOM (OIF), and the Global War on Terrorism (GWOT). These networks form the backbone of what is envisioned to become a global infrastructure, allowing the U.S. to share information rapidly with coalition partners worldwide, in support of local, regional, and global combined operations.

The CENTRIXS network is both network-centric and web-centric, using a combination of readily available Commercial-Off-The-Shelf (COTS) and Government-Off-The-Shelf (GOTS) solutions to reduce implementation costs while providing a robust, innovative approach to warfighting communications. CENTRIXS is designed to be a global, interoperable, interconnected, inexpensive, and easy-to-use information sharing system. Global, interoperable and interconnected means that all combatant commands, national agencies, and foreign partner nations will be able to communicate with each other in a seamless manner. Inexpensive means that even nations with limited fiscal means can still participate in

---

[141] http://www.centrixs.smil.mil/

the information sharing effort. In addition, easy-to-use means that even the most inexperienced users can learn and use CENTRIXS in a very short time, and with minimal technical training. CENTRIXS is installed in partner nation countries and their military forces after they establish their specific requirements and relate the requirements to the supporting U.S. Combatant Commands.

CENTRIXS implementation first focused on fielding core collaboration services (i.e., Email with and without attachments, web-browser-based data access, and file sharing). Other required services to include tactical, near-real-time data access, have been enabled as the network has matured. To the extent possible, CENTRIXS will subsume and consolidate existing stove-piped coalition networks as part of a single, unified system. The basic mission of CENTRIXS is to support the secure, sharing and exchanging of intelligence and operations information through reliable communications connectivity, data manipulation, and automated processes.

CENTRIXS services provide Combatant Command (COCOM) commanders with: Common and consistent situational awareness of the battlefield Common Operational Picture (COP). CENTRIXS uses existing communications infrastructures, such as the Secret Internet Protocol Routed Network (SIPRNet), whenever possible. The system employs NSA (National Security Agency)-Approved Type-1 encryption devices and uses an encrypted channel between AOR locations (strategic or tactical) to a Forward Point of Presence (FPoP) or Command Headquarters site.

The Joint Interoperability Test Command (JITC) is in the process of conducting a global Interoperability Certification on CENTRIXS. The JITC will directly observe or simulate testing and utilize any available test data from Developmental and/or Operational Tests conducted by any of the Combatant Commands (COCOMs) or the Program Office. Upon completion of testing, JITC will conduct a thorough analysis of all the test data gathered during the CENTRIXS Interoperability Test Phase. An Interoperability Certification Evaluation Test Report will be provided to the DISA Multinational Information Sharing (MNIS) Joint Program Office (JPO) upon completion of the analysis. The JITC maintains a CENTRIXS test network, which includes a command headquarters configuration and various deployable configurations and also maintains various NSA-approved type I encryptions devices, including the KG-175 TACLANE and KIV-7, and a data link simulator, which facilitates simulation testing of the live CENTRIXS network in a lab environment. This test network is also used to test Information Assurance Vulnerability Alerts (IAVA) for the DISA MNIS JPO, prior to enterprise-wide deployment.

Atlantic, Barents, Baltic and Arctic Regions (ABBA)

The US forces intend to set up a CENTRIXS-based Federated Mission Network (FMN) compliant and over-arching information domain for the Atlantic, Barents, Baltic and Arctic regions (ABBA enclave). This would provide the possibility of an encrypted (comparable to US-SECRET) communication between the Baltic Sea States and intends to introduce ABBA as a standardized information environment "Baltic Sea" in addition to the unclassified Sea Surveillance Cooperation Baltic Sea (SUCBAS). It can be expected that ABBA will base on CENTRIXS experiences and base on its hierarchical rather than federated architecture and technology.

Maritime Organization for West and Central Africa (MOWCA)

The <u>Maritime Organization for the West and Central Africa (MOWCA)</u>[142] was established in May 1975 (Charter of Abidjan) as the Ministerial Conference of West and Central African States on Maritime Transport (MINCONMAR). The name was changed to MOWCA as part of reforms adopted by the General Assembly of Ministers of Transport, at an extraordinary session of the Organization held in Abidjan the Republic of Cote d'Ivoire from 4-6 August 1999. Objective of MOWCA is to serve the regional and international community for handling all maritime matters that are regional in character.

The focus of MOWCA policy is on the following:

- Encouraging participation of the private sector in West/Central Africa in ship operation particularly in coastal shipping, by way of ownership/chartering of tonnage and forging co-operation/partnerships between regional operators and foreign shipping companies operating to the sub-region;
- Development of coastal shipping networks and establishment of feeder systems to connect hub and spoke ports - Establishment of an effective multimodal transport system for the sub-region;
- Port development and facilitation with particular reference to achieving a cost effective / faster ship turnaround times and creating special berths and conditions for landlocked countries, coastal /feeder shipping in MOWCA port;
- Strengthening of service-oriented shippers' councils to effectively protect and represent the users or the demand side of the shipping industry;
- Strengthening of the regional maritime academies of Abidjan, Accra and the Nigerian Academy of Oron to provide training at all levels of the maritime, fishing and petroleum industry, including exchange of teaching personnel, provision of facilities for sea-training for cadets;
- Establishment of national transport observatories to be coordinated by a regional observatory based in the Secretariat, to generate an up-to-date, uniform, computerized data base for the entire sub-region;
- Maritime safety and environmental protection with regards to creating effective contingency plans for pollution prevention/curtailment in member states, establishment of reception facilities for the discharge of waste from tankers, enhancement of efficiency of maritime administrations, the implementation of flag state control measures and regional Memorandum of Understanding (MOU) on port state control.

---

[142] http://www.amssa.net/framework/mowca.aspx

MOWCA unifies 25 countries on the West and Central African shipping range (inclusive of five landlocked countries). These countries comprise of 20 coastal states bordering the North and South Atlantic Ocean, and to explain the maritime link for landlocked countries the ports of the Ocean interfacing countries provide the seaborne trade of those that are landlocked.

What is of special interest for MOWCA is that together the countries within the remit of the organization in 1998 generated an estimated 247million tons of cargo which represented 4.8% of world cargo, 95% of which was seaborne. MOWCA has also identified a number of problems at the sub regional level relating to the cost-effectiveness of shipping services these are; availability of shipping space, frequency of sailings, level of freight rates, competitiveness and survival of national/regional operators, efficiency of seaports, safety of cargo/ships, inland transportation networks, availability of coastal shipping services, efficiency of multi-modal transport systems and trade facilitation, protection of shippers interests, and the special case of landlocked countries.

The overarching decision-making body for MOWCA is a General Assembly of Ministers of Transport of Member States meeting at ordinary sessions every two years and at extraordinary sessions if necessary.

The MOWCA Secretary-General coordinates the following three specialized Units for MOWCA these represent the ports, shippers and operators. Countries affiliated with MOWCA are the Port Management Association of West and Central Africa (PMAWCA), the Union of African Shippers Councils (UASC), the Association of African Shipping Lines (ANSL), Angola, Benin, Burkina Faso, Cape Verde, Cameroon, Central African Republic, Chad, Côte d'Ivoire, Democratic Republic of the Congo, Equatorial Guinea, Gabon, Ghana, Guinea, Guinea Bissau, Liberia, Mali, Mauritania, Niger, Nigeria, Republic of the Congo, Sao Tome & Principe, Senegal, Sierra Leone, The Gambia and Togo.

Maritime Security Center Horn of Africa (MSCHOA)

The Maritime Security Center Horn of Africa (MSCHOA)[143] aims to provide a service to mariners in the Gulf of Aden, the Somali Basin and off the Horn of Africa and is situated in the Operational Headquarters (OHQ). MSCHOA is an initiative originated in EU NAVFOR SOMALIA with close co-operation from industry.

MSCHOA is a Coordination Centre dedicated to safeguarding legitimate freedom of navigation in the light of increasing risks of pirate attack against merchant shipping in the region, in support of the UN Security Council's Resolutions (UNSCR) 1814, 1816 and 1838. It provides 24-hour manned monitoring of vessels transiting through the Gulf of Aden, whilst the provision of an interactive website enables the Centre to communicate the latest anti-piracy guidance to industry, and for shipping companies and operators to register their vessels' movements through the region.

Through close dialogue with shipping companies, masters and other interested parties, MSCHOA will build up a picture of vulnerable shipping in these waters and their approaches. The Centre, which is manned by military and merchant navy personnel from several countries will then coordinate with a range or military forces operating in the region (notably EU NAVFOR) to provide support and protection to mariners.

There is a clear need to protect ships and their crews from illegitimate and dangerous attacks, safeguarding a key global trade route. To do this effectively, MSCHOA needs to know about merchant vessels approaching, transiting or operating in the region. The website was created by International Maritime Bureau, UKMTO and Fairplay, however due to the BREXIT the Maritime Security Center Horn of Africa (MSCHOA) moved to the Maritime Information Cooperation and Awareness Center (MICA Center) in BREST (FRA).

The Login offers ship owners, ships Masters and agents the facility to register their details securely with MSCHOA, update positions of their vessels and receive information and guidance designed to reduce the risk of pirate attacks. MSCHOA has been set up by the EU as part of a European Security and Defence Policy initiative to combat piracy in the Horn of Africa. This work commenced with the establishment of EU NAVCO in September 2008.

The Coordination Cell working in Brussels established linkages with a broad cross section of the maritime community and provided coordination with EU forces operating in the region. In November 2008, the Council of the European Union took a major step further by setting up a naval mission – EU NAVFOR ATALANTA – to improve maritime security off the Somali coast by preventing and deterring pirate attacks and help safeguard merchant shipping in the region. From mid-December 2008, an EU Naval Task Group, supported by maritime patrol aircraft, will be operating in the region.

---

[143] http://www.mschoa.org

Mercury

Mercury was developed as an urgent operational requirement to satisfy information exchange requirements between the various stakeholders of the Maritime Security Center – Horn of Africa (MSCHOA), the UK Maritime Trade Operations (UKMTO) and in EU NAVFOR Somalia – Operation ATALANTA.

As the unclassified EU Counter Piracy Coordination system it has proved a pragmatic and successful approach and replaced the former, more limited FEXWEB as the system of choice for EU, NATO, Coalition and other actors in the AOO of Operation ATALANTA. Mercury offers a degree of information protection by means of the https secure protocol.

Ship Security Reporting System (SSRS)

The Ship Security Reporting System (SSRS)[144] is an counter-piracy service that enhances the effectiveness of existing Ship Security Alert Systems by providing a link from a ship sending an alert direct to the Maritime Security Center – Horn of Africa (MSCHOA) and UK Maritime Trade Operations (UKMTO) and onward to EU NAVFOR Somalia – Operation ATALANTA and associated participating naval forces responsible for maritime security in the Gulf of Aden and off the Somali coast.

The SSRS service continuously monitors alerts transmitted by the SSAS onboard subscribed vessels. The system compares the position of the vessel sending the alert to the published MSCHOA / UKMTO Area of Operation and, establishing that the vessel is within the area, routes the alert, which includes the vessel name, report date/time stamp, associated latitude / longitude, and also the IMO number, MMSI, and freeboard and speed, to MSCHOA and UKMTO. Once they confirm the alert is genuine, the updated information is disseminated through the Mercury system which is viewed by all participating task forces.

The alert is copied to the registered Company Security Officer (CSO) as confirmation that it has been disseminated and to provide the CSO with relevant contact details, should he have additional information to communicate.

In the event of a piracy attack, SSAS activation supplements the main piracy reporting procedures set out in Best Management Practices for Protection against Somalia Based Piracy. If the crew find themselves under duress and unable to make or receive telephone calls, the SSRS service is an effective, covert method of automatically alerting the military assets faster than via the SSAS alone. However, to maximize the speed of response, it is advised that, whenever possible, the SSAS should be activated in concert with a telephone call to UKMTO.

The SSRS has been designed to integrate seamlessly with existing SSAS equipment installed on all commercial vessels of 500GT and above engaged on international voyages (defined in the ISPS Code, SOLAS XI-1/6).

---

[144] https://www.ssrs.org

Maritime Surveillance (MARSUR) Networking

2005 the EU Defense Ministers tasked the European Defense Agency (EDA) to create a maritime network using the existing naval and maritime information exchange systems of the Member States. On 14. September 2006 EDA launched the first Maritime Surveillance project (MARSUR I) and under the Integrated Development Team (IDT) with the MARSUR Working Group.

On 30. June 2011 the EDA project terminated in the Distinguished Visitors Day (DV-Day) at the Diamond Center, in Brussels, Belgium, with a demonstration to Maritime Surveillance Stakeholders, at first connecting six Member States of the EU. 30. June 2011 marks the end of the first project of the European Defence Agency (MARSUR I) and the start of the MARSUR Networking LIVE PHASE.

1. July 2011 was the start of the operational use of the first MARSUR Capability developed during the EDA project with the MARSUR Exchange System (MEXS 1.0) by Spain, Finland, France, Italy, Sweden and United Kingdom.

On 27. October 2011 the 15 initial participating Member States (pMS) Belgim, Cyprus, Germany, Spain, Finland, Greece, France, Ireland, Italy, Lithuania, Croatia, the Netherlands, Poland, Portugal, Sweden and United Kingdom and Northern Ireland signed the MARSUR Networking Technical Arrangement (TA) which marked the start of the MARSUR Operational LIVE PHASE under the lead of the new established MARSUR Management Group (MMG). On 11. October 2012 Bulgaria, Latvia and Norway joined the Technical Arrangement (TA) raising the number to 18 participating Member States (pMS) in the MARSUR LIVE PHASE. Since 2011 the MARSUR Networking is closely related to the European context, however independently driven by the MARSUR Members under the TA.

The MARSUR DEVELOPMENT PROJECT (01. October 2012 - 30. October 2014) was to improve the functionalities and add new requirements in MARSUR Networking with the newly developed state-of-the-art technology and replaced the obsolete solution of the former EDA project. This second MARSUR project (MARSUR II) brought enhanced software development with a MARSUR Exchange System (MEXS 2.0), Mobile MARSUR Exchange System (MEXS Mobile), MARSUR User Interface (MARSUR-UI), MARSUR Test Tool and a MARSUR Management Tool.

The new developed capability was presented at the EURONAVAL in Paris, France on 28. October 2014. As the third MARSUR project under EDA (MARSUR III) it was initiated as Adaptive Maintenance for the support of the software components in 2017. The EU Satellite Center joined in 2017, EDA in 2018, both without voting rights on new MARSUR members, Malta joined in 2018 and Slovenia wrote a note of accession in 2017. The Project Arrangement (PA) Adaptive Maintenance (MARSUR III) started in 2017 the third project (MARSUR III) with the contributing Member States (cMS) of BEL, CYP, DEU, ESP, FIN, FRA, GBR, GRC, ITA, IRL, NDL, NOR, PRT, SWE in order to improve the current software with new developments until 2020.

Maritime Surveillance Networking (MARSUR Networking) aims at the information exchange of the participants under the Technical Arrangement (TA) for the MARSUR LIVE PHASE in the EU by:

- Aspiring to share part of their surveillance-information of the Maritime Domain with each other.
- Aiming at establishing policies, procedures and technical arrangements for their needs in the area of maritime surveillance.

The aim of MARSUR LIVE PHASE is to share Maritime Surveillance information and further develop the existing concept, technical solutions and procedures as established by the demonstration phase. The decision to supply information into the system would be a responsibility of the participants, and is as such, the information is owned by the providing participant. Further dissemination of the information might be limited by the originator of the information.

The main objectives of the MARSUR are to contribute, pursuant to national laws and regulations, to security, safety and protection of the environment in the maritime domain and improve maritime situational awareness, produce and share information from the respective maritime picture, improve interoperability and co-operation between EU military and civilian maritime authorities and other international maritime actors.

The aim of MARSUR LIVE PHASE is to share Maritime Surveillance information and further develop the existing concept, technical solutions and procedures as established by the demonstration phase. The decision to supply information into the system would be a responsibility of the participants, and is as such, the information is owned by the providing participant. Further dissemination of the information might be limited by the originator of the information.

The main objectives of the MARSUR LIVE PHASE are to contribute, pursuant to national laws and regulations, to security, safety and protection of the environment in the maritime domain and improve maritime situational awareness, produce and share a maritime picture, improve interoperability and co-operation between EU military and civilian maritime authorities and other international maritime actors.

The European Defence Agency (EDA) was in 2011 no signatory member of the Technical Arrangement (TA) and did not become a member of the MARSUR Management Group (MMG) until 2018. The agreement on the accession of EDA in the MARSUR community was as observer with no voting rights and at no charge/fees. EDA does not take part in the operational activities of the MARSUR LIVE Phase. Further European Costal Countries have been agreed for extension of the MARSUR Networking i.e. Albania, Croatia, Denmark, Estonia, Montenegro, and Romania.

The MARSUR Virtual Private Network (MARSUR VPN) for connecting the Maritime Operation Centers of the Members reached its Initial Operational Capability (IOC) on 30. June 2011, after the EURONAVAL in Paris, France the IOC for the second software version was declared on 28. October 2014 using the MARSUR Virtual Private Network (VPN) as the Network Layer for nine MARSUR Core Services. The Network Layer, the MARSUR Virtual Private Network (VPN) is established between the national routers, which connects the hard- and software installations for the national MARSUR Exchange Nodes-Nodes (National MEXS-Node).

MARSUR Networking aims at connecting all available national services from National Maritime Surveillance Systems (NMSS) of the MARSUR Member States via the nine core services. The track streaming is mainly thought for members using the MARSUR User Interface (MARSUR-UI) until they implemented their National Maritime Surveillance System (NMSS). Nations without a NMSS connected - or for interim use during development of a national MEXS-Node for NMSS implementation - can therefore still participate on the ADVANCED LEVEL.

However, if not one member would connect a National Maritime Surveillance System (NMSS) feed, the logical consequence would be that there would be no track stream available for any maritime picture, no tracks could then be visible on the MARSUR User Interface. Therefore the sharing of national data is of imminent importance for the benefit of the whole community and the enhancing of the overall MSA in Europe and beyond.

The MARSUR Networking has two participation levels, the first one simply providing Point-of-Contacts for operational and technical information exchange via commercial e-mail, fax and phone, the second level on using the operational VPN for secured information exchange.

MARSUR Basic Level (Mandatory)

- Exchange of national Point of Contact information related to maritime situational awareness, with the intention of manually sharing maritime surveillance information. Basic level will be mandatory for the Participants. *Note: Initial Operation Capability (IOC) on Basic Level is 24/7 since 27. October 2011.*
- The Basic Level is mandatory for all MARSUR Networking Member States using commercial telephone, e-mail and fax conducted by the MOWG POCs or respective MOC (24/7) with a weekly rotating schedule. The Basic Level is also the fallback solution in case of technical failure of the National Maritime Surveillance System (NMSS) and/or the national MEXS-Node.

The MARSUR Networking in the mandatory BASIC LEVEL based on a 24/7-Point-of-Contact List of all Member States MOCs with phone and fax numbers and e-mail-addresses is in Full Operational Capability (FOC) since 27. October 2011. The weekly Senior POC rotates between the nations and weekly role calls and exercises have been conducted by the MOWG since then.

The common chat rooms are provided by one nation in a technical failover procedure in order to ensure 24/7 operation; same accounts for a time server failover sequence.

<u>Advanced Level</u>

- Mutual automated exchange of information obtained through national analysis of information, where permitted by national law and any applicable agreements or arrangements. The Advanced Level requires implementation of a National Maritime Surveillance System (NMSS) to the MARSUR Release forming a national MARSUR Exchange System, the national MEXS-Node. The MARSUR User Interface (MARSUR-UI) may be used as interim solution.
- The Advanced Level in MARSUR Networking is described as the information exchange using the National Maritime Surveillance System (NMSS) via national MARSUR Exchange System (MEXS-Node) connected to other MARSUR Member States MEXS-Nodes. Nations with no NMSS implemented to a national MEXS-Node can make use of the MARSUR User Interface (MARSUR-UI) via national MEXS-Node connected to other MARSUR Member States MEXS-Nodes.

<u>The Maritime Surveillance (MARSUR) Networking participants are:</u>

1) Belgium
2) Bulgaria (2012)
3) Cyprus
4) Germany
5) Greece
6) Spain
7) Finland
8) France
9) Ireland
10) Italy
11) Latvia (2012)
12) Lithuania
13) Netherlands
14) Norway (2012, first non-EU Member in MARSUR)
15) Poland
16) Portugal
17) Sweden
18) United Kingdom of Great Britain and Northern Ireland (No changes due to BREXIT)
19) EU Satellite Center (EU SatCen, 2018, Non-Voting rights on new members)
20) Malta (2018)
21) European Defence Agency (EDA, 2018, Non-Voting rights)
22) Slovenia (Note of accession in 2017)

Maritime Surveillance (MARSUR) Capability

Maritime Surveillance (MARSUR) Capability is the software of the MARSUR Release Versions and its documentation of the MARSUR Cooperation. The MARSUR Interface Technology is closely related to the one used in the Sea Surveillance Cooperation Baltic Sea (SUCBAS) Interface Technology all origination back to TIDE Sprint Specifications (FMN).

The MARSUR concept and architecture is based on the European Union Network Enabled Capability (EUNEC), NATO Network Enabled Capability (NNEC), the concept of the Sea Surveillance Cooperation Baltic Sea (SUCBAS) and the Common Information Sharing Environment (CISE).

Network-Enabled Capability, or NEC, is originally the name given to the United Kingdom long term intent to achieve enhanced military effect through the better use of information systems. NEC was envisaged as the coherent integration of sensors, decision-makers, effectors and support capabilities to achieve a more flexible and responsive military. This is intended to make commanders better aware of the evolving military situation and better able to react to events through communications.

Network-Enabled Capability is further related to the US concept of network-centric warfare (NCW), which at the time was described as "translating an information advantage into a decisive war fighting advantage". This was later renamed "*network-centric operations*" (NCO), to encompass activities such as peacekeeping. Out of these efforts came the initiative for the Technology for Information, Decision and Execution Superiority (TIDE) Sprint.

The basic ideas of NEC/NCW, following the NATO NEC (NNEC) and European EDA NEC (EUNEC) and the widely available specifications and standards from the TIDE Sprint build the sole technical foundation for the Maritime Surveillance (MARSUR) Networking, here the software and documentation under the MARSUR Release, familiar with the technology used in the Sea Surveillance Cooperation Baltic Sea (SUCBAS) and many international and national projects and developments of the nation's participating in the TIDE Sprints.

National MARSUR Capability Implementation is using the MARSUR Release as the common interface software to connect to a national implementation or a graphical viewer; the MARSUR User Interface. The implementation of the MARSUR Release to the National Maritime Surveillance System (NMSS) and/or the MARSUR User Interface (MARSUR-UI) is a prerequisite for a national MARSUR Exchange Node (National MEXS-Node) to achieve an Initial Operational Capability (IOC). As an alternative the NMSS can be adapted to the standards and specification used in MARSUR.

The MARSUR Release provides a common standardized interface for a development of a national node, the national implementation forming a national MARSUR Exchange System (MEXS) and a graphical user interface.

The MARSUR Capability in the releases 1.0/2.0 up to 3.x are the steadily enhanced software solutions for technical implementation including the related documentation for:

- MARSUR Exchange System (MEXS)
- MARSUR Exchange System Mobile (MEXS Mobile)
- MARSUR User Interface (MARSUR-UI)
- MARSUR Test Tool (MTT)
- MARSUR Management Tool (MMT)
- MARSUR DEVELOPMENT Documentation

The required changes on the Intellectual Property Rights (IPR) later brought a renaming into:

- MARSUR Exchange System 2.0 in new version MEXS 3.x
- MARSUR Exchange System Mobile (MEXS Mobile) in new version MEXS Mobile 2.x
- MARSUR Test Tool (MTT) as new version the MARSUR Client Test Utility (MCTU).
- MARSUR Management Tool in new version the MARSUR Server Management Utility (MSMU)
- MARSUR User Interface (MARSUR-UI) as unsolved IPR not renamed.

A fully functional national MEXS-Node is an implemented National Maritime Surveillance System (NMSS) enabling the automated information exchange with all nine core services via the MARSUR Virtual private Network (MARSUR VPN) using secure E-Mail, Chat, Notifications, Collaboration Tools and the additional available information in each NMSS.

All MARSUR Networking Core Services can only be tested within the MARSUR Networking VPN using the National MEXS-Nodes including the implementation of the National Maritime Surveillance System (NMSS) and/or a MARSUR User Interface (MARSUR-UI). The MARSUR-UI can also act as emergency or backup solution in case NMSS failure.

Only the development of the MARSUR Release into a fully functional national MEXS-Node is enabling a MARSUR MS information exchange using all nine core services on the MARSUR ADVANCED LEVEL. The fully functional national MEXS-Node is a sole national responsibility.

Any national MEXS-Node can add the installation of a MARSUR User Interface (MARSUR-UI) providing manual track exchange, secure E-Mail, Chat, Notifications, Collaboration Tools. However, without the available services of the NMSS any MARSUR MS can provide a limited contribution to the MARSUR Networking. The MARSUR-UI and a MARSUR D2 Release form a partial national MEXS-Node with only limited functionality.

The interface protocols used in MARSUR Development derive to a large extend from the TIDE Sprint, are also used in SUCBAS and various national projects/systems. They will be part of the NATO Project TRITON and have been submitted in CISE (EUCISE2020) creating therefore a family of interfacing standards.

The concept in MARSUR, SUCBAS as well as in CISE is the implementation of *"Smart Agents"* running inside the national systems. Information provided to MARSUR Networking via services is therefore nationally pre-processed, correlated/fused and evaluated. Each partner decides for what needs and will be shared, but no pressure or force due to obligation exists.

Each nation is responsible for its national maritime picture/RMP. Missing or searched information can be send or retrieved via Queries on the connected other systems, or using E-Mails, Chats, Notifications, etc., all linked via a national MEXS-Node. The distribution of an entire national picture is no longer required or only activated in specific operational situations, reducing the workload (bandwidth requirement).

In the case a single or several nations have no national situational picture/RMP connected to a national MEXS-Node the MARSUR Exchange System (MEXS) can be used to share a track-streaming service for the other nations as consumers. However, at least one nation does have to provide a track-streaming service from its NMSS implementation in the MARSUR VPN for other national MEXS-Nodes to be consumed or military units with Mobile MEXS-Nodes connected.

Technical specifications and standards in MARSUR and other international and regional communities benefit in projects and technical realization from the knowledge exchange and the work conducted in the TIDE Sprint.

In order to enter an operational status, the MARSUR Capability needs to be integrated with the National Maritime Surveillance Systems (NMSS) in the Member States. See Development of MARSUR Networking Core Services.

The challenge is to harmonize two independent project cycles:

1. The International Project Cycle under the MARSUR Project Arrangement
2. The National Implementation Project Cycle

The national implementation project cycle involves national procurement agencies and the operational and technical military organizations while the ministerial representatives mainly handle the MARSUR Project Arrangement.

Standard Implementation Cycle of a project for Centralized Services Implementation is the realization of an application, or execution of a plan, idea, model, design, specification, standard, algorithm, and policy, in a project. The above relative simple standard implementation cycle does unfortunately not apply fully to the MARSUR Networking, it refers more to the process of the national integration of the national systems/services towards the MARSUR Release forming a fully functional MARSUR Exchange System (MEXS). After a new MARSUR Release development cycle above the same applies for the national interface, the national implementation.

Implementation of the MARSUR Capability refers to the process of the national integration of the national systems/services towards a MARSUR Release forming a fully functional MARSUR Exchange System (MEXS).

In the IT Industry, implementation refers to post-sales process of guiding a client from purchase to use of the software or hardware that was purchased. MARSUR Capability Implementation refers to the process of guiding the MARSUR Member States national Maritime Operations Centers (MOC) under the national task of the use of the software (MARSUR D2 Release) and supports the national adaption of the national systems/services towards this software with national projects.

Just as the MARSUR DEVELOPMENT project included requirements analysis, scope analysis, customizations, systems integrations, user policies, user training and delivery, the same steps need to follow on the national implementation process which adds a second time line for and overall declaration of operational IOC.

The national MARSUR Capability Implementation is the core to reach any operational stage in MARSUR Networking. It involves any international EDA project for MARSUR Networking under a Project Arrangement (PA) and will be finished with the at MARSUR Member States participating being electronically connected with the National Maritime Surveillance System (NMSS).

The MARSUR Release is the common interface software and documentation needed to connect to a national implementation or the use with a graphical viewer (MARSUR User Interface). As an alternative, the NMSS can be adapted to the standards and specification used in MARSUR.

The implementation of the MARSUR Release to the National Maritime Surveillance System (NMSS) and/or the MARSUR User Interface (MARSUR-UI) is a prerequisite for a national MARSUR Exchange Node (National MEXS-Node) to achieve FOC.

As one of very few federated/decentralized operational networks, the connections are point-to-point, this means relative from and to each national MEXS-Node versus the view from the other MEXS-Node implementations towards the respective node (send/receive) and each corresponding service.

The MARSUR Networking Tests provide an overview of the functionalities available on the MARSUR Networking Virtual Private Network (MARSUR Networking VPN) by all MARSUR Member States in the MARSUR ADVANCED LEVEL on the nine MARSUR Networking Core Services:

1) Electronic Mail (E-Mail)
2) Real-time transmission of text messages (Chat)
3) Voice-over-Internet Protocol (VoIP)
4) File Transfer
5) Notifications
6) Video
7) Track Streaming
8) Tactical Drawing
9) Query

Software implementations involve several professionals that are relatively new to the knowledge that is based on business analysts, technical analysts, solutions architects, and project managers and developers.

To implement a system successfully, a large number of inter-related tasks need to be carried out in an appropriate sequence. Utilizing a well-proven implementation methodology and enlisting professional advice can help but often it is the number of tasks, poor planning and inadequate resourcing that causes problems with an implementation project, rather than any of the tasks being particularly difficult. Similarly, with the cultural issues it is often the lack of adequate consultation and two-way communication that inhibits achievement of the desired results.

Since the federated concept approach has so far only been set into real world environments very few projects i.e. in the Sea Surveillance Cooperation Baltic Sea (SUCBAS) and the Maritime Surveillance Networking (MARSUR Networking) there are no standardized procedures for implementing the comprehensive Overarching MARSUR Architecture.

The nine MEXS Core Services Track Exchange, Notification, File Transfer, Query, Tactical Drawing, Video, Chat, E-Mail and VOIP run on virtual servers on the Operating System CentOS. The MEXS Core Services are standardized towards all other MEXS-Nodes connected, towards the national site the countries have to implement the services bi¬ directional to the National Maritime Surveillance System (NMSS).

As interim solution until the NMSS services are up and running via a national Implementation the MARSUR User Interface (MARSUR-UI) provides the operator with a possibility of viewing information from services of already connected NMSS, Collaboration Tools like E-Mail, Chat, Notifications, File Exchange and also with a limited capability of track interaction without a own NMSS in place.

The MARSUR Exchange System (MEXS) and the newly designed MARSUR User Interface started operations in the MARSUR LIVE PHASE in 2014 and the MARSUR Capability was handed over under the Common Security and Defence Policy (CSDP) to the embodied in the Leadership of the European Union Military Staff (EUMS).

The participants of the MARSUR LIVE PHASE have their systems active and regularly establish communications, exchanging virtual data to test the system (ADVANCED LEVEL) or exchange information in the BASIC LEVEL (regular phone, fax, e-mail), which also is the fallback solution in case of system-failures.

Technical - and in consequence financial - difficulties to achieve integration with the national systems vary considerably among the participant nations. Some nations that belong to the SUCBAS community and Spain have certain commonalities between their national systems and MARSUR Exchange System due to the similar standards and specifications used.

In summary, an MARSUR IOC status still serves a low operational objective, until the National Maritime Surveillance Systems (NMSS) are connected to the national MEXS-Nodes.

However, the timeline of each national implementation varies to a great amount in the Member States.

An important factor is the established knowledge about the technology used federated networking which is established with the national technical and operational Subject Matter Experts (SME) participation in the MARSUR Technical Working Group (MTWG), the MARSUR Operational Working Group (MOWG), and especially by the SME of administrators, system architects, developers and other SMEs from the national projects initiated to provide the national implementation to the MARSUR Capability.

These SME act as main force-multiplier in the overall development in the MARSUR Networking, any MARSUR project development, and the independent national project process. This human factor cannot be valued high enough in all international projects, any cooperation or organization. For example the second Project Arrangement (MARSUR II) was signed on 11. October 2012 with the following kick-off on 12. April 2013. The overall project duration was 759 calendar days, with 235 weekends and holidays, the total development time resulted in only 524 working days.

In Phase 1 the delivery of the MARSUR D2 Release 2 software and documentation and the following software demonstration was postponed by only one week, however the MTWG Subject Matter Experts (SME) had only 40 working days in preparation for the main software verification and validation. The last software release from contractor was delivered three weeks prior to the Verification Event, which left limited time for installation and prior testing. The national technicians had to concentrate therefore their efforts on the MEXS 2.0 and the MARSUR User Interface.

The ever first MARSUR Maintenance & Development Workshop (MDWS) ever by contractor took place on the 11.-12. September 2013. The training was to provide information on setup, installation and configuration of the MARSUR Release products. The workshop conducted according to project plan, unfortunately 21 days prior to the delivery of the MARSUR D2 Release.

One Lesson Learned is to leave a project planning up to the proposals of the contractors. Project Arrangement (PA) should contain the project start and end date, possibly the interim milestones with their content. To impress a pre-established a detailed project plan on the tender proposals for contractors reduces the flexibility of the contractor as well as of the participating Member States for own adaptive planning and adjustments within the framework of Mile Stones and Project Plan.

The main Lessons Learned from the decentralized approaches in the Sea Surveillance Cooperation Baltic Sea (SUCBAS), the Maritime Surveillance Networking (MARSUR Networking), or the Common Information Sharing Environment (CISE), is the benefit of information exchange on the knowledge building from the participation of identical national SMEs.

The productivity of the national MTWG POCs and their highly skilled SME (administrator, technicians, consultants), who submitted their knowledge into the project, paved the way to the completion of the project. These experts build the basis of all national expertise for systems integration over the years to follow after any MARSUR project has long ended.

Mediterranean Coast Guard Functions Forum (MedCGFF)

The purpose of the Mediterranean Coast Guard Functions Forum (MedCGFF) is the exchange of ideas about Maritime Safety and Security issues in the Mediterranean Sea in order to improve international cooperation and facilitating an appropriate networking.

The Mediterranean Coast Guard Functions Forum (MedCGFF) is a non-binding, voluntary, independent and non-political forum bringing together representatives from institutions and agencies with related competencies in coast guard functions in the Mediterranean to facilitate multilateral cooperation on a wide range of issues such as maritime safety, security and environmental protection activities as well as the potential partnership for their application, seeking solutions to common problems and issues confronting participating countries by sharing expertise and best practices in a cooperative and consensual manner.

Members are Albania, Algeria, Bosnia Herzegovina, Croatia, Cyprus, Egypt, France, Greece, Israel, Italy, Lebanon, Libya, Malta, Monaco, Montenegro, Morocco, Palestine, Slovenia, Spain, Tunisia, Turkey and from the European Union DG MARE. Observers are Belgium, Bulgaria, Canada, Denmark, Estonia, Finland, Georgia, Germany, Iceland, Ireland, Latvia, Lithuania, Netherlands, Norway, Poland, Portugal, Romania, Russia, Sweden, Ukraine, United Kingdom of Great Britain and Northern Ireland, and the United States of America.

General objectives of the MedCGFF are to promote a Forum for the adequate level of discussion, exchanging information on juridical, technical and operational experiences within the framework of current law, and in close relationship with existing international and regional organizations specialized in maritime matters that enable to enhance maritime safety, security and environmental protection activities as well as partnership that allow to find solutions to avoid regional risks and threats in the Mediterranean Basin.

MedCGFF intends to promote international efforts that enhance the safety and security of the maritime commons while preserving freedom of the seas for legitimate purposes. Success cannot be achieved by any one country acting unilaterally, but requires a partnership of nations willing to maintain a strong, united international front. The conferences are attended by the representatives of administrations of participating countries and by the representatives of the European Commission, EMSA, REMPEC, DG MARE, EFCA, FRONTEX, a.o..

The conference adopt concrete suggestions and conclusions for the purpose of future regional cooperation and exchange of data among Mediterranean countries and organization of necessary trainings with the support of European Union.

The EU funded the SAFEMED IV project supported the organization and the hosting of the 1. Mediterranean Coast Guard Functions Forum Secretariat Meeting of 2018. The 2. Maritime Coast Guard Functions Forum Secretariat Meeting 2018 was hosted by EMSA with the participation of the Secretariat Members of France and Morocco (Co-Chairs 2018-2019), Turkey (Chair 2017) and the EMSA and FRONTEX representatives as observers. The main objective of the meeting was to work on the proposed draft agenda for the Plenary Meeting in Marseille in June 2018. A structure comprising of an opening session and 3 additional sessions (Policy, Maritime Surveillance and Environment) was agreed. In addition, the possibility of organizing a round table with the three Agencies (EMSA, EFCA and FRONTEX) was discussed and the forum SAFEMED IV renewed its commitment to support the participation of ENP SOUTH Neighbors countries' participants to the plenary conference.

The conferences consist of several panels with following topics:

- Data exchange building,
- Maritime safety and prevention of marine pollution,
- Monitoring and inspection of fisheries,
- Consideration of the situation related to immigration crisis and smuggling
- Organization of regional exercises and cooperation of institutions with coast guard responsibilities.

Specific objectives of the MedCGFF are:

- To enhance cooperation and information sharing among member states and third countries in their effort to detect, monitor, deter and intercept transnational maritime threats to the global safety, security, economy and environment.
- To develop a more effective cooperation among member states and third countries of the Mediterranean Sea in matters related to maritime safety, security and environmental protection.
- To create inter-institutional connections oriented to facilitate stakeholders in their contribution for the best use of all different capabilities at national level.
- To contribute to develop cooperative disaster response plans to assist member states and third countries in preventing natural and man-made disasters, mitigation and recovery.
- To contribute to promote common education and training standard.

North African Port Management Association (NAPMA)

The North African Port Management Association (NAPMA) includes Mauritania, Morocco, Algeria, Tunisia, Libya, Egypt and Sudan and was created by the Economic Commission for Africa of the United Nations in 1974 at the conclusion of the Transport Minister Conference held in Alexandria from 17. to 21. of June 1974. Headquarter is in Casablanca, Morocco.

NAPMA is also member of the Pan African Association for Port Co-operation, (PAPC), the International Association of Ports and Harbors (IAPH) and have close relationship to the Port Management Association for West and Central Africa (PMAWCA), Port Management Association for East and South Africa (PMAESA) and also to the Arab Sea Port Federation (ASPF). NAPMA has also working relations to the African Section of CEDA International and recently to the Canaries Islands/Africa Section of ICHCA International Limited.

The North African Port Management Association (NAPMA) aims to improve, coordinate and standardize port operations, materials and services of members with a view to increase efficiency of their harbors in relation to ship movement and other forms of transport.

NAPMA has relations with 4 non-governmental organizations and 4 inter-governmental organizations and is also referred to as Port Management Association of North Africa (PMANA), the association des administrations portuaires de l'Afrique du Nord. Statutes adopted 17-21 June 1974, Alexandria (Egypt); amended 18-22 Nov 1974, Tunis (Tunisia), 15-16 Dec 1993, Casablanca (Morocco).

North Atlantic Coast Guard Forum (NACGF)

The objective of the North Atlantic Coast Guard Forum (NACGF)[145], which was formed in 2007, is to increase cooperation amongst member countries on matters related to maritime safety and security in the region.

At the inaugural meeting it was agreed that the NACGF would be non-binding, voluntary, neither policy, nor regulatory-oriented and would operate within existing legal frameworks. It was agreed that this forum would not duplicate work undertaken in other international fora. The following are member countries of the NACGF: Belgium, Iceland, Portugal, Canada, Ireland, Russia, Denmark, Latvia, Spain, Estonia, Lithuania, Sweden, Finland, The Netherlands, United Kingdom and Northern Ireland, France, Norway, United States, Germany, Poland.

The annual forum provides an opportunity for Coast Guard representatives from twenty countries to discuss mutual priorities related to maritime safety and security, two further annual meetings are held by:

- Subject matter experts (March);
- Coast Guard leaders (September)

The Coast Guards are the national point of contact for the NACGF. Chairmanship rotates per year among the participating countries and an annual theme is decided upon. There are seven working groups covering different themes (fishery, migration, SAR, etc. ...).

---

[145] http://www.ccg-gcc.gc.ca/NACGF

North Atlantic Treaty Organization (NATO)

The North Atlantic Treaty Organization (NATO)[146], also called the North Atlantic Alliance, is an intergovernmental military alliance based on the North Atlantic Treaty, which was signed on 4 April 1949. The organization constitutes a system of collective defense whereby its member states agree to mutual defense in response to an attack by any external party.

Like any alliance, NATO is ultimately governed by its 28-member states. However, the North Atlantic Treaty and other agreements outline how decisions are to be made within NATO. Each of the 28 members sends a delegation or mission to NATO's headquarters in Brussels, Belgium. The senior permanent member of each delegation is known as the Permanent Representative and is generally a senior civil servant or an experienced ambassador (and holding that diplomatic rank). Several countries have diplomatic missions to NATO through embassies in Belgium.

Together, the Permanent Members form the North Atlantic Council (NAC), a body which meets together at least once a week and has effective governance authority and powers of decision in NATO. From time to time the Council also meets at higher level meetings involving foreign ministers, defense ministers or heads of state or government (HOSG) and it is at these meetings that major decisions regarding NATO's policies are generally taken. However, it is worth noting that the Council has the same authority and powers of decision-making and its decisions have the same status and validity, at whatever level it meets. France, Germany, Italy, the United Kingdom and the United States are together referred to as the Quint, which is an informal discussion group within NATO. NATO summits also form a further venue for decisions on complex issues, such as enlargement.

The meetings of the North Atlantic Council are chaired by the Secretary General of NATO and, when decisions have to be made; action is agreed upon on the basis of unanimity and common accord. There is no voting or decision by majority. Each nation represented at the Council table or on any of its subordinate committees retains complete sovereignty and responsibility for its own decisions.

NATO's headquarters are located in Haren, Brussels, Belgium, where the Supreme Allied Commander also resides. Belgium is one of the 28 member states across North America and Europe, the newest of which, Albania and Croatia, joined in April 2009. An additional 22 countries participate in NATO's Partnership for Peace program, with 15 other countries involved in institutionalized dialogue programs. The combined military spending of all NATO members constitutes over 70 percent of the global total. Members' defense spending is supposed to amount to 2 percent of GDP.

NATO was little more than a political association until the Korean War galvanized the organization's member states, and an integrated military structure was built up under the

---

[146] http://www.nato.int

direction of two US supreme commanders. The course of the Cold War led to a rivalry with nations of the Warsaw Pact, which formed in 1955. Doubts over the strength of the relationship between the European states and the United States ebbed and flowed, along with doubts over the credibility of the NATO defense against a prospective Soviet invasion—doubts that led to the development of the independent French nuclear deterrent and the withdrawal of France from NATO's military structure in 1966 for 30 years. After the fall of the Berlin Wall in 1989, the organization was drawn into the breakup of Yugoslavia, and conducted its first military interventions in Bosnia from 1992 to 1995 and later Yugoslavia in 1999. Politically, the organization sought better relations with former Warsaw Pact countries, several of which joined the alliance in 1999 and 2004.

Article 5 of the North Atlantic treaty, requiring member states to come to the aid of any member state subject to an armed attack, was invoked for the first and only time after the 11 September 2001 attacks, after which troops were deployed to Afghanistan under the NATO-led ISAF. The organization has operated a range of additional roles since then, including sending trainers to Iraq, assisting in counter-piracy operations and in 2011 enforcing a no-fly zone over Libya in accordance with U.N. Security Council Resolution 1973. The less potent Article 4, which merely invokes consultation among NATO members, has been invoked five times: by Turkey in 2003 over the Iraq War; twice in 2012 by Turkey over the Syrian Civil War, after the downing of an unarmed Turkish F-4 reconnaissance jet, and after a mortar was fired at Turkey from Syria; in 2014 by Poland, following the Russian intervention in Crimea; and again by Turkey in 2015 after threats by the Islamic State to its territorial integrity.

NATO has twenty-eight members, mainly in Europe and North America. Some of these countries also have territory on multiple continents, which can be covered only as far south as the Tropic of Cancer in the Atlantic Ocean, which defines NATO's "area of responsibility" under Article 6 of the North Atlantic Treaty. During the original treaty negotiations, the United States insisted that colonies like the Belgian Congo be excluded from the treaty. French Algeria was however covered until their independence on 3 July 1962. Twelve of these twenty-eight are original members who joined in 1949, while the other sixteen joined in one of seven enlargement rounds. Few members spend more than two percent of their gross domestic product on defense, with the United States accounting for three quarters of NATO defense spending.

From the mid-1960s to the mid-1990s, France pursued a military strategy of independence from NATO under a policy dubbed "Gaullo-Mitterrandism". Nicolas Sarkozy negotiated the return of France to the integrated military command and the defense Planning Committee in 2009, the latter being disbanded the following year. France remains the only NATO member outside the Nuclear Planning Group and unlike the United States and the United Kingdom, will not commit its nuclear-armed submarines to the alliance.

The Partnership for Peace (PfP) program was established in 1994 and is based on individual bilateral relations between each partner country and NATO: each country may choose the extent of its participation. Members include all current and former members of the

Commonwealth of Independent States. The Euro-Atlantic Partnership Council (EAPC) was first established on 29 May 1997, and is a forum for regular coordination, consultation and dialogue between all fifty participants. The PfP program is considered the operational wing of the Euro-Atlantic Partnership. Other third countries also have been contacted for participation in some activities of the PfP framework such as Afghanistan.

The European Union (EU) signed a comprehensive package of arrangements with NATO under the Berlin Plus agreement on 16 December 2002. With this agreement the EU was given the possibility to use NATO assets in case it wanted to act independently in an international crisis, on the condition that NATO itself did not want to act—the so-called "*right of first refusal*". It provides a "*double framework*" for the EU countries that are also linked with the PfP program.

Additionally, NATO cooperates and discusses their activities with numerous other non-NATO members. The Mediterranean Dialogue was established in 1994 to coordinate in a similar way with Israel and countries in North Africa. The Istanbul Cooperation Initiative was announced in 2004 as a dialog forum for the Middle East along the same lines as the Mediterranean Dialogue. The four participants are also linked through the Gulf Cooperation Council.

Political dialogue with Japan began in 1990, and since then, the Alliance has gradually increased its contact with countries that do not form part of any of these cooperation initiatives. In 1998, NATO established a set of general guidelines that do not allow for a formal institutionalization of relations, but reflect the Allies' desire to increase cooperation. Following extensive debate, the term "*Contact Countries*" was agreed by the Allies in 2000.

By 2012, the Alliance had broadened this group, which meets to discuss issues such as counter-piracy and technology exchange, under the names "partners across the globe" or "*global partners*". Australia and New Zealand, both contact countries, are also members of the AUSCANNZUKUS strategic alliance, and similar regional or bilateral agreements between contact countries and NATO members also aid cooperation. In June 2013, Colombia and NATO signed an Agreement on the Security of Information to explore future cooperation and consultation in areas of common interest; Colombia became the first and only Latin American country to cooperate with NATO.

NATO's military operations are directed by the Chairman of the NATO Military Committee, and split into two Strategic Commands (Allied Command Transformation (ACT) and Allied Command Operations (ACO)) which are commanded by a senior US officer and a senior French officer assisted by a staff drawn from across NATO. The Strategic Commanders are responsible to the Military Committee for the overall direction and conduct of all Alliance military matters within their areas of command.

Each country's delegation includes a Military Representative, a senior officer from each country's armed forces, supported by the International Military Staff. Together the Military Representatives represent the Military Committee, a body responsible for recommending to NATO's political authorities those measures considered necessary for the common defense of the NATO area. Its principal role is to provide direction and advice on military policy and strategy. It provides guidance on military matters to the NATO Strategic Commanders, whose representatives attend its meetings, and is responsible for the overall conduct of the military affairs of the Alliance under the authority of the Council.

Like the Council, from time to time the Military Committee also meets at a higher level, namely at the level of Chiefs of Defense, the most senior military officer in each nation's armed forces. Until 2008 the Military Committee excluded France, due to that country's 1966 decision to remove itself from NATO's integrated military structure, which it rejoined in 1995. Until France rejoined NATO, it was not represented on the defense Planning Committee, and this led to conflicts between it and NATO members. Such was the case in the lead up to Operation Iraqi Freedom. The operational work of the Committee is supported by the International Military Staff.

The NATO Command Structure evolved throughout the Cold War and its aftermath. An integrated military structure for NATO was first established in 1950 as it became clear that NATO would need to enhance its defenses for the longer term against a potential Soviet attack. In April 1951, Allied Command Europe and its headquarters (SHAPE) were established; later, four subordinate headquarters were added in Northern and Central Europe, the Southern Region, and the Mediterranean.

From the 1950s to 2003, the Strategic Commanders were the Supreme Allied Commander Europe (SACEUR) and the Supreme Allied Commander Atlantic (SACLANT). The current arrangement is to separate responsibility between Allied Command Transformation (ACT), responsible for transformation and training of NATO forces, and Allied Command Operations (ACO), responsible for NATO operations worldwide.[149] Starting in late 2003 NATO has restructured how it commands and deploys its troops by creating several NATO Rapid Deployable Corps, including the EUROCORPS, the 1. German/Dutch Corps, Multinational Corps Northeast, and NATO Rapid Deployable Italian Corps among others, as well as naval High Readiness Forces (HRFs), which all report to Allied Command Operations.

In early 2015, in the wake of the War in Donbass, meetings of NATO ministers decided that Multinational Corps Northeast would be augmented so as to develop greater capabilities, if thought necessary to prepare to defend the Baltic States, and that a new Multi-National Division Southeast would be established in Romania. Six NATO Force Integration Units would also be established to coordinate preparations for defense of new Eastern members of NATO.

The NATO Maritime Situational Awareness MSA) Program states MSA as an enabling capability which seeks to deliver the required Information Superiority in the maritime environment to achieve a common understanding of the maritime situation in order to increase effectiveness in the planning and conduct of operations.

The Military Committee (MC) endorsed the NATO MSA Concept on 14 January 2008. Council agreed that the MSA Concept should be pursued further by the NATO Military Authorities (NMAs) by establishing a clear vision of the required capability and developing a comprehensive, integrated implementation plan. The MC divided the development of MSA into three stages:

- Precursor Phase - leading to provision of the MSA Concept Development Plan
- Phase I - Development and Validation
- Phase II - Implementation of NATO MSA

Maritime Situational Awareness (MSA) evolved in the end of 2006 from NATO Maritime Domain Awareness (MDA), previously US concept, following the NATO RIGA Summit 2006. As a result of the event, HQ SACT was tasked by International Military Staff (IMS) to develop MSA Definition & Scope; Allied Command Transformation (ACO) in turn gained responsibility for capturing Information Requirements for MSA.

HQ SACT was officially tasked in 2007 to develop the BI-SC Draft NATO MSA Concept after approval of the MSA working Definition & Scope. In 2008 the MC approved the NATO MSA Concept (NATO Restricted). The North Atlantic Council (NAC) agreed that the MSA Concept should be pursued further by the NATO Military Authorities (NMAs) by establishing a clear vision of the required capability and developing a comprehensive, integrated implementation plan.

The MSA Concept Development Plan was provided by ACT in 2008 as a result of combined efforts of ACT, Allied Command Operations (ACO) MCC Northwood (today MARCOM), MCC Naples with ACT in the lead. The plan carried out in several phases was based on the analysis in the areas Doctrine, Legal aspects, Material & Technologies, Organization, Facilities, Training and Interoperability.

The MC approved the Maritime Situational Awareness (MSA) Concept Development Plan (CD Plan) in 2008 and authorized SCs, with ACT in the lead, to implement it. The MC approval and the following NAC notation of MSA CD Plan provided military validation for release of resources for the envisaged work in MSA Phase-I (Development and Validation) in 2008.

The NATO MSA Concept is closely related with NATO MSO (Maritime Security Operations) and with NATO MSO paper that identifies "MSA as one of the means by which the ends of MSO can be achieved and hence maritime security maintained". Most of the nations have different Concepts and Plans to achieve MDA/MSA, i.e. in the National Maritime Domain Awareness (MDA) Architecture Plan of the United States of America representing the federal design for information sharing and safeguarding.

Under the NATO MSA Program and MSA Concept were various areas addressed in general, as technical specifications, NATO products. The NATO MSA Concept Development Plan calls for seven inter-dependent studies (doctrine, legal, organization, training, material & technology, facilities and interoperability) to be performed in parallel, with the details of the material & technology study as well as the relationships and dependencies on other studies:

- MSA Material & Technology Study
- MSA Study Deliverables
- MSA Doctrine and Organization Study
- MSA Legal Study
- MSA Interoperability
- MSA Training
- NATO MSA Experimentation

The Maritime Situational Awareness (MSA) Material & Technology activities are part of NATO's larger MSA program. The MSA technology work started in 2006 to better support Maritime Component Command (MCC) Naples in the execution of Operation Active Endeavour (OAE). After the approval of the NATO MSA Concept and NATO MSA Concept Development Plan, the technology activities were embedded in the larger NATO MSA program. MSA study deliverables concentrate on the more technical aspects.

The aim of the legal study is to inform the Maritime Component Commands (MCC) of any legal or policy-related constraints impacting their ability to obtain the information needed for baseline MSA. The study is to incorporate the MCC-determined data set into the tabletop experiment and test the extent to which legal considerations affect operators' ability to obtain information about vessels and include a scenario that will allow the Legal Study team to test data received from each NATO country. The desired scenario would be one that traces a ship through the Baltic Sea, North Sea, North Atlantic Ocean, and Mediterranean Sea.

The Legal study was to demonstrate to the Military Committee the current information sharing practices within the alliance for MSA data and to contact individuals/operating centers determined by the nations to be their national MSA point of contact and request information. It needed to identify areas in which information sharing currently works well and areas in which it could be improved or enhanced through the use of legal agreements.

Maritime Situational Awareness Interoperability Study to demonstrate use of interoperability with external sources of MSA data covers references:

- Appendix A - MSA Interoperability Scoping Document
- Appendix B - MSA Interoperability Profile
- Appendix C - MSA Enterprise Architecture

The Allied Maritime Strategy (AMS) is the overarching strategy for Maritime Situational Awareness and Maritime Security in NATO. AMS sets out, in full consistency with the Strategic Concept, the ways that maritime power could help resolve critical challenges facing the Alliance now and in the future, and the roles - enduring and new - that NATO forces may have to carry out in the maritime environment in order to contribute to the Alliance's defense and security and to promote its values.

These roles capitalize upon the ability of maritime forces to provide a spectrum of strategic options to the Alliance, and include appropriate contributions to: Deterrence and collective defense, Crisis management, Cooperative security, and Maritime Security.

History shows that NATO has always been in the business of transformation and the end of the Cold War and the conflict in the Balkans are recent illustrations of this fact. During the period from 1989 to 1991 there were many momentous events that occurred in Europe confirming that the Cold War was coming to an end. In November of 1990. At the end of the Cold War, the world's attention was shifted to the conflict in the Balkans. This conflict clearly highlighted the need for NATO to be able to adapt more quickly and effectively in order to tackle new post-Cold War challenges. At the Ministerial Meeting of the North Atlantic Council in June 1996, it was stated that *"we are determined to … develop further our ability to carry out new roles and mission relating to conflict prevention and crisis management and the Alliance's efforts against the proliferation of weapons of mass destruction and their means of delivery, while maintaining our capability for collective defense."*

Entering the 21st Century, the threats to NATO are not what they once were and the Alliance continues to transform in order to respond to the new threat environment. NATO Secretary General Lord Robertson said, "*NATO must change radically if it is to be effective*" and "*Modernize or be marginalized*". At the Prague Summit in November 2002, the Heads of State and Government from the member countries of the North Atlantic Alliance resolved to approve the Prague Capabilities Commitment (PCC) as part of the continuing Alliance effort to improve and develop new military capabilities for modern warfare in a high threat environment; to implement all aspects of the PCC as quickly as possible; to improve capabilities in the identified areas of continuing capability shortfalls; to create a NATO Response Force (NRF) consisting of a technologically advanced, flexible, deployable, interoperable and sustainable force to more effectively respond to the new threats of the 21st century and also to provide a means for focusing and promoting improvements in the Alliance's military capabilities.

On June 19th 2003, the Allied Command Transformation (ACT), as NATO's forcing agent for change, assumed the lead in the effort to build through innovation and collaboration the "*right capabilities*" in support of the NRF. The importance of the NRF was reinforced at the Ministry of Defense Seminar at Colorado Springs in October 2003 when the US Secretary of Defense Donald Rumsfeld stated that the seminar "*…highlighted the need for that Response Force to have capabilities that are agile, swift and lethal, so that this wonderful alliance of ours can respond quickly and effectively to rapidly unfolding crises*".

In today's complex heterogeneous world, military operations cover a wide variety of missions making it difficult to pre-design and pre-integrate Information Technology (IT) solutions. In the past, the Alliance and its member nations have developed numerous capabilities supporting particular functional services, for example maritime, air, ground, logistics and intelligence. Most of these capabilities are inherently powerful but have consistently failed to integrate horizontally between functional services to provide dominant and effectual combined and joint capabilities. Flexible and innovative IT capabilities must in the future support the full dimension of horizontal and vertical integration of functional services to counter any potential threat in ever-changing environments.

There are multiple communication networks and system used by NATO and its members to support its exercises and operations. The NATO Tidepedia list alone 60 portals on systems, networks and development activities which all exceed the possibilities to be described in detail.

The NATO Standardization Agency (NSA) mission is to initiate, coordinate, support and administer standardization activities conducted under the authority of the Committee for Standardization (CS). The NSA is also the Military Committees lead agent for the development, coordination and assessment of operational standardization.

The NATO Communications and Information Agency (NCI Agency) was established on 1 July 2012 as a result of the merger of the NATO Consultation, Command and Control Agency (NC3A), the NATO ACCS Management Agency (NACMA), the NATO Communications and Information System Services Agency (NCSA), the ALTBMD Program and elements of NATO HQ. The establishment of the Agency is part of a broader NATO reform. The new NCI Agency "connects forces, NATO and Nations" - it is NATO's IT and C4ISR provider, including cyber and missile defense. Some of the most important networks and systems:

- Battlefield Information Collection and Exploitation Systems (BICES)
- Crisis Response Operations in NATO Operating Systems (CRONOS), which is a system of interconnected computer networks used by NATO to transmit classified information at the level of NATO Secret.
- Combined Federated Battle Laboratories Network (CFBLNet), which is a wide area network connecting the US, the UK, Canada, Australia, New Zealand, six NATO countries and Sweden for sharing research and development information.

Active Endeavour

Active Endeavour or Operation Active Endeavour (OAE) is NATO's only Article 5 operation on anti-terrorism. It aims to demonstrate NATO's solidarity and resolve in the fight against terrorism and to help deter and disrupt terrorist activity in the Mediterranean. Under Operation Active Endeavour, NATO ships are patrolling the Mediterranean and monitoring shipping to help deter, defend, disrupt and protect against terrorist activity. The operation evolved out of NATO's immediate response to the terrorist attacks against the United States of 11. September 2001 and, in view of its success, is being continued.

Allied Command Operations (ACO)

Allied Command Operations (ACO) is one of the two strategic commands of the North Atlantic Treaty Organization (NATO), the other being Allied Command Transformation (ACT). The headquarters and commander of ACO is Supreme Headquarters Allied Powers Europe (SHAPE) and Supreme Allied Commander Europe (SACEUR), respectively.

Under ACO, there are two joint force operational headquarters and several single service commands:

- Allied Joint Force Command Brunssum (JFCBS), Netherlands
- Allied Joint Force Command Naples (JFCNP), Italy

Single-service commands:

- Allied Air Command (AIRCOM) at Ramstein, Germany
- Allied Land Command (LANDCOM) at Izmir, Turkey
- Allied Maritime Command (MARCOM) at Northwood, United Kingdom

Other commands:

- Naval Striking and Support Forces NATO (Strike Force NATO, STRIKFORNATO) at Oeiras, Portugal
- NATO Communication and Information Systems Command (NCISG) at Mons, Belgium

Maritime Command (MARCOM)

Allied Maritime Command (MARCOM) is the central command of all NATO maritime forces and the Commander MARCOM is the prime maritime advisor to the Alliance. MARCOM is responsible for planning and conducting all NATO maritime operations. Currently, this includes the Alliance's Counter Piracy Operation OCEAN SHIELD, as well as its Counter-Terrorism Operation ACTIVE ENDEAVOUR in the Mediterranean. MARCOM also has to ensure it is capable at all times of contributing to potential maritime operations. This requires the highest level of readiness, of awareness of the maritime environment and it also requires the headquarters to maintain a constant dialogue with key maritime stakeholders.

The Allied Maritime Command (MARCOM) has the Maritime Situational Awareness (MSA) tasks and the operational responsibility for the security patrols in the NATO Area of Responsibility. In November, Exercise Trident Juncture 18, the largest NATO Live Exercise since the 1980s, concluded after 14 intensive days. Additionally, Operation Sea Guardian (OSG) concentrated maritime security efforts in the Eastern Mediterranean for Focused Operations (FOCOPS).

Allied Command Transformation (ATO)

The Allied Command Transformation (ACT) is a military command of the North Atlantic Treaty Organization (NATO), formed in 2003 after restructuring. It was intended to lead military transformation of alliance forces and capabilities, using new concepts such as the NATO Response Force and new doctrines in order to improve the alliance's military effectiveness.

The command's headquarters is in Norfolk, Virginia, in the United States. HQ SACT itself is organized into a command group, the Transformation Directorate, the Transformation Support Directorate, National Liaison Representatives, the Partnership for Peace Staff Element and Reservists responsible to HQ SACT.

The Transformation Directorate is headed by the Deputy Chief of Staff (DCOS) Transformation who acts as the Supreme Allied Commander, Transformation's (SACT) Director for guidance and coordination of the activities of his or hers Directorate Transformation, divided in two divisions: Implementation and Capabilities. Within the full scale of SACT's transformational responsibilities the Deputy Chief of Staff (DCOS) Transformation assists the Chief of Staff (COS) in the execution of his or her duties with emphasis on deliverables to the Alliance Military Transformation Process in order to enhance NATO's operational capabilities and to meet NATO's future requirements.

The Implementation Division, led by Assistant Chief of Staff (ACOS) Implementation, is responsible for guidance and coordination of the activities of two Sub-Divisions, Joint Education and Training (JET) and Joint Experimentation, Exercises and Assessment (JEEA), as well as providing guidance for the Joint Warfare Centre (JWC) and Joint Analysis Lessons Learned Centre (JALLC), in their efforts to enhance training programs, to path on what does this mean? breaking concept development and experimentation, to develop effective programs to capture and implement lessons learned and to press on common standards. This division probably is there some doubt? serves as NATO's link point to the annual U.S.-led Coalition Warrior Interoperability Demonstration.

The Capabilities Division, led by Assistant Chief of Staff (ACOS) Capabilities, is responsible for guidance and coordination of the activities of three Sub-Divisions: of Strategic Concepts, Policy and Interoperability (SCPI); Future Capabilities, Research and Technology (FCRT) and Defence Planning (Def Plan) in their efforts to staff Capabilities, Concepts and Development products.

Reflecting NATO as a whole, ACT has a presence on both sides of the Atlantic. Before the deactivation of United States Joint Forces Command, the two organizations were co-located, and indeed shared a commander for some time. There is an ACT command element located at SHAPE in Mons, Belgium. ACT's major subordinate commands are the Joint Warfare Centre (JWC) in Stavanger, Norway; the Joint Force Training Centre (JFTC) in Bydgoszcz, Poland; and the Joint Analysis and Lessons Learned Centre (JALLC) in Monsanto, Portugal.

Under a customer-funded arrangement, ACT invests annually about 30 million Euros into research with the NATO Communications and Information Agency (NCIA) to support scientific and experimental programs.

One of the important achievements in concept and development is the TIDEPEDIA, the "*NATO Wikipedia*", providing forums for information exchange and discussions in several environments for the NATO members, Partnership-for-Peace participants, academia and industry. TIDEPEDIA has websites with limited access for events like the TIDE Sprint, CWIX, NATO Hackathon a.o. providing a fruitful ground for future work with ACT.

Allied Worldwide Navigational Information System (AWNIS)

The <u>Allied Worldwide Navigation Information System (AWNIS)</u>[147] is used in the NATO Shipping Center (NSC) collates, coordinates and communicates navigational safety and security information to merchant shipping and military authorities within an operational area. AWNIS delivers assurance to military commanders and merchant mariners against the additional risks to safety and security of Navigation that are associated with maritime operations. It is responsible for both classified and unclassified information. Wherever the military impinges on the maritime there is a requirement for AWNIS.

The NATO Shipping Centre (NSC) is part of Allied Maritime Command and the link between NATO and the merchant shipping community. Permanently manned by NATO, the NSC is the primary point of contact for the exchange of merchant shipping information between NATO's military authorities and the international shipping community.

NATO and the Maritime Warfare School, HMS COLLINGWOOD, offer a two year course for NATO officers (both regular and reserve) of any specialization, undertaking the appointment of Staff Officer (AWNIS) or Safety of Navigation Information Coordinator (SONIC) in a NATO Command. This course is also open to civilian whose job involves an understanding of AWNIS in both crisis and non-crisis situations.

The course contains:

- Introduction to AWNIS
- AWNIS Publications
- AWNIS Concept
- Worldwide Navigation Warning System (WWNWS)
- Maritime Safety Information (MSI)
- SafetyNet
- Navtex
- Classified and Unclassified Messages
- Q-route Management and Q-Routes Design
- Totes and Logs
- OPTASK AWNIS
- Navigation Information Cycle
- Mine Danger Area Management
- Wrecks

---

[147] http://www.shipping.nato.int/Pages/NCAGS.aspx

Planning Board for Ocean Shipping (PBOS)

The NATO AC/271 (PBOS) chronological series contains the records of the Planning Board for Ocean Shipping (PBOS)[148]. The North Atlantic Council approved the establishment of the Planning Board for Ocean Shipping (PBOS) in May 1950 with its acceptance of the *"International Working Group Report on Establishment of a Planning Board for Ocean Shipping in the North Atlantic Treaty Organization"* (C-R-4/5).

The initial mission of PBOS was the planning for an emergency organization which would be able to coordinate the most efficient use of shipping resources in wartime. A small number of technical working groups were set up in 1950 to facilitate this planning work. This effort resulted in the Defense Shipping Authority (DSA), which was to have the authority to pool all allied ocean going merchant vessels for the purpose of maintaining a steady flow of food, supplies, military equipment and armed forces personnel.

The DSA was to carry out this role according to the civil and military shipping priorities set by the Alliance. After the first two years of existence, during which it met three times, PBOS met once a year for a plenary to discuss a wide range of issues related to shipping and emergency planning. PBOS membership consisted of national representatives from all Allied nations with merchant marine fleets, and the Board was supported by a permanent secretariat headquartered in London. The chair of PBOS alternated each year between an American official and a British official, corresponding to the nation hosting the plenary meeting.

In 1997 the United States agreed to take over responsibility for the secretariat, and the London office was closed on 31 March 1997. To fully explore the relevant issues and to draw on national experts, PBOS over the years set up a number of different working and study groups, including the PBOS Bunker and Tanker Committee; PBOS Coasting Committee; PBOS Working Committee on Heavy Lifts; PBOS Communications Working Party; PBOS Working Committee on Troopship Standards; PBOS Special Working Committee - SIDESTEP Sub-Group; PBOS Joint Technical Committee; PBOS FALLEX 62 Sub-group; PBOS CIVLOG 65 Planning Team; PBOS Container Study Group; PBOS D and L Card Study Group; PBOS Defense Shipping Authority Plans Review Study Group; and the PBOS Sealift Procurement Study Group.

In addition, PBOS created joint groups with the Planning Board for European Inland Surface Transport (PBEIST), the Joint PBOS/PBEIST Coastal Shipping Group, and with the Petroleum Planning Committee, the Joint PBOS/PPC Bunkering Study Group and the Joint PBOS/PPC Oil and Tanker Sub-Group. These different subordinate groups generally produced reports which were forwarded to PBOS for review and acceptance.

---

[148] http://archives.nato.int/planning-board-for-ocean-shipping-2

NATO Centers of Excellence (COE)

At the 2002 Prague Summit it was decided that NATO should change its military structures and concepts, and acquire new types of equipment to face the operational challenges of the new millennium. Thus NATO's military command structure was reorganized with focus on becoming a leaner and more efficient organization. And as certain warfare areas and expertise were not reflected any more, Nations offered NATO to build up and keep this expertise concentrated in *"Centers of Excellence"* (COEs).

The idea for NATO COEs originated in MC 324/1, "The NATO Military Command Structure," dated 14. May 2003. The Military Committee refined this idea into MCM-236-03; "MC Concept for Centers of Excellence (COE)" dated 04 Dec 2003. Once the idea and the concept were firmly established, the accreditation criterion was defined. In 2004, IMSM-0416-04, *"NATO COE Accreditation Criteria"* was agreed on and the first NATO COE was formally accredited on 01 Jun 2005.

The common definition was formulated, as *"A COE is a nationally or multi-nationally sponsored entity, which offers recognized expertise and experience to the benefit of the Alliance, especially in support of transformation."* A COE is not part of the NATO Command Structure (NCS), but forms part of the wider framework supporting NATO Command Arrangements (NCA).

The concept of the NATO Center of Excellence allows the Alliance to maintain and improve selected competences in several key military domains. They also contribute to share those skills across the whole NATO Alliance as well as partner Nations. The COE community has now grown to 23 accredited Centers raising the full potential of their capacity to the benefit of NATO.

There are many steps in the COE establishment process. COEs generally start either as an idea from a NATO Nation, or as the offer of an established (multi-) national entity by an Alliance Nation. It culminates with the accreditation by the Military Committee (MC) which is forwarded for approval of the North Atlantic Council (NAC). At the same time the COE can be activated by the NAC as a NATO Military Body and thus hold International Status. Much dialogue and teamwork is required throughout the process starting with a close coordination between the Framework Nation (FN) and the HQ SACT Transformation Network Branch (TNB).

In return the expertise available to Allies and cost effectiveness make the centers critical enablers towards NATO efforts to prepare for the future operating environment, here also for the Maritime Entities. For a complete coverage all COEs are listed, the one with mainly maritime expertise emphasized.

Pillars of the support of the NATO Centers of Excellence and principles are basically they are no cost to NATO, they conform to NATO procedures, doctrines and standards while they are no duplication off existing assets. Relationships with Strategic Commands through

Memorandum of Understanding agreements and relationships with partners are supported and encouraged.

The role of ACT is to work on the outputs of Transformation, in order to improve NATO's posture. This will enable NATO to deliver strategic military effects to fulfill the three core tasks defined in the Strategic Concept:

- Collective Defense,
- Crisis Management and
- Cooperative Security

All the Centers of Excellence can provide tangible actions in expanding six focus areas in support of NATO's Transformation in order to develop a robust foundation based on six focus areas:

- Command and Control,
- Logistics and Sustainment,
- Collective Training,
- Partnership,
- Manpower, and
- Capabilities.

The aim of the Centers of Excellence is to provide opportunities to enhance education and training; improve interoperability and capabilities; assist in doctrine development and/or test and validate concepts through experimentation. The makeup and characteristics of the COEs are therefore unique to the work of each one with key points of interest:

- 1-17 Sponsoring Nations (SN)
- 25 of 28 NATO nations participate in COEs
- Manning from 4 to 95 posts
- Operating costs per position: 6 to 25K €
- Total manning is approximately 1139 billets/877 filled for 23 accredited COEs
- SNs are encouraged not to populate a COE at the expense of NATO billets in the NCS

The number of NATO COEs is consistently growing. Through the MC Concept and the NATO accreditation criteria, COEs have proven to be a successful and enduring model for strong multinational solutions. As a result, a robust network of COEs, which are nationally or multi-nationally managed and funded and are open for participation by all member states, supports the NCA.

The Headquarter Supreme Allied Commander Transformation (HQ SACT) acts on behalf of Allied Command Transformation (ACT) and Allied Command Operations (ACO) as the strategic staff authority with overall responsibility for all COEs and manages the ACT-COE staff level relationships. ACT performs this coordination on three distinct levels:

- The overall coordination of COEs and their processes is accomplished by the HQ SACT TNB. Main processes under TNB cognizance include COE establishment, accreditation, NATO RFS and periodic assessments.
- The functional coordination of products and services offered by COEs is accomplished by an ACT SME in the specialty of the COE.
- The strategic link between the COE and NATO is assisted by an appointed Flag or General Officer (FOGO). Each COE is paired with an ACT FOGO who will act as a champion within the command. The intent is for these champions to improve the NATO RFS process, to better align the work of each COE with NATO priorities, and to provide enhanced visibility for ACT activities. ACT designated COE/FOGO pairing below.

Each year HQ SACT coordinates the submission of NATO Requests for Support (RFS) to the COE Programs of Work (POW). The overall aim of the process is to optimize the use of the COEs. The milestones established align the timelines between the centers and NATO planning, forming the COE POW development cycle.

The RFS are prioritized lists submitted by NATO to each COE. The COE POW development cycle is coordinated between NATO customers and COEs, and is the primary tool for generating NATO RFS. The NATO RFS, together with inputs from Sponsoring Nations and other entities are presented to the COE Steering Committees for approval as the following year's POWs.

In 2016 the COEs received 818 RFS, an increase of 218 RFSs or 36.3% over 2015. They accepted 601 RFSs. The reasons for not accepting all of the requests were due to limitations to the amount of COE resources (Funding, People etc.) available to meet demand and many requests were poorly explained by requestors.

Analysis and Simulation Center for Air Operations (CASPOA)

Main task of the Center d'Analyse et de Simulation pour la Preparation aux Operations Aeriennes (CASPOA)[149] is to train individuals for present and future NATO air operations in combined and joint environments. CASPOA also functions as the NATO Department Head for AirC2 Systems Education & Training.

As an Air C2 Subject Matter Expert, CASPOA is experimenting with new concepts and doctrine, as well as educating and training personnel to plan, task and control air operations. Moreover, capitalizing on the collection of lessons identified, it feeds the NATO Lessons Learned process, enhancing the efficiency and effectiveness of the Joint Force Air Component structures.

By demonstrating excellence in providing training solutions for NATO requirements, CASPOA earned an unconditional Quality Assurance certification by ACT in December 2013. All eight CASPOA courses offered to NATO are included in NATO's Education and Training Opportunities Catalog (ETOC) and are "NATO Approved" by ACT.

Center of Excellence Defense Against Terrorism (COE DAT)

The mission of the Centre of Excellence Defense Against Terrorism (COE DAT) is to assist NATO, Nations, Partners and other bodies by supporting NATO's capability development process, effectiveness, and interoperability by providing comprehensive and timely expertise on Defense Against Terrorism.

---

[149] http://wise.defensenns.aouv.fr/WISE/CASPOA

Center of Excellence for Operations in Confined and Shallow Waters (COE CSW)

The <u>Center of Excellence for Operations in Confined and Shallow Waters (COE CSW)</u>[150] is working in a cramped, congested and contested operational environment that is especially characterized by extraordinary complexity, interaction, surprise, speed, disguise plus diversity of actors. Furthermore, the rapid as well as unpredictable change of conditions and circumstances, including the frequent shift of tactical advantage from one side to another, are typical.

With these attributes, the area of Confined and Shallow Waters (CSW) constitute an extremely challenging littoral battlespace, which affects the freedom of movement and action by specific geographical and geophysical factors as well as manifold threats and risks. On the other side, CSW also offers a broad range of possibilities and opportunities for military operations. Principally being a maritime sphere, CSW is a theatre of operations also being significantly affected by the other military domains (land, air, space and cyber). Consequently, the greatest possible joint interaction takes place in CSW involving all major military components and services.

The COE CSW mission is to provide joint and combined subject matter expertise in the range of operations in confined and shallow waters in order to support the Alliance, the COE CSW Participants and other Customers, thus contributing to NATO transformation and enhancing the overall Alliance's interoperability.

The mission of the of the COE CSW is to provide Joint and Combined Subject Matter Expertise in the range of operations in confined and shallow waters (CSW) for NATO and the nations participating in the COE CSW in order to advance future developments and in particular to support NATO Transformation. Their Latin slogan is *"Nemo solus satis sapit!"* (English *"Nobody alone is clever enough"*).

Initiated by a *"Request for Support"*, usually from a NATO entity or a participating nation, the COE CSW contributes to NATO Transformation by continuous activities such as participation in various working groups or supporting exercises as well as by working on specific projects related to its range of expertise. The latter have a clear end state and date; this could for example be a research task, which is concluded by a final report. In conjunction with its activities and projects, COE CSW plans and hosts a number of events such as conferences, symposia, and workshops. The Steering Committee approves the annual "Program of Work" detailing the major activities and projects of COE CSW.

One of the most important items in the newer maritime history is the Request for COE CSW support to facilitate the kick-off meeting, the *"Maritime Security Regimes Roundtable (MSR Roundtable)"*, in order to initiate global MSR cooperation. At that time there was no cooperation between the regions and the Maritime Security Regimes, which was seen a key

---

[150] http://www.coecsw.org

factor for the future when making the seas more safe and secure. The Cooperation Maritime Surveillance Networking (MARSUR Networking) submitted this request on 10. MARCH 2014 to the COE CSW.

The contributing nations and organizations in the Multi-national Experiment (MNE) 7 together with the COE CSW and the Combined Joint Operations from the Sea Center of Excellence (CJOS COE) defined in their *"Enterprise Implementation Proposal and MSR Manual"* a Maritime Security Regime (MSR) as a group of states and/or organizations acting together, with an agreed upon framework of rules and procedures, to ensure security within the Maritime Domain. COE CSW in Kiel, Germany, the CJOS COE in Norfolk, United States of America, and the Maritime Security Center of Excellence (MARSEC COE) in Marmaris, Turkey, have dedicated their work to the various fields in Maritime Security and MSA.

The COE CSW initiated/organizes four major events:

- Kiel Conference
- Annual Discipline Conference
- Maritime Security Regimes Roundtable
- Conference on Operational Maritime Law

Center of Excellence for Cold Weather Operations (COE CWO)

The NATO Center of Excellence for Cold Weather Operations (COE CWO) acts as the main provider and coordinator of expertise and capabilities in the area of Cold Weather Operations in NATO. The COE CWO will provide NATO and Partner nations the necessary competence in order to operate under Arctic, sub-Arctic and Cold Weather conditions. This is done through utilizing the full spectrum of competence in the Norwegian Armed Forces, coordinated with other nations competence in the Cold Weather environment.

The COE CWO is not one geographical location, but a concept that encompasses training areas, infrastructure, manpower, skills and knowledge drawn from all four services in the Norwegian Armed Forces, the Norwegian Defense Research Establishment, the NATO and PfP-Nations Cold Weather community of Interest and Civilian Academia.

Civil-Military Cooperation Center of Excellence (CIMIC COE)

The mission of the Civil-Military Cooperation Centre of Excellence (CIMIC COE) is to assist NATO, Sponsoring Nations (SN) and other military and civil institutions/organizations in their operational and transformation efforts in the field of Civil-Military Interaction (CMI)/CIMIC:

- providing innovative and timely advice and subject matter expertise (SME),
- providing specialized training and education,
- contributing to the lessons learned processes,
- supporting the development of existing and new concepts, policy and doctrine.

Combined Joint Operations from the Sea Center of Excellence (CJOS COE)

The mission of the Combined Joint Operations from the Sea Center of Excellence (CJOS COE)[151] is the work in conjunction with Commander, U.S. Fleet Forces, where the CJOS COE provides a focus for the Sponsoring Nations and NATO in improving allied ability to conduct combined joint operations from the sea in order to ensure that current and emerging global security challenges can be successfully tackled.

Vision is to become the pre-eminent source of innovative specialist advice and recognized expertise on all multinational aspects of combined joint operations from the sea in support of the sponsoring nations, NATO, and other allies.

The Combined Joint Operations from the Sea Centre of Excellence (CJOS COE) was established in May 2006, to provide a focal point for Joint Maritime Expeditionary Operations expertise for allied nations. With 13 nations represented, CJOS COE is the only Centre of Excellence in the United States, and is one of 20 NATO accredited Centers worldwide, representing a collective wealth of international experience and expertise, and best practices.

Independent of the NATO Command Structure, CJOS COE draws on the knowledge and capabilities of Sponsoring Nations, United States Fleet Forces, and neighboring U.S. Commands to promote "*best practices*" within the Alliance. CJOS COE also plays a key role in aiding NATO's transformational goals, specifically those focused on maritime-based joint operations. We enjoy close cooperation with Allied Command Transformation (ACT), other maritime COEs, NATO Joint Forces Commands, various national commands, academia.

With almost 30 permanent staff and 20 USN reservists, CJOS COE is highly flexible and responsive to its customers' needs. The Centre has worked to reduce consultations between COE staff and key exports within individual Sponsoring Nation and other customers setting up "*Focal Points of Contact*" that put the Centre directly into contact with their Subject Matter Experts.

In many situations, Combined Joint operations 'from the sea' will present a viable, even the preferred option. The seas will continue to play a vital role in the dynamic and uncertain situations of the modern world, and provide a unique capability to match the pace and to reflect the tone of diplomatic activity. A large number of countries continue to rely on an unhindered use of the sea for their security, prosperity and wellbeing. Sea-based operations reduce the political complications and military risks of deploying forces (and their logistic support) for extended periods on land, whilst enabling the movement of the base of operations across 75% of the earth's surface.

Operating from sophisticated platforms at sea takes advantage of Allied supremacy against Transnational Terrorists in this environment and turns the asymmetry in our favor. If

---

[151] http://www.cjoscoe.org

Combined and Joint military power is to be effectively projected in this area, all aspects must be continuously re-examined and revitalized to incorporate new concepts and technologies. This coordination and harmonization of effort requires a focused approach across all Services and functional components, and through all levels of command.

The US FLEET FORCES COMMAND (USFFC) possesses a number of attributes and provides opportunities within the joint maritime expeditionary operating area that merit the establishment of a Combined and Joint Operations from the Sea Center of Excellence:

- forefront of US Navy transformational activities;
- major stakeholder in numerous US joint and combined exercises;
- responsible for the training and employment of a large number of assets, from Strike Carriers and Amphibious forces, to frigates and destroyers;
- designated US CTF Commander; thus his staff make-up provides strong joint and combined focus and experience.

The naming of the U.S.-Commander of U.S. Fleet Forces prior to anything else is the program. The Combined Joint Operations from the Sea Center of Excellence (CJOS COE) list no slogan, however the Latin slogan "*Non Sibi Sed Patriae*" (English "*Not self but country*") is inscribed over the entrance to the Chapel at the U.S. Naval Academy in Annapolis, Maryland.

Command and Control Center of Excellence (C2 COE)

The Command and Control Centre of Excellence (C2 COE) in Utrecht, The Netherlands, will support NATO, nations and international institutions/organizations with subject matter expertise on Command and Control. The main level of interest is C2 at the operational level. To achieve this mission, the C2COE's aim is to be a principal source of expertise focusing on specific areas in the domain of Command and Control in order to best support the Transformation of NATO. The NATO C2COE will:

- Network with Sponsoring Nations, ACT, ACO and other international institutions/organizations.
- Contribute C2-expertise to the operational communities, catalyzing C2.
- Contribute to the NATO mission.
- Attend NATO exercises in support of JWC and SHAPE, with a special interest in Federated Mission Networking.
- Organize events to promote social networks and spread knowledge and expertise with NATO and nations.
- Publish documents related to C2.

With the NATO Pillars of Support in mind, the NATO C2COE has defined three Focus Areas, in which it intends to provide subject matter expertise:

- C2 Processes and Structures
- Information & Knowledge Management
- Human Factors (incl. Leadership)

Cooperative Cyber Defense Center of Excellence (CCDCOE)

The NATO Cooperative Cyber Defense Centre of Excellence (CCDCOE) based in Tallinn, Estonia is a NATO-accredited knowledge hub, think-tank and training facility. The international military organization focuses on interdisciplinary applied research and development, as well as consultations, trainings and exercises in the field of cyber security.

The Czech Republic, Estonia, France, Germany, Greece, Hungary, Italy, Latvia, Lithuania, the Netherlands, Poland, Slovakia, Spain, Turkey, the United Kingdom and the USA have signed on as Sponsoring Nations. Austria and Finland have joined the Centre as Contributing Participants - the status available for non-NATO nations. The Centre is funded and staffed by the aforementioned nations.

The vision of the Center is to be the main source of expertise in the field of cooperative cyber defense by accumulating, creating, and disseminating knowledge in related matters with NATO, NATO nations and partners. The primary aim of the Centre's POW is to support work on the most pressing cyber defense issues facing the Alliance, nations and partners. The Centre will focus on the following areas of expertise by collecting, maintaining and disseminating knowledge, skills and best practices:

- Policy, Strategy and Doctrine
- National and International Law
- Terminology and Standards
- Technical and Tactical Environment

Mission is to enhance the capability, cooperation and information sharing among NATO, NATO nations and Partners in cyber defense by virtue of education, research and development, lessons learned and consultation.

Research into the cyber information exchange practices and the organizational and technical means used to assure situational awareness. Studies include possible enforcement measures available across NATO countries to issue and enforce mandatory guidelines in crises situations in order to maintain the secure operation of the cyber domain. The Tallinn Manual 2.0 is the follow-on to the successful Tallinn Manual on the International Law Applicable to Cyber Warfare. Publication is forthcoming from Cambridge University Press in the second half of 2016.

Both publications aim to offer guidance on applying existing international norms to the cyber arena, consist of black letter rules with commentary and are based on the consensus of an international group of legal experts. Tallinn Manual 2.0 will expand the scope of the original piece to so-called peacetime international law, addressing incidents that state frequently face.

Counter-Improvised Explosive Devices Center of Excellence (C-IED COE)

The Counter-Improvised Explosive Devices Centre of Excellence (C-IED COE) mission is to provide subject matter expertise in order to support the NATO Alliance, its partners and the international community in the fight against Improvised Explosive Devices (IED), collaborate to increase the security of Allied Nations and troops deployed in theatres of operations and reduce or eliminate the threats from IEDs used by terrorists or insurgents.

Five goals are:

- Establish the C-IED COE as an organization capable of establishing relationships with organizations to exchange information to Attack the Network (AtN) and Counter Threat Network (CTN) operations.
- Establish the C-IED COE as the training and education subject matter experts (SME's) for C-IED related activities.
- Establish the C-IED COE as the lessons learned coordinator for C-IED related activities.
- Become one of the focal points for the generation of modern defense technology military capabilities in C-IED related technologies, capabilities and knowledge.
- Achieve a robust capability to communicate and exchange IED and C-IED information with other organizations.

The Long Term Vision of the CIED COE is to have a solid capability to co-operate with military and civilian international organizations and industry to increase security of Nations and troops deployed in theatres of operations, to assist in reducing or eliminating the threat of attacks with IEDs and to provide information to national and international organizations to combat the IED as a weapon of strategic influence.

Counter Intelligence Center of Excellence (CI COE)

The decision to form the Counter Intelligence Center of Excellence (CI COE) has been made following tensions between the Western military alliance and the Russian Federation. The unit will be located in southern suburbs of Kraków. The CI COE is not accredited or in accreditation Process.

Poland and Slovakia signed a memorandum to open the NATO center a year ago in October. The memorandum stipulates the involvement of 10 NATO member states in installing structures on the territory of Poland. Germany is to send among other things training officers.

Crisis Management and Disaster Response Center of Excellence (CMDR COE)

The mission of the Crisis Management and Disaster Response Centre of Excellence (CMDR COE) is to act as the catalyst for improvement of NATO, Nations and Partners capabilities in crisis and disaster response operations through collaborative partnerships.

Energy Security Center of Excellence (ENSEC COE)

The mission of the Energy Security Centre of Excellence (ENSEC COE) is to assist Strategic Commands, other NATO bodies, nations, partners, and other civil and military bodies by supporting NATO's capability development process, mission effectiveness, and interoperability in the near, mid and long terms by providing comprehensive and timely subject matter expertise on all aspects of energy security.

NATO ENSECCOE acts to be the recognized hub of knowledge and expertise in Energy Security within NATO, being a unique platform of cooperation and partnership in this area.

Explosive Ordnance Disposal Center of Excellence (EOD COE)

The mission of the Explosive Ordnance Disposal Centre of Excellence (EOD COE) is to support and enhance the NATO transformation and operational efforts in the field of EOD.

Vision is to be NATO's leading agent in the preparation of technologically advanced, interoperable and well trained EOD experts capable to support and to enable entire spectrum of Alliance operations.

The EOD COE is designed to perform tasks in support of:

- Lessons Learned and Analysis
- Training and Education
- Concept and Doctrine Development and Experimentation
- Standardization
- Technology Development

Human Intelligence Center of Excellence (HUMINT COE)

The Human Intelligence Centre of Excellence (HUMINT COE) provides the highest quality NATO HUMINT- focused services and products in response to the requirements and needs of the NATO Command Structure, NATO Force Structure, NATO Nations, and, when feasible, Partner Nations. The HCOE is the focal point of HUMINT expertise within NATO and is the spearhead of all major HUMINT initiatives within the Alliance.

Joint Air Power Competence Center (JAPCC)

Joint Air Power Competence Centre (JAPCC) is NATO's catalyst for the improvement and transformation of Joint Air and Space Power; delivering effective solutions through independent thought & analysis.

The JAPCC, as a team of multinational experts, is to provide key decision makers effective solutions on Air and Space Power challenges, in order to safeguard NATO and the Nations' interest.

Joint Chemical Biological Radiological Nuclear – Defence Center of Excellence (JCBRN COE)

Nuclear states are striving to modernize their capabilities, while new nuclear states appear on the threat table. Weapons of mass effect are under continued international surveillance, but limited tactical use by states, rogue regimes, or ideologically-driven non-state actors is increasingly likely and happening.

The Joint Chemical Biological Radiological Nuclear - Defence (JCBRN COE) is to support NATO and its military transformation in the field of CBRN defense through and in support of HQ SACT, assist Sponsoring Nations, other NATO nations and other Customers in their CBRN defense-related efforts.

Maritime Security Center of Excellence (MARSEC COE)

The Maritime Security Center of Excellence (MARSEC COE)[152] in Maris, Turkey is both a center for academic research as well as a (multinational) hub for practical training in the field of maritime security, along with relevant domains (Maritime Trade, Energy Security, Maritime Environment, Maritime Resources, Public Health, Maritime Transport-Logistic). MARSEC COE strives to achieve the necessary collaboration amongst stakeholders from government, industry, academia and private sector.

MARSEC COE sees Maritime Security with different dimensions, including but not limited to Maritime Situational Awareness (MSA), Law enforcement, maritime safety, maritime environment, maritime science & technology, maritime trade & economy, maritime law and public health. Therefore, in national terms, Maritime Security can only be achieved by a "whole of government" approach. Maritime Security downgraded to a Maritime Interdiction Operation (MIO) with a defense-minded mentality is covering only % 10 of Maritime Security. The idea of a Maritime Security Center of Excellence (MARSEC COE) stemmed from the need for coordination and de-confliction among governmental and interagency organizations as well as defense and law enforcement entities towards a more secure maritime environment.

One of the most important lessons Turkey has learned through the recently conducted Maritime Security Operation- MSO is the significant role Maritime Situational Awareness (MSA) plays in countering maritime risks and threats. Another equally important lesson is the fact that, MSA can only be achieved effectively through working together with other regional maritime security organizations and civilian agencies, employing all the other instruments of national power and thus enabling a *"whole of government approach"*.

The Allied Maritime Strategy (AMS) sets out, in full consistency with the Strategic Concept, the ways that maritime power could help resolve critical challenges facing the Alliance now and in the future, and the roles-enduring and new-that NATO forces may have to carry out in the maritime environment in order to contribute to the Alliance's defense and security and to promote its values. These roles capitalize upon the ability of maritime forces to provide a spectrum of strategic options to the Alliance, and include appropriate contributions to:

- Deterrence and collective defense,
- Crisis management,
- Cooperative security, outreach through partnerships, dialogue and cooperation,
- Maritime Security.

NATO launched the Smart Defense Initiative in 2011 that aims at providing cost-effective

---

[152] http://www.dgmm.tsk.tr/

solutions for capability development. This was very appropriate and timely taking into account the recent worldwide financial and the shrinking defense budgets.

This Initiative, mainly aims at collective prioritization of the nations' defense requirements, role specialization among nations and the multinational projects developments to this end, has been welcomed and fully supported by the nations. Turkey fully supports this initiative and encourages all Allies to foster it in NATO. Maritime security is suitable area for cooperation with partners.

Piracy directly affects the crew of merchant fleets and company owners in a negative way. Today, NATO, CTF-151, EU (ATALANTA) and several countries (such as Japan, India, China, RF, South Korea, New Zealand and Australia) are actively taking part in activities conducted against piracy in the Gulf of Aden and off the coast of Somalia. However, a common platform in which solutions could be found and acted upon is needed in implementing a maritime cross–functional inter-agency approach, healing the wounds of the maritime sector, as well as easing the workload of defense and security forces.

MARSEC COE is another approach to Maritime Security based on Multi-national cross–functional-interagency co-operation and work program of includes courses, workshops and exercises providing an ideal platform for cross-functional inter-agency cooperation including Counter Piracy Training in Mediterranean before deployment of maritime forces.

Military Engineering Center of Excellence (MILENG COE)

The Military Engineering Centre of Excellence (MILENG COE) mission is to enable the development of Sponsoring Nations (SN) and Alliance MILENG capability and interoperability, in order to enhance the effectiveness of MILENG support to NATO and other operations.

The intent is to deliver the mission at the Strategic, Operational and Tactical levels, encompassing, but not limited to, all aspects of MILENG as encapsulated in MC Policy for MILENG and its supporting doctrine documents.

The Director MILENG COE is the Principle Advisor to SACT for MILENG, the Deputy Chairman of MCLSB MILENG WG, the Chairman of MILENG DPAG (to provide SME in support of N DPP) and Member of the NSJEC Advisory Board.

The MILENG COE provides the secretariat for the annual NATO Senior Joint Engineer Conference (NSJEC), is Department Head for the MILENG Training and Education Discipline and is a permanent member of the NATO C-IED Task Force.

MILENG is a Functional Area in support of operations covering the shaping, improving and protecting the physical operating environment, coordinated by an Engineer staff. The MILENG COE therefore is engaged in all aspects covered by the MILENG Functional Area, like Environmental Protection, Infrastructure and Energy Efficiency.

Military Medicine Center of Excellence (MILMED COE)

The mission of the NATO Centre of Excellence for Military Medicine (MILMED COE) is to support and assist the Strategic Commands, other NATO bodies, nations and other civil and military organizations by supporting the transformation of the Alliance and thereby improving medical support to operations and to provide subject matter expertise in the following areas:

- Medical training, exercises and evaluation leading to certification
- Medical Lessons Learned
- Standards development and custodianship
- Deployment related health surveillance

Moreover, the COE provides relevant knowledge, expertise, and best practices to the NATO community to enhance and develop the provision of effective, sustainable and ethical full spectrum health services at best value to the Allies.

Since its founding 5 years ago, the MILMED COE has become a hub of military medical expertise and a focal point of knowledge, providing invaluable training, education and deployment health surveillance capabilities, lessons learned databases and concept development support. The COE has excellent working relations with all relevant NATO bodies and working groups operating in the field of military medicine. These achievements are achieved by the NATO and civilian university accreditation of the COE's courses, with the permanent support of the Sponsoring and Partner Nations, the exercise series Vigorous Warrior, frequent requests for medical evaluation of multinational medical units, and the continuous support of the Committee of the Chiefs of Military Medical Services in NATO (COM EDS).

Military medicine is a discipline that is strongly connected to civilian medical developments. The MILMED COE, as a knowledge center, remains an open institution that links military and civilian medicine together by collecting up-to-date medical knowledge and expertise from both communities, continuing to cooperate with the widest possible range of partners.

Military Police Center of Excellence (MP COE)

The NATO Military Police COE (MP COE) mission is to enhance the capabilities of NATO and PfP-Nations, foster interoperability and provide subject matter expertise on MP activities in accordance with the Alliance's Strategic Concept. The NATO MP COE aim is to support the transformation and enhancement of NATO's Military Police capabilities and to contribute to the NATO Readiness Action Plan in the field of Military Policing. Our vision is to offer Military Police expertise and experience to the benefit of the Alliance, especially in support of enhancing those areas that are underdeveloped in the NCS, thereby helping to fill capability shortfalls with the aim to contribute to the Capability Development (CP), the NATO Defense Planning Process (NDPP), Smart Defense (SD), Connected Forces Initiative (CFI) and Framework for Future Alliance Operations (FFAO).

Modeling and Simulation Center of Excellence (M&S COE)

The NATO Modeling and Simulation COE (M&S COE) is dedicated to the promotion of M&S in support of operational requirements, training and interoperability. The Centre will act as a catalyst for transformation through the involvement of NATO, governments, academia, industry, operational and training entities, by improving the knowledge of NATO and Nation's M&S professionals, promoting cooperation between Nations and organizations through the sharing of M&S information, contributing to the development of new M&S concepts and standards, and serving as an international source of expertise.

Mountain Warfare Center of Excellence (MW COE)

The establishment of the NATO Mountain Warfare Centre of Excellence (NATO MW COE) contributes to the transformation and adaptation of the Alliance' capabilities which enable its forces to better operate in the mountain environment. The NATO MW COE incorporates a professional core that will ensure and develop the subject matter expertise to meet the requirements of mountain warfare challenges. This will enhance the ability of individuals and military units to engage in mountain warfare, as well as the Alliance's interoperability to operate in mountainous environment.

The mission of the NATO MW COE is to assist NATO member countries, partners, other countries and international organizations, in order to enhance mountain warfare capabilities through the following core areas: Development of mountain warfare-specific doctrine and tactics. The vision of the NATO MW COE is to be the hub for mountain warfare expertise in the NATO community.

Naval Mine Warfare Center of Excellence (NMW COE)

The Naval Mine Warfare Centre of Excellence (NMW COE) focuses on Education and Training, Doctrine and Standardization, Analysis and Lessons Learned, Concept Development and Experimentation in order to deliver the best possible support in NMW to NATO, Sponsoring Nations and Partners.

While the NMW COE has retained its function as an education and training institution for the Belgian and Netherlands Navies, the Framework Nations provide ACT, other NATO entities and Nations prioritized access to the services and support of the NMW COE.

Stability Policing Center of Excellence (SP COE)

The mission of the NATO Stability Policing Centre of Excellence (SP COE)[153] is to be an internationally recognized focal point and a hub of expertise for a Community of Interest in the field of Stability Policing, which is a set of police related activities intended to reinforce or temporarily replace the indigenous police of an unstable area in order to contribute to the restoration and/or upholding of the public order and security, rule of law, and the protection of human rights (AJP 3-22 - Ratification Draft).

To this end, the SP COE acts as a prime mover to increase the contribution to the Stabilization and Reconstruction efforts of the Alliance in unstable scenarios, providing NATO with a unique tool to fill the capability gap in the area of Stability Policing.

The SP COE is committed to help the Alliance, the Sponsoring Nations and PfP-Nations to enhance and transform their capabilities, procedures and functions in order to meet potential and future security challenges in line with NATO's three declared core tasks of collective defense, crisis management and cooperative security, in the framework of the Smart Defense, Connected Force Initiative and Framework for Future Alliance Operations.

---

[153] http://www.nspcoe.org

Strategic Communications Center of Excellence (STRATCOM COE)

Strategic communication is an integral part of the efforts to achieve the Alliance's political and military objectives. The NATO Strategic Communications COE (STRATCOM COE), based in Riga, Latvia, contributes to improved strategic communications capabilities within the Alliance and Allied nations.

The mission of the NATO StratCom COE is to contribute to the Alliance's communication processes in order to ensure that it communicates in an appropriate, timely, accurate and responsive manner on its evolving roles, objectives and missions. The NATO StratCom COE provides comprehensive analyses, timely advice and practical support to the Alliance, designs programs to advance doctrine development, and conducts research and experimentation to find practical solutions to existing challenges. Its strength is built by multinational and cross-sector participants from the civilian and military, private and academic sectors and usage of modern technologies, virtual tools for analyses, research and decision-making.

NATO StratCom COE will continue to support NATO institutions and nations on build-up of Strategic Communications understanding and mind-set. The ambition is to be part of the NATO Strategic Communications doctrine development process; therefore we will continue to conduct research and training. NATO StratCom COE will participate in NATO StratCom training events and exercises.

NATO Science and Technology Organization (STO)

Science & Technology (S&T) in the NATO context is defined as the selective and rigorous generation and application of state-of-the-art, validated knowledge for defense and security purposes. S&T activities embrace scientific research, technology development, transition, application and field-testing, experimentation and a range of related scientific activities that include systems engineering, operational research and analysis, synthesis, integration and validation of knowledge derived through the scientific method.

In NATO, S&T is addressed using different business models, namely a collaborative business model where NATO provides a forum where NATO Nations and partner Nations elect to use their national resources to define, conduct and promote cooperative research and information exchange, and secondly an in-house delivery business model where S&T activities are conducted in a NATO dedicated executive body, having its own personnel, capabilities and infrastructure.

The mission of the NATO Science & Technology Organization (STO)[154] is to help position the Nation and NATO is S&T investments as a strategic enabler of the knowledge and technology advantage for the defense and security posture of NATO Nations and partner Nations, by conducting and promoting S&T activities that augment and leverage the capabilities and programs of the Alliance, of the NATO Nations and the partner Nations, in support of NATO its objectives, and contributing to NATO's ability to enable and influence security and defense related capability development and threat mitigation in NATO Nations and partner Nations, in accordance with NATO policies.

The total spectrum of this collaborative effort is addressed by six Technical Panels who manage a wide range of scientific research activities, a Group specializing in modelling and simulation, plus a Committee dedicated to supporting the information management needs of the organization.

- Applied Vehicle Technology (AVT)
- Human Factors and Medicine (HFM)
- Information Systems Technology (IST)
- NATO Modelling and Simulation Group (NMSG)
- System Analysis and Studies (SAS)
- Systems Concepts and Integration (SCI)
- Sensors and Electronics Technology (SET)

These Panels and Group are the power-house of the collaborative model and are made up of national representatives as well as recognized world-class scientists, engineers and

---

[154] https://www.sto.nato.int/

information specialists. In addition to providing critical technical oversight, they also provide a communication link to military users and other NATO bodies.

The scientific and technological work is carried out by Technical Teams, created under one or more of these eight bodies, for specific research activities which have a defined duration. These research activities can take a variety of forms, including Task Groups, Workshops, Symposia, Specialist in Meetings, Lecture Series and Technical Courses.

The Science & Technology Organization (S&T) works in close cooperation with the NCIA and the NSA.

NATO SECRET Wide Area Network (NSWAN)

There are actually several NATO SECRET Networks and Services, one of the most important one being the Crisis-Response Operations NATO Open Systems (CRONOS) as a mission related classified network, where the Web Information Services Environment (WISE) enables CRONOS users to design, develop and manage their own web sites, applications, content and security permissions in real time. Analysts working within NATO should be familiar with WISE because it is used by all NATO commands and HQs to build their web sites and to act as a tool for information management and knowledge management. WISE provides a plethora of tools (called objects) to help build web pages and some of these may be useful for data collection.

The Web Information Services Environment (WISE) is a web application framework containing a collection of predefined applications suitable for building structured web sites built on an Open Source web application server (ZOPE). WISE supports PC platforms including Windows NT, 2000, XP and 2003 as well as the HPUX platform.

NATO engineers from the former ACLANT System Support Center (ASSC) originally developed WISE Portal technology as a web interface for NATO's Maritime Command and Control Information System (MCCIS) using Open Source software from ZOPE Corporation. Since its inception, WISE utility has expanded to support network-enabled requirements throughout NATO. Each new release has incorporated feedback from WISE users through innovative applications and improved business practices.

WISE software is easy to obtain and install, providing a proven web capability for knowledge sharing, collaboration, decision support, planning and mission execution. WISE empowers users to create and manage a custom website using only a standard PC and web browser. WISE has proven its reliability and value as an information management tool at home or in the field. WISE recently demonstrated superior performance, supporting NRF forces in Athens during the Distinguished Games, as well as in exercises such as Allied Warrior, Allied Reach and MNE-4.

An experimental version of WISE has been absorbed into the BRITE baseline. At some point, this version of WISE was referred to as WISE 1.3 and WISE 2.0 which caused a lot of confusion. This is no longer the case and the term WISE is no longer used in the BRITE project. Over time, BRITE has become completely independent and today only the history books reveal the original relationship.

Aside from this several Mission related NATO SECRET Networks (NSWANs, i.e. the NATO General Purpose Segment Communications System (NGCS), the NATO Initial Data Transfer System (NIDTS) or CRONOS) exist other enclaves, in agencies and institution, which are either not connected to the main network or through a diode like for the Maritime System Interface Services, which provides the interfaces to external data sources. There is a dedicated module to control and monitor each physical interface with a data source.

NATO Shipping Center (NSC)

The NATO Shipping Center (NSC)[155] is the link between NATO and the merchant shipping community. Permanently manned by NATO, the NSC is the primary point of contact for the exchange of merchant shipping information between NATO's military authorities and the international shipping community.

The NSC is part of Allied Maritime Command (MARCOM) Headquarters at Northwood, GBR, and NATO's Point-of-Contact with the Maritime Community and primary advisor to merchant shipping regarding potential risks and possible interference with maritime operations. In addition to its role in Operations OCEAN SHIELD and ACTIVE ENDEAVOUR, the NSC supports NATO, national and multinational operations and exercises with two specialist capabilities:

- Naval Cooperation and Guidance for Shipping (NCAGS)
- and Allied Worldwide Navigation Information System (AWNIS).

In order to both reduce conflict of interest between military and merchant shipping, and to enhance safety and security on sea, NATO developed and implemented the concept of Naval Cooperation and Guidance for Shipping (NCAGS).

NATO maintains its own NCAGS website where NCAGS personnel can gain access to publications, news, calendars etc. This is a password protected site for the NCAGS community only. NATO has released a non-classified publication which covers the relationship with civilian shipping as: *"NAVAL COOPERATION AND GUIDANCE FOR SHIPPING (NCAGS) – GUIDE TO OWNERS, OPERATORS, MASTERS AND OFFICERS"*.

The NSC, embedded in NATO MARITIME COMMAND HQ (Northwood) currently supports military operational planning and execution and serves as a conduit between NATO and the maritime community for on-going operations (Counter-Terrorism Operation Active Endeavour (OAE) and Counter-Piracy Operation Ocean Shield (OOS)) in order to share information on maritime risks with the merchant shipping industry.

However, as NATO areas of interest transition from regional to global, and its planning efforts shift from on-going operations to contingencies, the NSC must similarly clarify its role as an enduring organization with a strategic vision that addresses MARCOM's overarching objectives and provides purpose in its close relationship with the wider community. This vision document outlines the strategic plan for the NSC and delineates its role, areas of interest and engagement plan.

Based on the current situation, voluntary or mandatory reporting schemes may be imple-

---

[155] https://shipping.nato.int/nsc

mented regarding entering, sailing in and leaving defined areas. The normal vessel reporting format is called the Format Alfa, as found in the ATP 2, Guide to Masters. By submitting the Format Alfa, you will make your presence in the area known to the navy. This will enhance the navy's ability both to assist you and to avoid interference between naval and merchant shipping.

Requirements regarding information to be included in the Format Alfa can and will differ dependent on the schemes set for the specific area. In general the Format Alfa includes:

- Ship Name
- Flag
- IMO Number
- Inmarsat Telephone Number
- Time (UTC) and Position
- Course
- Passage Speed
- Freeboard
- Cargo
- Destination and Estimated Time of Arrival
- Last Port, Departure Date and Time (UTC)
- Additional Ports, ETA and ETD Dates and Times (UTC)
- Start Suez Canal Transit, Date and Time UTC (if applicable)

The North Atlantic Treaty Organization (NATO) has been engaged in counter-piracy missions off the Horn of Africa since October 2008 when forces from Operation Allied Provider which provided protection to World Food Program vessels also helped deter acts of piracy. This mission was followed by Operation Allied Protector and since August 2009 by Operation Ocean Shield, which has been extended until December 2012. Operation Ocean Shield entails at-sea counter piracy operations using naval forces from the two Standing NATO Maritime Groups (SNMG1 and SNMG2). These operations include deterring, disrupting and protecting against pirate attacks, rendering assistance to ships in extremis as required, and actively seeking suspected pirates and preventing their continued activity through detention, seizure of vessels and property.

The NATO Shipping Centre (NSC) provides the commercial link with NATO's Maritime Forces and is NATO's primary contact point with the maritime community and is used to communicate and coordinate with other military actors engaged in counter-piracy operations. The NSC website provides piracy warnings and alerts with positional information on sighted pirates. In addition, the website informs the commercial shipping community on Best Management Practices (BMP) through online content, workshops and the possibility of personal contact. The NSC disseminates among other information about piracy incidents daily piracy updates and weekly assessments on their website. Also, the NSC provides a piracy alert map that can be found on their website.

Naval Cooperation and Guidance for Shipping (NCAGS)

Naval Cooperation and Guidance for Shipping (NCAGS)[156] is the NATO cooperation with the merchant shipping. Maritime trade is of fundamental strategic interest to nations and their economic well-being depends on freedom of movement on the seas. Military operations at sea will frequently involve, or have some impact, on merchant shipping and likewise merchant shipping may affect military operations.

In order to both reduce conflict of interest between military and merchant shipping, and to enhance safety and security on sea, NATO developed and implemented the concept of NCAGS. NCAGS provides the interface between military operations and merchant shipping. This interface involves the provision of military cooperation, guidance, advice, and assistance to merchant shipping. The NCAGS capability is employed to enhance the safety of participating merchant ships in the operations area while supporting military objectives.

---

[156] http://www.shipping.nato.int/Pages/NCAGS.aspx

Planning Board for Ocean Shipping (PBOS)

The NATO Planning Board for Ocean Shipping (PBOS) plans for the provision of transportation of persons and goods by sea in crisis and conflict within and beyond NATO's area of responsibility. The Maritime Administration (MARAD) is the United States representative to NATO's Planning Board for Ocean Shipping. Additionally, MARAD's Office of National Security Plans provides the Secretariat for PBOS and the Associate Administrator for National Security currently serves as the Chairman of PBOS.

PBOS was established by the NATO North Atlantic Council (NAC) in 1950. It is one of nine NATO Planning Boards and Committees (PB&Cs) responsible in peacetime for coordinating and monitoring National and NATO arrangements for civil emergency preparedness and crisis management. The Planning Board is responsible for developing and maintaining plans for civil shipping support to the Alliance in crisis and war.

PBOS planning takes into account the international character of merchant shipping and seeks to facilitate access to worldwide shipping. Its planning responsibilities include planning for the provision of shipping resources to support military lift requirements through appropriate shipping crisis management arrangements, and planning for the availability of marine war risks insurance for merchant ships supporting the alliance.

PBOS plans for the use of merchant shipping in crises or wars affecting the interests of the Alliance. All other sealift activities in peacetime are solely a national responsibility.

Standing NATO Maritime Groups (SNMG1 and SNMG2)

NATO has two Standing Maritime Groups (SNMG1 and SNMG2) and two Standing Maritime Mine Countermeasure Groups (SNMCMGs). The groups provide the alliance with a continuous maritime capability for NATO Response Force (NRF) operations, non-NRF operations and other activities in peacetime and periods of crisis and conflict. The primary role of these Forces, as standing elements of their respective NRF, is the full integration and participation in the NRF, providing maritime support to operations.

Other missions that are applicable to both SNMGs and SNMCMGs include establishing alliance presence, demonstrating solidarity, conducting routine diplomatic visits to member, partner and non-NATO countries, supporting transformation and providing a variety of maritime military capabilities to ongoing missions.

Standing NATO Maritime Group 1 (SNMG1) is one of NATO's standing naval maritime immediate reaction forces. Prior to 1 January 2005 it was known as Standing Naval Force Atlantic (STANAVFORLANT). The group was also briefly called the Standing NATO Response Force Maritime Group One.

SNMG1 consists of 4 to 6 destroyers and frigates. The force operates, trains and exercises as a group, providing day-to-day verification of current NATO maritime procedures, tactics and effectiveness. Ships are usually attached to the force for up to six months, on a rotating basis. Units of one nation do not necessarily relieve ships of the same nation. The force commander and the staff are appointed for one year, with the force commander rotating among the participating nations.

Standing NATO Maritime Group 2 (SNMG2) is a NATO standing maritime immediate reaction force. Prior to 1 January 2005 it was known as Standing Naval Force Mediterranean (STANAVFORMED or more colloquially as SNFM).

SNMG2 is a multinational, integrated maritime force – made up of vessels from various nations that are part of NATO, training and operating together as a single team – that is permanently available to NATO to perform a wide range of tasks, from participating in exercises to crisis response and operational missions. Usually the force is employed in the Mediterranean Sea but, as required, will be available anywhere NATO requires it to deploy.

SNMG2 carries out a continuous program of operational training and conducts port visits to know and get known in many ports in and out of the Mediterranean, in NATO and non-NATO nations. These include ports in nations that are part of the Partnership for Peace, Mediterranean Dialogue and the Istanbul Cooperation Initiative programs.

The composition of SNMG2 varies depending on the current contributions of nations, but generally consists of 4–8 frigate or destroyer type ships and one oiler or support ship. Command of the force rotates in one year intervals among participating countries. The commander of SNMG2 until 2013 reported to the Commander of Allied Maritime Command

Naples, one of the two component commands of Allied Joint Force Command Naples.

Composition of the force varies as naval units are provided by NATO contributing nations on a rotational basis while command of the force rotates among them. Nations normally contributing to the group include Canada, Germany, Greece, Italy, The Netherlands, Spain, Turkey, United Kingdom and United States. Other NATO nations have also occasionally contributed.

Tidepedia

Tidepedia is the online information repository for the NATO TIDE community. Tidepedia is modelled after Wikipedia, an Internet open source encyclopedia collaboratively written by volunteers. Members of the TIDE community can and are strongly encouraged to create and edit articles in this repository. Over time, this repository will become the collective knowledge base for the community. Tidepedia is hosted by Allied Command Transformation (ACT), driven by an Open Source Wiki engine called Media Wiki, and all closely in approach related to the Wikipedia available on the internet with access on request.

In the past, the Alliance and its member nations have developed numerous capabilities for particular communities of interest (e.g. maritime, air, ground, logistics and intelligence). Most of these capabilities are inherently powerful but often fail to integrate horizontally with other functional services to provide dominant and effectual joint/combined capabilities. Furthermore, today's IT solutions do not necessarily dovetail with human processes such as knowledge gathering and decision making. It is increasingly essential to improve the interface between human and automated processes to maximize our abilities and to facilitate other transformational initiatives.

The Allied Command Transformation (ACT) as NATO's forcing agent for change, supports Transformation as a means to achieve Information, Decision and Execution superiority expressed by the TIDE concept.

1. TIDE is a life style (oversensitive males can call it a concept or initiative) whereby the emphasis is placed on sharing, reusing and collaborating.
2. TIDE is not a product itself but TIDE encourages product owners to step outside their traditional boundaries to improve the reuse value of their products.
3. TIDE emphasizes the power of modern technology while addressing critical human factors such as Usability, Usefulness, and Understanding.
4. TIDE aims to rapidly improve operational capabilities through iterative processes based on horizontal and vertical integration of existing and emerging products.
5. TIDE stimulates peaceful coexistence of NATO and National programs and services.
6. TIDE embraces network enabled principles and service oriented architectures.
7. TIDE is not about ownership or (hostile) takeovers. Stakeholders are simply encouraged to do business differently to ensure better cohesion across the board.

By capitalizing on NATO Network Enabled Capability (NNEC) developments, reusing existing capabilities as the foundation and by focusing on human cognitive processes, TIDE aims to quickly provide a joint Command, Control, Communication, Computer, Intelligence, Surveillance and Reconnaissance (C4ISR) capability. The TIDE Consortium will seek to break down current stovepipe barriers and place strong focus on rapid reaction forces and effects-based planning, decision-making and execution.

In 2003, ACT endorsed the Technology for Information, Decision and Execution Superiority (TIDE) concept. In an effort to provide the *"right capabilities"*, TIDE supports the following business practices:

- Collect data in real time through intelligence, surveillance and reconnaissance activities.
- Transform raw data into human understandable information in real time enhancing Information Superiority.
- Deliver the right information in the right format to the right person at the right place and the right time.
- Deploy cognitive science principles to allow humans to more rapidly gain knowledge and understanding.
- Deploy decision support tools to all command levels to ensure Decision Superiority.
- Support mission execution and subsequent monitoring to seek Execution Superiority.

The goal of TIDE is to contribute to the achievement of the Alliance to transform current forces and capabilities into forces capable of rapid reaction to any situation with the ability to achieve information superiority, decision superiority and execution superiority.

The purpose of the TIDE initiative is to rapidly improve the IT capabilities of the NATO Alliance by reusing existing systems/components and by steering current and future projects toward greater openness and cooperation in a common framework. The TIDE concept does not seek to take over these products or projects but rather to create and foster an environment to achieve transparent integration between services provided by the disparate systems/components. It was therefore decided to use the notion of a loosely coupled consortium of the willing to bring stakeholders together and move this initiative forward.

During and after the Cold War, it was quite common to specify, design and engineer Communication and Information Systems (CIS) along military service or functional area lines. This practice followed the approach taken for military organizations, doctrine, tactics and procedures where each service or functional area managed and developed their programs in isolation. With the introduction of joint warfare in the 1990's, it became clear that these traditional boundaries had to be eliminated to achieve horizontal integration of multi-dimensional capabilities into a powerful Joint Force. All aspects of a capability (typically referred to as: Doctrine, Organization, Training, Material, Leadership, Personnel, Facilities and Interoperability (DOTMLPFI) must be synchronized across traditional boundaries to ensure a powerful well-integrated force.

One of the major drawbacks of structuring CIS capabilities along service or functional areas lines is the overlap between systems and the struggle in achieving operability. It is quite common for existing CIS systems to each have their own mapping, relational database and message processing capabilities. Since all these implementations are slightly different, it is almost impossible to achieve common situational awareness (i.e. a Common Operational Picture (COP)), have common collaboration tools or provide joint planning and execution capabilities. Another important consequence of the existing business model is the significant Operations & Maintenance (O&M) cost due to overlapping, duplicating or conflicting parts of delivered systems. Harmonization of the capabilities in a Joint context is therefore a necessity to support emerging requirements and to reduce support costs wherever possible.

Today's threats are substantially different from a decade ago. We no longer face a large predominately static opponent that calls for relatively simple (but large scale) countermeasures supported by predefined CIS solutions in clearly defined regions. Today, we face an asymmetric threat from smaller but equally lethal groups at a global scale. Crisis scenarios are harder to predict requiring a much more flexible approach to capability specification, design and implementation. Most envisioned operations require relationships with new partners like non-governmental organization or non-member nations that go far beyond the traditional NATO borders. These new relationships will require new interfaces and in some cases a complete redesign of the CIS infrastructure (e.g. information security).

The NATO Network Enabled Capability (NNEC) initiative is starting to address the required capabilities in a network centric alliance. The word capability should be seen in the larger DOTMLPFI context and not just as material (e.g. hardware and software). The word Network does not just refer to the notion of computer networks but the art of networking which includes the integration of automated and human processes and collaboration. NNEC will have a significant impact on the way CIS capabilities are created and used.

Network Enabled Capability (NEC) is also the name given to the United Kingdom Ministry of Defense intent to achieve enhanced military effects through the better use of information systems towards the goal of *"right information, right place, right time - and not too much"*. NEC is envisaged as the coherent integration of sensors, decision-makers, effectors and support capabilities to achieve a more flexible and responsive military. In this future vision commanders will be better aware of the evolving military situation and will be able to react to events through voice and data communications. NEC is a long-term change program. The UK NEC is related to the US concept of Network-Centric Warfare (NCW), which at the time was described as *"translating an information advantage into a decisive warfighting advantage"*.

More recent US thinking has moved towards the vision of network-centric operations (NCO) which reflects that not all military activity is warfighting, for example peacekeeping. NEC is related to the Australian concept of Ubiquitous Command and Control (UC2), which includes not only network enabled capability, but network enabled intent, and network enabled awareness. UC2 extends the "*networking position*" of NEC and NCW, to include positions on decision devolution; embracing the ubiquity of available decision makers; a future of computing in achieving that ubiquity; the necessary human-computer integration in decision making; decentralization of intent and physical dispersion; social coordination protocols to unify intent, capability and awareness; and management levels to bound behaviors. In Sweden the armed forces are being transformed in a closely related program labeled Network Based Defense, while the European Union used in the EDA the term EURONEC or EU NEC, where the first prototype was the MARSUR Network 2011-2014, with the MARSUR Network in the LIVE PHASE starting 2014 is the first operational decentralized network in the European Union and Norway connecting the National Maritime Operation Centers.

However, in a constant flow new military concepts are emerging to deal with this new environment. Concepts like effects-based planning and execution are gaining momentum. Effects Based Planning (EBP) and Effects Based Operations (EBO)aim to achieve the maximum intended effect using the least amount of resources while constantly avoiding unintended side effects. The Effects Based Approach to Operations (EBAO) goes one step further and addresses the need to maximize the effectiveness of all instruments of power including Political, Military and Economic. These concepts will have a significant effect on CIS capabilities and the way they interact with human processes.

To support these new missions, architectures and concepts, it is recommended to adopt a component-based approach to CIS systems. Each component should provide a clearly defined capability (or service) and should be reusable in a number of environments and scenarios. Components are like Lego™ blocks in the sense that in themselves they are not significant but combined with other components, they quickly turn into powerful capabilities in support of a variety of modern NATO missions. A component view of the overall required capabilities can help to remove traditional service or functional area barriers, to lower acquisition risk due to the smaller size, to avoid duplication of capability and to lower acquisition and O&M cost though reuse.

Transition from a service/functional area approach to a component-based approach requires a new framework that allows the identification of requirements and the grouping of these requirements into component packages for implementation. Following is a proposal for such a framework in support of emerging missions and concepts.

The TIDE Framework presented here provides a mechanism to categorize the components in a network-enabled environment. To support effects-based doctrine, the collection of information system components can in the simplest form be seen as a black box representing all DOTMLPFI aspects of the capability taking raw data as an input and producing the intended effects as an output.

The conceptual framework was developed within the TIDE (Technology for Information , Decision and Execution Superiority) initiative and describes how Network Enabled Capabilities will transform raw data into intended effects and how they aid in achieving NATO's

Transformation Goals and Objectives.

Major components of the conceptual framework are:

1. The ENGINE. To enable all components in a network enabled environment to communicate, a robust physical infrastructure must be made available as the foundation or core of the enterprise. The Expandable NATO Global Infrastructure for a Networked Environment (ENGINE) provides this part of the capability. Often the ENGINE is referred to as a global grid by some NATO member nations. The ENGINE is a network of networks that provides a plug-and-play infrastructure to achieve end-to-end connectivity (including common services) between stakeholders (providers and consumers) in a networked enabled enterprise.

2. The Information Sphere. The information sphere contains the sum of all information required by a network-enabled enterprise, derived from the NATO Global Grid, implemented as a collection of logical information busses or virtual databases. Information sphere content must be accessible to all stakeholders, regardless of their location. The information sphere is not a tangible capability or component but rather a set of standards, agreements, models and protocols.

3. The Collectors. Collectors are the set of all automated, semi-automated and manual sensor, collecting, mining, analyzing, parsing, processing, translating, aggregating, fusing and correlating capabilities/ processes required to achieve information superiority in a network enabled enterprise. Collectors transform raw data into validated and recognized information.

4. The Decision Makers. Decision makers are the set of all automated, semi-automated and manual awareness, assessment, prediction, simulation, planning and decision making capabilities/processes required to achieve decision superiority in a network enabled enterprise.

5. The Effectors. Effectors are the set of all automated, semi-automated and manual political, economic, military and information capabilities/processes required to achieve execution superiority in a network enabled enterprise.

6. The Human Process. The human process comprises human thinking and acting – dependent on culture, personal education and experience, training attitudes, political will and views. Where human beings are involved in network-enabled enterprise they will influence and alter automated generated assessments, predictions and proposals. Using human experience and skills supported by automated processes provides an additional component on the way towards achieving decision superiority and desired effects. However, it might also be a factor leading to less qualified results and therefore has to be taken into consideration. The human processes in this framework span the collector, decision maker and effector columns highlighting the fact that human processes should not be seen in isolation but need to be dovetailed with automated processes and capabilities.

It is important to recognize that in a complex environment that includes the Alliance, its member Nations, partner Nations, non-military national agencies and non-governmental organizations, no single player has control over all aspects of the presented framework. The piece owned by NATO is very small and the alliance heavily depends on capabilities provided by partners to support its missions. Collaboration with national, industrial and agency partners is therefore key in the success of effects-based network-oriented missions. Achieving transparency at the security, information definition and interoperability levels is required to enable full plug and play of components from different stakeholders. NATO's role as standardization coordinator is vital in this sense.

To deliver information system capabilities to the alliance, an iterative multi-level spiral approach has been recommended by Allied Command Transformation (ACT). This approach brings together transformational thinking with traditional processes like defense planning, capital investment and operational maintenance ensuring accelerated delivery of more integrated solutions to the operational customers.

In the strategic spiral, political guidance is translated into NATO investment activities and national force commitments. The Defense Requirement Review (DRR) process must be synchronized with the NATO Security Investment Program (NSIP) to ensure that high level political ambitions are translated properly into complementary national and NATO actions.

The strategic spiral operates on a 2-year cycle and seeks to achieve political, operational, and resource endorsement for a capability vision. There is a need for a strategic information system operational perspective to be used as a framework for the development and follow-up of Capability Packages (CPs). This document would describe at the highest conceptual level what operational capability is required by what time. This would ensure that individual CPs and at a lower level projects within CPs, effectively contribute to a balanced and achievable capability. The "strategic perspective" is an operational framework that would not need approval by the resource community, but could be used in support of the development of the Mid-Term Resource Plan (MTRP). This could also be used to seek Nations' endorsement (consensus building) of the overall shortfall being addressed and a high level view of the required scope. The CP approval should then fulfill the requirement to gain Nations' approval for programming and project definition as normal.

Future, revised information system CPs should: be concise and simple; address a limited set of consolidated, joint and therefore integrated requirements; focus on a timeframe of limited duration (2-3 years) but could be for longer timeframe; and be revised frequently to reflect the dynamic operational environment (NRF cycle/requirements) and pace of technological change. This will contribute to more accurate program level cost estimation and accommodate the flexibility required in a transformational/spiral business approach. The CP revision process should be loosely coupled to the periodic ministerial meetings which establish revised priorities for NATO capability.

ACP should address the total requirements for a Capability Area and clearly reflect interdependency between Capability Planning and Force Planning processes including alignment of component delivery cycles. Availability of maintenance funding against a relatively static ceiling within the Alliance is a significant constraint in capability planning. Introduction of a life cycle view (including NATO manpower costs) as a critical element of support to forces will help to present a more integrated perspective to Nations.

In the capability review and screening process, there is discontinuity between political decisions/targets and the resources available to implement these agreements. Early notification of the military/political authorities must sometimes be achieved when implementation of a decision is realistically unachievable. Execution of the Strategic Spiral involves the broadest, most extensive effort. This spiral must execute first in order to provide the necessary guidance to the subsequent spirals.

In the present ACT transformational concept, Capability Package (CP) development is a serial process of Concept Development, Research & Technology (R&T), and Experimentation solution development followed by CP generation. In reality, CP development, submission, and approval must occur concurrently with these events or else information system delivery becomes a lengthy and inflexible process.

One of the major aims of the new transformational process is to close the C4ISR capability gap between NATO forces and selected Nations by achieving interoperable, compatible standards. Developmental interaction with Nations will include dialogue below NATO HQ (delegations, missions, IMS) level.

Although feasible, Capability Development under the NATO Securities Investment Program (NSIP), based on re-use of national products has historically often been confronted with difficulties. Development must generally involve an element allowing national industries an opportunity to compete and provide products.

Models exist in which products developed through laboratory and operational experimentation can be concurrently (1) fielded for operational use (with satisfactory support structure) and (2) successively formalized using industrialization and implementation resources made available through NSIP.

Substantial funding is programmed through existing CPs for Bi-SC AIS development in four categories: Consolidation, Core IS, Functional IS, and Architectural and Planning.

Development should be based on smaller, more manageable steps which provide for early delivery of a partial capability. This allows for a phased approach in which out-year increments can be scoped to address emergent needs and lessons learned from use of initial increments.

There are precedents where implementation of multi-year, task order based level of effort, and/or product-deliverable, and, technical services contracts requiring selected special personnel qualifications were funded by NSIP, e.g. for ACCS implementation support. In HQ SACT's view this would provide a more effective developmental environment. It is the Staff's view that, based on the above precedents, this could be achievable as long as the principles of scope definition; competitive bidding; and clear, definable output related to

CPs are included. As long as this clarity is achieved, it will be possible to closely phase both specification and implementation of projects.

Use of this approach should offer as much visibility and control to the Nations as possible through periodic reporting and milestone achievement. Such an approach should be presented to the IC in principle prior to actual submission of funding requests. The approach should be proposed within the context of 2261, also making use of the best value procurement that was recently agreed for a trial period. The TOPFAS phase II project is an example where a best value contract approach will be proposed in the near future. Once this example is validated, precedent can be used for future experimentation based solutions in AIS.

Under NSIP rules, the responsibility for the procurement method is with the Host Nation (HN). It is therefore essential that any implication of spiral development on the procurement aspect has the full support of the HN (NC3A) and that all aspects are fully documented by the HN. The IS is aware of initiatives being undertaken by NC3A in that area, and early coordination between SACT, NC3A and IS in the area is essential.

Allocation of increased resources for operational/military experimentation (SACT initiative) against a relatively static MBC Budget ceiling, may reduce the funding availability to support delivered Capabilities (affordability issue). An overall balance will need to be maintained in the allocations of resources between experimentation and actual operations.

The scope of experimentation includes multiple events of limited duration across the spectrum of experimentation types: e.g. concept development, laboratory experiments, limited field experiments, full live exercises. Experimentation will involve multiple partners (National tech centers, RTO, NC3A, etc.).

There are precedents to the funding of experimentation projects within NSIP, either as an integral part of the project(s), or as a separate project as long as the tenets of authorization are met: clear definition of scope and deliverables, competitive bidding, how deliverables will be used in achieving ultimate capability.

Late-stage validation of proposed solutions, development of designs and specifications, and feasibility studies are examples of transformational-like activities which have been funded in the past. However, NSIP cannot be used for concept definition and early exploratory experimentation.

Another significant issue that requires resolution is the need to bridge the gap between successful demonstration of an experimental baseline and the formalization (industrialization) of the solution. Today, this time gap can be a number of years and since operational expectations were raised during experimentation, a supported "leave behind" option must be evaluated. It is recommended to create a small ACT lead Support Entity that can provide limited operational support for the experimental baseline.

Technology for Information, Decision and Execution Superiority (TIDE)

Technology for Information, Decision and Execution Superiority (TIDE), former Transforming technology towards Information Decision superiority and Execution superiority, has the aim in NATO for pre-designed and pre-integrated Information Technology (IT) solutions as flexible and innovative IT capability to counter any potential threat in ever-changing global environments. TIDE could be soon renamed into Think-tank for Information, Decision & Execution Superiority.

TIDE is an enabling and iterative initiative that was launched in 2003 and endorsed by the first Supreme Allied Commander Transformation (SACT), ADM Giambastiani as a multi-national think-tank initiative for C4ISR technology & standards exploration, architecture development and human factors improvements. By reaching out to a diverse audience, the TIDE initiative explores pragmatic innovations that can assist NATO and its partners in quickly delivering highly effective yet cost efficient solutions.

TIDE Sprints are open source/information events where operational users, scientists, engineers, and managers get together to share information & ideas, synchronize activities, and develop transformational concepts.

The purpose of the TIDE initiative is to bring together C4ISR stakeholders (operational users, engineers, scientists, managers) from various NATO and National IT programs to improve interoperability and to support the transition to highly-networked capabilities through a bottom-up community driven approach. The TIDE initiative focuses on building human networks and is driven by a very simple question:

"How can we expect information systems to talk to each other if the people behind them have never communicated?[157]*"*

The TIDE community meet twice a year during think-tank meetings called TIDE Sprints. The spring Sprint is always in Europe and the fall Sprint is always in the USA. A Sprint is a focused "*doing*" session where people actively search for solutions to stated problems. "Death by PowerPoint" is not allowed at TIDE Sprints and all participants are required to contribute and network. ACT sets the agenda based on community input and moderates/facilitates all sessions.

The term "*Sprint*" was chosen to reflect the fact that we want people and their programs to rush together for one week and through intense dialogue rapidly solve interoperability issues. The term Sprint is frequently used in the open source community for the same purpose. Between Sprints (a period affectionately called the "*drift phase*"); online tools such as Wikis, Forums and other Social Media like tools are available at on Tidepedia to sustain the effort at a lower level of intensity.

Nations are fully engaged in TIDE activities and today many of them use TIDE standards

---

[157] Johan Goossens

and products for National purposes. For example, Finland, Germany, Italy, Poland, Portugal, Spain, Sweden, Turkey and even the United States use TIDE standards and/or products for their National Common Operational Picture (COP) and Maritime Situational Awareness (MSA) projects. Finland has even introduced TIDE standards in the EU for its maritime program.

Based on the Framework for Collaborative Interaction (FFCI), industry has become a major player since 2010. Examples of industry interaction include Cloud Computing, Mobile Computing and Maritime Situational Awareness. Over the years, the TIDE community has examined emerging academic and industrial standards and evaluated today's practical use. As a result of this work, they have compiled a list of practical solutions that are submitted for inclusion in the NATO Interoperability Standards and Profiles (NISP) publication.

TIDE Sprints

The Think-Tank for Information Decision and Execution (TIDE) Sprints, or TIDE Sprints, bring together the extended TIDE Community (NATO and Nations) to discuss and rapidly evolve TIDE concepts. As one of NATO's premier 'think-tank' events TIDE Sprint improves Alliance and Partner Nations interoperability by formulating concepts, developing roadmaps and exploring specifications that turn ideas into actionable outcomes. During this TIDE Sprint, experts from Alliance and Partner Nations, Military, Academia and Industry collaborate, exchange ideas and innovate to reach common solutions to many of the most pressing interoperability challenges faced by NATO. TIDE Sprint is part of a continuum of events that will improve interoperability, experts from Alliance and Partner Nations use them to 'fast-track' interoperability challenges to improve Federated Mission Networking (FMN) toward 'day zero' readiness. TIDE Sprint encourages scientists, engineers, managers and operators from National Governments, Military, Industry and Academia to collaborate and solve problems, they can do this because:

- They understanding of command and control (C2) and where interoperability needs improving.
- They can identify innovative opportunities offered by technology to overcome complex problems.
- They can share knowledge, actively participate and contribute in an environment that celebrates open-mindedness.
- They recognize a unique opportunity to improve interoperability between all Alliance and Partner Nations CIS.

TIDE Sprint is solving tomorrow's interoperability challenges, today as Scientists, Engineers, Managers and Operators take forward in a single week what would otherwise take many months to achieve.

The TIDE Sprint is steadily growing and transforming itself, just the 30. TIDE Sprint Autumn 2017 i.e. included twelve tracks that addressed detailed CIS interoperability issues that may be operational, technical, architectural or even 'human' in nature. The outcomes aligned roadmaps, develop concepts and affect priorities for developing and testing specifications.

During the daily Plenary Sessions, participants engaged with senior experts from NATO, Industry and National Governments on challenges surrounding Adaptive Manufacturing, the benefits of software virtualization and will have an opportunity to take part in a NATO Industry Forum (NIF) event to examine approaches to federating Cyber Defence. This makes the TIDE Sprint to a fruitful and wide ranged information platform for all participants and NATO itself.

The TIDE Sprint is self-funded through conference fees, all Nations/NATO command/Agencies/Industry pay for the own participation. The C2 Interoperability Deliverable (3_20) budget is only used to fund ACT travel and minor upkeep of the reference facility.

In the past, the NATO Alliance and its member nations have developed numerous capabilities for particular communities of interest (e.g. maritime, air, ground, logistics and intelligence). Most of these capabilities are inherently powerful but often fail to integrate horizontally with other functional services to provide dominant and effectual joint/combined capabilities. Furthermore, today's IT solutions do not necessarily dovetail with human processes such as knowledge gathering and decision making. It is increasingly essential to improve the interface between human and automated processes to maximize our abilities and to facilitate other transformational initiatives.

The TIDE has several event series like the Maritime Information Services (MIS) Conferences, TIDE Sprint Conference Tracks and TIDE related events and groups like i.e. the NATO Coalition Warrior Interoperability Exercise (CWIX), the NATO Network Enabled Capability (NNEC, dormant), the FMN Capability Planning Working Group (CPWG), the C3B Interoperability Profiles Capability Team (IP CaT) or the Mission Thread forum.

<u>Maritime Information Services (MIS) Conferences Working Groups include:</u>

- Maritime Information Services Technical Working Group (MTWG)
- Maritime Community User Group, former the MCCIS User Group (MCUG)
- MCCIS Life Cycle Working Group (MCCIS LCWG)
- NATO Project TRITON
- Maritime Information Services (MIS) Sub-Working Groups

<u>TIDE Sprint Tracks:</u>

- Operational Command & Control (OC2)
- Command & Control (C2) of the IE
- Communications
- Cyber Defence (CD)
- Data Centric Security (DCS)
- Enterprise Architecture (EA)
- Federated Mission Networking (FMN)
- Logistics
- Maritime Situational Awareness (MSA)
- Modelling & Simulation (M&S)
- Service Management (SM)
- Technology
- Open Source Intelligence (OSINT)

The Allied Command Transformation (ACT) as NATO's forcing agent for change, supports Transformation as a means to achieve Information, Decision and Execution superiority expressed by the TIDE concept. By capitalizing on NATO Network Enabled Capability (NNEC) developments, reusing existing capabilities as the foundation and by focusing on human cognitive processes, TIDE aims to quickly provide a joint Command, Control, Communication, Computer, Intelligence, Surveillance and Reconnaissance (C4ISR) capability. Therefore each track covers a specific area in C4ISR.

Operational Command and Control (OC2)

The Operational Command and Control (OC2) Track provides attendees with a wide array of operational and technically focused C2 updates, soliciting audience feedback, discussion, and support in ongoing initiatives. For the 30th edition of TIDE Sprint, the OC2 Track featured i.e. two themes, first focused on the end user by providing briefings and working sessions on current and future C2 technology solutions. Next was the OC2 Track with its attention to the capability developers by providing opportunities to better understand and influence NATO's Interoperability Standardization Program. Feedback received during these sessions was directly feed the ongoing development of key white papers and concepts.

Command and Control of the Information Environment (C2 of the IE)

Command and Control of the Information Environment (C2 of the IE) encompasses all of the aspects that have been discussed in previous TIDE Sprint Open Source Information (OSINT) and Social Media tracks. The C2 of the IE track is now aligned in such a way that is structured towards current developmental thinking within ACT so that there is direct cross over with other activities. Essentially C2 of the IE covers 4 areas that have concepts, policies or research and experimentation in various developmental stages within NATO that coalesce in this common goal. This TIDE Sprint track will be an important contributor to this effort. The 4 areas of C2 of the IE are as follows:

1. Publicly Available Information (PAI): The policies, sources, federation and tools that provide an assessment of PAI. This includes but is not limited to, Information Environment Assessment (IEA) and OSINT.
2. Strategic Awareness: The persistent, comprehensive and anticipatory network of collaborating experts assessing and understanding the IE for decision making at the Strategic level. This is categorized by building an understanding in 4 subject areas, that of the physical environment, the PAI environment, traditional intelligence capabilities and resilience of the NATO Alliance.
3. Strategic Communications: this requires an understanding the IE for planning, synchronizing and delivering effects. This includes Information Operations, Psychological Operations and Public Affairs and is the essential non-lethal element of synchronized effects.
4. Measuring Strategic Effects (MSE): PAI, Strategic Awareness and Strategic Communications support of MSE criteria and methodologies.

This track aims to gain a broader understanding across the community of the current work underway, identify gaps in the current construct and investigate potential conceptual and technical solutions associated with those gaps. Essentially this is about what the Alliance can do better within the Information Environment.

Communications

These events provides an excellent opportunity for the CPWG Communications Syndicate to gather in person and work on specific tasks and products. The 30. TIDE Sprint was the first of such occasions and was used to finalize the syndicate products in the Draft FMN Spiral 3 Specification.

As another follow-up from the work at the previous TIDE Sprint, the need to continue to work collectively on *"Ubiquitous Communications"* was recognized - which was then the major theme of the track. By now the concept proposal is finalized and that means that the TIDE participants can initiate concept development with the creation of the first draft.

Furthermore, the Communications track provides participants information about the latest developments in major communications programs in NATO, in the communications roadmap for Federated Mission Networking, in Protected Core Networking and the status of the respective STANAG, and feedback from the communications focus area in CWIX.

Cyber Defense

At the Warsaw summit, Allies reaffirmed NATO's defensive mandate and recognized Cyberspace as a domain of operations. This will improve NATO's ability to protect and conduct operations across all domains and maintain NATO's freedom of action and decision, in all circumstances. The 'operationalization' of the Cyberspace will come with deep modifications of existing mindsets, processes and structures and will entail the development or acquisition of new relevant assets. The Cyber Defense track deals with the answers to some of the many questions raised by this new Cyber Domain from a military, industrial or academic perspective.

The focus at this Tide Sprint will be on, but not limited to, the federation aspect of the defense of cyberspace, education and training of personal and finally to achieve contributions to a concept for Federated Cyber Defense and Federated Cyber Training. In order to do so, comparisons to organizations from outside that are in close coordination with NATO will be presented, their approach on training and federation in cyber defense is discussed. Furthermore the challenges within a federated cyberspace domain to assure Mission assurance will be scrutinized by been addressed, discussed and brought into line with the Federated Mission Networking requirements.

Data Centric Security

Data-centric security architecture rather than focusing on network perimeter defence focuses on securing access to the data itself. Data-centric security (DCS) effectively bridges the divide between IT security technology and information sharing. The DCS track aims at delivering a DCS Collaborative Capability Development Framework for the Alliance.

In addition to being vulnerable to insider threat, Network-Centric Security presents two information sharing challenges. First, it lacks an inherent capability to discriminate each information asset based on its protection requirement. Consequently, all information assets in the network are assumed to require the same level of protection. Second, Network-Centric Security protection is fix that is it lacks the ability to adapt to operational conditions.

Previous coalition discussions have demonstrated the potential DCS represents in addressing the first challenge. Indeed the coalition efforts in developing STANAG 4774 Confidentiality Metadata Label Syntax and STANAG 4778 Metadata Binding Mechanism address the need to differentiate information asset in order to apply the proper level of protection. However, little discussion has taken place on the second challenge; protecting information assets at a level commensurate with operational conditions.

TIDE Sprint 2/2017 DCS track addresses both challenges. Using CWIX 2017 DCS Interoperability Challenges the track will investigate STANAGs 4774 and 4778 technical implementation issues. Also using the result of CWIX 2017 DCS Table Top exercise the track reviews the Concept Proposal for *Dynamic Information Protection*." The Track is to follow-up on the TIDE Mini-Sprint on DCS Interoperability - Crypto Considerations findings and reviews the initial result from the MC tasking to develop a strategy for implementing a Data Centric Security Approach.

Enterprise Architecture (EA)

The recognition of the importance and maturity of Enterprise Architecture (EA) discipline is constantly growing within NATO Enterprise and the Alliance. The wider use and adoption of the architectures leads to the next generation challenges including but not limited to: federation environments, SMC and Cyber aspects, coherency among policies and directives, training provision, enterprise requirements management. The EA Track focuses on those development areas and identify best practices and lessons learned that could be applied immediately or in the near future.

In particular this edition of EA track will address the C3 Taxonomies 3.0 as a coherency enabling tool. Additionally the results from the last, and planning for the upcoming NATO Hackathon are reviewed and the enterprise architects support all the other tracks by providing EA expertise.

Federated Mission Networking (FMN)

Today, Federated Mission Networking (FMN) is the main NATO initiative regarding coalition interoperability. Inspired by a new way of thinking, trust and willingness to become interoperable and to achieve operational readiness and effectiveness, to-date 34 FMN Affiliates including six Partner Nations are gathering under the FMN umbrella, with other nations and organizations already having indicated their interest to join. Not far from reaching a steady state with a standing FMN Framework been established and a FMN Governance and Management structure in place, now FMN can be considered operational with the NRF 2017 being assessed FMN Spiral 1 Specification compliant. Meanwhile, the next FMN Spiral 2

Specification was developed awaiting approval and the development of the follow-on FMN Spiral 3 Specification has started.

The FMN Track during TIDE Sprint 02-2017, in light of the FMN Vision and the FMN Spiral Specifications Roadmap, aims at providing a status update on FMN-related developments since Spring 2017 and to discuss hot-topics in support of current and upcoming FMN Framework activities (e.g. the FMN Spiral 3 Specification development).

The FMN Track strives to continue sharing a common FMN knowledge in an open discussion forum in dedicated FMN Track sessions, while at the same time intend to increase cooperation and engagement with the other COIs represented at TIDE Sprint by deliberately reaching out to the Tracks through participation of SMEs to contribute to COI focused discussions in the context of FMN.

It is also important to notice that, beyond the FMN track, some "*domain*" tracks sessions, even if not joint, address FMN concepts or activities (for example, Technology track with the FMN core Services syndicates).

Logistics

The Logistics Track will provide communities of interest across NATO, industry and academia the opportunity to come together to discuss the potential of emerging technologies for future NATO logistics capabilities. The development of a persistent, agile, scalable and resilient NATO logistics capability is one of ACT's Focus Area of effort. The Logistics Track is in support of the Focus Area and feed the NATO Future Sustainment Concept.

The Logistics Track clearly understands itself as an idea-creating track and a driver for innovation. During the autumn 2017 event the focus was on 3D printing with the keynote presentation and panel discussion. FMN specifications for Logistics, Modelling & Simulation (M & S), Autonomous Systems, Big Data focusing on the newly introduced topic Logistics Intelligence will stimulate lively discussions and create ideas to improve the alliance's logistical capability.

Maritime Situational Awareness (MSA) Track

The most significant areas of naval interest up to 2030 will remain the defense of states (including maritime economic zones and stand-off deterrence) and protection of trade routes. Maritime distribution networks will be challenged with the aim to deny natural resources to states. This includes maritime freedom of navigation and commerce (e.g. pirates, undersea robots and sea mines). Potential adversaries already extend their reach beyond the littorals to blue water. They will seek to further increase their technical capabilities.

National and international efforts to control lawbreakers exploiting the freedom of the seas require the cooperation of navies. This requires a federated MSA which links directly to a Federated Mission Networking (FMN) approach to sharing of MSA data and information, since FMN provides the specifications to ensure interoperability. Improved MSA interoperability is part of NATO's increased interest in developing efficient capabilities to meeting existing and future challenges in the global commons. The MSA track will review the FMN Service Instructions for MSA as well as the MSA Interoperability Profile to migrate it into the FMN structure.

To exploit the advances in new maritime technologies for manned and unmanned water vehicles, both in deep water and the littorals, Nations should obtain the latest surface and subsurface maritime technologies including anti-submarine detection, and underwater warfare capabilities. While new technologies may provide tools to further improve interoperability, the disproportionate rates of technological development amongst Alliance Nations could lead to compatibility issues. This track covers especially the unmanned autonomous underwater technology, but will also discover links to autonomous technology in other domains and identify risks related to interoperability.

The amount of data gained by a wide variety of sources is huge. Systems dealing with these large data sets are complex and require a highly structured development process as well as sophisticated testing procedures. Modelling and Simulation Tools may support not only development and standalone testing, but will also change the way interoperability can be achieved and proven. In a combined discussion with M&S this track looks for ways of making better use of M&S. The track also provides a forum to exchange knowledge and experiences about existing systems, new technologies, and identifying further needs of experimentation.

Modelling & Simulation Track

The primary aim for this Tide Sprint is to advance on the understanding and scope of the FMN objective for spiral 4: *"To enable automated exchange of simulation data in support of multiple operational, education and training processes"*. Other aims are to continue our support to the logistic community to identify gaps that may be addressed using M&S; to define a process to capture M&S concepts and plan for experimentation and to advance in the achievement of the M&S as a Service approach (MSaaS).

Modelling & Simulation (M&S) is one of key enablers to enhance both operational and cost effectiveness across NATO, Allies and partners. M&S has been a success story for NATO in cross-cutting and inspiring new ways of thinking regarding support to operations, capability development, mission rehearsal, training, education and procurement. The NATO M&S vision as defined is to *"Exploit M&S to its full potential across NATO and the Nations to enhance both operational and cost effectiveness"*. (NATO M&S Master Plan, 14 September 2012, AC/323/NMSG(2012)-015).

Five objectives were identified for realizing the vision: Establish a common technical standard framework (VV&A) to foster a true common interoperability understanding and reuse of models, data and simulations across the Alliance; Provide coordination and common services to increase cost-effectiveness in a budget-constraint situation; New models and simulations development projects should be co-operative in nature and follow a common;

development process by reusing and networking existing capabilities whenever practical; Employ simulations to enhance NATO, member Nations, and partner operations in all application areas, utilizing a distributed simulation environment; Incorporate technological advances increasing functional capabilities, performance and overall M&S effectiveness.

Service Management Track

The IT Service Management Track is for Subject Matter Experts to discuss IT Service Management policy implementation within a NATO-led, partnership environment. Of particular interest are matters related to IT Service Management interoperability and its role in Federated Networks, both nationwide and between different service providers. It will highlight the impacts of shifting to a service-oriented approach and emphasis the need to take a life cycle view of interoperability.

Furthermore, the IT Service Management Track provides participants information about the latest developments in major NATO and national projects, in the IT Service Management Roadmap for Federated Mission Networking, in the SMC Roadmap, and feedback from IT Service Management testing in FMN Focus Area at CWIX.

Technology Track

The Technology Track gathers software engineers from multiple NATO organizations, the Nations, industry and academia to discuss, evolve and validate federation interoperability architectures and specifications.

During the TIDE Sprint participants established FMN Core Services Syndicate, with the initial goals to review and update the FMN Roadmap for Core Services, and plan 2018 activities to develop and validate required Service Instructions for Spiral 2 and 3. The track also explores opportunities and benefits in NATO environment of some emerging technologies: Internet of Things (IoT) and Containers.

The results from the track are incorporated in current and future FMN Spiral Specifications with the aim to improve the interoperability between systems in a federated environment.

Maritime Information Services Conference (MISC)

The Maritime Information Services Conference (MISC) is today the governing conference for all MSA activities related to NATO Maritime Command and Control Information System (MCCIS)[158] and Increment 1 + MSA as well as Increments 2-3 in the project TRITON, the MSA Demonstrator and MSA Prototype derived from the Baseline for Rapid Iterative Transformational Experimentation (BRITE).

The Maritime Situational Awareness Prototype (MSA Prototype) was/is a BRITE-Installation in CC MAR NORTHWOOD and in CCMAR NEAPLES (closed 2013) in order to maintains an maritime picture of commercial shipping (WHITE SHIPPING) and determine contacts of significant of significant interest to improve quality of decisions to act. This MSA operational prototype was developed to capture requirements, but was also used in Operations Active Endeavor and Ocean Shield to support MCCs.

Maritime Functional Services (MFS) are included into the set of processes followed within HQ SACT leading to delivery of capabilities for NATO and NATO Nations is a category of automated information services designed to support operational decision making by maritime commanders. In view of the important changes to NATO's operational objectives and processes in recent years, a new approach to the development and delivery of information services is required.

To guide this approach, a NATO-wide Information Services Strategy is being developed under the NATO C3 Organization. Maritime FAS developments will be compliant with: NATO Network Enabled Capability (NNEC) and therefore with NNEC Feasibility Study; NATO Architecture Framework (NAF); FUMARSER (Future Maritime Services) or Future Maritime Information Services (FMIS) in the project TRITON as part of Capability Package 9C0107 (CP 9C0107), whose core will be MSA enabling capability tools based on components coming from maritime related ongoing projects.

The mix up of the terms Maritime Functional Services (MFS), FUMARSER (Future Maritime Services), Future Maritime Information Services (FMIS) or the Maritime Information Services (MIS), defined as the functional services in the maritime domain, happens when people can't keep their own terminology clear, because it all aims at the same, to develop TRITON.

Out of these different approaches the Maritime Information Services Conference (MISC) was formed governing the Maritime Community User Group (MCUG), heading the MCCIS Technical Working Group (MTWG), the MCCIS User Group (only represented by Belgium, Denmark, Germany, the Netherlands, Norway and Poland), and Maritime Information Service Trainers Sub Working Group (MTSWG).

---

[158] Maritime Command and Control Information System (MCCIS) is since the 1990s NATO's existing principal maritime C2 system which is also used as the national maritime C2 exchange system by 22 NATO nations and basic C2 national system in some NATO nations (DEU, POL, PRT, DNK a.o.). This system is today obsolete and is to be replaced through the project TRITON developments.

FMIS (Future Maritime Information Services) is a generic term encompassing the maritime C2 capability to be delivered incrementally under Capability Package 9C0107 (CP 9C0107, Functional Services for Command & Control of Operations) and beyond. Initial FMIS operating capability will incorporate enhanced versions of the functionality provided through MCCIS and the MSA operational prototype validated in Phase I.

Subsequent levels of FMIS operating capability, to be delivered under the first increment, will deliver Mine Warfare Decision Making tools and Amphibious Warfare functionality. FMIS will interface with the remaining C2 elements of the Capability Package (Ref.: IMPLEMENTATION OF MARITIME SITUATIONAL AWARENESS (MSA) WITHIN FUTURE MARITIME INFORMATION SERVICES (FMIS) - IMSWM-0347-2011).

The Future Maritime Information Services (FMIS), Future Maritime Information Services (FMS) or former Future Maritime Functional Area Services (FUMARSER) cover information services in respect to the project TRITON and Maritime Situational Awareness (MSA). A prioritized list of possible candidates and/or mission types out of the NATO MSA Projects for FMIS are e.g.:

5. enhancement of NATO capabilities in the area of Maritime Situational Awareness (MSA)
1. employing, among others, components from the following main products like NSCIMA; BRITE; MCCIS; WISE;
2. the Naval Cooperation & Guidance of Shipping (NCAGS)
3. Allied Worldwide Navigational Information System (AWNIS)
4. Maritime Counter Terrorism Warfare (MCTW) Services
5. Maritime Environmental & Pollution Warfare (MEPW) Services
6. Maritime Illegal Immigration Warfare (MIIW) Services
7. Maritime Anti-Drugs Warfare (MADW) Services
8. RMP (part of the NATO COP project <u>NCOP</u>)
9. Maritime Operational Planning
10. Maritime Force Protection
11. Maritime Data Bases
12. Merchant Shipping Information Services
13. Maritime SAR Operations
14. Maritime Analysis of Exercises
15. Maritime Training
16. Computer Assisted Exercises (CAX)
17. Tactical Decision Aids (TDA's)
18. Maritime Special Operations
19. Maritime Air Operations Services
20. Mine Warfare (MW) Services
21. MEDAL (Mine Warfare)
22. MCM EXPERT (Mine Warfare)

23. DARE (Mine Warfare)
24. Water Space Management (WSM)
25. Anti-Surface Warfare Services (ASUW)
26. Maritime Land Attack Operations Services
27. Anti-Air Warfare Services (AAW)
28. Electronic Warfare (EW) Services
29. Amphibious Warfare (AW) Services
30. Anti-Submarine Warfare (ASW) Operations

Baseline for Rapid Interactive Transformational Experimentation (BRITE)

The Baseline for Rapid Interactive Transformational Experimentation (BRITE) is first of all a NATO Allied Command Operation (ACT) sponsored program, second it is an experimentation framework which allows for the rapid implementation of new ideas and capabilities to support experimentation. BRITE has been developed as part of the TIDE (Technology for Information, Decision and Execution superiority) initiative and is intended to rapidly improve the IT capabilities of the NATO Alliance by reusing existing systems / components and by steering current and future projects towards greater openness and cooperation in a common framework.

The experimentation capability development activity is built around a Service Oriented Architecture (SOA) framework which separates information, presentation, interoperability and business logic. Each of these is implemented using a modular architecture which allows application components to be added to them. The real benefit of BRITE comes when it is installed on a network with other TIDE compatible systems. TIDE compatible systems will discover each other on the network and work together to provide a richer information environment. The ability to discover, acquire and exploit this information is the purpose of BRITE.

BRITE can very easily and dynamically integrate the information from multiple sources and display it in tabular and graphical format. The ability to discover, acquire and exploit information from various sources is the purpose of BRITE. It can be used in isolation on a network as it provides complete capability from the acquisition to the presentation of information to improve situational awareness and support the decision making process. But, the real benefit of BRITE comes when BRITE is installed on a network with other TIDE compliant systems. TIDE compliant systems will discover each other on the network and work together to provide a richer information environment.

<u>BRITE had two main development phases:</u>

- First phase from 2005 to 2010, were the primary objective was development of BRITE prototype. Team of 8 (almost full time) engineers. 4 civilians (1 x A3, 3 x A2) & 4 contractors The 4 civilians were also requested to work on other activities (mainly, MCCIS and WISE support), but main priority was BRITE.
- Second Phase from 2010 to 2013, when the objectives were split between maintenance and stability development. Team of 1 A3, 1 A2 and different development teams provided by national companies (Spain, Germany, Italy). Outcome were several different versions of BRITE existed which NATO ACT attempted to merge into a single product.

BRITE is i.e. used as step II of the Maritime Safety & Security System (MSSIS) of the US-American Department of Transportation to limit and pre-evaluate the huge amount of available AIS contacts prior to a manual feed into the obsolete Maritime Command and Control

System (MCCIS), the Joint Common Operational Picture (JCOP), the Joint Situational Awareness Sharing Tool (JSAST) and in the Maritime Situational Awareness (MSA). BRITE Baselines according to NATO APFL are Version 1.1.2. (2010), 2.0/2.1 (2011) and is continuously updated.

The four development teams introduced their own technologies and products. The version of MSA BRITE handed over to NCIA (on 1st April 2013) included over 150 different technologies (The software installer had 15 technologies alone). ACT informed NCIA of 2 known significant problems: Having more than 100k tracks in the AIS DB would cause some serious slowdown of the system Having more than a certain number of concurrent tracks arriving within a second would cause performance issues and eventually result in crashes and more than 900 other documented issues were listed in the MSA issue tracker.

The MSA Prototype of BRITE was used in MCC NEAPLES/MARCOM NORTHWOOD with the O&M funding for MSA BRITE approved in mid 2013 with 160,000 Euros. (50% of what had been requested). ACO directed NCIA to only support the MSA BRITE prototype for MARCOM and focus should be directed at supporting the current version rather than introducing any new features. Funding was enough to hire a dedicated BRITE support developer, while the system management responsibilities were resumed by the MCCIS System Manager. Many of the ~150 technologies used by BRITE were no longer supported by vendors and BRITE was very complex and suffered from poor stability, i.e. the Live system at MARCOM (MSA Prototype) crashed at least once per day.

BRITE MSA patch (2.1.2) was released in September 2014, which included support for additional AIS messages. Extra tracks showed that the virtual machine (VM) platform in MARCOM was not adequate.

The NATO Shipping Centre at MARCOM signed User Acceptance Testing for the updated BRITE Version 2.2 in March 2015 and the MSA BRITE approved by NCIA Change Management in September 2015. The new hardware significantly improved performance in Version 2.2 and the dedicated servers showed huge improvement in stability of the product. Version 2.1.2 on VM platform processed around 50% of all data received. Version 2.2 running on dedicated hardware processed 99% of all data received.

The potential impact of increasing AIS feed into BRITE was obvious, as the introduction of a new Global AIS data feed was expected to increase the work load demand on the MSA BRITE prototype by 500%. It had not been possible to simulate this potential increase in capacity in a test environment. This increase would take data input to MSA BRITE Prototype well beyond the limits that were known at the time of the handover from ACT to NCIA.

This envisioned way-ahead as seen in 2016 was a significant increase of use of the MSA BRITE Prototype, which would cause the system to become unstable/less responsive. NCIA C2SL had made strong recommendations that any new AIS feed should support functionality that could split the feed into multiple smaller feeds and include the ability to alter the frequency at which data was sent. The multiple feed option would allow to have separate geographic feeds for different areas of the globe (e.g. one each for Mediterranean, North

Atlantic, Baltic Sea etc.). If the prototype would not support the full global feed, then this would allow us to disable feeds that were considered lower importance. A possibility was also to introduce multiple MSA servers to process each feed separately.

Unless the MSA BRITE prototype was tested against any potential new global coverage feed, NCIA C2SL would not be able to offer any guarantee that BRITE would be ready to process any planned new global AIS data feed. If the supplier of the new feed is able to provide multiple AIS feeds and allow the ability to throttle the speed at which data is sent, then there is a higher probability that the prototype would be able to cope with any new global feed.

Since 2014, NCIA C2SL has requested an additional 160,000 Euros to increase the level of manpower to support the MSA prototype. This increase in manpower would significantly increase the level of possible support. The increased funding for BRITE was declined in 2014, 2015 and 2016. The additional funding was again requested in the 2017 O&M budget and had been given priority by ACO J6.

In the following the MSA BRITE prototype became much more stable and powerful than the one inherited in 2013, however the MSA BRITE funding levels were only sufficient to provide day to day support for MSA BRITE administrators in MARCOM and to implement some stability improvements. NCIA C2SL couldn't guarantee the continued reliability and stability of the MSA BRITE prototype in case the capacity levels would be significantly increased in an uncontrolled manner. Introduction of a more agile development and release cycle assisted by virtual hosting technologies and changes to the current release process was to improve customer satisfaction.

BRITE usage increased continuously in the NATO MSA community:

- MARCOM has purchased extra AIS data feeds to increase the number of tracks and data within the system.
- BRITE 2.2 still had known limitations and not as stable as it could be.
- BRITE 2.3 was under development for release in mid 2016. (RFC created on 18th May 2016)
- BRITE 2.3 focuses on improving stability and performance and fixing noted security vulnerabilities.
- BRITE 2.3 is removing many of the unnecessary technologies tools that were included in the BRITE 2.1 version as many of these were no longer supported and include potential security threats.
- BRITE Beta Version 2.3 was hosted on Microsoft Cloud to allow for quicker feedback from user community.

Bi-Strategic Command Automated Information System (Bi-SC AIS)

The Bi-Strategic Commands Automated Information System (Bi-SC AIS) is one of the key elements of NATO CIS Capabilities, which includes a number of strategic sub-systems and Functional Services to be implemented under a Program. Objective of the Bi-SC AIS is to provide NATO commands with effective and integrated core (common to all users) and functional (specific to staff functions) services within both static and deployed contexts.

The range of functional services for supporting the operational commanders includes the C2 capabilities for Air, Land, Maritime, Special Operations Forces (SOF), Joint Operations, Logistics and Intelligence, including interfaces to NATO national systems and PfP-Nations, Non-Governmental Organizations (NGO), EU and UN, as required, over the period 2004-2012. The strategic NATO Consultation, Command and Control (C3) capability includes the following closely related and interdependent systems/program:

- the Automated Information Systems (AIS) the NATO General Purpose Communications Systems (NGCS);
- the Satellite Communication Post 2000 (SATCOMP2K).

The BI-SC AIS is one of the largest NATO common funded programs resulting from the Washington Summit in June 1999 and reinforced by the Prague summit in November 2002. More than 120 projects funded by a dozen Capability packages, implemented by several Host Nations, will contribute to establish the foundation of the NATO Automated Information Services. Core and Functional Services will provide the necessary capabilities for both NATO C2 and supporting staff work to both the static and the deployable environments.

A key objective of the Bi-SC AIS is to achieve the implementation of Core and Functional Services in an integrated and coherent way on the same desktop. The Bi-SC AIS Program Implementation Strategy described the overall concept and key objectives. The management structure is based on a 3-pillar concept:

- the System Management Authority (SMA), carried out by the Strategic Commands;
- the Implementation Management Authority (IMA), carried out by NC3A; and
- the Operational Support Authority (OSA), carried out by NC3A.

The Bi-Strategic Command Automated Information System (Bi-SC AIS) is the information system which supports the Command and Control business process when exercised in a NATO Command in either a static or deployed situation. The Bi-SC AIS is a series of Information System nodes which are connected together by the NATO Global Communication System (NGCS). The Bi-SC AIS is a customer of the NGCS which provides communications also for other NATO systems.

The Bi-SC AIS is logically one information network but physically is divided into several networks operating at different security levels. In order to simplify operation and to reduce operating costs, the various networks follow a common architecture and are constructed to the maximum extent possible from identical hardware and software components.

On 24 April 2016, the implementation and configuration of the NATO Information Portal (NIP) pilot site (SHAPE) was completed, and IOC was achieved. This milestone represents the first important step towards reaching the full roll-out of all Bi-Strategic Commands (Bi-SC) NIP sites. NIP basic functionalities are now available online.

The NATO Information Portal (NIP) project is a web-based application that enables Bi-SC Automated Information System (AIS) network users to publish and maintain web content across commands and specific communities of interest. The AIS Information Portal also serves as an important element of the emerging NATO Network Enabled Capability (NNEC). As a key functionality to enable the achievement of NNEC, the progressive evolution of portal technology will contribute to the pace of reaching an overall net-centric capability.

The NIP offers dynamic web pages that pull last available content through automatic queries from the "authoring" space instead of static textual pages. The NIP will provide new functionalities such as improved web hosting capabilities, improved federated search, web interfaces to data handling systems (DHS) and tasker tracker enterprises (TTEs). It will also provide social computing/networking functionalities and:

- standardize and centralize a Bi-SC Information and Knowledge Management tool; the initial scope of the project is the Operational Network (ON), while the new project P100 will mirror same NIP functionalities on the Private Business Network (PBN) in 2017-2020;
- offer NATO Functional Area Services (FAS) service-oriented, on-demand web hosting capabilities and a multi-tenant environment; the initial scope of the project is the delivery of a web hosting and multi-tenant environment, until a new project from CP150 enhances interoperability online services in accordance with the STANAG 4774 and the STANAG 4778 (2018-2020 time period, following the IKM Tools Program's roadmap).

The Intelligence Functional Services are i.e. combining the:

- ACO Open Source System for Intelligence (AOSS-INTEL)
- Battlefield Information Collection and Exploitation System (BICES)
- Counter-Improvised Explosive Device Toolbox (C-IED Toolbox)
- Intelligence Toolbox
- Intelligence Training Management Software (ITMS)
- Joint Operations/Intelligence Information System (JOIIS)
- Request for Information Management System (RFIMS)
- I2 Analyst's Notebook and I2 Analyst's Workstation
- I2 ChartReader

Battlefield Information Collection and Exploitation System (BICES)

Battlefield Information Collection and Exploitation System (BICES) is a joint US/NATO initiated network to integrate current and future intelligence networks (N2). BICES is intended to coordinate and exploit battlefield intelligence gathering among all NATO commands and participating nations. BICES supports faster and more accurate detection and assessment of the military situation of foreign countries.

For this purpose, knowledge from all available news sources is gathered effectively to support national N2 distribute the information to users within NATO. The function is to enable with an encrypted exchange of proprietary owned information between departments of the military within the NATO nations and thereby ensure the ownership of the providing party in further distribution of the data. BICES is the only way to access the Linked Operations-Intelligence Centers Europe (LOCE), one reason BISEC is a network connecting already a large number of nations and MSA stakeholders.

Maritime Command and Control System (MCCIS)

The Maritime Command and Control Information System (MCCIS) provides Maritime Functional Services in NATO's command and control environment on the Bi-SC Automated Information System (Bi-SC AIS).

MCCIS grew out of the Alpha CCIS program that delivered a modern maritime C2 capability in the early 1990's to SACLANT (Norfolk) and CINCIBERLANT (Oeiras, PO). The Alpha CCIS/MCCIS program developed strong ties with national programs. A close relationship with the US Navy has existed since 1992 to ensure operational interoperability and technology synchronicity. The power of MCCIS was quickly discovered by other NATO commands/Nations and full proliferation to all corners of the Alliance (including 16 nations) occurred over the next 10 years. Today, the MCCIS product and community contribute a high quality Recognized Maritime Picture (RMP) to NATO's situational awareness and Common Operational Picture (COP). The vertical cooperation and integration between National and NATO commands is well developed. MCCIS has been aggressively enhanced over the last 12 years based on a spiral development methodology. Sixteen upgrades (5 major) have been delivered to meet emerging operational requirements. Modern technologies have been inserted along the way and NATO's de-facto standard web portal (WISE) resulted from the MCCIS program. It should be argued that MCCIS is one of NATO's earliest forward-looking transformational programs.

The "M" in front of CCIS is a misnomer as MCCIS has always had a substantial Joint capability, however it was first used during the exercise Strong Resolve in 1998, afterwards MCCIS was providing the COP to NATO commanders during the security operations for the Olympic Games in Athens.

The network centric capabilities of MCCIS and WISE enabled consumption of this COP are provided throughout NATO's Wide Area Network (WAN). MCCIS is highly interoperable with other NATO and National systems. Since the beginning MCCIS proves to be capable of exchanging Joint and Combined information (e.g. Land, Air, Maritime, Space, Intel, Logistics and Environmental) with numerous external systems in the Coalition Warrior Interoperability Demonstration (NATO CWID, today the CWIX).

MCCIS has been managed on a Bi-SC basis since 1997 under an agreed Protocol and NATO Nations have made substantial contributions at the life cycle management, operational and technical user level. The MCCIS life cycle process and methodology is modern, iterative, responsive, decentralized and flexible. It has been a major reason for the success of MCCIS as a product in NATO and National environments. Continued implementation of this methodology is a key element for future system operation. The current MCCIS approach is highly recommended for future capability management that aims to quickly deliver solutions to counter asymmetric threats in complex and rapidly evolving Joint, Coalition and Non-Governmental environments.

With the advent of a major restructuring within the NATO Military Authorities to establish

Allied Command Transformation (ACT) and extend the responsibility of Allied Command Operations (ACO) to encompass all NATO operations, new Terms of Reference were assigned to each Strategic Commander. This reorganization has precipitated the realignment of specific responsibilities for system management. In the case of MCCIS, the requirement was established, as expressed in the reference D. letter, to transfer management authority for the fielded, operational MCCIS baseline to SHAPE J6 exclusively and responsibility for in-service support to NCSA. As part of this transfer process, enhancements to selected elements of the baseline are required to facilitate overall system stability and accommodate the implementation of architectural changes consistent with the Bi-SC AIS Reference Architecture. As a result of the MCCIS transfer of responsibility from ACT to ACO & NCSA an MCCIS Stabilization and Enhancement Project is being implemented.

MCCIS 5.2 and MCCIS 5.3 have been delivered by ACT but MCCIS 6.0, subcontracted to Northrop Grumman as established in MCCIS Stabilization and Enhancement Project, was delivered since 2008 under NC3A MCCIS Stabilization Project Management responsibility until its replacement. MCCIS has been enhanced over the last 12 years based on a spiral development methodology. Nearly twenty upgrades have been delivered to meet emerging operational requirements, five of them major ones. Modern technologies have been inserted along the way and NATO's de-facto standard web portal (WISE) resulted from the MCCIS program, and was base for the MSA BRITE prototype.

However, the support for the HP rp34xx (PA-RISC architecture) hardware platforms discontinued and software support for the associated HP-UNIX operating system version (HP-UX 11iv3 11.31 PA RISC) will end in late 2020. The HP Integrity rx2800 i4 Itanium server platform is no longer produced. Thus, only the HP Integrity rx2800 i6 server platform was left available as a hardware platform for MCCIS, while the software support for the associated HP-UNIX operating system version (HP-UX 11iv3 11.31 Itanium) will end in late 2025. In 2018 the Approved Fielded Product List (AFPL) included MCCIS Versions 6.2, 6.3, 6.3.1, 6.4, 6.4.1 and 6.4.2 approved for use in the NS-WAN by the NATO Security Advisory Board (NSAB). The latest MCCIS release in 2018 was MCCIS Version 6.4.2.

MCCIS Version 6.5 is planned to support both the classified ADatP-3 Baseline 13.1 (APP-11 (C)) and Baseline 15 (APP-11 (D)) in the future. MCCIS Version 6.5 / 6.5.1 will be the last MCCIS Version with new operational capabilities supported by the NCIA due to the end of the contract in 2021, afterwards only Patch-Versions will be released. After the OS OTHT-Gold Version (2000 Rev. D NATO UNCLASS) the US developed a OS OTHT-GOLD Version (Revision14) possibly to be integrated into the US national GCCS-M system. This Integration would also have direct impact on MCCIS as approximately 70% of the NATO situational picture is delivered by the US military through the Radiant Mercury Network using a GCCS-M system.

Harmonization between the different NATO and US as well as the EU and the worldwide used standards is imminent for the future of character oriented information exchange in NATO and beyond.

TRITON

NATO Project TRITON is to replace the obsolescent NATO Maritime Command & Control System (MCCIS) and to deliver Maritime C2 capabilities to the NATO Command Structure, NATO Headquarters and Operation Centers, including the Joint Warfare Center, Joint Forces Training Center (JFTC), respectively the subjects in the NATO Communications Information Agency (NCIA), NATO Command, Control and Information System School (NCCIS) and Training Centers.

TRITON is funded in the NATO Security Investment Program, part of the Capability Package CP09107 with the Functional Services for Command & Control of Operations; also the Future Maritime Information Service (FMIS), formerly known as Future Maritime Functional Areal Services (FUMARSER). Project TRITON is 1 of 13 Functional Services (FS) to support the Planning, Conduct, Analysis and Evaluation Operations and Exercises. Project Triton (NSIP) will deliver Maritime Information Services to replace MCCIS and additionally MSA and Mine Warfare functionality. The common funded part of this project will be executed under CP 9C0107 at an estimated cost of EUR M13 (est. 2016). TRITON development will take into account the MCCIS CLASSIC (until Version 6.0) and TIDE with BRITE and other software as services in the development.

Project TRITON is intended not only to replace the Maritime functionality that was provided by MCCIS, but also to incorporate all emerging MSA requirements (NATO BRITE MSA) as well. With this in mind and close participation from the MSA community, namely the Maritime Components Commands (MCCs), including the Naval Cooperation and Guidance for Shipping (NCAGS) cells from the operational side and the MSA Integrated Project Team (IPT) on the technology and interoperability side, the Maritime Command and Control Reference Architecture (MC2RA) was developed.

MCCIS, BRITE and TRITON are subjects under the bi-annual held Maritime Information Services Conference (MISC) with representatives from Nations and NATO entities (Allied Command Transformation, Allied Command Operations, NC3A and NCSA) and addressed issues related to current and future Maritime C2 Information Services. From an operational and transformation perspective the MISC focuses the provision of Maritime C2 tools, Maritime Situational Awareness, and the associated sense making and decision making tools to enable effective C2. The Conference addressed issues related to: the operation and maintenance of MCCIS, particularly the hardware baseline, Project TRITON (the NATO funded implementation program to replace MCCIS) and associated requirements capture; the future Bi-SC Maritime C2 Capability Development plan, CWIX and the associated Maritime Aspects; and the Multi-National Maritime Information Service project (MNMIS).

The first time a Full Operational Capability (FOC) form TRITON was projected for June 2013 since the MCCIS End-of-Life-Time was December 2013. The project had to deal with the strains of NATO procurement and slow process reality, the time line was continuously changed with an estimated IOC of all components only for the first Increment in 2020 and

FOC of Increment 3 in 2027. Due to the delay the MCCIS Life-Cycle Extension required several extensions now terminated accordingly. The current Maritime Information Services in TRITON are:

- Maritime Command & Control Information System
- Maritime Situational Awareness Prototype (BRITE)
- Mine Warfare Planning Tools (with MCM Expert and DARE)

Triton will have a Mission Partner Environment Information Service (MPE IS) in order to implement national services into the NATO environment. Project TRITON will further implement ADatP-3 Baseline 15, however the new time table pushed the timelines further into the future:

- Initial Operational Capability (IOC) of Increment 1 estimated in November 2020
- Enhanced Operational Capability (EOC) estimated beginning 2021
- Enhanced Operational Capability (EOC) of Increment 2 estimated end of 2024
- Full Operational Capability (FOC) of Increment 3 estimated end of 2027

North Pacific Coast Guard Forum (NPCGF)

The North Pacific Coast Guard Forum (NPCGF)[159] was initiated by the Japan Coast Guard in 2000 as a venue to foster multilateral cooperation through the sharing of information on matters related to combined operations, exchange of information, illegal drug trafficking, maritime security, fisheries enforcement, illegal migration, and maritime domain awareness. The current membership includes agencies from Canada, China, Japan, South Korea, Russia, and the United States.

The first Forum was held in Tokyo in 2000 and has followed an alternating semi-annual cycle of technical experts and principals meetings since. Between 2000 and 2005, meetings were sponsored by Japan, Russia, United States, South Korea, Canada and Japan.

The forum has had success in documenting best practices from the member countries in areas of illegal drug trafficking, maritime security, fisheries enforcement and illegal migration, has a web-based information exchange system, and has published a manual for combined operations.

The cooperation fostered in the forum was successfully tested during a tabletop exercise in Victoria, British Columbia, Canada in March 2004 involving the tracking of a fictitious vessel suspected of carrying weapons of mass destruction as it transited the international waters off each member country's coast. During the summer of 2005, the forum planned and executed actual at-sea combined operations were conducted to enforce Illegal, Unregulated, Unreported (IUU) Fisheries regulations in the North Pacific Ocean. Member nations noted the success of the 2005 combined operations and have continued them in recent years.

The success of the NPCGF encouraged the United States to take the lead on establishing a similar forum for north Atlantic countries. In 2005 and 2006 the US Coast Guard worked with dozens of European countries and Canada to advance the idea and in August 2007 a plenary session will be held in Sweden to finalize the details of that organization.

---

[159] https://en.wikipedia.org/wiki/North_Pacific_Coast_Guard_Agencies_Forum

Pacific Islands Forum (PIF)

The Pacific Islands Forum (PIF)[160] is an inter-governmental organization that aims to enhance cooperation between the independent countries of the Pacific Ocean. It was founded in 1971 as the South Pacific Forum, but in 1999, the name was changed to Pacific Islands Forum in order to be more inclusive of the Forum's Oceania-spanning membership of both north and south Pacific island countries, including Australia. PIF is an observer at the United Nations.

The mission of the Pacific Islands Forum is *"to work in support of Forum member governments, to enhance the economic and social well-being of the people of the South Pacific by fostering cooperation between governments and between international agencies, and by representing the interests of Forum members in ways agreed by the Forum"*. Its decisions are implemented by the Pacific Islands Forum Secretariat (PIFS), which grew out of the South Pacific Bureau for Economic Co-operation (SPEC). As well as its role in harmonizing regional positions on various political and policy issues, the Forum Secretariat has technical programs in economic development, transport and trade. The Pacific Islands Forum Secretary General is the permanent Chairman of the Council of Regional Organizations in the Pacific (CROP).

Australia and New Zealand are generally larger and wealthier than the other countries that make up the rest of the forum, Australia's population is around twice that of the other 17 members combined and its economy is more than five times larger. They are significant aid donors and big markets for exports from the other countries. Military and police forces as well as civilian personnel of Forum states, chiefly Australia and New Zealand, have recently been part of regional peacekeeping and stabilization operations in other states, notably in Solomon Islands (2003-) and Nauru (2004-2009), under Forum auspices. Such regional efforts are mandated by the Biketawa Declaration, which was adopted at the 31st Summit of Pacific Islands Forum Leaders, held at Kiribati in October 2000.

In September 2011, the U.S. territories of American Samoa, Guam and the Northern Mariana Islands were granted observer status in the Pacific Islands Forum, while in September 2016 the French territories of French Polynesia and New Caledonia were granted full membership.

From 5.-7. August 1971, the first meeting of the South Pacific Forum was initiated by New Zealand and held in Wellington, with attendants of the following seven countries: the President of Nauru, the Prime Ministers of Western Samoa, Tonga and Fiji, the Premier of the Cook Islands, the Australian Minister for External Territories, and the Prime Minister of New Zealand. It was a private and informal discussion of a wide range of issues of common concern, concentrating on matters directly affecting the daily lives of the people of the islands of the South Pacific, devoting particular attention to trade, shipping, tourism, and

---

[160] https://www.forumsec.org

education. Afterwards this meeting was held annually in member countries and areas in turn. In 1999, the 30th South Pacific Forum decided to transform into Pacific Islands Forum, with more extensive and formal way of discussion and organization. Immediately after the forum's annual meeting at head of government level, the Post Forum Dialogue (PFD) is conducted at ministerial level with PFD development partners around the world.

In August 2008, the Forum threatened to suspend Fiji if the latter did not commit to holding a general election by March 2009. Subsequently, at a special leaders' meeting of the Pacific Islands Forum, held in Papua New Guinea in January 2009, Forum leaders set a deadline of 1 May, by which date Fiji must set a date for elections before the end of the year. Fiji rejected the deadline. Consequently, on May 2, Fiji was suspended indefinitely from participation in the Forum with immediate effect.

The Chair of the Pacific Islands Forum and Premier of Niue, Toke Talagi, described the suspension as "*also particularly timely given the recent disturbing deterioration of the political, legal and human rights situation in Fiji since April 10, 2009*". He described Fiji as "*a regime which displays such a total disregard for basic human rights, democracy and freedom*" which he believed contravened membership of the Pacific Islands Forum. Talagi emphasized, however, that Fiji had not been expelled and that it would be welcomed back into the fold when it returned to the path of "*constitutional democracy, through free and fair elections*".

The 2009 suspension of Fiji marked the first time that a country had been suspended from the Pacific Islands Forum in the history of the then 38-year-old organization. Following the Fijian general election of 17. September 2014, the Forum lifted the suspension of Fiji on 22. October 2014.

The Pacific Island Countries Trade Agreement (PICTA) aims to establish a free-trade area between 14 of the Pacific Islands Forum countries. As of 2013, it had been signed by 12 states, here Cook Islands, Fiji, Kiribati, Micronesia, Nauru, Niue, Papua New Guinea, Samoa, Solomon Islands, Tonga, Tuvalu, and Vanuatu. All of the signing states have ratified the treaty, with the exception of Micronesia. As of March 2008, six countries had announced that domestic arrangements had been made enabling them to trade under the agreement, here Cook Islands, Fiji, Niue, Samoa, Solomon Islands, Vanuatu. PCTA has not been signed by either Palau or the Marshall Islands.

After the trade agreement goes into force, countries commit to removing tariffs on most goods by 2021. As of April 2008, The Forum Island Countries are also negotiating an Economic Partnership Agreement (EPA) with the European Union. It is important to note that the PICTA discussed here covers only the trade of goods. At the Forum Island Leaders Meeting held in Rarotonga, Cook Islands on 28. August 2012, nine members signed the Pacific Island Countries Trade Agreement Trade in Services (PICTA TIS). As of April 2008, there is an ongoing negotiation to design and agree on a protocol to include trade in services and the temporary movement of natural persons (a broader concept than the GATS's Mode 4).

The Office of the Chief Trade Adviser was established on 29. March 2010 to provide independent advice and support to the Pacific Forum Island Countries (FICs) in the PACER Plus trade negotiations with Australia and New Zealand.

An *"open skies"* policy has been under work by a number of nations. The Pacific Islands Air Services Agreement or PIASA would allow member nations to have more access for their airlines to other member countries. To date there have been ten signatories, Cook Islands, Kiribati, Nauru, Niue, Papua New Guinea, Samoa, Solomon Islands, Tonga, Tuvalu and Vanuatu, while only six have ratified the agreement. These six are Cook Islands, Nauru, Niue, Samoa, Tonga and Vanuatu.

At the 19.–20. August 2008 Pacific Islands Forum meeting in Niue, the leaders discussed Pacific Plan priorities including, *"fisheries, energy, trade and economic integration, climate change and transport, in addition to information and communication technology, health, education, and good governance."* Leaders also discussed the impacts of climate change and adopted the Niue Declaration on Climate Change. Restoration of democratic governance in Fiji was discussed as were consequences should the interim government fail to meet established deadlines. The Pacific Region Infrastructure Facility was launched 19 August 2008 to provide up to A$200 million over four years to help improve infrastructure in Kiribati, Samoa, Solomon Islands, Tonga, Tuvalu and Vanuatu.

The United Nations announced that it would partner with Samoa to develop an Inter-Agency Climate Change Centre to help Pacific island nations combat the impacts of climate change in the region. In the 2013 forum, the Marshall Islands, supported by all other Pacific nations, claimed compensation from the United States for the nuclear tests conducted on the islands during the 1940s and 1950s. There has been a call from within both the Australian and New Zealand business communities to extend the CER (Closer Economic Relations) to other Pacific island nations, moving towards a single market and allowing the free movement of people and goods. A Pacific Union has been theorized as the next step of the forum.

In September 2016, the Pacific Islands Association of Non-Governmental Organizations (PIANGO) regional network, encouraged member states' leaders to include in the organization's agenda the issue of human rights violations in West Papua.

Pan-African Association for Port Cooperation (PAPC)

The <u>Pan-African Association for Port Co-operation (PAPC)</u>[161] is a federation of the three sub-regional port Associations in Africa. Since its establishment through its biannual flagship PAPC Conference, the federated body has promoted continent-wide forum for ports and the maritime business community to meet, share experiences and deliberate on the challenges and prospects of the port, maritime transport and trade sector of the continent.

As cooperation for port trade, operations and efficient service delivery information sharing is being enhanced through the establishment and works of the three sub-regional associations of port authorities, namely the:

- Port Management Association of West and Central Africa (PMAWCA)
- Port Management Association of East and Southern Africa (PMAESA)
- and the Union of Port Administrations of Northern Africa (UAPNA)

Individually, these three ports Associations established since 1972, worked and did achieve significant milestones in their respective regions, however, there was the need to harmonize these exchanges and create a continental pool of port Authorities that could ensure that effectively, bad lessons in one sub-region were not repeated in another region and that good lessons and best practices in one sub-region could be very efficiently and harmoniously shared and replicated all over the continent. It was this dream that led to the establishment in 1999 of the Pan African Association for Ports Cooperation (PAPC), which is a federation of PMAWCA, PMAESA and UAPNA.

In July 1999, the first meeting of the ad-hoc committee for the creation of the African ports association was convened by ECA, with the participation of all three sub-regional port associations (PMAWCA, UAPNA and PMAESA), in Addis Ababa (Ethiopia). The committee reviewed the draft Constitution for the Association, and agreed on its name, the Pan-African Association for Port Co-operation (PAPC).

The draft constitution, soon supplemented by the Rules of Procedure of the Association, was reviewed and adopted by each of the three sub-regional associations, on the occasion of their respective Council Meetings.

The 2nd meeting of the ad-hoc committee was held in Conakry (Guinea), in April 1999, in conjunction with the PMAWCA Council Meeting. The Conakry Declaration, adopted by the committee, approved the Constitution, and paved the way for the final meeting of the ad-hoc committee, which would launch the PAPC.

The 3rd and final meeting of the ad-hoc committee was held in Alexandria (Egypt), in

---

[161] http://agpaoc-pmawca.org

conjunction with the Council Meeting of UAPNA, in November 1999. The Alexandria Resolution celebrated the launching of the activities of the PAPC, and confirmed the leading role played by PMAWCA in the establishment of the PAPC by appointing the then Chairman of PMAWCA as spokesperson of PAPC, pending election of the organs of the Association.

Port Management Association of Eastern and Southern Africa (PMAESA)

The Port Management Association of Eastern and Southern Africa (PMAESA) was first established as the Port Management Association of Eastern Africa, in Mombasa, Kenya, on April 1973, under the auspices of the United Nations Economic Commission for Africa (ECA), following a recommendation made at a meeting of the African Ministers in charge of transport, held in Tunisia in February 1971.

PMAESA is a non-profit, inter-governmental organization made up of Port Operators, Government Line Ministries, Logistics and Maritime Service Providers and other port and shipping stakeholders from the Eastern, Western and Southern African and Indian Ocean regions. The PMAESA Secretariat, based in Mombasa, Kenya was established to coordinate the activities of the Association.

PMAESA's primary objective is to strengthen relations among member ports with a view to promoting regional cooperation and subsequently regional integration. The Association offers an appropriate framework for exchange of information and ideas among members and to create an enabling environment whereby member can interface with one another in the port, transport and trade arenas.

PMAESA also works towards improving conditions of operation and management of ports in its region of coverage with a view to enhance their productivity.

PMAESA seeks to maintain relations with other port authorities or associations, regional and international organizations and governments of the region to hold discussions on matters of common interest. Maritime safety and protection of the marine environment, transit transport, port operations issues such as port statistics, the public sector-private sector partnership, communication, the cruise industry and regional cooperation are PMAESA's main areas of activity.

Over the years, PMAESA has strengthened its relationships with the other African port management associations: the North African Ports Association (UAPNA), and the Port Management Association of West and Central Africa (PMAWCA). This co-operation established the Pan-African Association for Port Co-operation (PAPC). PMAESA has tried to forge links among its member countries so as to fulfill the ambitions of the Association.

Port Management Association of West and Central Africa (PMAWCA)

The Port Management Association of West and Central Africa (PMAWCA) is a sub-regional intergovernmental economic organization, made up of twenty two regular member Ports and Nine Associate members including landlocked Countries and Maritime Organizations, all located along the West Coast of the Africa Continent extending from Mauritania to Angola. The Association was established during the historic inaugural Assembly of October 1972 in Freetown (Sierra Leone) by the United Nations Economic Commission for Africa (UNECA).

PMAWCA is a specialized organ of the Maritime organization for West and Central Africa (MOWCA). The first statutes were signed by the representative of Nine (9) member ports and the Secretary General of ECA.

Objectives are to:

- Contribute towards the improvement, coordination and harmonization of port and harbor activities, services and infrastructure in the West and Central African sub region so as to increase the effectiveness of their services to ships and other means of transportation;
- Help, in relationship with similar Port Organizations or concerned Governments, to strengthen cooperation among member Ports, in a manner that will encourage the development of their activities,
- Establish and maintain relationship with transport enterprises, institutions, associations, Governmental or international Organizations in order to have a closer look at problems facing the members;
- Establish a forum for the exchange of ideas among member Ports to freely discuss their common problems, so as to contribute to regional integration.

PMAWCA is made up of twenty two (22) regular member Ports and Nine (9) associate members including landlocked Countries and Maritime Organizations, all located along the West Coast of the Africa Continent Stretching from Mauritania to Angola covering a coast line of about 9,400km on the Atlantic Ocean.

PMAWCA also has observer members located in Europe. These memberships comprise of Anglophone, Francophone and Lusophone Speaking Countries thus, PMAWCA's official languages are English, French and Portuguese.

The Port Management Association of West and Central Africa (PMAWCA) together with the Union of Port Administrations of Northern Africa (UAPNA)and the Port Management Association of West and Central Africa (PMAWCA) build the Pan-African Association for Port Cooperation (PAPC).

Union of Port Administrations of Northern Africa (UAPNA)

The Union of Port Administrations of Northern Africa (UAPNA) is a regional intergovernmental economic organization for the ports of Northern Africa and part of the Pan African Association for Ports Cooperation (PAPC).

The ports of Africa are upgraded to accommodate bigger vessels, but poor road and railway infrastructure have hindered their growth. The Union of Port Administrations of Northern Africa (UAPNA) play a role in the Africa Integrated Marine Strategy (AIMS).

The Union of Port Administrations of Northern Africa (UAPNA) together with the Port Management Association of West and Central Africa (PMAWCA) and Port Management Association of East and Southern Africa (PMAESA) build the Pan-African Association for Port Cooperation (PAPC).

Port State Control (PSC)

Port State Control (PSC)[162] is responsible for the inspection of foreign ships in national ports to verify that the condition of the ship and its equipment comply with the requirements of international regulations and that the ship is manned and operated in compliance with these rules.

Nine regional agreements on port State control - Memoranda of Understanding or MoUs - have been signed: Europe and the north Atlantic (Paris MoU); Asia and the Pacific (Tokyo MoU); Latin America (Acuerdo de Viña del Mar); Caribbean (Caribbean MoU); West and Central Africa (Abuja MoU); the Black Sea region (Black Sea MoU); the Mediterranean (Mediterranean MoU); the Indian Ocean (Indian Ocean MoU); and the Riyadh MoU. The United States Coast Guard maintains the tenth PSC regime.

---

[162] http://www.imo.org/en/OurWork/MSAS/Pages/PortStateControl.aspx

Red Operativa de Cooperación Regional de Autoridades Maritimas de las Americas (ROCRAM)

Red Operativa de Cooperación Regional de Autoridades Maritimas de las Americas (ROCRAM[163]) is the Operative Network for Regional Co-operation among Maritime Authorities of South America, Cuba, Mexico and Panama Is an informal regional organization, for Maritime Authorities to interact at different levels of cooperation, by means of fluent, open and permanent communications.

The Assembly is the governing body of ROCRAM and gathers these Maritime Authorities every two years. Its tasks are to analyze and support the achievements made over the last term, to approve the Network's biannual agenda and to discuss general political matters, as well as to assign the Maritime Authority that will fulfill the duties of General Secretariat of the Operative Network, SECROCRAM. The objective of the Secretariat, which rotates every two years, is to coordinate activities of the Maritime Authorities according to the goals and resolutions adopted during the meetings.

The ROCRAM Maritime Strategy includes the guidelines, to which members adhere, and the tasks and actions necessary to coordinate proceedings regarding maritime safety and security, training and certification of seafarers, protection of the environment, maritime legal and facilitation aspects, and all matters that each organization considers appropriate to ensure the decision making process regarding the effective implementation of IMO instruments. The current strategy has been established for the 2005-2010 period.

---

[163] http://www.rocram.net

Organization of American States (OAS)

The Organization of American States (OAS)[164] is a continental organization that was founded on 30. April 1948, for the purposes of regional solidarity and cooperation among its member states. Headquartered in the United States capital Washington, D.C., the OAS's members are the 35 independent states of America.

The notion of an international union in the New World was first put forward by Simón Bolívar who, at the 1826 Congress of Panama (still being part of Colombia), proposed creating a league of American republics, with a common military, a mutual defense pact, and a supranational parliamentary assembly. This meeting was attended by representatives of Gran Colombia (comprising the modern-day countries of Colombia, Ecuador, Panama and Venezuela), Peru, Bolivia, The United Provinces of Central America, and Mexico but the grandly titled *"Treaty of Union, League, and Perpetual Confederation"* was ultimately ratified only by Gran Colombia. Bolívar's dream soon floundered with civil war in Gran Colombia, the disintegration of Central America, and the emergence of national rather than New World outlooks in the newly independent American republics. Bolívar's dream of American unity was meant to unify Hispanic American nations against external powers.

The pursuit of regional solidarity and cooperation again came to the forefront in 1889–1890, at the First International Conference of American States. Gathered together in Washington, D.C., 18 nations resolved to found the International Union of American Republics, served by a permanent secretariat called the Commercial Bureau of the American Republics (renamed the International Commercial Bureau at the Second International Conference in 1901–1902). These two bodies, in existence as of 14. April 1890, represent the point of inception to which the OAS and its General Secretariat trace their origins.

At the Fourth International Conference of American States (Buenos Aires, 1910), the name of the organization was changed to the Union of American Republics and the Bureau became the Pan American Union. The Pan American Union Building was constructed in 1910, on Constitution Avenue, Northwest, Washington, D.C..

In the mid-1930s, U.S. President Franklin Delano Roosevelt organized an inter-American conference in Buenos Aires. One of the items at the conference was a *"League of Nations of the Americas"*, an idea proposed by Colombia, Guatemala, and the Dominican Republic. At the subsequent Inter-American Conference for the Maintenance of Peace, 21 nations pledged to remain neutral in the event of a conflict between any two members. The experience of World War II convinced hemispheric governments that unilateral action could not ensure the territorial integrity of the American nations in the event of external aggression. To meet the challenges of global conflict in the postwar world and to contain conflicts within the hemisphere, they adopted a system of collective security, the Inter-

---

[164] http://www.oas.org and https://en.wikipedia.org/wiki/Organization_of_American_States

American Treaty of Reciprocal Assistance (Rio Treaty) signed in 1947 in Rio de Janeiro.

The Ninth International Conference of American States was held in Bogotá between March and May 1948 and led by United States Secretary of State George Marshall, a meeting which led to a pledge by members to fight communism in the western hemisphere. This was the event that saw the birth of the OAS as it stands today, with the signature by 21 American countries of the Charter of the Organization of American States on 30 April 1948 (in effect since December 1951). The meeting also adopted the American Declaration of the Rights and Duties of Man, the world's first general human rights instrument.

As of 31. January 2014, there are 69 permanent observer countries including the four countries with territories in the Americas, Denmark, France, the Netherlands, and the United Kingdom; as well as the European Union.

The transition from the Pan American Union to OAS would have been smooth if it had not been for the assassination of Colombian leader Jorge Eliécer Gaitán. The Director General of the former, Alberto Lleras Camargo, became the Organization's first Secretary General. The current Secretary General is former Uruguayan minister of foreign affairs Luis Almagro.

In the words of Article 1 of the Charter, the goal of the member nations in creating the OAS was *"to achieve an order of peace and justice, to promote their solidarity, to strengthen their collaboration, and to defend their sovereignty, their territorial integrity, and their independence."* Article 2 then defines eight essential purposes:

- To strengthen the peace and security of the continent.
- To promote and consolidate representative democracy, with due respect for the principle of non-intervention.
- To prevent possible causes of difficulties and to ensure the pacific settlement of disputes that may arise among the member states.
- To provide for common action on the part of those states in the event of aggression.
- To seek the solution of political, judicial, and economic problems that may arise among them.
- To promote, by cooperative action, their economic, social, and cultural development.
- To eradicate extreme poverty, which constitutes an obstacle to the full democratic development of the peoples of the hemisphere.
- To achieve an effective limitation of conventional weapons that will make it possible to devote the largest amount of resources to the economic and social development of the member states.

Over the course of the 1990s, with the end of the Cold War, the return to democracy in Latin America, and the thrust toward globalization, the OAS made major efforts to reinvent itself to fit the new context.

OAS stated priorities include today the following:

- Strengthening democracy: Between 1962 and 2002, the Organization sent multinational observation missions to oversee free and fair elections in the member states on more than 100 occasions. The OAS also works to strengthen national and local government and electoral agencies, to promote democratic practices and values, and to help countries detect and defuse official corruption.
- Working for peace: Special OAS missions have supported peace processes in Nicaragua, Suriname, Haiti, and Guatemala. The Organization has played a leading part in the removal of landmines deployed in member states and it has led negotiations to resolve the continents' remaining border disputes (Guatemala/Belize; Peru/Ecuador). Work is also underway on the construction of a common inter-American counter-terrorism front.
- Defending human rights: The agencies of the inter-American human rights system provide a venue for the denunciation and resolution of human rights violations in individual cases. They also monitor and report on the general human rights situation in the member states.
- Fostering free trade: The OAS is one of the three agencies currently engaged in drafting a treaty aiming to establish an inter-continental free trade area from Alaska to Tierra del Fuego.
- Fighting the drugs trade: The Inter-American Drug Abuse Control Commission was established in 1986 to coordinate efforts and cross border cooperation in this area.
- Promoting sustainable development: The goal of the OAS's Inter-American Council for Integral Development is to promote economic development and combating poverty. OAS technical cooperation programs address such areas as river basin management, the conservation of biodiversity, preservation of cultural diversity, planning for global climate change, sustainable tourism, and natural disaster mitigation.

The Organization of American States is composed of an Organization of American States General Secretariat, the Permanent Council, the Inter-American Council for Integral Development, and a number of committees.

1. The General Secretariat of the Organization of American States consists of six secretariats.
2. Secretariat for Political Affairs
3. Executive Secretariat for Integral Development
4. Secretariat for Multidimensional Security
5. Secretariat for Administration and Finance
6. Secretariat for Legal Affairs
7. Secretariat for External Relations

The various committees of the Organization of American States include:

- The Committee on Juridical and Political Affairs
- The Committee on Administrative and Budgetary Affairs
- The Committee on Hemispheric Security
- The Committee on Inter-American Summits Management and Civil Society Participation in OAS

The General Assembly is the supreme decision-making body of OAS. It convenes once every year in a regular session. In special circumstances, and with the approval of two-thirds of the member states, the Permanent Council can convene special sessions.

The Organization's member states take turns hosting the General Assembly on a rotating basis. The states are represented at its sessions by their chosen delegates: generally, their ministers of foreign affairs, or their appointed deputies. Each state has one vote, and most matters, except for those for which the Charter or the General Assembly's own rules of procedure specifically require a two-thirds majority, are settled by a simple majority vote.

The General Assembly's powers include setting the OAS's general course and policies by means of resolutions and declarations, approving its budget and determining the contributions payable by the member states, approving the reports and previous year's actions of the OAS's specialized agencies, and electing members to serve on those agencies.

All 35 independent nations of the Americas are members of the OAS. Upon foundation on 5. May 1948, there were 21 (22) members: Argentina, Bolivia, Brazil, Chile, Colombia, Costa Rica, Cuba, Dominican Republic, Ecuador, El Salvador, Guatemala, Haiti, Honduras, Mexico, Nicaragua, Panama, Paraguay, Peru, United States, Uruguay, and Venezuela.

The later expansion of the OAS included Canada and the newly independent nations of the Caribbean. Members with later admission dates (sorted chronologically): Barbados (member since 1967), Trinidad and Tobago (1967), Jamaica (1969), Grenada (1975), Suriname (1977), Dominica (1979), Saint Lucia (1979), Antigua and Barbuda (1981), Saint Vincent and the Grenadines (1981), Bahamas (1982), Saint Kitts and Nevis (1984), Canada (1990), Belize (1991), and Guyana (1991).

Although Canada has been a founding member of the League of Nations in 1919 and has joined international organizations since that date, it chose not to join the OAS when it was first formed, despite its close relations with the United States. Canada became a Permanent Observer in the OAS on 2 February 1972. Canada signed the Charter of the Organization of American States on 13 November 1989 and this decision was ratified on 8. January 1990.

In 2004–2005, Canada was the second largest contributor to the OAS, with an annual assessed contribution representing 12.36 percent of the OAS Regular Budget (US$9.2 million) and an additional $9 million in voluntary contributions to specific projects. Shortly after joining as a full member, Canada was instrumental in the creation of the Unit for the Promotion of Democracy, which provides support for the strengthening and consolidation of democratic processes and institutions in OAS member states.

The current government of Cuba was excluded from participation in the Organization under a decision adopted by the Eighth Meeting of Consultation in Punta del Este, Uruguay, on 31 January 1962. The vote was passed by 14 in favor, with one against (Cuba) and six abstentions (Argentina, Bolivia, Brazil, Chile, Ecuador, and Mexico).

This meant that the Cuban nation was still technically a member state, but that the current government was denied the right of representation and attendance at meetings and of participation in activities. The OAS's position was that although Cuba's participation was suspended, its obligations under the Charter, the American Declaration of the Rights and Duties of Man, etc. still hold: for instance, the Inter-American Commission on Human Rights continued to publish reports on Cuba's human rights situation and to hear individual cases involving Cuban nationals. However, this stance was occasionally questioned by other individual member states.

Cuba's position was stated in an official note sent to the Organization *"merely as a courtesy"* by Minister of Foreign Affairs Dr. Raúl Roa on 4 November 1964: *"Cuba was arbitrarily excluded ... The Organization of American States has no juridical, factual, or moral jurisdiction, nor competence, over a state which it has illegally deprived of its rights."*.

The reincorporation of Cuba as an active member regularly arose as a topic within the inter-American system – for instance, it was intimated by the outgoing ambassador of Mexico in 1998, but most observers did not see it as a serious possibility while the present government remained in power. Since 1960, the Cuban administration had repeatedly characterized the OAS as the *"Ministry of Colonies"* of the United States of America. On 6 May 2005, President Fidel Castro reiterated that the island nation would not *"be part of a disgraceful institution that has only humiliated the honor of Latin American nations."* After Fidel Castro's recent retirement and the ascent of his brother Raúl to power, this official position was reasserted. Venezuelan President Hugo Chávez promised to veto any final declaration of the 2009 Summit of the Americas due to Cuba's exclusion.

On 17 April 2009, after a *"trading of warm words"* between the administrations of U.S. President Barack Obama and Cuban leader Raúl Castro, OAS Secretary General José Miguel Insulza said he would ask the 2009 General Assembly to annul the 1962 resolution excluding Cuba.

On 3 June 2009, foreign ministers assembled in San Pedro Sula, Honduras, for the OAS's 39th General Assembly, passed a vote to lift Cuba's suspension from the OAS. The United States had been pressuring the OAS for weeks to condition Cuba's readmission to the group on democratic principles and commitment to human rights. Ecuador's Foreign Minister Fander Falconí said there will be no such conditions. *"This is a new proposal, it has no conditions of any kind,"* Falconí said. *"That suspension was made in the Cold War, in the language of the Cold War. What we have done here is fix a historic error."* The suspension was lifted at the end of the General Assembly, but, to be readmitted to the Organization,

Cuba will need to comply with all the treaties signed by the Member States, including the Inter-American Democratic Charter of 2001. A statement issued by the Cuban government on 8 June 2009 stated that while Cuba welcomed the Assembly's gesture, in light of the Organization's historical record "*Cuba will not return to the OAS.*"

On 26 April 2017, Venezuela announced its intention to withdraw from the OAS. Venezuelan Foreign Minister Delcy Rodríguez said that President Nicolás Maduro planned to publicly renounce Venezuela's membership on 27 April 2017. It would take two years for the country to formally leave. During this period, the country does not plan on participating in the OAS.

Inter-American Committee against Terrorism (CICTE)

Despite the political differences and tensions within the Organization of American States (OAS) the threats of international terrorism enabled important decisions by the participants of the cooperation.

Following the recommendations of the Second Inter-American Specialized Conference on Terrorism (November 1998), the Organization of American States (OAS) General Assembly created the Inter-American Committee Against Terrorism (CICTE)[165] with the objective of fostering cooperation among OAS member states to prevent, combat, and eliminate terrorism in the Hemisphere. CICTE fosters increased cooperation and coordination among member states through training and the exchange of information among specialists and political leaders/decision-makers working together to strengthen hemispheric solidarity and security.

The first regular session of CICTE was held in Miami, Florida in October 1999, which developed a plan of work. While no sessions took place in 2000 or 2001, the events of 11. September 2001 brought renewed focus to the Inter-American efforts to confront terrorism. In 2002, the OAS Secretary General established a secretariat within the General Secretariat to support CICTE and appointed an Executive Secretary in October 2002 to direct the operations of the CICTE secretariat. The CICTE Secretariat implements the member-state approved Work Plan of the Committee.

Another key milestone in 2002 was the adoption of the Inter-American Convention against Terrorism, signed on June 3rd by 30 Member States at the OAS General Assembly in Bridgetown, Barbados, and which entered into force in July 2003. This Convention has now been ratified by 24 member states and signed by 33, and is regarded as providing the legal structure for cooperation among OAS member states in the fight against terrorism, with CICTE as the main multilateral vehicle for promoting that cooperation and facilitating implementation of the Convention.

The nature, powers, and functions of CICTE include:

1. The Inter-American Committee on Terrorism (CICTE) shall be an entity established by the General Assembly of the Organization of American States (OAS) in conformity with Article 53 of the Charter of the Organization, which shall enjoy technical autonomy. It will be composed of the competent national authorities of all the member states and be governed in the exercise of its functions by the provisions of Article 91.f of the Charter.
2. In the exercise of its functions, CICTE shall promote the development of inter-American cooperation on the basis of international conventions on this matter and the

---

[165] http://www.cicte.oas.org

Declaration of Lima to Prevent, Combat, and Eliminate Terrorism. It shall be empowered to encourage, develop, coordinate, and evaluate implementation of the Plan of Action of Lima, the recommendations of the Meeting of Government Experts to Examine Ways to Improve the Exchange of Information and Other Measures for Cooperation among Member States to Prevent, Combat, and Eliminate Terrorism, as well as the recommendations contained in this Commitment.

3. CICTE will provide assistance to member states requesting it, in order to prevent, combat, and eliminate terrorism, while promoting, in accordance with the domestic laws of the member states, the exchange of information and experiences with the activities of persons, groups, organizations, and movements linked to terrorist acts as well as with the methods, sources of finance and entities directly or indirectly protecting or supporting them, and their possible links to other crimes.

4. In order to ensure an adequate exchange of information on the issue of illicit trafficking in arms, munitions, explosives, materials, or technology capable of being used to perpetrate terrorist acts or activities, CICTE will coordinate with the Consultative Committee established by the 1997 Inter-American Convention against the Illicit Production of and Trafficking in Firearms, Ammunition, Explosives, and Other Related Materials.

5. CICTE will hold at least one annual session. During its first session, CICTE will draw up its work schedule designed to implement the following guidelines:

a) To create an inter-American network for gathering and transmitting data via the competent national authorities, designed to exchange the information and experiences referred to in paragraph 3, including the creation of an inter-American database on terrorism issues that will be at the disposal of member states.

b) To compile the legal and regulatory norms on preventing, combating, and eliminating terrorism in force in member states.

c) To compile the bilateral, subregional, regional, or multilateral treaties and agreements signed by member states to prevent, combat, and eliminate terrorism.

d) To study the appropriate mechanisms to ensure more effective application of international legal norms on the subject, especially the norms and provisions contemplated in the conventions against terrorism in force in states parties to those conventions mentioned in paragraph xiv of this Commitment.

e) To formulate proposals designed to provide assistance to states requesting it in drafting national antiterrorist laws.

f) To devise mechanisms for cooperation in detecting forged identity documents.

g) To devise mechanisms for cooperation among competent migration authorities.

h) To design technical cooperation programs and activities for training staff assigned to tasks related to preventing, combating, and eliminating terrorism in each of the member states that request such assistance.

6. The above-mentioned guidelines do not preclude the possibility of CICTE carrying out other activities should the General Assembly so determine.
7. With the acquiescence of the competent authorities, CICTE may establish mechanisms for coordinating with other competent international entities, such as INTERPOL.

In light of these and subsequent resolutions of the organs of the Inter-American system, CICTE began holding regular annual sessions as a forum for discussion and decision-making on counter terrorism issues, measures, and cooperation. Member States designate a competent national authority, a principal representative, alternate representatives, and advisors. Member States also appoint one or more National Points of Contact with competence in the field of prevention and elimination of terrorism to serve as the principal liaison among governments of the Member States and with the CICTE Secretariat for developing cooperation programs.

Member States often issue a "*declaration*" at each CICTE Regular Session laying out their priorities on a specific counterterrorism topic. At these meetings, the Member States also approve by consensus the CICTE Secretariat's Annual Work Plan, which seeks to address Member State security vulnerabilities that can be exploited not only by terrorists, but also for transnational criminal purposes.

CICTE has emerged as a key player in the Inter-American system supporting the "*Multidimensional Security*" pillar of the OAS, one of four pillars guiding the work of the OAS. CICTE works in the areas of maritime security, document security, tourism security, security for major events, legislative assistance for the financing of terrorism, aviation security, immigration and customs controls, cyber security, supply chain security, critical infrastructure protection and emerging threats preparation.

In 2013, the CICTE Secretariat conducted 113 activities, training courses, and technical assistance missions which benefited more than 4181 participants in five thematic areas: border control, critical infrastructure protection, counterterrorism legislative assistance and terrorist financing, strengthening strategies on emerging terrorist threats (crisis management), and international cooperation and partnerships.

The United States has been a major contributor to CICTE's training programs and has directly provided funding and expert trainers for capacity building programs focused on maritime security, aviation security, travel document security and fraud prevention, cybersecurity, counterterrorism legislation, and efforts to counter terrorist financing.

Regional Cooperation Agreement on Combating Piracy an Armed Robbery against Ships in Asia (ReCAAP)

The <u>Regional Cooperation Agreement on Combating Piracy and Armed Robbery against Ships in Asia (ReCAAP)</u>[166] is the first regional multilateral government-to-government agreement to promote and enhance cooperation against piracy and armed robbery in Asia and has the status of an international organization with local staff and overseas representatives from member countries. 20 States have become Contracting Parties to ReCAAP, which was recognized as an international organization on January 2007 and is the first regional government-to-government agreement to promote and enhance cooperation against piracy and armed robbery in Asia.

The ReCAAP Information Sharing Center (ReCAAP ISC) in Singapore was established under the ReCAAP Agreement, and was officially launched in Singapore on 29. November 2006. It was formally recognized as an international organization on 30. January 2007. The roles of ReCAAP ISC include exchanging information among Contracting Parties on incidents of piracy and armed robbery supports capacity building efforts of Contracting Parties, and for cooperative arrangements.

The roles of the ReCAAP ISC are to:

- serve as a platform for information exchange with the ReCAAP Focal Points via the Information Network System (IFN), facilitate communications and information exchange among participating governments to improve incident response by member countries; analyses and prove accurate statistics of the piracy and armed robbery incidents to foster better understanding of the situation in Asia,
- facilitate capacity building efforts that help improve the capability of member countries in combating piracy and armed robbery in the region, and
- cooperate with organizations and like-minded parties on joint exercises, information sharing, capacity building program, or other forms of cooperation, as appropriate, and agree upon among the Contracting Parties.

The ReCAAP ISC facilitates exchange of information among the ReCAAP Focal Points through a secure web-based Information Network System (IFN). Through this network, the ReCAAP Focal Points are linked to each other as well as the ReCAAP ISC operating on a 24/7 basis, and are able to facilitate appropriate responses to incident.

ReCAAP has recently opened the INTERPOL Global Complex for Innovation (IGCI), a cutting-edge research and development facility for the identification of crimes and criminals,

---

[166] http://www.recaap.org

innovative training, operational support and partnerships.

The agency receiving the incident report will manage the incident in accordance to its national policies and response procedures, and provide assistance to the victim ship where possible. The agency will in turn, inform their ReCAAP Focal Point which will submit an incident report to the ReCAAP ISC and its neighboring Focal Points including external participants in the form of states such as Malaysia and also as organizations like the Asian Ship-owners Forum (ASF), Baltic and International Maritime Council (BIMCO), EU, IFC, ICS, INTERPOL, IMO, Singapore Shipping Association. ReCAAP is further officially linked with the maritime centers in Yemen (SANAA Ceneter), Kenya and Tanzania.

However; while Australia, Bangladesh, Brunei, Cambodia, China, Denmark, India, Japan, Korea, Laos, Myanmar, the Netherlands, Norway, Philippines, Singapore, Sri Lanka, Thailand, UK, USA, Vietnam have signed the official agreement, neither Malaysia nor Indonesia and other parties have officially joined in ReCAAP which limits the information exchange.

Information Network System (IFN)

The Regional Cooperation Agreement on Combating Piracy and Armed Robbery against Ships in Asia (ReCAAP) uses the secure web-based Information Network System (IFN)[167]. Through this network, the ReCAAP Focal Points are linked to each other as well as the ReCAAP ISC on a 24/7 basis, and are able to facilitate appropriate responses to incident.

The Information Network (IFN) System is a 24-hour, secure, web-based information system that supports the ISC in the collection, organization, analysis and sharing of piracy and armed robbery information among ReCAAP member countries. It links up the ReCAAP Information Sharing Centre (ISC) in Singapore with the designated Focal Points of ReCAAP member countries to enable the dissemination and exchange of information. These Focal Points (e.g. national coast guard, maritime authority) are designated by each individual ReCAAP member country as the points of contact for the ISC.

When a piracy or sea robbery incident occurs, the victim reports to the closest Focal Point which would respond to the incident as well as submit an incident report to the ISC through the IFN. The ISC would be able to communicate with this Focal Point and other Focal Points via the IFN, as well as disseminate alerts to other Focal Points. The ISC would also post messages on the ISC website to alert the shipping community to the incident.

The key features that the IFN provides are a 24/7 Information Management enabling the expeditious sharing of information on piracy and armed robbery incidents among ReCAAP member countries, no matter what time of day an incident occurs. This feature is particularly critical when an incident reported by one member country requires the immediate assistance of other member countries, such as in the tracking of hijacked vessels. Second is are analytical and reporting tools enabling the ISC to easily identify patterns and trends in pirate behavior.

Such insights can help member countries improve enforcement and preventive measures as well as the shipping community avoid and deter attacks. The secured system ensures only authorized users issued with user ids and passwords can access the network. The system also makes use of encrypted connections and firewalls to ensure the security and integrity of information captured.

The IFN System also facilitates interaction between Focal Points of ReCAAP member countries through the use of on-line collaboration tools such as an electronic forum, document exchange and news alerts.

As probably the first international organization ReCAAP developed applications for Android and iPhones for download in the online-store which allow a monitoring and alerting by registered users even from onboard the vessels.

---

[167] https://www.mot.gov.sg

Russia (RUS)

Russia has always been a potent maritime entity and global player. It is the largest country in the world in terms of area but unfavorably located in relation to major sea lanes of the world, Despite its size, much of the country lacks proper soils and climates, either too cold or too dry for agriculture. Lake Baikal is the deepest lake in the world and is estimated to hold one fifth of the world's fresh water. The website Global Firepower (GFP)[168] listed in 2018 a total of 136 countries in the Global Firepower database ranking Russia second after the USA and before China and India.

Russia is approximately 1.8 times the size of the USA located at the Arctic Ocean, extending from Europe (the portion west of the Urals) to the North Pacific Ocean and bordering Azerbaijan, Belarus, China (southeast), China (south), Estonia, Finland, Georgia, Kazakhstan, North Korea, Latvia, Lithuania (Kaliningrad Oblast), Mongolia, Norway, Poland (Kaliningrad Oblast), and Ukraine.

According i.e. to a study by the Center for Strategic & International Studies[169], Northern Europe, and specifically the Baltic and Norwegian Seas, has been the site of increasingly provocative and destabilizing Russian actions. The country's use of a range of military, diplomatic, and economic tools to undermine the North Atlantic Treaty Organization (NATO) and its allies highlights the need to monitor and understand Russian activity. The region is characterized by complex factors like unique geographic features, considerable civilian maritime traffic, the presence of advanced Russian and Western military capabilities, and strategic proximity to Russia and the Kola peninsula, home to the Russian Northern Fleet. While the Norwegian and Baltic Seas do differ in key ways, they are linked by the emerging risk posed by Russia's long-range strike capabilities.

Responding to Russian challenges across the competitive space requires a deep understanding of the Northern European maritime environment. Maritime Domain Awareness (MDA), defined by the United States as the effective understanding of anything associated with the maritime domain that could impact the security, safety, economy, or environment of a nation or region, is an exceptionally broad concept.

At its core, MDA has three functions in Russia, the collection of raw data, the analysis of that data, and the action of disseminating information to and coordinating among the different components of the framework. In order to provide security in Northern Europe, NATO and its allies must use MDA frameworks to understand and respond to the challenges above, on, and underneath the sea, as well as the surrounding land environment. While some constructive work has been done to address the evolving Russian threat, NATO and its partners must make changes to their current MDA capabilities to evolve alongside

---

[168] https://www.globalfirepower.com
[169] Kathleen H. Hicks, Senior Vice President; Henry A. Kissinger Chair; Director, International Security Program, Center for Strategic & International Studies, https://www.csis.org/analysis/contested-seas

with it.

Russia presents three challenges of particular concern to the MDA efforts in Northern Europe, the maritime hybrid warfare, electronic and cyber warfare capabilities, and long-range strike systems.

1. Maritime Hybrid Warfare—The Russian military is experienced and effective in its use of hybrid warfare, seen in Syria, Crimea, and Northern Europe. The ambiguity possible in the maritime domain lends itself well to this strategy. Russia uses three specific approaches in this realm: deception through different types of vessels including civilian ships, deniable forces like the amphibious and light infantry that easily navigate the complex Baltic and Norwegian Seas, and the country's well-developed and diverse force for seabed warfare.
2. Cyber and Electronic Warfare—Russia's advanced EW capabilities have the potential to hinder information gathering and dissemination methods, which are both vital functions of MDA. These capabilities are challenging for military personnel but potentially devastating in civilian contexts, especially as civilian networks and technology (like GPS) are far less secure.
3. Long-Range Strike Capabilities—New challenges for NATO and Northern European partners have emerged with Russia's development of a long-range precision strike complex. The weapons, now being mounted on new and existing Russian naval vessels, give these vessels the option to stay in the Barents or White Seas and strike targets across Northern Europe. This, combined with air force capabilities based on the Kola Peninsula and in Kaliningrad, presents threats unlike any seen by NATO before. These capabilities require NATO and its partners to use MDA frameworks to monitor launch platforms across the domain.

A weakness of the original US MDA and Maritime Situational Awareness (MSA) concepts is that many of the associated capabilities and frameworks are focused on civil maritime issues. Given the global proliferation of advanced military capabilities, like anti-ship cruise missiles, NATO and its partners require a holistic understanding of the maritime environment that focuses on everything from civil maritime actions to high-end military operations and even issues associated with the maritime environment.

A key implication of the heightened maritime threat environment is the need to improve the integration of and attention to undersea aspects of MDA. Antisubmarine warfare (ASW), a traditional strength of Western naval intelligence and operations, has atrophied since the end of the Cold War. Today, Russian submarines with conventional long-range missiles pose a threat to NATO. ASW must be integrated with MDA to address these concerns. Comprehensive understanding of the undersea realm should extend beyond ASW. Russia's amphibious special forces and combat swimmers threaten more than just military targets, including civilian vessels and undersea cables, which are an integral part of MDA. ASW technology can be useful in countering these and other threats.

In the Norwegian Sea, the biggest challenge for NATO is detecting advanced ultra-quiet submarines. This issue is sharpened by dramatically depleted stockpiles of sonar-buoys, a constant need for increasingly advanced sonar-buoy technology, and an American unwillingness to share highly classified information about the undersea domain. NATO would benefit from an apparatus like the ASW Operations Centers (ASWOC), used most prominently during the Cold War to streamline ASW operations. Integration of platforms is a challenge in the Baltic Sea as well, largely because Sweden and Finland are not NATO states, making data sharing challenging.

Frameworks like Sea Surveillance Co-Operation Baltic Sea (SUCBAS) and the Maritime Surveillance (MARSUR) project facilitate the work of regional states to address these issues but more must be done. Additionally, NATO monitoring of the Baltic region is largely domain specific and suffers from not examining the maritime domain holistically. The alliance and its partners should also act to focus on resiliency to continue to operate in the face of jamming and non-kinetic attacks from Russia.

The key to enhancing MDA capabilities in Northern Europe is the integration of frameworks across the maritime domain. Cooperation between NATO states and allies is vital to understanding the complex environment. The CSIS study team has identified seven recommendations of particular importance:

- Create a Baltic Sea MDA analytic center at the Baltic Maritime Component Command (BMCC) at Rostock, Germany;
- Empower a small analytic team at the BMCC to focus on maritime hybrid issues;
- Develop a training course for military intelligence officers on best practices for Baltic Sea MDA analysis;
- Create a classified Baltic Sea data environment that can incorporate both NATO and partner states;
- Develop a multinational operational framework for the Baltic Sea;
- Integrate subsurface sensors and antisubmarine warfare into a comprehensive MDA framework; and
- Acquire significant stockpiles of advanced sonar-buoys and associated acoustic processing systems.

Collectively, it will be necessary to enhance security in Northern Europe by closing identified gaps and ensuring capabilities for collection, analysis, and action in MDA to counter balance Russian influence and activities, whereas Russia keeps its cooperation and organizations limited to national interests.

Russia has environmental issues with air pollution from heavy industry, emissions of coal-fired electric plants, and transportation in major cities, industrial, municipal, and agricultural pollution of inland waterways and seacoasts, deforestation; soil erosion, soil contamination from improper application of agricultural chemicals, scattered areas of sometimes intense radioactive contamination, groundwater contamination from toxic waste, urban solid waste management, abandoned stocks of obsolete pesticides.

Russia is in international environmental agreements party to Air Pollution, Air Pollution-Nitrogen Oxides, Air Pollution-Sulfur 85, Antarctic-Environmental Protocol, Antarctic-Marine Living Resources, Antarctic Seals, Antarctic Treaty, Biodiversity, Climate Change, Climate Change-Kyoto Protocol, Desertification, Endangered Species, Environmental Modification, Hazardous Wastes, Law of the Sea, Marine Dumping, Ozone Layer Protection, Ship Pollution, Tropical Timber 83, Wetlands, Whaling, signed, but not ratified is Air Pollution-Sulfur 94.

Globalnaya navigatsionnaya sputnikovaya Sistema (GLOSNASS)

Globalnaya navigatsionnaya sputnikovaya sistema (GLONASS)[170] or "Global Navigation Satellite System" (Russian: ГЛОНАСС; Глобальная навигационная спутниковая система), is a Russian space-based satellite navigation system operating in the radio navigation-satellite service. It provides an alternative to the US-GPS and is the second navigational system in operation with global coverage and of comparable precision.

Manufacturers of GPS navigation devices say that adding GLONASS made more satellites available to them, meaning positions can be fixed more quickly and accurately, especially in built-up areas where the view to some GPS satellites may be obscured by buildings. It is also more suitable for use in high latitudes (north or south).

Development of GLONASS began in the Soviet Union in 1976. Beginning on 12. October 1982, numerous rocket launches added satellites to the system until the constellation was completed in 1995. After a decline in capacity during the late 1990s, in 2001, under Vladimir Putin's presidency, the restoration of the system was made a top government priority and funding was substantially increased. GLONASS is the most expensive program of the Russian Federal Space Agency, consuming a third of its budget in 2010.

By 2010, GLONASS had achieved 100% coverage of Russia's territory and in October 2011, the full orbital constellation of 24 satellites was restored, enabling full global coverage. The GLONASS satellites' designs have undergone several upgrades, with the latest version being GLONASS-K2, scheduled to enter service in 2019. By 2040 a group of communications and navigational satellite is announced to be deployed. The task also includes the delivery to the Moon of a series of spacecraft for orbital research and the establishment of a global communications and positioning system.

The main contractor of the GLONASS program is Joint Stock Company Reshetnev Information Satellite Systems (ISS Reshetnev, formerly called NPO-PM). The company, located in Zheleznogorsk, is the designer of all GLONASS satellites, in cooperation with the Institute for Space Device Engineering (ru:РНИИ КП) and the Russian Institute of Radio Navigation and Time. Serial production of the satellites is accomplished by the company PC Polyot in Omsk.

Over the three decades of development, the satellite designs have gone through numerous improvements, and can be divided into three generations: the original GLONASS (since 1982), GLONASS-M (since 2003) and GLONASS-K (since 2011). Each GLONASS satellite has a designation 11F654 and the military "Cosmos-NNNN" designation.

The system requires 18 satellites for continuous navigation services covering the entire territory of the Russian Federation, and 24 satellites to provide services worldwide. The

---

[170] https://en.wikipedia.org/wiki/GLONASS

GLONASS system covers 100% of worldwide territory, however, on 2. April 2014 the system experienced a technical failure that resulted in practical unavailability of the navigation signal for around 12 hours and on 14.–15. April 2014 nine GLONASS satellites experienced a technical failure due to software problems.

On 19. February 2016 three GLONASS satellites experienced a technical failure, here the batteries of GLONASS-738 exploded, the batteries of GLONASS-737 were depleted, and GLONASS-736 experienced a station keeping failure due to human error during maneuvering. GLONASS-737 and GLONASS-736 are expected to be operational again after maintenance, and one new satellite (GLONASS-751) to replace GLONASS-738 is expected to complete commissioning in early March. The full capacity of the satellite group was expected to be restored after the launching of two new satellites and maintenance of two others.

The GLONASS accuracy is up to 2.8 meters, in comparison with GPS, which have accuracy up to 5 meters. Civilian GLONASS used alone is therefore slightly less accurate than GPS. On high latitudes (north or south), GLONASS' accuracy is better than that of GPS due to the orbital position of the satellites.

Some modern receivers are able to use both GLONASS and GPS satellites together, providing greatly improved coverage in urban canyons and giving a very fast time to fix due to over 50 satellites being available. In indoor, urban canyon or mountainous areas, accuracy can be greatly improved over using GPS alone. For using both navigation systems simultaneously, precision of GLONASS/GPS navigation definitions were 2.37–4.65 meters.

Actions were undertaken to expand GLONASS's constellation and to improve the ground segment to increase the navigation definition of GLONASS to an accuracy of 2.8 meters by 2011. The latest satellite design, GLONASS-K has the ability to double the system's accuracy once introduced.

As of early 2012, sixteen positioning ground stations are under construction in Russia and in the Antarctic at the Bellingshausen and Novolazarevskaya bases. New stations will be built around the southern hemisphere from Brazil to Indonesia. Together, these improvements are expected to bring GLONASS' accuracy to 0.6 m or better by 2020. The setup of a GLONASS receiving station in the Philippines is under negotiation.

Scandinavia

Norway (NOR), Denmark (DNK), Sweden (SWE), Finland (FIN) and Iceland (ISL)

Scandinavia, including Norway, Denmark, Sweden, Island and Finland, is by geographical position, coastal area, and economy a harmonious group of Maritime Entities with strategic importance in the Baltic region. The Scandinavian countries have strong historical, cultural, and linguistic ties – the later in exception of Finland - and are all in a strategic position for the whole region in Northern Europe, but especially for the Baltic Sea.

Norway is bordering the North Sea and the North Atlantic Ocean, west of Sweden, with about two-thirds mountains, some 50,000 islands off its much-indented coastline. Norway has a strategic location adjacent to sea lanes and air routes in North Atlantic; one of the most rugged and longest coastlines in the world.

Denmark is bordering the Baltic Sea and the North Sea, on a peninsula north of Germany (Jutland); also includes several major islands (Sjaelland, Fyn, and Bornholm) and has together with Norway strategic access to the North Sea, Skagerrak, Kattegat, and the Baltic Sea. Denmark, Norway and Sweden control the Danish Straits (Skagerrak and Kattegat) linking Baltic and North Seas.

Sweden is geographically located along Danish Straits linking Baltic and North Seas and in a strategic position between Norway, Denmark and Finland. Finland has a long boundary with Russia with Helsinki being the northernmost national capital on European continent. Finland is bordering the Baltic Sea, Gulf of Bothnia, and Gulf of Finland, controlling the access between Sweden and Russia.

Iceland is in a strategic location between Greenland and Europe; westernmost European country; Reykjavik is the northernmost national capital in the world; more land covered by glaciers than in all of continental Europe.

Finland has environmental issues with air pollution from manufacturing and power plants contributing to acid rain, water pollution from industrial wastes, agricultural chemicals, habitat loss threatens wildlife populations.

Finland is in international environmental agreements party to Air Pollution, Air Pollution-Nitrogen Oxides, Air Pollution-Persistent Organic Pollutants, Air Pollution-Sulfur 85, Air Pollution-Sulfur 94, Air Pollution-Volatile Organic Compounds, Antarctic-Environmental Protocol, Antarctic-Marine Living Resources, Antarctic Treaty, Biodiversity, Climate Change, Climate Change-Kyoto Protocol, Desertification, Endangered Species, Environmental Modification, Hazardous Wastes, Law of the Sea, Marine Dumping, Marine Life Conservation, Ozone

Layer Protection, Ship Pollution, Tropical Timber 83, Tropical Timber 94, Wetlands, Whaling

Denmark has the highest percentage of arable land for any country in the world and therefore environmental issues with air pollution, principally from vehicle and power plant emissions, nitrogen and phosphorus pollution of the North Sea, drinking and surface water becoming polluted from animal wastes and pesticides. Sweden has environmental issues with acid rain damage to soils and lakes, pollution of the North Sea and the Baltic Sea.

Denmark is in international environmental agreements party to Air Pollution, Air Pollution-Nitrogen Oxides, Air Pollution-Persistent Organic Pollutants, Air Pollution-Sulfur 85, Air Pollution-Sulfur 94, Air Pollution-Volatile Organic Compounds, Antarctic Treaty, Biodiversity, Climate Change, Climate Change-Kyoto Protocol, Desertification, Endangered Species, Environmental Modification, Hazardous Wastes, Law of the Sea, Marine Dumping, Marine Life Conservation, Ozone Layer Protection, Ship Pollution, Tropical Timber 83, Tropical Timber 94, Wetlands, Whaling

Norway is in international environmental agreements party to Air Pollution, Air Pollution-Nitrogen Oxides, Air Pollution-Persistent Organic Pollutants, Air Pollution-Sulfur 85, Air Pollution-Sulfur 94, Air Pollution-Volatile Organic Compounds, Antarctic-Environmental Protocol, Antarctic-Marine Living Resources, Antarctic Seals, Antarctic Treaty, Biodiversity, Climate Change, Climate Change-Kyoto Protocol, Desertification, Endangered Species, Environmental Modification, Hazardous Wastes, Law of the Sea, Marine Dumping, Ozone Layer Protection, Ship Pollution, Tropical Timber 83, Tropical Timber 94, Wetlands, Whaling.

Sweden is in international environmental agreements party to Air Pollution, Air Pollution-Nitrogen Oxides, Air Pollution-Persistent Organic Pollutants, Air Pollution-Sulfur 85, Air Pollution-Sulfur 94, Air Pollution-Volatile Organic Compounds, Antarctic-Environmental Protocol, Antarctic-Marine Living Resources, Antarctic Treaty, Biodiversity, Climate Change, Climate Change-Kyoto Protocol, Desertification, Endangered Species, Environmental Modification, Hazardous Wastes, Law of the Sea, Marine Dumping, Ozone Layer Protection, Ship Pollution, Tropical Timber 83, Tropical Timber 94, Wetlands, Whaling.

Iceland has environmental issues with water pollution from fertilizer runoff and inadequate wastewater treatment and is in international environmental agreements party to Air Pollution, Air Pollution-Persistent Organic Pollutants, Biodiversity, Climate Change, Climate Change-Kyoto Protocol, Desertification, Endangered Species, Hazardous Wastes, Kyoto Protocol, Law of the Sea, Marine Dumping, Ozone Layer Protection, Ship Pollution, Transboundary Air Pollution, Wetlands. Whaling, signed, but not ratified are Environmental Modification and Marine Life Conservation.

Scandinavian countries and Baltic Sea nations are cooperating more than ever to ensure safety, security and sovereignty, but it is a complex environment, with threats seen and unseen on, over and under the sea. Ensuring the safety and security of maritime activities requires close cooperation and collaboration among stakeholders. Shipping on the Baltic Sea is substantial and has grown steadily over the past decade. More than 2,000 large ships are underway at any given time on the Baltic, including a significant amount of container traffic and transport of oil. While each Baltic Sea country has its own sovereignty and national interests, maritime safety and security are a shared necessity, and best accomplished by means of mutual cooperation. The Baltic Sea also has been through two world wars, with many sea mines that still pose a danger, as well as a significant amount of unexploded ordnance and even chemical weapons that were dumped, some indiscriminately, in its waters; approximately 20 percent of as estimated 170 thousand mines are still on the seabed.

Half of Finland's international trade is with Baltic Sea neighbors. The Swedish port of Gothenburg, at the entrance to the Baltic, is the commercial transportation hub not only for Sweden, but for all of Scandinavia. Sweden is completely dependent on a functioning navigation system for its entire supply chain. 72 percent of all goods coming or going to Sweden pass through a Swedish port. The Baltic Sea is also very important for Russia as 40% of its trading goes through the Baltic Sea, including an increasing transportation of Russian crude oil through the Baltic Sea. There is trade under the sea, too, with numerous pipelines and cables. The Baltic Sea North Stream gas pipeline directly links Russia and Germany, it plays an important role in ensuring reliable supplies of Russian natural gas.

Perhaps the best example of such cooperation is between the Nordic neighbors of Sweden and Finland with the establishment of SUCFIS (Sea Surveillance Cooperation Finland Sweden) in the early 2000s. It began as an effort to notify each other about impending military maneuvers, and grew into a mechanism for the Swedish and Finnish national maritime surveillance systems to create some transparency and exchange data between their autonomous systems to add value to their maritime situational awareness, and ensure environmental and economic security. SUCFIS utilizes the combined sensors for sea surveillance in order to establish one continuous, stable and functional recognized maritime picture (RMP), and has evolved to cross border operations using each other's nations' vessels for sea surveillance.

SUCFIS established the model and led the trend towards coordination of networks and sharing of information through their bilateral agreement, which has been expanded to include other partners in various parts of the world.

Based on the success of the SUCFIS model, a broader coalition was established in 2009 called Sea Surveillance Cooperation Baltic Sea (SUCBAS), aimed at improving the exchange of ship positions, tracks, identification data, chat and images, which today includes Finland, Sweden, Denmark, Germany, Estonia, Latvia, Lithuania, Poland and the United Kingdom. It is not a new system, but rather a mutual support structure, a network connecting existing systems in each country. While each maintains its own capability, they are able to share appropriate information with each other to the extent that their national conditions allow. Today, all nine SUCBAS countries are networked together to maintain a common operating picture (COP) for multinational maritime situational awareness.

Based on the example of SUCFIS and SUCBAS, and the experience of the participating countries, the European Union initiated the Maritime Surveillance project (MARSUR) in 2006 to connect the existing national naval and maritime information exchange systems to reduce duplication of effort better make use of data and information to support safety and security.

Interoperability is achieved through the adoption of common standards for reliable distribution of data between the autonomous national systems. By sharing available track history, notifications and alerts, each nation can enhance their own recognized maritime picture (RMP).

From a military point of view, it is hard to defend against air, surface and sub-surface threats in the Baltic Sea. Together, Finland and Sweden have far more than 100,000 islands and islets. Because of the currents and varying depths, bottom topography, water temperatures and salinity, open ocean solutions are considered to be less successful. There is a common interest there for all the Baltic Sea nations to maintain freedom of navigation and keep sea lines of communications in the Baltic Sea and the Baltic Straits open. This created a need for the Baltic Sea nations to tailor their military forces adapted to the environmental, economic and military challenges.

While all the stakeholders in the Baltic Sea region rely on a safe and secure maritime environment, a revanchist Russia is raising concerns for Baltic Sea nations, and prompting closer coordination and collaboration. The dramatic video of Russian tactical aircraft buzzing a U.S. Navy destroyer in international waters was just one example of Russia's military bold but unprofessional, reckless and dangerous operations in peace time. The annexation of Crimea and the Russian supported insurgents in Eastern Ukraine have made the former Warsaw Pact members or occupied states wary. Russia has increased activities at and over the Baltic Sea, including surveillance, tests of weapon and sensors, trials and more complex exercises. There are constant numerous reports of suspected submarine sightings within the Swedish Archipelago near Stockholm and the Finnish territorial waters.

All Baltic Sea nations (except Russia) and other NATO-nations are frequently conducting national and international combined mine clearance operations and mine warfare exercises. The Baltic Sea nations conduct own weapon testing and training, making the Baltic Sea a crowded playground. This that creates situations where there easily could be misunderstandings or mistakes, and a risk for escalation if there is something happening, foremost in the air where you have fast developing and close situations, but also on the surface.

Denmark, Norway and Island are part of NATO and EU, Sweden and Finland are not part of NATO, but are part of the EU and joined the Partnership-for Peace (PfP) initiative in NATO. The United States presence in the Baltic Sea Region is crucial and has a clear threshold effect.

Knowing the environment is critical in naval warfare, especially as it relates to the capabilities and limitations of sensors, and an adversary's systems, as well. Nowhere is this truer than the Baltic Sea. Radar ducting is common. Parts of the sea freeze solid each winter.

When that ice melts, it creates a current of fresh water. Currents have different temperatures at different depths, with multiple and distinct boundary layers which deflect sonar. The Baltic Sea also has a much lower average salinity than the ocean itself, but is notorious for salt pockets where the salinity is much greater than surrounding waters, creating a layer that sound waves can bounce off.

Some fishermen have learned from hard experience where the mines must be avoided. Mines may not pose a threat as long as they remain undisturbed in the sea floor. But fishermen might haul up an unknown object in their nets, and even bring it ashore and discard it, where it could be exposed to heat and air. Some of the old munitions are filled with deadly chemicals that are still very toxic. There is a new military occupation field growing for "*Historical Ordnance Disposal*" (HOD OPS).

In 2014, Sweden and Finland started the development and training to establish a bilateral standing naval task group, the Swedish-Finnish Naval Task Group (SFNTG), with an initial operational capability for maritime surveillance in 2017 and with full operational capability to provide protection of shipping by 2023.

The cooperation between the two countries also aims at an increased level of interoperability between the Swedish and Finnish Air Forces with the capacity for joint operations, common base operations and common command and control (C2) capability, together with the development of a combined Finnish-Swedish Brigade Framework. There have even been discussions between Finland and Sweden as well as other Baltic Nations to collaborate in other areas, such as education, logistics and procurement, with the benefits of sharing costs and resources.

Barents Watch

BarentsWatch[171] is a collaboration between government agencies and research institutions working to collect, develop and share knowledge of coastal and marine areas close to Norway, however the area extents from Denmark in the south, to Greenland in the west, the North Pole in the north and Novaya Zemlya in the east.

BarentsWatch collects, develops and shares information about Norwegian coastal and marine areas. Ten ministries and 29 administrative agencies and research institutes are our partners. The office is situated in Tromsø, and the actual development work is carried out by commercial suppliers through public procurement.

The program provides the basis for better cooperation, professional development and sharing of information, both for public agencies, trade and industry, and the public. BarentsWatch provides a comprehensive networking platform, collecting and sharing existing data, databases such as The Shared Resources Register[172] and services such as the nationwide Wave Forecast[173].

BarentsWatch is subject to the Ministry of Transport and Communications, and the Norwegian Coastal Administration is responsible for the implementation of the program. This is done through the Norwegian Coastal Administration's *Centre for the Sharing of Sea and Coast Information* (BarentsWatch Center).

Technical solutions from BarentsWatch are set to help an analysis unit at the Vardø vessel traffic service (VTS) center in northern Norway with exposing illegal activity on the high seas. The new unit is a collaboration between the Directorate of Fisheries and the Norwegian Coastal Administration (NCA), with two staffers from each agency receiving and executing assignments.

Its primary job will be to analyze available information on vessel activities and movements in order to identify illegal fishing and cargo shipments. While its responsibility is restricted to areas with Norwegian interests, the unit can also accept assignments from other agencies, both national and international. In addition to work for the NCA and the fisheries directorate, the unit will collaborate with the joint armed forces headquarters (FOH) and the Norwegian Coast Guard.

Other potential partners include Norwegian Customs and Excise, the Directorate of Taxes, the police, the National Criminal Investigation Service (Kripos), the National Authority for Investigation and Prosecution of Economic and Environmental Crime in Norway (Økokrim), the Norwegian Environment Agency (NEA) and the Norwegian Maritime Authority (NMA).

---

[171] https://www.barentswatch.no/en/
[172] https://www.barentswatch.no/en/articles/the-service-shared-resources-register/
[173] https://www.barentswatch.no/en/waveforecast/

Services for end users are presented in the BarentsWatch Portal[174]:

- Fishinfo[175] service provides relevant map information from Norwegian authorities and download files for use in a chart plotter.
- Wave forecast[176] for particularly vulnerable areas and stretches along the Norwegian coast.
- Fishhealth[177] as weekly overview including salmon lice, diseases and countermeasures down at locality level.
- Fishery activity[178] as historical overview of fishing facilities.
- Maps[179] for creating and sharing your own maps and geographical information.
- Overview of ports[180] for viewing and searches in Norwegian ports based on their facilities.
- Polar lows[181] for receiving polar low forecasts and warnings by SMS or E-Mail. The season spans from October to May.
- Saltstraumen[182] for automatically updated forecast for the Saltstraumen tidal current the coming 48 hours. Saltstraumen is a small strait with one of the strongest tidal currents in the world.

The technical system and portal of Barents Watch is named after the Barents Watch Entity, however it does covers a much wider area than just the Barent Sea. By coordinating information and developing new services based on the combination of data, BarentsWatch will disseminate a better factual basis and more comprehensive picture of the activities in, and condition of, our seas and coastal areas.

The system will make relevant information and services more easily accessible for authorities, decision-makers and general users. This will simplify access to and ensure the exchange of public information.

An open part of BarentsWatch shall be an information portal available to everyone. This web site was launched in 2012, and is developed incrementally. The portal has information about topics such as the climate and environment, marine resources, oil and gas, maritime transport and maritime law, among other things. There are also map services, an overview of ports, and news from the 25 partners.

---

[174] https://www.barentswatch.no/en/
[175] https://www.barentswatch.no/en/fishinfo/
[176] https://www.barentswatch.no/en/waveforecast/
[177] https://www.barentswatch.no/en/fishhealth/
[178] https://www.barentswatch.no/en/fisheryactivity/
[179] https://kart.barentswatch.no/?epslanguage=no
[180] https://www.barentswatch.no/en/ports/
[181] https://www.barentswatch.no/en/polar-low/
[182] https://www.barentswatch.no/en/saltstraumen/

C2SöC

The Swedish Navy uses the system C2SöC in the two sea surveillance centers in Muskö and Gothenburg and one standby-center in Visby. The Swedish Navy cooperates closely with the Swedish Coast Guard, which is using its own surveillance system SJÖBASIS, however both systems have the same core by the company SAAB.

The approach is similar to the system modularity used by the Finish Coast Guard, Navy and Customs, which are the main cooperation partners together with Norway and Denmark as well as Poland, Lithuania and Latvia in the Baltic.

Criads

Criads is a operational system for Norwegian coastal radars. It can export several formats, among them US OS-OTH Gold used in NATO. A simplified OS-OTH Gold track store with support for the parts of the specification used by Criads was first created to convert OS-OTH Gold to the NATO Friendly Force Information (NFFI) Format. This track store was connected to a TCP socket in Criads providing for an OS-OTH Gold feed. The track store used a mediation service to translate OS-OTH Gold to NFFI, and exposed the NFFI track as an NFFI SIP3 request/response service. This service was then publish/subscribe-enabled by a connection from the JBridge.

The JBridge would invoke the NFFI track service at regular intervals, and publish the response using WS-Notification via WSMG. This approach is best used when dissemination of tracking information at regular intervals is desired, and an existing request/response NFFI service is already available. Exporting NFFI tracks automated is a feature to be added.

MDA-Tool

The MDA-Tool is the Situational Awareness Suite used in the Danish Navy and Coast Guard. Based on the Danish software the company Navicon provided the SUCBAS Stand-Alone-Client within the Sea Surveillance Cooperation Baltic Sea (SUCBAS) from the start in 2007.

The Navicon MDA service provides the ability to acquire, maintain, and share information between all stakeholders inside an organization or between multiple organizations. The advanced tools of the MDA service enables the operating organization to be proactive in its assignments and have a risk balanced approach towards all identified activities. This is supported through:

- A high degree of service scalability.
- Advanced graphical display and tailored symbology.
- Multi-source information collection.
- Sharing of value added information with other stakeholders.
- Ability to communicate across communication channels with different bandwidth characteristics.
- Automated risk assessment of vessel transits.
- Display of information secured by filtering mechanisms.
- Advanced alerting services, flexible and adaptable to suit operational needs.
- Advanced anomaly detection functionality.
- A highly configurable Graphical User Interface.
- Automated process management and quality assurance.
- Automated risk assessment of vessel transits.

Standard MDA service sub modules for management of information applicable to vessel contacts include:

- Current vessel transit and journey details.
- Historical information including possible vessel incidents and/or detainments.
- Information offered by external or legacy subsystem and public internet sites.
- IHS Maritime Sea-Web.
- Law enforcement.
- Search And Rescue.
- Incident management and analysis.
- Maritime environmental protection.
- Automated Risk Assessment.
- Exercise of Sovereignty.
- Detection of abnormal behavior.

The MDA Service is adaptable and configurable to meet the requirements of different organizations. The MDA Service is extensible and supports integration with most legacy systems. The central core is the tracking/GIS centered engine around which different standard or specially developed sub modules may be integrated.

The core functionality of the MDA service is provision of situational awareness by displaying tracked vessels in their real time position on an electronic map. The MDA service takes input from connected sources (typically radar, AIS, LRIT and AIS, but in mobile configurations also GPS, Speed Log and Wind) and the situational display is automatically updated by these sources. Supplementary to the automated data sources, tracks may also be driven by manual input performed by an operator.

Depending on data sources available, the amount of tracks may be immense. To help the operator maintaining focus on vessels of interest, the MDA service features advanced build-in filtering functions that may force display (or suppression) of tracks representing certain vessels based on their type, cargo, size, behavior or combinations hereof. Inside the Maritime Situational Display the symbology supports:

- Suppressing tracks representing vessels that are considered trivial.
- Special encoding of track symbols with colors, outline, and decorations representing availability of external information.

The GUI of the MDA service applies to the Windows standard featuring both a menu as well as a tool bar. Otherwise the GUI is organized around a number of panels that may be selected, displayed, sized and minimized at the discretion of the individual operators.

The central panel of the MDA service GUI is the Chart panel catering for a layered display of the maritime situation. Through the Chart panel the operator performs most of the interactions with the service including:

- Track- and transit management.
- Management of Alarm Zones.
- Geo-whiteboarding.
- Exchange of information with collaborating authorities.

When a vessel is selected somewhere in one of the MDA service panels, immediately all other open panels displaying vessel related information shift their focus to the selected vessel displaying the attributes they hold applicable to the selected vessel.

For the operator to find specific information, comprehensive search functionality is available, as well are various lists (incidents, transits, and AIS) for searchable overviews. While the vessel itself may be of interest, it is more often the specific voyage that is of special concern. Following this all vessels tracked by MDA service may be assigned a transit.

The transit is a case file holding all information relevant to the specific voyage of the vessel. Once the vessel has finalized its journey the transit is archived. When the vessel then at a later date revisits the area, the archived transit may be recalled and its content used in evaluating the current vessel risk profile and the measures needed against the vessel.

Assignment of transits to tracked vessels may happen automatically when the vessel fulfills certain configurable criteria or be manually initiated by an operator. Supplementary to the transit database the MDA service maintains a vessel database where all information applicable to the ship itself is maintained. While information may be stored manually the vessel database may also inherit information from connected services.

Transit risk assessment is an important issue. The likelihood that a vessel becomes involved in a dangerous situation during its voyage through the AOR, and balancing the effort that authorities shall assign to monitor and investigate individual vessels in the most cost effective way.

When the MDA Service initiates a new transit for a vessel the service will also automatically generate a risk profile for the vessel. The risk profile will be automatically be updated during the voyage of the vessel through the AOR. The parameters which influence the risk profile are configurable and are easily tailored to meet specific customer requirements.

To define if actions are needed against a specific vessel, a vessel risk profile must be established. Important information in this context is the historic activities of the vessel. Supplementary to management of transit and vessel specific information, the MDA service supports recording of vessel incidents and hails performed against the vessel.

Incident records may comprise activities like violation of navigational rules, lack of seaworthiness, groundings, collisions, pollution detainments etc. Hail records may originate from routine hails as well as special hails performed because the vessel has exercised dubious or illegal behavior. Like the transit and vessel records all the incident and hail records applicable to a certain vessel may be retrieved instantly any time needed.

The MDA service is a comprehensive tool for both small stand-alone configurations as well as large integrated center solutions. A prime capability of the service is its ability to support collaboration both inside the host organization as well as with partner authorities outside.

The MDA service is a truly network centric solution enabling maritime surveillance centers, remote coastal lookout stations and mobile units like maritime patrol vessels and aircraft to act as one coherent unit:

- Mutually enhance their commonly share domain awareness.
- Automatically share task orders and other communication.

While data and information applicable to the maritime situation are overwhelming, and human resources available to evaluate these are often scarce, the risk exist that dubious or illegal activities may go on undetected. To mitigate such risks the MDA service features comprehensive tools that can dig deep into the maritime situation and highlight abnormal activities. The MDA service caters for a number of tools that enables the surveillance organization to be proactive.

In terms of collaboration with external partner organizations the MDA service offers real time information exchange tools like:

- Information Discovery: enabling partners to share database information in a controlled environment.
- Notification Sharing: providing the ability for partners to mutually alert each other of suspicious activity.
- Geo-whiteboarding: allowing operators to draw symbols, areas, and other graphics on a com- mon chart shared across the network in a chat-like fashion.

The tool box includes an advance route monitoring service that from big data analysis automatically identifies ships that operate out of the *"normal pattern"*. This can be ships of a specific type and/or size that do not behave as other vessels of their category. The toolset enables automated detection of dubious activities that unfolds over time (i.e. if a ships gradually over hours reverts is course).

The MDA service toolset enables opera- tors to generate, customize and edit the rule set for detecting abnormal events. Detected abnormal activities may be spanned as alerts to all operators or only to those subscribing to events of a special nature. The MDA Service offers and supports a variety of different tasks. One of these is Search and Rescue, where the service:

- calculates search areas and recommends search patterns, and
- monitors the movement of alerted SAR units.

The MDA service includes a IAMSAR compliant set of tools for calculation of search areas and search patterns based on parameters including: last known position (datum), search object characteristics, wind and current parameters, sea state, visibility, SAR unit capabilities, and probability of detection. The toolset supports automated distribution of datum, search areas and search patterns to the electronic chart plotter of the involved SAR units (vessels, dinghies and aircraft), eliminating the possibility error in communication while expediting the information transfer.

SAR operations often involves a lot of participants and it may prove difficult for the decision makes to maintain overview of the entire situation, especially in the initial phase where information often is incomplete . In the event that a SAR unit assigned to a maritime calamity does not show movement within the a pre- defined lead time, the MDA service spans an alert to all involved parties.

With potentially multi thousands contacts inside the AOR, process management and quality assurance are key issues. The MDA Service track colour schema and outline supports an easy overview of the target process status, identifying the targets that need operator attention and those that are fully processed. Build in functionality guides the operator through the steps of target processing.

Inside the MDA service the maritime situation is displayed in a multi layer environment, where the different layers may be switched on and off individually. The MDA service supports a large number of different electronic chart formats including shape KML, and S57. Charts may be imported from services on the Internet like for instance Google Earth, NOAA or satellite images from NASA and weather services.

User defined back ground layers can be e.g. outline of the national territorial waters, fishing areas, wildlife sanctuary areas, and military restricted areas. Vessel tracks are displayed in separate layers that are configurable to best meet the requirements of the opera- tor. Parameters that may be tailored by the individual user includes: track size, text font, label content and outline, tooltips, course and speed vector, and history trail.

All layers are individually selectable, and additional layers can be added on request, while other layers include:

- Management of Alarm Zones.
- Fleet management included distributions of e.g. Rescue stations SAR Helo pads.
- Camera integration.
- Geo-whiteboarding enabling multiple operators to collaborate using maps as frameworks.

Depending on the configuration of the implementation the different sub modules forming the MDA service, including track services and database services, may be deployed as server side modules. Because of the modularity of the MDA service the product will easily adapt to any organization or business inside the maritime domain ranging from a stand-alone service on a single ruggedized Laptop for small craft installations or PC, over a singular MSA center solution where multiple operators collaborate using multiple clients connected to shared servers, to multiple center solutions where the business obligations are distributed among different geographically distributed sites.

Norwegian Command and Control Information System (NORCCIS)

Norwegian Command and Control Information System (NORCCIS) is a system for command and control in military combat developed by developed by the Norwegian Armed Forces in collaboration with the Norwegian defense industry. The Defense Logistics Organization is responsible for the system. The system has been in use since 1992 and has benefited throughout the Norwegian armed forces. In addition, there are some installations of the system in NATO context. The Norwegian Special Forces Command were also equipped with the system when they were in Afghanistan in 2002-2003.

The system provides an overview (common combat picture - common operational picture) based on various sensors such as radar, aircraft monitoring and real-time video stream. The system also has full overview of its own strengths and has its own logistics functions. The system also has messaging features according to NATO standard.

NORCCIS II is an upgraded version which can export its tracks using several formats, the NATO Friendly Forces Information (NFFI) included. However, the NFFI support in NORCCIS II follows an older specification using NFFI over TCP, and not the Web service interface from the current NFFI SIP3. This was solved by creating an NFFI TCP client that was connected to the configured NORCCIS II export socket. The tracks pushed over this socket were NFFI 1.3 formatted, i.e., using the same schema as SIP3 does. Creating a Web service wrapper was the next task, as even if both NORCCIS II and Criads are fundamentally different systems, the Web services offered from the wrapper used the exact same interface.

The JBridge was used in this case as well and in the same manner as for Criads. Being able to reuse functionality in this manner, due to the use of common interfaces, is one of the main strengths of the SOA approach. This, in turn, leads to increased operational agility.

MARIA Geo Development Toolkit (MARIA GDK)

The Norwegian MARIA Geo Development Toolkit (MARIA GDK) was introduced in 2003 in a demonstration connecting the Norwegian Command and Control System II (NORCCIS II), Maria, NATO MCCIS and XOMail (Tactical military message handling system) and other components using the Common Object Request Broker Architecture (CORBA) and the Control of Agent Based Systems (CoABS).

MARIA Geo Development Toolkit (MARIA GDK) is for development of high performance GIS applications. With a modern microservice based architecture, well designed APIs and flexible styling mechanisms, MARIA GDK provides all you need to build scalable world class systems for real-time situational awareness and decision making as a map application developed by Teleplan Globe used in several Norwegian Command and Control systems.

Maria GDK has a collection of building blocks that can be used by a developer to assemble a tailored GIS application. The GDK is designed to work well with WPF applications and encourage the MVVM development pattern.

MARIA GDK supports high performance tracking and geo-fencing with a wide variety of map data formats (using GDAL/OGR or FME) and visualization standards like MGCP and S-52 (ENC). An automated data processing toolchain greatly improves map production pipelines for advanced users.

The tracking functionality is highly optimized for receiving, processing and visualizing massive amounts of real-time tracking data (more than 100.000 track updates per second), including aggregation of history and statistics along with rule based geo-fencing.

MARIA GDK has a flexible draw object functionality built-in to support for Mil Std 2525C, and flexible functionality for definition and visualization of custom draw objects. GeoJSON can be used for simple map overlays.

MARIA GDK has support for fast terrain analysis functions, elevation profiles, radio propagation and radio coverage calculations and visualization, uses Microsoft .NET technology, and works particularly well with WPF client applications using the MVVM development pattern. However, the MARIA services can support almost any kind of client technology via platform independent REST interfaces.

Sealion

Sealion is the Finnish Navy C4ISR system as follow-up of the former MEVAT system and has several derivates in the Maritime Awareness Tactical Information System (MATIS) all developed by the company Navielektro. Sealion and MATIS were from the start designed to facilitate federated information sharing with a robust and scalable technology.

Sealion and MATIS excels in peer-to-peer, inter-agency and multinational information sharing solutions and supports a very wide variety of information exchange interfaces, such as military datalinks, NATO TIDE and numerous legacy military information exchange mechanisms. They also supports Cursor on Target-based information sharing. Operators can use the same, single solution for mission planning and surveillance as well as for tactical-, targeting-, intelligence- and information exchange activities. The system may be deployed on everything from rugged tablets to fixed tactical operator desks and cloud based environments.

The system has been designed using service oriented architecture, in order to offer maximum efficiency and versatility - especially when it comes to collaboration. MATIS provides high availability and redundancy for the most critical functions, ensuring stable operation in even the most challenging environments. Surveillance data is presented as a UCOP (User Defined Common Operational Picture), providing operators with a comprehensive picture of the surveillance area, helping operators assess available mission resources.

Sealion and MATIS are modular, scalable and based on open, service oriented architecture, supporting collaboration between various authorities. The component-based build allows software components to be added according to the needs of the customer, including third-party extensions. This design is efficient and flexible, while also ensuring affordable life-cycle costs. As development can continue after deployment, Sealion and MATIS can be expanded and adapted to suit the changing needs of any maritime domain.

The user interface is built around the Joint Tactical Mapping (JMAP) Tactical Display Framework. JMAP is a software platform capable of performing real-time visualization tasks for operationally demanding applications, as well as managing safety and security based assets. Sealion and MATIS have already been deployed within several inter-agency national and multinational information sharing domains, and the JMAP software platform is widely used for both civilian and military situational awareness.

Sealion and MATIS provide integration to a very wide variety of sensors, and can be used to control sensors and present sensor information as part of the situation picture. The system is also capable of controlling various sensor systems through middleware services. Sensors can be displayed as part of the situation picture, so users need only select which camera source to use. The video image is presented either as an embedded video window, or using a separate display located at the operator working position. Integrated camera sleuthing is also possible, so the chart view and sensors can also be set to track objects on chart.

All of the data collected (including i.e. AIS track data, radar data, weather data and manual data) is fused into a UCOP (User-defined Common Operational Picture). Operators are presented with a single, comprehensive picture of the entire surveillance area, enabling them to quickly assess the situation at hand. Any operator from any workstation can access the situation picture at any time. This allows operators to jump in when swift action is required, shortening response times.

Track location, status and other factors that have an impact on situational awareness are displayed in real- time, on chart. Weather information can be incorporated in the situational picture for the purposes of predicting conditions on the field.

Sealion and MATIS provide the operator with a set of advanced, automated tools that alert the operator regarding various safety and security related conflicts that may arise within the situation picture. Performance of surveillance tasks is further ameliorated through intuitive window layouts and controls.

Collaboration capabilities are integrated with the geospatial view, offering a direct presentation of any position or information that is related to an object on the chart. This allows users of the system to receive and transmit various messages, and exchange data with different actors or allied services in a secure and reliable fashion.

Sealion and MATIS can predict track movements based on track history and voyage information. This allows users to quickly check the heading of a track, recognizing hazardous situations ahead of time. Track can also be linked to routes within the system, allowing track movement simulation according to routes.

Sealion and MATIS performs automatic route selection as part of the fusion process, using an integrated route management system. The system assigns routes to tracks within the situation picture based on the current location and available voyage information of the track, taking into account information such as the origin, destination, course and draught of the track. This information is compared to the network of available routes in order to determine which routes vessels will use to get to their destination.

Sealion and MATIS can be deployed with a hosted application server. They run on Linux, Windows and most cloud based systems, both private and public. They can also be supplied in a hardened operating system solution. Data link functionality is delivered using JREP-C, with an interface to a Data Link Processor (DLP). Tactical data can be exchanged between units using only the DLP, which supports data links such as Link 16 and Link 22. Support mechanisms for network enabled-capability (NEC) are built into the system, so NEC may be cost-effectively achieved even in low bandwidth environments without expensive military data links. Our system allows federated information sharing without a king of the hill type solution, creating truly network enabled capability.

From the Network Logging System operators can log almost all of the information within a control center. Stream-based logging services log real-time data for replay purposes, while database logging stores large amounts of data that may later be queried and analyzed using set search criteria.

Logged information can be presented and replayed in our Tactical Display Framework in completely synchronized fashion. Audio, plots, tracks, radar video, RDF strobes and more are played back so that the logged situation is completely indistinguishable from live data. Our system is also capable of creating electronic videos, printouts and screenshots of the replayed data, in order to facilitate sharing and reviewing of important data.

Sealion and MATIS support the distribution of weather, gyro and own ship's position data, which make out an important part of tactical operations. The data is distributed throughout the system, in real-time. They are also capable of receiving and presenting track data from ARPA navigation radars.

Anomaly detection is an integral part of Sealion/MATIS, enabling users to detect and prevent collisions, grounding as well as illegal activity within their area of responsibility. The MATIS applications utilize intelligent decision aids to analyze statistical track movements and user-defined rules, automatically generating alerts and responses for anomalous conditions, including intruder detection, unexpected vessel behavior, and other safety or security hazards.

A significant feature of Sealion and MATIS is the Navielektro Track Fusion and Management System (NETFMS), a multi-sensor integration and correlation for detecting and identifying potential sea-based threats utilizing inputs from maritime surveillance radars, AIS receivers, GPS-based self-reporting devices and other external data sources. Information services (databases), targeting and geospatial integration are also a central part of the system. Tactical track information is collected from various sources, and this information is used to produce an accurate, coherent Single Integrated Picture for presentation, control, and dissemination. The NETFMS system is capable of high-level fusion and has been deployed in operational use at the Finnish Navy, Coast Guard and Vessel Traffic Management Systems.

In today's demanding environment with multiple actors, the need to collaborate in realtime is constantly increasing, therefore the integrated collaboration capabilities may be accessed directly from the operator workstation. Collaboration is integrated with the geospatial view, offering direct presentation of any position or information related to an object on chart. In addition to exchange of position, static and voyage related data, MATIS contains a Military Message Handling System (MMHS) that enables authorized users to receive and transmit various messages.

Military Message Handling System (MMHS) is a profile and set of extensions to X.400 for messaging in military environments. It is NATO standard STANAG 4406 and CCEB standard ACP 123. It adds to standard X.400 email support for military requirements such as mandatory access control (i.e. Classified/Secret/Top Secret messages and users, etc.). In particular it defines a new message format, P772 that is used in place of X.400's interpersonal message formats P2 (1984 standard) and P22 (1988 standard). The derivate German COI

Specific MSA MMHS is i.e. capable of supporting the following formats:

- AIS SRM
- SMS
- EMAIL/SMTP & IMAP
- TETRA/SDS
- OS-OTH GOLD
- ADaTP-3

Sealion is far more than MATIS integrating sensors and partners on a national information platform. Today the modular system are able to provide far more functionalities as the standard NATO MCCIS up to Water Space Management (WSM) required for submarine operations. The participation of the Finnish SME in the NATO TIDE Sprints and CWIX together with the integration of the new technologies into national systems provided for one of the most sophisticated modular software suites for maritime surveillance.

The Finnish language is one of the most difficult in the world to learn and articulate, some city names hardly fit the street signs. Therefore the original Finnish name of MATIS did no suit any international commercial ambitions. In reference to the company CEO's first name a proposal was made to name it *"Maritime Awareness Tactical System (MATS)"*[183]. This was then changed into Maritime Awareness Tactical Information System (MATIS).

---

[183] © by the Author

Sitaware Suite

The SitaWare Suite originates from the company Systematic in Denmark. It provides all essential Command & Control and Battle Management capabilities right out of the box, including the all-important interoperability capabilities that allow nations to exchange battlespace information with coalition partners. Components are:

- Sitaware Headquarters
- Sitaware Frontline
- Sitaware Edge

SitaWare Headquarters is a powerful and scalable C4I system (Command, Control, Communications, Computers, & Intelligence) that offers an accessible, easy-to-use yet comprehensive toolset that can significantly increase operational flexibility and reduce deployment time while providing extensive interoperability capabilities.

SitaWare Frontline is a breakthrough in battle control software that offers clarity, simplicity and high performance that addresses real world command and control challenges at the frontline. It has been designed by and for frontline commanders in tough environments, where a clear operational view with rapid updates of friendly force tracking (FFT) is absolutely essential and has been built with deployment and in-theatre management in mind.

SitaWare Edge is a simple, lightweight, easy to use Android software designed for the dismounted domain. SitaWare Edge is part of the SitaWare Suite, which is compliant with the latest standards, making it fully interoperable with other systems, coalition partners and NATO member countries. When deployed in an operational theatre, SitaWare Edge can be used with SitaWare Frontline and SitaWare Headquarters to provide a unified C2 system from the tactical edge to the highest levels of command.

SitaWare Edge is specifically designed for the dismounted commander at the tactical edge. Its operational simplicity provides the commander with fast and clear friendly force tracking (FFT) picture, tactical situation and latest intelligence overlays thereby enhancing force protection. With the ability to exchange tactical command layers, complex maneuvers can be quickly illustrated on the map and shared with subordinates via tactical data communication, thus saving valuable voice time and speeding up operation execution.

SitaWare Edge uses SitaWare Tactical Communication (STC) - a network protocol optimized for low bandwidth, high latency tactical radio data communication supporting IP as well non-IP radios. STC is shared between SitaWare Edge (dismounted), SitaWare Frontline (mounted) and SitaWare Headquarters systems, thus establishing networked command and control and shared situational awareness all the way from higher headquarters right down to the dismounted commander at the tactical edge.

SitaWare Edge is part of the SitaWare suite of C2 products supporting a wide range of contemporary interoperability standards, making it fully interoperable with NATO and coalition partners.

SitaWare products are used in many military core systems, however they have been mainly used in land operations with centered functionalities. In order to provide the required maritime capabilities as i.e. NATO Water Space Management (WSM) for submarines and certain message text formats for automated message exchange or air functionalities, the company envisioned new developments.

Once the suite will cover a wider band of capabilities for air, land and maritime forces it could be one of the first joint platforms for harmonized systems approach in programs and project for procurement. This would in exchange possibly minimize the vast number of different systems today needed in creation of a Common Operational Picture (COP).

SJÖBASIS

The Swedish Coast Guard is on patrol along the entire Swedish coastline, at first using the Sjöbasis surveillance system to display information about all maritime traffic in the area and providing support for advanced analyses, such as those required for planning different types of rescue operation, and to be used by all of Sweden's civil authorities and shared with agencies in other countries.

The Swedish Coast Guard's task to conduct maritime surveillance includes assuming responsibility for or supporting other authorities in surveillance, crime prevention activities, and control and inspection. In its day-to-day activities the Swedish Coast Guard conducts independent maritime surveillance activities for a number of authorities, such as the Swedish Police and Swedish Customs. In some areas the authorities have reached agreement on specific measures for co-ordination, such as:

- <u>Swedish Coast Guard Maritime Clearance (SMC)</u>: Set up in July 2003 as a national point of contact for the shipping industry with responsibility for receiving and reviewing notifications in advance with regard to border control and maritime security.
- Maritime Intelligence Center (MUC): Set up with the Swedish Police and Swedish Customs in July 2004, with the aim of combating organized and cross-border criminality within the maritime sector.
- Fisheries Competence Center (FKC): Set up with the Swedish Board of Fisheries in April 2005. Provides support for daily inspection and control activities, conducts risk analyses and prepares background data.
- Notification in advance for shipping is a joint system for receiving all notification data that must be submitted before arrival at a Swedish port. Developed in collaboration with the Swedish Maritime Authority and Swedish Customs.
- Collaboration with the Swedish National Police Board's National Task Force: The Swedish Coast Guard provides boat operators for the National Task Force, and owns and manages the boats used by the Task Force. The cooperation agreement was signed in 2007.
- International Cooperation: Oil and chemical discharges, the shipping industry and organized crime recognize no borders, and require international collaboration. Examples of such organizations are the Baltic Sea Region Border Control Cooperation (BSRBCC), which deals with maritime surveillance and maritime information exchange, the Task Force on Organized Crime in the Baltic Sea Region, working against cross-border, organized crime, the North Atlantic Coast Guard Forum (NACGF), which is a partnership between North America and Europe, and HELCOM, which is responsible for environmental observation and protection. There is

also a partnership between Sweden and the USA on coast guard matters and security research. There are also various authorities operating within the EU, such as EMSA, CFCA and Frontex.

The Swedish Coast Guard, together with the civil authorities and in cooperation with the Swedish Armed Forces, has determined the civil authorities' requirements for real time maritime data and maritime information. Real time maritime data may consist of information about vessels and other objects moving on the sea. Maritime information contains additional information that Swedish authorities might need, such as a vessel's owner, crew and cargo, maps and weather information, or suspicions of crimes committed. In order to satisfy the needs of the civil authorities, the Swedish Coast Guard has developed a system for the co-ordination of sea-based information between authorities: SJÖBASIS.

However, when the Headquarter in Stockholm was commissioned, SJÖBASIS lacked a number of functions with the potential to simplify work for the Coast Guard. The development of a new monitoring system (SJÖBASIS 2) made use of the latest technological developments to simplify the vitally important work of the Coast Guard, both at sea and in the air in the two sea surveillance centers in Muskö and Gothenburg.

Most user switched until autumn 2016 to the SJÖBASIS Generation 2 which recieves data from various sources to pinpoint vessel positions and monitor weather conditions and geographical data. Large amounts of information are processed in real time and presented in a new dynamic user interface. Minimizing the need for a manual, despite the system's complex analytical functions, was one of several key goals.

SJÖBASIS was developed in two initial stages with information from:

Stage 1

- Swedish Coast Guard
    - Information and decision-making support system, KIBS
    - System for research vessel permits
- Swedish Maritime Authority
    - Vessel reporting system, FRS
    - Swedish AIS system
    - HELCOM AIS system
- Swedish Meteorological and Hydrological Institute
    - Weather forecasts
    - Weather observations
- Swedish Environmental Protection Agency
    - VIC Natur (Nature reserves)
- Swedish Maritime Authority & Lantmäteriet (Swedish Mapping, Cadastral and Land Registration Authority)
    - Map databases

- Swedish Board of Fisheries
  - Vessel Monitoring System (VMS)
- Swedish Armed Forces
- Real time maritime data
- Organizational administrative resources
- National resource list

Stage 2

- All authorities
  - Image database
  - Additional information
  - National resource list. Operational resources from all authorities concerned.
- Swedish Board of Fisheries
- Register of fishing vessels
- Register of professional fishing licenses
- LANDBAS (fisheries control database)
- Swedish Coast Guard
  - Routes for aircraft
- Police
  - Boats disappeared or being searched for
- MSB
  - RIB and WEB-RIB
- Swedish Maritime Authority
  - Register of vessels
  - Register of seamen
  - Vessel inspection system
- SGU
  - Environmental chemical sediment database
  - Marine geological map database
  - Database for measurement lines and tests
- Swedish Customs
  - Register of traffic permits
- SMHI
  - SeaTrackWeb

The Swedish Coast Guard cooperates closely with the Swedish Navy, which is using its own surveillance system C2SöC, however both systems have the same core by the company SAAB.

Sea Surveillance Cooperation Baltic Sea (SUCBAS)

The <u>Sea Surveillance Cooperation Baltic Sea (SUCBAS)</u>[184] has its origin in the Sea Surveillance Co-operation Finland- Sweden (SUCFIS) that commenced during the beginning of the 3rd millennium. The spirit of SUCFIS is that that multinational co-operation is considered the best way to ensure efficient Maritime Situational Awareness.

SUCBAS aims at improving the information exchange between its members and therefore enhance the Maritime Security Environment in being beneficial to the Maritime Safety, Security, Environmental and Economic matters by sharing knowledge in sea surveillance between the relevant authorities of the participating nations.

In 2008 an initiative was taken by the two countries Finland and Sweden, already cooperating in the Sea Surveillance Cooperation Finland-Sweden (SUCFIS), to enlarge the co-operation in order to encompass all countries around the Baltic Sea. A seminar was hosted in September 2008 at the Berga Castle in Sweden to boost the idea, which was attended by most North European countries.

During the first SUCBAS Conference hosted by Finland at the House of Estates in Helsinki on the 4. of March 2009, the countries Finland, Sweden, Denmark, Germany, Estonia and Lithuania signed the SUCBAS Letter of Intent (LoI). Building upon the spirit and lessons learned from SUCFIS, SUCBAS commenced operations on the 2. of April 2009 and on the 24. of September 2009 the SUCBAS co-operation was joined by Poland and Latvia. During the SUCBAS seminar held in the International Maritime Museum in Hamburg, the United Kingdom became the ninth member of SUCBAS on 12. March 2015.

Until recently collaboration between the different services operating in the maritime domain in the Baltic has been fragmented. The maritime rescue service in one nation might share information with a maritime rescue service of another nation, likewise would the navy of one nation share information with navy of another nation. However until the implementation of SUCBAS only in very rare cases would information be shared across service across borders.

During the first time of SUCBAS operations, sharing of information was confined to manual exchange of reports only, but since then all nine SUCBAS countries have implemented automated solutions to support establishment of sustainable multinational Maritime Situational Awareness in connecting their national systems to the SUCBAS Network.

The SUCBAS network to share maritime information across borders and across services has been implemented into national Maritime Situational Awareness. Today MSA information is continuously shared between the participating parties benefitting at the same time maritime safety, maritime rescue, maritime assistance, maritime environmental protection,

---

[184] http://sucbas.org/

maritime security and law enforcement in the Baltic Sea region.

The technical concept is based on the approach of interaction of existing sources and procedures using common standards and distribution principles with a design having low impact on the autonomous national systems. Sharing services depending on national and international agreements, laws and regulations with priority given to locally and nationally available track history, notifications and alerts from well managed and sustainable data bases will enhance the individual nation Recognized Picture (RMP).

In recognition of the fact that responsibility for of maritime surveillance, maritime safety, maritime security, the maritime environment and maritime law enforcement are implemented differently in each country, SUCBAS information can be shared among national governmental institutions with a maritime responsibility regardless if these are civil or military at the discretion.

The main objectives of SUCBAS are to:

- enable the exchange of maritime information and enhancing MSA.
- support and deepen the overall cooperation between countries involved.
- enhance maritime safety and security in the Baltic region.
- support national authorities with responsibilities regarding environmental hazards.
- support authorities conducting maritime law enforcement and border control.

Similar to other Maritime Entities, SUCBAS is not limited to monitoring AIS transmissions of commercial vessels. In support of risk evaluation the SUCBAS participants continuously share information on all type of vessels in their area of responsibility and evaluate issues like:

- ships records of dubious or criminal activities,
- previous port calls,
- cargo being transported,
- navigational behavior observed,
- seaworthiness and
- civil and military intelligence.

Based on the continuous evaluation performed the SUCBAS participants will identify the vessels which is considered to pose extraordinary risk and therefore should be kept under a close surveillance.

The risk posed by the individual ships travelling through the Baltic waters is continuously adjusted though compilation of all different the pieces of information provided by the different national SUCBAS participants. Many different organizations on the national and internationally level are engaged in maritime safety, security, environmental law enforcement activities and provision of maritime situational awareness. It is imperative than no single organization or single country is capable of establishing sustainable maritime situational awareness on its own. In the worlds common quest to safeguard the maritime environment it is imperative that nations and organization work together. SUCBAS welcomes cooperation with other initiatives supporting maritime situational awareness, and in support hereof SUCBAS utilize commonly agreed concepts and terminology when sharing information.

Within the SUCBAS cooperation the couplings between the participating nations are very loose. The cooperation entails no command structure and all participants corporate on an equal level and all participants decide what parts of their national information base they will share with the rest of the community. While other co-operations mandate that participants invest in a common IT infrastructure SUCBAS, like MARSUR, utilizes the infrastructure already implemented by the participating parties.

SUCBAS Organization:

- Steering Board,
- Coordination Group,
- Technical Group ,
- Operational Group,
- and Security Group (initiated on request)

The highest level is the Steering Board composed of mostly flag officers appointed from each participating state, followed by the Co-ordination Group responsible for overall SUCBAS operations and the further development of the multinational collaboration. Future projects requiring decisions are subject to the Steering Board.

The three sub-groups contain the respective representative and the national Subject Matter Experts (SME) from each member state, are responsible for development and maintenance of SUCBAS concepts of operations, operational procedures, standards and specifications, and automated interoperability solutions.

The Chair of each Group and the Steering Board is annually rotated between all SUCBAS countries while the Senior POC for the operational networking is rotating on a weekly bases.

SUCBAS distinguishes two levels of operation. SUCBAS Level 1 cooperation comprises:

- Exchange of points of contact responsible for national maritime surveillance, and
- Manual exchange of maritime contacts of special surveillance interest though a daily report via e-mail supplemented by public telephone as applicable.

SUCBAS Level 1 information exchange comprises sharing of data information and knowledge that is not classified higher than MILITARY UNCLASSIFIED/AUTHORITY EYES ONLY and Level 1 cooperation is mandatory for all SUCBAS participants.

SUCBAS Level 2 cooperation is an enhancement to that of Level 1 comprising:

- Online automated sharing of maritime situational awareness data, information and knowledge through reciprocal access to each other's databases,
- Online sharing of sensor such as radar generated tracks and electro-optical video and images, supplemented via chat services.

SUCBAS Level 2 information exchange comprises sharing of data information and knowledge that is not classified higher than MILITARY UNCLASSIFIED/AUTHORITY EYES ONLY. In Level 2 all data, information and knowledge sharing between the participating authorities takes place though Virtual Private Network tunnels over the public Internet.

SUCBAS Level 3 cooperation is envisioned as an enhancement to that of Level 2 comprising online automated sharing of maritime situational awareness data, information and knowledge which level of classification may exceed MILITARY UNCLASSIFIED/AUTHORITY EYES ONLY.

SUCBAS Level 3 sharing of information comprise discretionary access control to sensitive information safeguarding that certain pieces of information is only shareable among the limited share of SUCBAS participants (organizations and/or individuals) specially authorized hereto. In support of sharing data, information and knowledge that may be military classified an envisioned SUCBAS Level 3 entails utilization of appropriate certified military encryption equipment and the definition on which classification shall be used within the layer of a restricted second SUCBAS Network, which will also be more cost intensive.

SUCBAS Level 3 would require a second, parallel classified network to the existing SUCBAS UNCLASS and national entry points with the connection of the respective national classified system as well as an agreed and common crypto device and classification level.

Since not all SUCBAS members are neither in NATO or the EU and the lack of defined SUCBAS Classification the SUCBAS Level 3 is a long term investigation currently set dormant. Similar to MARSUR, the cooperation SUCBAS is also missing a common organization similar to the NCIA and NSA in NATO to ensure common funding and project arrangements.

SUCBAS is one of the world's first federation-of-systems in MSA, built on common standards and procedures. SUCBAS is using a standardized Sea Surveillance Cooperation Baltic Sea (SUCBAS) Interface Technology related to Federated Mission Networks (FMN) in NATO. As the "*mother of federated military systems*[185]" SUCBAS thrives to implement further the developments in FMN.

---

[185] ... when considering Maritime Surveillance (MARSUR) Networking the "*father of federated military systems*"... (Joachim Beckh)

Sea Surveillance Cooperation Baltic Sea (SUCBAS) Interface Technology

The Sea Surveillance Cooperation Baltic Sea (SUCBAS) Interface Technology is used in SUCBAS Level 2 and Level 3 cooperation and based on the so called TIDE (Technology for Information, Decision and Execution superiority) initiative originally developed NATO ACT (Allied Command Transformation).

TIDE is developed for use when heterogeneous organizations come together to cooperate. The TIDE concept acknowledges cooperation where there is no central organization that handles central infrastructure. All cooperation is done by voluntary cooperation on agreed standards.

Within SUCBAS and TIDE services are based on ontologies as an alternative to information models creating a more open environment for information sharing. An entity can be described by one or several ontologies at the same time. In a simplified example a vessel can be described by using two ontologies: *"Vessel"* and *"Position Report"*. Static information like ship name and nationality is contained in the *"Vessel"* ontology and dynamic information like Ship's position is enclosed in the *"Position Report"* ontology. The common element for both ontologies linking the information together in this case is a Ship-ID.

A vital part in a dynamic environment is how services are found. In SUCBAS the common strategy is using multicast DNS2 (mDNS). The technology is based on DNS with records A and MX. In these records information about services is stored (DNS Service Discovery). In the picture below Party A is searching for services. Party B and Party C have already registered services locally on their machines. When party A broadcast a call searching for services Party B and Party C responds with their matching result.

The TIDE Information Discovery concept utilized within the SUCBAS cooperation evolved out of the requirement to discover, extract, and make useful, information from emerging and legacy systems. To achieve the desired effect the participating systems must implement the Information Discovery Protocol. This protocol is designed to be scalable, and small and easy to implement.

The Information Discovery concepts are implemented in two distinct delivery mechanisms:

- Request-Response implementation is a web services (http) based on Representational State Transfer (REST6) style calls, and
- Subscribe-Publish implementation is based on the streaming services provide by XMPP.

The standards used in the Sea Surveillance Cooperation Baltic Sea (SUCBAS) Interface Technology are closely related to the standards used in the Maritime Surveillance Networking (MARSUR Networking) Interface Technology and are based on the Federated Mission Network (FMN) Track in TIDE.

Sea Surveillance Cooperation Finland Sweden (SUCFIS)

Finnish and Swedish Forces had been cooperating for a long period of time. After the „Hoburg" Project 1997 followed the first study to exchange information between FIN MEVAT and SWE STRIMA systems. In the following year the bilateral „Loviisa" exercises started annually between Finland and Sweden and in 2000 a plan for advanced exercises and the development of the „Loviisa" test system was initiated. During 2001 the Swedish Defense Forces and Finnish Navy were assigned to start developing the sea surveillance exchange with the existing systems and in October 2001 was the Kick-off meeting, with the first operational and technical group meeting in January 2002.

The aim of the project under the Sea Surveillance Cooperation Finland-Sweden (SUCFIS) was to develop and implement information exchange between Finnish and Swedish Sea surveillance systems. Main objectives of SUCFIS are to:

- Improve maritime situational awareness in the Baltic Sea, Gulf of Finland and Gulf of Bothnia
- Improve interoperability between Sweden and Finland
- Support maritime safety
- Support the control of environmental hazards
- Support authorities conducting border control according to Schengen agreement

In January 2003 followed the Demonstration Phase already including Technical tests, Operator training and Operator exchange and in October 2003 the development of the interface and its implementation with tests started and the Standard Operation Procedures (SOP) were drafted. A total of 30 working group meetings, a four man-years work, had created SUCFIS.

SUCFIS is exchange of information concerning surface maritime picture with attributes and properties. The respective parties are responsible to notify each other on the following incidents or information (classified "*FIN/SWE CONFIDENTIAL*"):

- Information about the SUCFIS status
- Malfunction of sea safety systems
- Incidents (Start/End)
- Environmental disasters
- Observed pollution at sea
- Mines and dangerous objects for maritime traffic
- Exceptional weather conditions
- Presence of third party military vessel

The information about Finnish and Swedish Navy vessels can be exchanged when specially agreed. The national control centers in Sweden in Muskö and in Finland in Korppoo are authorized to make the decisions on releasing the position and information on Swedish and Finnish navy vessels. The national RMP is feed with the relevant information from the Swedish SjöCny and Finnish MEVAT system, both been updated to new system versions, but keep independent under national authority. Prior the Swedish Maritime Surveillance System Safir SeaOps was interfaced to Finland.

The SUCFIS Steering Group (SG) is the coordinating element of the SUCFIS. The working groups (WG) control the progress of the SUCFIS and takes action if needed. The working groups prepare and present the updates of documents for the Steering Group according the orders from the Steering group or gathered lessons learned. The working groups are responsible to execute the decisions made by the Steering Group, which will forward issues out of their authority to national Navy or Armed Forces HQ, while the SUCFIS working groups are responsible for making conclusions of the lessons learned and proposals for development to be presented to the SUCFIS Steering group.

The Sea Surveillance Cooperation Finland-Sweden (SUCFIS) aims at creation and use of a bilateral tool for analyzing data and improves the quality of the national RMP under MSA and the national experience of the experts supported the development of FMN in NATO as well as in SUCBAS and MARSUR. The Sea Surveillance Cooperation Finland Sweden (SUCFIS) is the foundation stone of the Sea Surveillance Cooperation Baltic Sea (SUCBAS) and the Maritime Surveillance Networking (MARSUR Networking) and also basis for the work in the NATO Multinational Experiment 5 (MNE 5).

SUCFIS with a common interface system between the two national Sea Surveillance Systems of Finland and Sweden, referred to as Sea Surveillance Information System (SSIS), is the root for the federated networks of systems in Maritime Situational Awareness.

Sea Surveillance Information System (SSIS)

Sea Surveillance Information System (SSIS) is the interface between the Finnish control center and the Swedish Maritime Operation Centers in the Sea Surveillance Cooperation Finland Sweden (SUCFIS) and not to be seen as an independent system.

Only the respective national fused or merged track contacts are exchanged from the national system in a push principle, therefore the original source is no longer recognizable. The track quality depends on the sensors selected for the track information, i.e. a contact in the Swedish area and system might be supplied by a Finnish radar track.

SUCFIS and the Sea Surveillance Information System (SSIS) are operational since 2006 for exchange of track and other sensor data. It serves mainly military but also environmental purposes. The interface connects the respective national systems in the context of departmental responsibility, the information can be forwarded to other authorities, i.e. the METO database (Tri-Maritime Environment Authority Operations) in Finland.

Seychelles (SYC)

The Seychelles are a maritime stronghold, however due to geographical location and limited resources the country requires to be integrated in cooperation with other neighboring countries. The Seychelles are an archipelago in the Indian Ocean, northeast of Madagascar, and is the smallest African country including 155 islands with a water supply depending on collection of rainwater.

The Seychelles are in international environmental agreements party to Biodiversity, Climate Change, Climate Change-Kyoto Protocol, Desertification, Endangered Species, Hazardous Wastes, Law of the Sea, Marine Dumping, Ozone Layer Protection, Ship Pollution, Wetlands.

Regional Fusion and Law Enforcement Center for Safety and Security at Sea (REFLECS3),

Regional Fusion and Law Enforcement Center for Safety and Security at Sea (REFLECS3) was former the Seychelles Regional Anti-Piracy Prosecutions Intelligence Coordination Center (RAPPIC), which played an integral role in anti-piracy and countering drug trafficking in the Indian Ocean Region.

RAPPICC announced in 2013 that the activities of the organization would expand to include combating other forms of maritime crime. The transition of the original RAPPICC concept towards a broader remit addressing other types of serious cross-border crime affecting the region, such as human trafficking, drugs smuggling, environmental crime and, of course, piracy.

The REFLECS3 is the transition from a purely piracy focused theme to a broader mission which focuses on enhanced maritime security in the region and building on our successful program for counter-piracy, commencing a wider engagement with regional partners to help support the fight against transnational organized crime (TNOC).

This new mandate which was authorized by the multi-nation Steering Group, the independent body which provides strategic direction to the center, and takes the form of three interlinked missions:

- the Transnational Organized Crime Unit (TOCU),
- a Maritime Trade Information Sharing Center (MTISC)
- and a Local Capacity Building Coordination Group (LCBCG).

Singapore (SGP)

The Republic of Singapore is by geographical position and economical power one of the most influential entities in its region. It has a strategic position between Malaysia and Indonesia, as focal point for Southeast Asian sea routes and consists of about 60 islands, the by far the largest is Pulau Ujong. The area of interest of the Republic of Singapore has been intensely hit by piracy and terrorism which resulted in a wide range of national and international activities.

Singapore is in international environmental agreements party to Biodiversity, Climate Change, Climate Change-Kyoto Protocol, Desertification, Endangered Species, Hazardous Wastes, Law of the Sea, Ozone Layer Protection, Ship Pollution.

Information Fusion Center (IFC)

Established on 27 April 2009 by the Republic of Singapore Navy, the Information Fusion Center (IFC)[186] is a regional Maritime Security (MARSEC) information-sharing center with wide ranging international links and geographic area from the Gulf of Bengal, Malacca Strait to the South China Sea.

IFC aims to facilitate information sharing and collaboration between partners to enhance maritime security. Through the speedy sharing of information, it facilitates timely and effective responses from partner countries for MARSEC incidents. The links to 65 agencies in 35 countries, with 16 International Liaison Officers (ILOs) from 15 countries, 82 ILOs representing 23 states (Australia, Brunei, Cambodia, Canada, Chile, China, France, Greece, India, Indonesia, Italy, Japan, Malaysia, Myanmar, New Zealand, Pakistan, Peru, Philippines, South Africa, Thailand, UK, USA, Vietnam) provide a wide cooperation area with thirteen Memorandum of Understandings, two Technical Arrangements and a Standing Operational Procedure with the Malacca Straits Patrol.

Therefore IFC has played a role in resolving various MARSEC incidents; it has for example provided timely situational updates on ships hijacked by pirates in the Gulf of Aden to facilitate better operational decisions. In Nov 2012, through the IFC's real-time updates, the Vietnam Peoples' Navy and Vietnam Marine Police (now renamed Vietnam Coast Guard) were also able to localize a hijacked Malaysia-flagged tanker, the MV ZAFIRAH, in the South China Sea and arrested the perpetrators.

To support the Search and Locate (SAL) operations for the missing the MH 370 flight, the

---

[186] https://www.mindef.gov.sg

IFC first consolidated a situation picture of the SAL operation in South China Sea and Malacca Strait. The details of the SAL operation, including assets deployed and search sectors where available, were then shared among the various ILOs and Operation Centers that were linked to the IFC. With the shift of the search to the Southern Corridor, the IFC also engaged commercial ships transiting the Indian Ocean through specific IFC advisories to more than 330 shipping companies to report sightings or nil sightings to the IFC. This was to create awareness for all the partners, and also to assist the SAL coordinators, who could take into account the relevant information to decide the allocation of resources for their subsequent searches.

The IFC also conducts capacity-building activities such as international information sharing exercises and MARSEC workshops, for example, the biennial Maritime Information Sharing Exercise (MARISX) and the annual Regional Maritime Security Practitioner Course.

IFC generates unclassified reports on positions of tracks from AIS and/or LRIT, IMO Number, Images and Videos of the tracks, notifications and warnings for distribution to the partners with a possible limited further distribution.

The system surveillance is divided into:

- piracy and armed robbery
- IUU fishing
- smuggling
- terrorism
- illegal immigration and human trafficking
- maritime incidents
- natural disasters
- arms smuggling

In addition to the IFC Singapore established the national maritime security framework on an inter-ministry policy level and the Maritime Security Task Force located in the Changi Command and Control Center, which also houses the Singapore Maritime Security Center (SMSC) as national platform for law enforcement agencies sharing a common situation picture including harbor system, AIS and VTS.

The ASEAN ILOs in IFC also serve as the Permanent Secretariat of the ASEAN Navy Chiefs' Meeting. As the Permanent Secretariat, the IFC facilitates and monitors the development of new MARSEC initiatives among ASEAN Navies. The IFC also hosts maritime information sharing portals such as the ASEAN Information Sharing Portal and the Regional Maritime Information eXchange (ReMIX), which facilitates information sharing among ASEAN navies and Western Pacific Naval Symposium members, respectively. The backbone of the network is the Open & Analyzed Shipping Info-System (OASIS), also using information from the Malacca Straits Patrol Info-System (MSP IS).

Through ReMIX the IFC shares information with the 24 countries of the Western Pacific Naval Symposium (WPNS). They organize regular workshops and seminars to foster common practice. The IFC hosted a MARISX information sharing exercise with other OPCENs. There are links with 64 agencies in 34 countries in the EU (France, UK (UK-MTO), and via Italy into the Virtual-Regional Maritime Traffic Center V-RMTC).

Malacca Straits Patrol Info-System (MSP IS)

The Malacca Straits Patrol Info-System (MSP IS)[187] was introduced in the Information Fusion Center (IFC) to facilitate collaboration between Indonesia, Malaysia, Singapore and Thailand with bilateral links.

The launch and operational use of the MSP-IS provides Singapore a co-operative tool to enhance the situation awareness of Malacca Strait and its approaches. With the MSP-IS, air and sea assets deployed on scene can quickly pass information of an unfolding incident to all Monitoring and Action Agencies (MAA) on a real-time basis. This will enhance the shared situation awareness and facilitate co-ordination in term of responses.

The MSP-IS was set up in November 2006 for trial operations. Since then, the MAAs have been accessing it on a regular basis to share maritime information. The MSP-IS is connected to the Open & Analysed Shipping Information System (OASIS) and a Sense-Making Analysis and Research Tool (SMART).

Open & Analysed Shipping Info-System (OASIS)

Open & Analysed Shipping Info-System (OASIS)[188] is the node for the ASEAN Information Sharing Portal via the Regional Maritime Information eXchange (ReMIX) system both serving to enhance the situation awareness of Malacca Strait and its approaches. OASIS was developed by the Comprehensive Maritime Awareness Team of the Defense Science and Technology Agency (DSTA) of Singapore as a centralized system connected to open sources, information sharing partners, national agencies and commercial partners. The IFC is developing a new system to replace OASIS with improved interfaces.

OASIS maritime database collects and fuses the information received information from the distinct information system and shares information with the Coast Guard, the ship owners and sailing ships, enhances it via the Sense-making, Analysis and Research Tool (SMART), and allows for collaboration via the forum, chat and email. SMART generates a list of vessels of interest (VOI). Via OASIS the IFC provides a Maritime Picture (MP) and sends regular

---

[187] http://www.mindef.gov.sg
[188] http://www.mindef.gov.sg

reports.

The operators have via OASIS the functionalities for File-Transfer, E-Mail, Whiteboard, Voice, and Chat, however the system has an open interface architecture that can be customized to different needs, including the WEB based and a mobile version. The scalable architecture facilitates integration with other systems.

OASIS is a living repository that aids trending and analysis. It has grown to contain over 180,000 vessel tracks to aid investigation of maritime activities. Customized user profiles can be rapidly configured via OASIS to meet the information needs of different stakeholders. For example, the CMA system supports the efforts of the Malacca Strait Patrols (MSP) and the Regional Maritime Information Exchange (REMIX) which focus on maritime security in different areas. In order to manage the massive amounts of information, the Information Management engine controls the complex sharing requirements of each information source.

Regional Maritime Information Exchange (ReMIX) System

The Regional Maritime Information (ReMIX) System[189] was introduced by the Royal Singapore Navy in the Information Fusion Center (IFC) to facilitate inter-OPCEN collaboration on the multinational level via a portal as well as OASIS. The IFC has a partnership with ReCAAP-ISC in which they contribute to the 24/7 surveillance watch and ReMIX was basically developed under the auspices of the Western Pacific Naval Symposium (WPNS).

Sense-making, Analysis and Research Tool (SMART)

Sense-making, Analysis and Research Tool (SMART)[190] SMART was built to enable early detection of maritime threats. Working as part of the Open & Analysed Shipping Info-System (OASIS), it uses an innovative fusion technique to piece together vague or partial information like a jigsaw to present a coherent picture. SMART also connects the dots between real-time and archived data, thereby facilitating the investigation of vessels across time periods and identification of emerging trends.

---

[189] http://www.mindef.gov.sg
[190] http://www.mindef.gov.sg

Surveillance Picture (Surpic)

On 27 May 2005 the Surveillance Picture (Surpic)[191] web based tool for sharing and supporting a common maritime situation picture was launched in Batam, Indonesia, by senior officers from both navies, as a platform to enhance the interoperability and information exchange between the Republic of Singapore Navy (RSN) and the Indonesian Navy (TNI AL). Under the ambit of the Indonesia-Singapore Coordinated Patrol (ISCP), the SURPIC system allowed both navies to share a real-time sea situation picture of the Singapore Strait. This enhanced the coordination of operational responses and deployment of assets by the various maritime forces under the ISCP.

The agreement to launch SURPIC II, an enhanced version of SURPIC, was signed by the Singapore and Indonesian navies on 3 Dec 2009. SURPIC II was launched on 11 May 2012 as part of the program to commemorate 20 years of ISCP operations. The key enhancements in SURPIC II are as follows:

- Inclusion of Analytics. SURPIC II incorporates the Open and Analysed Shipping Information System (OASIS) and the Sense-Making Analysis and Research Tool (SMART) which enhances maritime situation awareness and sense-making respectively. OASIS is a shipping database which allows the tracking of vessels real-time when they traverse the straits of Malacca and Singapore. It is able to display both real-time and historical trend data in a common display screen. SMART complements the OASIS database by detecting anomalies in the behavior of ships transiting in the region, thus providing early warning to cue operational responses where necessary.
- Real-Time Chat Function and Translation. SURPIC II facilitates real-time exchange of information between the operators in Singapore and Indonesia through a group chat function. This facilitates daily operational updates of maritime information about the Singapore Strait. SURPIC II also provides real-time translation between Bahasa Indonesia and English for SURPIC II operators from Indonesia and Singapore when they are utilizing the chat function. This enhances the communication and interoperability between both navies.
- Enhanced Accessibility. SURPIC II is an internet-based platform, and will be highly mobile and accessible compared to its predecessor, SURPIC.

---

[191] https://www.mindef.gov.sg, not identical with SURPIC II application of the US Coast Guard Rescue Coordination Centers (RCCs) used to determine the best Amver asset to utilize in the event of a distress at sea.

South Africa (ZAF)

South Africa is by geographical location and economic power the largest Maritime Entities within the African continent. It includes Prince Edward Islands (Marion Island and Prince Edward Island), completely surrounds Lesotho and almost completely surrounds Swaziland, an lays in a strategic position at the southern tip of the continent of Africa. As in most African countries there is a lack of important arterial rivers or lakes, which requires extensive water conservation and control measures, growth in water usage is outpacing supply, there is pollution of rivers from agricultural runoff and urban discharge, air pollution resulting in acid rain in the big cities including soil erosion and desertification.

South Africa is in international environmental agreements party to Antarctic-Environmental Protocol, Antarctic-Marine Living Resources, Antarctic Seals, Antarctic Treaty, Biodiversity, Climate Change, Climate Change-Kyoto Protocol, Desertification, Endangered Species, Hazardous Wastes, Law of the Sea, Marine Dumping, Marine Life Conservation, Ozone Layer Protection, Ship Pollution, Wetlands, Whaling.

South African Maritime Safety Authority (SAMSA)

The South African Maritime Safety Authority (SAMSA) is a South African government agency established on 1 April 1998 as a result of the 1998 South African Maritime Safety Authority Act 5. as such it is responsible for the implementation of current International & National Regulations regarding the Maritime Industry as well as upon all recreational marine vessels within its jurisdiction.

SAMSA via the administration and/or management of all things marine related is in effect the governing authority and as such is required to investigate maritime accidents/incidents & to provide various marine related services both on behalf of Government as well as to Government.

It is subordinate to the Minister of Transport, who heads the Department of Transport. Despite it being a marine authority its head office is over 500 km away from the nearest ocean in Pretoria.

SAMSA administers the South African ship register. In July 2012 the authority acquired the former Antarctic supply vessel S. A. Agulhas as a training ship.

Southern Africa Development Community (SADC)

The predecessor of the Southern African Development Community (SADC)[192] was the Southern African Development Co-ordination Conference (SADCC), established in 1980 in Lusaka, Zambia. In 1992, Heads of Government of the region agreed to transform SADCC into the Southern African Development Community (SADC), with the focus on integration of economic development. SADC members are Angola, Botswana, DR Congo, Lesotho, Madagascar, Malawi, Mauritius, Mozambique, Namibia, Seychelles, South Africa, Swaziland, Tanzania, Zambia and Zimbabwe.

SADCC was formed to advance the cause of national political liberation in Southern Africa, and to reduce dependence particularly on the then apartheid era South Africa; through effective coordination of utilization of the specific characteristics and strengths of each country and its resources. SADCC objectives went beyond just dependence reduction to embrace basic development and regional integration. Hence, SADCC was formed with four principal objectives, namely:

- Reduction of Member State dependence, particularly, but not only, on apartheid South Africa;
- Forging of linkages to create genuine and equitable regional integration;
- Mobilization of Member States' resources to promote the implementation of national, interstate and regional policies; and
- Concerted action to secure international cooperation within the framework of the strategy for economic liberation.

The main objectives of the Southern Africa Development Community (SADC) are to achieve development, peace and security, and economic growth, to alleviate poverty, enhance the standard and quality of life of the peoples of Southern Africa, and support the socially disadvantaged through regional integration, built on democratic principles and equitable and sustainable development.

The objectives of SADC, as stated in Article 5 of the SADC Treaty (1992) are to:

- Achieve development and economic growth, alleviate poverty, enhance the standard and quality of life of the people of Southern Africa and support the socially disadvantaged through Regional Integration;
- Evolve common political values, systems and institutions;
- Promote and defend peace and security;
- Promote self-sustaining development on the basis of collective self-reliance, and the inter-dependence of Member States;

---

[192] http://www.sadc.int

- Achieve complementarity between national and regional strategies and programs;
- Promote and maximize productive employment and utilization of resources of the region;
- Achieve sustainable utilization of natural resources and effective protection of the environment;
- Strengthen and consolidate the long-standing historical, social and cultural affinities and links among the people of the Region.

Based on the SADC Objectives is the SADC Common Agenda, originates in Article 5 of the SADC Treaty (1992) as amended. The Common Agenda summarizes the key strategies and policies of the institution. Subsequently, the SADC institutional structure is consistent with the SADC Common Agenda and Strategic Priorities that it encapsulates. The same values are echoed in the Regional Indicative Strategic Development Plan (RISDP) and Strategic Indicative Plan for the Organ (SIPO).

The SADC Common Agenda is underpinned by a series of principles and policies, including:

- Promotion of sustainable and equitable economic growth and socio-economic development that ensures poverty alleviation with the ultimate objective of its eradication;
- Promotion of common political values, systems, and other shared values, which are transmitted through institutions that are democratic, legitimate and effective; and
- Promotion, consolidation and maintenance of democracy, peace and security.

In relation to Maritime Security the Southern Africa Development Community (SADC) approved in 2011 a Maritime Security Strategy, and Tanzania, Mozambique and South Africa signed an MoU on maritime security cooperation with SADC to ensure the free trade and security in their territorial waters which forms the SADC Maritime Entity including Democratic Republic of Congo, Angola, Namibia, South Africa, Mozambique, Tanzania, Madagascar, Mauritius, Seychelles, Botswana, Lesotho, Malawi, Swaziland, Zambia, and Zimbabwe.

Spain (ESP)

Spain is by geographical position, together with Portugal, coastal area, economic and military capability one of the larger Maritime Entities. Spain is bordering the Mediterranean Sea, North Atlantic Ocean, Bay of Biscay, and Pyrenees Mountains and the southwest of France. Its strategic location along approaches to Strait of Gibraltar and Spain also controls a number of territories in northern Morocco including the enclaves of Ceuta and Melilla, and the islands of Penon de Velez de la Gomera, Penon de Alhucemas, and Islas Chafarinas.

Spain is facing pollution of the Mediterranean Sea from raw sewage and effluents from the offshore production of oil and gas, water quality and quantity nationwide, air pollution, deforestation, and desertification.

Spain is in international environmental agreements party to Air Pollution, Air Pollution-Nitrogen Oxides, Air Pollution-Sulfur 94, Air Pollution-Volatile Organic Compounds, Antarctic-Environmental Protocol, Antarctic-Marine Living Resources, Antarctic Treaty, Biodiversity, Climate Change, Climate Change-Kyoto Protocol, Desertification, Endangered Species, Environmental Modification, Hazardous Wastes, Law of the Sea, Marine Dumping, Marine Life Conservation, Ozone Layer Protection, Ship Pollution, Tropical Timber 83, Tropical Timber 94, Wetlands, Whaling; signed, but not ratified is Air Pollution-Persistent Organic Pollutants.

Turkey (TUR)

Turkey as by geographical position in the Mediterranean, coastal area, economic and military capability. one of the larger Maritime Entities. Turkey is part of Southeastern Europe and Southwestern Asia (that portion of Turkey west of the Bosporus is geographically part of Europe), bordering the Black Sea, between Bulgaria and Georgia, the Aegean Sea and the Mediterranean Sea, between Greece and Syria.

The strategic location is controlling the Turkish Straits (Bosporus, Sea of Marmara, Dardanelles) that links the Black and Aegean Seas; this geographical position makes it to a important partner in NATO. 3% of Turkish territory north of the Straits lies in Europe and goes by the names of European Turkey, Eastern Thrace, or Turkish Thrace, the 97% of the country in Asia is referred to as Anatolia.

Turkey has high water pollution from dumping of chemicals and detergents, also air pollution, particularly in urban areas, and deforestation. Concern are for oil spills from increasing Bosporus ship traffic.

Turkey is in international environmental agreements party to Air Pollution, Antarctic Treaty, Biodiversity, Climate Change, Desertification, Endangered Species, Hazardous Wastes, Ozone Layer Protection, Ship Pollution, Wetlands. Signed, but not ratified is Environmental Modification.

Operation Black Sea Harmony (OBSH)

Operation Black Sea Harmony (OBSH) with the Operation Black Sea Harmony Centre for Permanent Coordination (OBSH CPC) is aiming at the security in the Black Sea under the Turkish Naval Forces since 1. March 2004. The operation has been supported by a task group consisted of frigates, corvettes, Turkish Type Patrol Boats, fast attack boats and submarines. Additionally, it was backed by helicopters, maritime patrol aircrafts and coast guard boats.

OBSH CPC was established on 10 October 2005 in order to set up information interchange as well as coordination among other Nations components attending Operation Black Sea Harmony. Recognized maritime picture attained by Operation Black Sea Harmony has been shared by NATO authorities and headquarters.

3 Black Sea countries including Russian Federation, Ukraine and Romania have officially responded to the invitation made by Turkey. Following the exchange of letters between commanders of Turkey and Russian Federation, Russian Federation was officially included in this operation with the exchange of notes between ministries of foreign affairs on 27 December 2006.

United Kingdom of Great Britain and Northern Ireland (GBR)

The United Kingdom of Great Britain and Northern Ireland (GBR) is a traditional maritime power due to is history, geographical location and economy in Europe. The British Royal Navy is the largest Western European Navy with the French Navy coming second; both nations also represent the only European Nuclear Military Powers.

The main island lies near the vital North Atlantic sea lanes, is only 35 km from France and linked by tunnel under the English Channel. Because of heavily indented coastline, no location is more than 125 km from tidal waters. The GBR are including the northern one-sixth of the island of Ireland, between the North Atlantic Ocean and the North Sea, northwest of France.

United Kingdom of Great Britain and Northern Ireland (GBR) are in international environmental agreements party to Air Pollution, Air Pollution-Nitrogen Oxides, Air Pollution-Persistent Organic Pollutants, Air Pollution-Sulfur 94, Air Pollution-Volatile Organic Compounds, Antarctic-Environmental Protocol, Antarctic-Marine Living Resources, Antarctic Seals, Antarctic Treaty, Biodiversity, Climate Change, Climate Change-Kyoto Protocol, Desertification, Endangered Species, Environmental Modification, Hazardous Wastes, Law of the Sea, Marine Dumping, Marine Life Conservation, Ozone Layer Protection, Ship Pollution, Tropical Timber 83, Tropical Timber 94, Wetlands, Whaling.

United Kingdom Maritime Trade Operations (UKMTO)

<u>United Kingdom Marine Trade Operations (UKMTO)</u>[193] is a Royal Navy capability with the principal purpose of providing an information conduit between military/security forces and the wider international maritime trade. UKMTO delivers timely maritime security information, often acting as the primary point of contact for merchant vessels involved in maritime incidents or travelling within an area of high risk (HRA).

The United Kingdom Marine Trade Operations (UKMTO) network was deployed to the Middle East in 2001 as part of the UK response to the 9/11 terrorist attacks in the USA, with the principal purpose of providing a point of contact with industry and information on security issues in the region. Since April 2007 the UKMTO has moved its focus towards Anti-Piracy and Maritime Security Operations and is now the primary point of contact for merchant vessels in case of a pirate attack. The UKMTO supports the industry Best Management Practices (BMP) and is listed in the BMP as the primary point of contact for merchant vessels in case of a pirate attack.

The International Maritime Bureau (IMB) receives information on these pirate attacks directly from the UKMTO (IMB Reports). Merchant vessels are strongly encouraged to send regular reports to UKMTO by the Voluntary Reporting Scheme, providing their position/course/ speed and ETA at their next port while transiting the region bound by Suez, 78°E and 10°S. The day-to-day interface between Masters and Naval/Military forces is provided by UKMTO which talk to merchant ships and liaise directly with the Maritime Security Centre – Horn of Africa (MSCHOA) and Naval Commanders at sea and ashore.

UKMTO also administers Voluntary Reporting Areas (VRAs), as detailed in a variety of Maritime Security Charts. These schemes are to enhance the security of merchant vessels and therefore vessels/masters/CSOs and Companies are encouraged to send regular reports, providing their position/course/speed and ETA at their next port whilst in transit.

In the event of an incident UKMTO is able to inform relevant regional authorities and warn and advise vessels in the near vicinity of the incident (See and Avoid process). The information is provided to the wider shipping industry, therefore providing ship owners and Masters with information that could affect their own company risk assessment in that transit.

UKMTO receives information from many organizations within the region including a variety of maritime operations centers and port authorities, this enables enhanced Maritime Situational Awareness (MSA) that is utilized in supporting the global maritime trade. All information received by UKMTO is strictly controlled in a secure information system and recognizes that the source and content of the information is often sensitive. It is important to note that whilst UKMTO liaises with military maritime operations, its role is not to coordinate a military response. UKMTO strives to understand the global maritime environment therefore

---

[193] https://www.ukmto.org/about-ukmto

directly supporting the maritime industry. It achieves this by knowledge of events that can impact the security, safety, environment and subsequently the safe navigation of the sea.

The Voluntary Reporting Area provides Maritime Security Guidance (MSG) to the mariner operating in the Voluntary Reporting Area (VRA). It receives reports and information on suspicious incidents from merchant shipping and shares that information with its regional, national contacts, as well as Industry and vessels operating in that area.

The Voluntary Reporting Area (VRA) as described on UKHO Chart Q6099. All vessels operating within the VRA are encouraged to register with the UKMTO. Registering establishes direct contact between the reporting vessel and UKMTO Once direct contact has been established the UKMTO is able to:

- Inform / warn the vessel reporting about maritime incidents which may affect that vessel.
- Receive reports of maritime security incidents and suspicious activity from the reporting vessel.
- Share information provided by the reporting vessel to the appropriate authorities within the region.
- Inform the reporting vessel about suspect vessels in the VRA.
- Assist with other information enquiries from the reporting vessel.

Vessels are also encouraged to report positional information daily while operating in the VRA (including when at anchor or berthed alongside) using one of the reporting forms as contained within this website.

UKMTO receives information from a large number of sources. In addition to the invaluable information received from vessels reporting from the VRA. UKMTO also receives information from many other organizations within the region including military maritime operations' centers and port authorities. From the information received and the maritime domain picture produced, UKMTO is able to provide direct warnings of maritime security incidents (Attacks, Hijacks, Robbery, Boarding) to reporting vessels, and wider Industry. UKMTO also provides more general warnings in the form of advisory notices to vessels concerning suspicious incidents .

UKMTO shares relevant information with appropriate authorities within states in the region. All information received is strictly controlled in a secure information system and recognizes that the source and content of the information is often extremely sensitive.

It is important to note that whilst UKMTO liaises with military maritime operations centers, its role is not to coordinate a military response.

National Maritime Information Centre (NMIC)

The National Maritime Information Centre (NMIC)[194] is the National Maritime Information Centre the United Kingdom of Great Britain and Northern Ireland established the a cross-government body with a center located at the Northwood Headquarters in Middlesex.

The NMIC brings together existing functions to provide improved maritime situational awareness and support to lead agencies in the event of emergency or crisis. This will allow a much better understanding of maritime safety and security risks and opportunities; information will be shared across Government, and to Industry, regional & international partners, and the public as appropriate.

The NMIC was established in order for citizens, business and government to ensure the full benefits of a clean, safe, sustainable, secure and resilient maritime domain, working together, at home and globally with partners, to understand and address the risks, to create a hostile environment for criminals and terrorists, and to seize opportunities in the maritime domain to enhance the overall maritime development, safety, security and resilience.

Recognizing the need for cross-government understanding of activity within the maritime domain, the 2010 Strategic Defense and Security Review established the NMIC which brings together information and intelligence provided by Border Force, the MCA, the police, the Armed Forces, the Foreign and Commonwealth Office, the Marine Management Organization, the National Crime Agency (Border Policing Command) and other agencies and, supported by international partnerships, incorporates additional global information to provide an united situational awareness of maritime activity in GBR and international waters.

The NMIC's focus is on the here and now rather than delivering the longer-term strategic analysis undertaken elsewhere in government. It provides real-time information to assess the impact of maritime activity to the GBR and contribute to decision making and tasks are to:

- Monitor and track maritime activity around the GBR and areas of national interest, and collate that data within a trusted environment,
- Analyze and share information for better understanding of maritime security issues,
- Act as a national focal point for regional and international partners on maritime domain awareness,
- Support government and industry decision making in times of need,
- Respond to government department and agency tasking to support their outputs on a case-by-case basis.

---

[194] http://www.nmic.org.uk

NMIC provides a hub for national maritime surveillance information and coordinate the input of maritime departmental data. This will provide a trusted, coherent, robust, accurate and up to date picture of maritime activity affecting the UK and its interests.

NMIC establishes a consolidated situational awareness picture to enable better identification and assessment of potential changes in the risks, as prioritized by the Maritime Security Oversight Group, to GBR maritime safety and security. The consolidated picture will feed into Government hubs to collate, analyses and provide advice on identified threats and vulnerabilities.

The NMIC advices on current and anticipated state of GBR maritime situational awareness enabling informed decisions. Support of the Government in times of crisis is provided with a coherent picture of activity in the maritime domain to the Cabinet Office Briefing Room.

The NMIC harmonizes Maritime Situational Awareness development in support of EU/NATO and other MSA networks like SUCBAS and MARSUR. It provide a single point of contact for EU and international partners developing similar situational awareness capabilities on a wider scale.

The Maritime Security Oversight Group (MSOG) is made up with key representatives of core departments, agencies and the Cabinet Office and is the senior-level decision making group for maritime issues. The purpose of the group is to provide strategic oversight and direction on all cross-cutting maritime security issues and programs. The group is responsible for the Maritime Security vision, strategic objectives and risks, reviewing them as circumstances require, and allocating priorities in order to use them as a framework to drive and coordinate day-to-day policy on cross government programs of work.

The Maritime Matrix Group is chaired by Home Office, will be held regularly to continue to oversee the progress on all Maritime Security work streams, including the development of policy proposals identified to address risks and issues in the maritime security domain. This group will maintain a milestone document and risk register to determine issues for MSOG discussion.

The NMIC Stakeholders Board is a cross-government working group with the responsibility of overseeing future plans for NMIC development, reporting progress to Maritime Security Oversight Group. It will also contribute progress against the NMIC Business Plan, demonstrating the contribution to, and traceability from, the Maritime Security Program.

NMIC has agreed information and data handling procedures with a number of partner organizations. Partners should be aware that information is the NMIC's most prized asset and instructions will always be followed to ensure the maintenance of trust and confidence.

The NMIC is not seeking to be involved in every maritime incident unless certain thresholds are met. When the NMIC is involved it will be in support of a nominated lead department to add value. In most incidents we expect to act as the information coordinating center using our own departmental representatives to ensure there is validity and transparency.

The primary objective of the NMIC is to support a nominated department in providing a satisfactory response. The NMIC is accountable to the MSOG who will assess the NMIC's success in the involvement of any event under a certain Threshold Level activated.

Should a maritime safety or security incident occur that falls under any of the following conditions, the NMIC should be notified:

- Multiple UK Government departments are involved or need to be involved.
- The reputation of the UK might be significantly impacted.
- There is/will be significant Parliamentary interest.

The impact level of the incident will be assessed against the published HMG Business Impact Levels (BIL). The HMG Information Assurance Standards are issued jointly by the Cabinet Office and CESG. The standards outline minimum measures that must be implemented by Departments and Agencies. There are currently six BIL Tables; the titles of each have been included below for information:

- Defense, International Relation, Security and Intelligence
- Public Order, Public Safety and Law Enforcement
- Trade, Economics and Public Finance
- Public Services
- Critical National Infrastructure
- Personal / Citizen

Since its inception in 2011, the NMIC has undertaken daily provision of data to support many maritime security operations and provided support to ensure the safety and security of the Olympic Games and during 2011 operations to protect civilians in Libya.

This strategy recognizes the added value provided by NMIC to maritime stakeholders across government and in industry since its inception. Within revised maritime security governance arrangements (Annex A) maritime stakeholders are considering how NMIC can be utilized to maximum effect to support the GBR's maritime security objectives.

INMARSAT

INMARSAT is a British satellite telecommunications company originating from the International Maritime Satellite Organization (INMARSAT), a non-profit intergovernmental organization established in 1979 at the behest of the International Maritime Organization (IMO) - the United Nations' maritime body - and pursuant to the Convention on the International Mobile Satellite Organization (IMSO), signed by 28 countries in 1976.

INMARSAT is offering today global mobile services including telephone and data services to users worldwide, via portable or mobile terminals which communicate with ground stations through twelve geostationary telecommunications satellites.

INMARSAT's network provides communications services to a range of governments, aid agencies, media outlets and businesses with a need to communicate in remote regions or where there is no reliable terrestrial network. The company is listed on the London Stock Exchange, is a constituent of the FTSE 100 Index and a financial (as well as technical) sponsor of Télécoms Sans Frontières.

The organization was created to establish and operate a satellite communications network for the maritime community. In coordination with the International Civil Aviation Organization in the 1980s, the convention governing INMARSAT was amended to include improvements to aeronautical communications, notably for public safety. The member states owned varying shares of the operational business.

In the mid-1990s, many member states were unwilling to invest in improvements to INMARSAT's network, especially owing to the competitive nature of the satellite communications industry, while many recognized the need to maintain the organization's older systems and the need for an intergovernmental organization to oversee public safety aspects of satellite communication networks.

In 1998, an agreement was reached to modify INMARSAT's mission as an intergovernmental organization and separate and privatize the organization's operational business, with public safety obligations attached to the sale.

In April 1999, INMARSAT was succeeded by the International Mobile Satellite Organization (IMSO) as an intergovernmental regulatory body for satellite communications, while INMARSAT's operational unit was separated and became the UK-based company INMARSAT Ltd. The IMSO and INMARSAT Ltd. signed an agreement imposing public safety obligations on the new company. INMARSAT was the first international satellite organization that was privatized. The INMARSAT head office is at Old Street Roundabout in the London Borough of Islington.

In 2005, Apax Partners and Permira bought shares in the company. The company was also first listed on the London Stock Exchange in that year. In March 2008, it was disclosed that U.S. hedge fund Harbinger Capital owned 28% of the company.

In 2009, INMARSAT completed the acquisition of satellite communications provider Stratos Global Corporation (Stratos) and acquired a 19-percent stake in SkyWave Mobile Communications Inc., a provider of INMARSAT D+/IsatM2M network services which in turn purchased the GlobalWave business from TransCore. Aside from its commercial services, INMARSAT provides global maritime distress and safety services (GMDSS) to ships and aircraft at no charge, as a public service.

Services include traditional voice calls, low-level data tracking systems, and high-speed Internet and other data services as well as distress and safety services. The most recent of these provides GPRS-type services at up to 492 kbit/s via the Broadband Global Area Network (BGAN) IP satellite modem the size of a notebook computer. Other services provide mobile Integrated Services Digital Network (ISDN) services used by the media for live reporting on world events via videophone.

The price of a call via INMARSAT has now dropped to a level where they are comparable to, and in many cases lower than, international roaming costs, or hotel phone calls. Voice call charges are the same for any location in the world where the service is used. Tariffs for calls to INMARSAT country codes vary, depending on the country in which they are placed. INMARSAT primarily uses country code 870.

Newer INMARSAT services use an IP technology that features an always-on capability where the users are only charged for the amount of data they send and receive, rather than the length of time they are connected.

The first (F1) and second (F2) of INMARSAT's most recent series of satellites, known as the "I4" satellites, were launched in June and November 2005. The third and final satellite (F3) was launched from the Baikonur Cosmodrome in Kazakhstan on the 18 August 2008. In addition to its own satellites, INMARSAT has a collaboration agreement with ACeS regarding handheld voice services.

Inmarsat has developed a series of networks providing certain sets of services (most networks support multiple services). They are grouped into two sets, existing and evolved services, and advanced services. Existing and evolved services are offered through land earth stations which are not owned nor operated by Inmarsat, but through companies which have a commercial agreement with Inmarsat. Advanced services are provided via distribution partners but the satellite gateways are owned and operated by Inmarsat directly.

There are three types of coverage related to each Inmarsat I-4 satellite, the Global Beam Coverage, Regional Spot Beam Coverage, and Narrow Spot Beam Coverage:

- Global Beam Coverage
- Each satellite is equipped with a single global beam that covers up to one-third of the Earth's surface, apart from the poles. Overall, global beam coverage extends from latitudes of −82 to +82 degrees regardless of longitude.
- Regional Spot Beam Coverage

- Each regional beam covers a fraction of the area covered by a global beam, but collectively all of the regional beams offer virtually the same coverage as the global beams. Use of regional the beams allow user terminals (also called mobile earth stations) to operate with significantly smaller antennas. Regional beams were introduced with the I-3 satellites. Each I-3 satellite provides four to six spot beams; each I-4 satellite provides 19 regional beams.
- Narrow Spot Beam Coverage
- Narrow beams are offered by the three Inmarsat-4 satellites. Narrow beams vary in size, tend to be several hundred kilometers across. The narrow beams, while much smaller than the global or regional beams, are far more numerous and hence offer the same global coverage. Narrow spot beams allow yet smaller antennas and much higher data rates. They form the backbone of Inmarsat's handheld (GSPS) and broadband services (BGAN). This coverage was introduced with the I-4 satellites. Each I-4 satellite provides around 200 narrow spot beams.

Advanced services are available via the Broadband Global Area Network (BGAN), a set of IP-based shared-carrier services:

- Broadband Global Area Network (BGAN) is for land line use. BGAN benefits from the new I-4 satellites to offer a shared-channel IP packet-switched service of up to 492 kbit/s (uplink and downlink speeds may differ and depend on terminal model) and a streaming-IP service from 32 up to X-Stream data rate (services depend on terminal model). X-Stream delivers the fastest, on demand streaming data rates from a minimum of 384 kbit/s up to around 450 kbit/s (service depend on location of user and terminal model). Most terminals also offer circuit-switched Mobile ISDN services at 64 kbit/s and even low speed (4.8 kbit/s) voice etc. services. BGAN service is available globally on all I4 satellites.
- FleetBroadband (FB) are maritime services based on BGAN technology, offering similar services and using the same infrastructure as BGAN. A range of Fleet Broadband user terminals are available, designed for fitting on ships.
- SwiftBroadband (SB) are aeronautical services based on BGAN technology and offers similar services. SB terminals are specifically designed for use aboard commercial, private, and military aircraft.

The "*BGAN M2M Family*" is a set of IP-based services designed for long-term machine-to-machine management of fixed assets:

- BGAN M2M was launched at the beginning of January 2012, will deliver a global, IP-based low-data rate service, for users needing high levels of data availability and performance in permanently unmanned environments. Ideally suited for high-frequency, very low-latency data reporting, BGAN M2M will prove extremely attractive for monitoring fixed assets such as pipelines and oil well heads, or backhauling electricity consumption data within a utility.

- IsatM2M is a global, short burst data, store and forward service that will deliver messages of 10.5 or 25.5 bytes in the send direction, to 100 bytes in the receive direction. The service is delivered to market via two partners - SkyWave Mobile Communications and Honeywell Global Tracking. Each has their own solutions to integrate the service into customers' infrastructure.
- IsatData Pro is a global satellite data service designed for two-way text and data communications with remote assets and has the capability to exchange large amounts of data quickly (To mobile: 10kBytes / From mobile: 6.4kBytes with typical delivery time at 15 sec.) This service is used in mission-critical applications and is used in everything from managing trucks, fishing vessels and oil & gas and heavy equipment, to text message remote workers and security applications. It is provided by SkyWave Mobile Communications Inc.

The INMARSAT company offers portable and fixed phone services:

- IsatPhone Pro is Inmarsat's own-designed and manufactured robust mobile satellite phone, offering clear voice telephony. It also comes with a variety of data capabilities, including SMS, short message emailing and GPS look-up-and-send, as well as supporting a data service of up to 20kbit/s.
- IsatPhone Link is a low-cost, fixed, global satellite phone service. It provides essential voice connectivity for those working or living in areas without cellular coverage and also comes with a variety of data capabilities.
- FleetPhone service is a fixed phone service ideal for use on smaller vessels where voice communications is the primary requirement or on vessels where additional voice lines are needed. It provides a low-cost, global satellite phone service option for those working or sailing outside cellular coverage.

Basis for the existing and the evolved technologies and services are:

- Aeronautical (Classic Aero) provides voice/fax/data services for aircraft. Three levels of terminals, Aero-L (Low Gain Antenna) primarily for packet data including ACARS and ADS, Aero-H (High Gain Antenna) for medium quality voice and fax/data at up to 9600 bit/s, and Aero-I (Intermediate Gain Antenna) for low quality voice and fax/data at up to 2400 bit/s. Note, there are also aircraft rated versions of Inmarsat-C and mini-M/M4. The aircraft version of GAN is called Swift 64 (see below).
- Inmarsat-B provides voice services, telex services, medium speed fax/data services at 9.6 kbit/s and high speed data services at 56, 64 or 128 kbit/s. There is also a

- 'leased' mode for Inmarsat-B available on the spare Inmarsat satellites. It will be closed in December 2016.
- Inmarsat-C effectively this is a "satellite telex" terminal with store-and-forward, polling etc. capabilities. Certain models of Inmarsat-C terminals are also approved for usage in the GMDSS system, equipped with GPS.
- Inmarsat-M provides voice services at 4.8 kbit/s and medium speed fax/data services at 2.4 kbit/s. It paved the way towards Inmarsat-Mini-M.
- Mini-M: provides voice services at 4.8 kbit/s and medium speed fax/data services at 2.4 kbit/s. One 2.4kbit/s channel takes up 4.8kbit/s on the satellite.
- GAN (Global Area Network) provides a selection of low speed services like voice at 4.8 kbit/s, fax & data at 2.4 kbit/s, ISDN like services at 64 kbit/s (called Mobile ISDN) and shared-channel IP packet-switched data services at 64 kbit/s (called Mobile Packet Data Service or MPDS, formerly Inmarsat Packet Data Service – IPDS). GAN is also known as "M4".
- Fleet is actually a family of networks that includes the Inmarsat-Fleet77, Inmarsat-Fleet55 and Inmarsat-Fleet33 members (The numbers 77, 55 and 33 come from the diameter of the antenna in centimeters). Much like GAN, it provides a selection of low speed services like voice at 4.8 kbit/s, fax/data at 2.4 kbit/s, medium speed services like fax/data at 9.6 kbit/s, ISDN like services at 64 kbit/s (called Mobile ISDN) and shared-channel IP packet-switched data services at 64 kbit/s (called Mobile Packet Data Service or MPDS - see below). However, not all these services are available with all members of the family. The latest service to be supported is Mobile ISDN at 128 kbit/s on Inmarsat-Fleet77 terminals.
- Swift 64 is similar to GAN, providing voice, low rate fax/data, 64kbit/s ISDN, and MPDS services, for private, business, and commercial aircraft. Swift 64 is often sold in a multi-channel version, to support several times 64kbit/s.
- Inmarsat D/D+/IsatM2M: Inmarsat's version of a pager, although much larger than terrestrial versions. Some units are equipped with GPS. The original Inmarsat-D terminals were one-way (to mobile) pagers. The newer Inmarsat-D+ terminals are the equivalent of a two-way pager. The main use of this technology nowadays is in tracking trucks and buoys and SCADA applications. SkyWave Mobile Communications is a provider of D/D+/IsatM2M satellite data services with its DMR and SureLinx series products. SkyWave also provides satellite tracking, monitoring and control capabilities through its GlobalWave MT series products.[39] Competing systems such as from SkyBitz only operate on the MSAT geostationary satellite over North America.
- MPDS (Mobile Packet Data Service) was previously known as IPDS, this is an IP-based data service in which several users share a 64kbit/s carrier in a manner similar

to ADSL. MPDS-specific terminals are not sold; rather, this is a service which comes with most terminals that are designed for GAN, Fleet, and Swift64.
- IsatPhone provides voice services at 4.8 kbit/s and medium speed fax/data services at 2.4 kbit/s. This service emerged from a collaboration agreement with ACeS, and is available in the EMEA and APAC satellite regions. Coverage is available in Africa, the Middle-East, Asia, and Europe, as well as in maritime areas of the EMEA and APAC coverage.

In August 2010, Inmarsat awarded Boeing a contract to build a constellation of three Inmarsat-5 satellites, as part of a US$1.2 billion worldwide wireless broadband network called Inmarsat Global Xpress. The three Inmarsat-5 (I-5) satellites will be based on Boeing's 702HP spacecraft platform. The first (Inmarsat 5-F1) was launched in December 2013, entering commercial service on 1. July 2014, the second was launched in February 2015 and the third in August 2015.

The satellites will operate at Ka-band in the range of 20–30 GHz. Each Inmarsat-5 will carry a payload of 89 small Ka-band beams which combined will offer global Ka-band spot coverage. In addition each satellite will carry six fully steerable beams that can be pointed at commercial or government traffic hotspots. According to Inmarsat, Global Xpress will deliver download speeds in excess of 60 Mbit/s to a 60 cm dish.

Inmarsat has announced plans to offer high-speed in-flight broadband internet on airliners using a system that integrates the use of Global Xpress and S-band service by using two antennas, the GX antenna on top of the plane for satellite backhaul over Ka-band frequencies and an S band antenna underneath for backhaul from ground stations.

In February 2011, Inmarsat announced that iDirect had been awarded the contract to provide both the ground segment and the core module that provides the key electronics in the new Global Xpress (GX) maritime terminals. During the course in 2016, Inmarsat introduced a series of market-specific, high-speed connectivity services powered by Global Xpress.

On 30. June 2008, the European Parliament and the Council adopted the European's Decision to establish a single selection and authorization process, the European S-band Application Process (ESAP), to ensure a coordinated introduction of mobile satellite services (MSS) in Europe. The selection process was launched in August 2008 and attracted four applications by prospective operators (ICO, Inmarsat, Solaris Mobile (now EchoStar Mobile), TerreStar).

In May 2009, the European Commission selected two operators, Inmarsat Ventures and Solaris Mobile, giving these operators "*the right to use the specific radio frequencies identified in the Commission's decision and the right to operate their respective mobile satellite systems*". EU Member States now have to ensure that the two operators have the right to use the specific radio frequencies identified in the Commission's decision and the right to operate their respective mobile satellite systems for 18 years from the selection decision.

The operators were to compelled operations within 24 months (May 2011) from the selection decision.

Inmarsat's S-band satellite program, called EuropaSat, will deliver mobile multimedia broadcast, mobile two-way broadband telecommunications and next-generation MSS services across all 27 member states of the European Union and as far east as Moscow and Ankara, by means of a hybrid satellite/terrestrial network. It is being built by Thales Alenia Space and was slated for launch in 2016.

Launched on 25. July 2013, Alphasat I-XL was carried into orbit by an Ariane 5 ECA rocket from the Guiana Space Centre, Europe's spaceport in Kourou, French Guiana. The satellite was built by Astrium using an Alphabus platform and weighed more than six tons at launch. The new-generation Alphasat I-XL was positioned at 25 degrees East to offer advanced mobile voice and data communications services across Europe, Africa and the Middle East using L-Band. It features a new generation digital signal processor for the payload, an 11-meter aperture AstroMesh antenna reflector, supplied by Astro Aerospace in Carpenteria, CA. Its design life is 15 years.

In addition, Alphasat will host four ESA-provided technology demonstration payloads: an advanced star tracker using active pixel technology, an optical laser terminal for geostationary to low-Earth orbit communication at high data rates, a dedicated payload for the characterization of transmission performance in the Q-V band in preparation for possible commercial exploitation of these frequencies and a radiation sensor to better characterize the environment at geostationary orbit.

United Nations (UN)

The <u>United Nations (UN)</u>[195] is an international organization founded in 1945. It is currently made up of 193 Member States. The mission and work of the United Nations are guided by the purposes and principles contained in its founding Charter. The name "*United Nations*", coined by United States President Franklin D. Roosevelt was first used in the Declaration by United Nations of 1. January 1942, during Second World War.

Each of the 193 Member States of the United Nations is a member of the General Assembly. States are admitted to membership in the UN by a decision of the General Assembly upon the recommendation of the Security Council.

The main organs of the UN are the General Assembly, the Security Council, the Economic and Social Council, the Trusteeship Council, the International Court of Justice, and the UN Secretariat. All were established in 1945 when the UN was founded.

The Secretary-General of the United Nations is a symbol of the Organization's ideals and a spokesman for the interests of the world's peoples, in particular the poor and vulnerable. The current Secretary-General of the UN, and the eighth occupant of the post, is Mr. Ban Ki-moon of the Republic of Korea, who took office on 1 January 2007. The UN Charter describes the Secretary-General as "*chief administrative officer*" of the Organization.

The Secretariat, one of the main organs of the UN, is organized along departmental lines, with each department or office having a distinct area of action and responsibility. Offices and departments coordinate with each other to ensure cohesion as they carry out the day to day work of the Organization in offices and duty stations around the world. At the head of the United Nations Secretariat is the Secretary-General.

The UN system, also known unofficially as the "*UN family*" and better termed the "UN Network", is made up of the UN itself and many affiliated programs, funds, and specialized agencies, all with their own membership, leadership, and budget. The programs and funds are financed through voluntary rather than assessed contributions. The Specialized Agencies are independent international organizations funded by both voluntary and assessed contributions.

The United Nations as Maritime Entity are per se the highest Maritime Authority with its specialized maritime agency, the International Maritime Organization (IMO), being responsible for the safety and security of shipping and the prevention of marine pollution by ships and various worldwide organizations, initiatives, and programs.

---

[195] http://www.un.org/en

International Maritime Organization (IMO)

The International Maritime Organization (IMO)[196] is the United Nations (UN) specialized agency with responsibility for the safety and security of shipping and the prevention of marine pollution by ships.

IMO is the global standard-setting authority for the safety, security and environmental performance of international shipping. Its main role is to create a regulatory framework for the shipping industry that is fair and effective, universally adopted and universally implemented.

In other words, its role is to create a level playing-field so that ship operators cannot address their issues by simply cutting corners and compromising on safety, security and environmental performance. This approach also encourages innovation and efficiency.

Shipping is a truly international industry, and it can only operate effectively if the regulations and standards are themselves agreed, adopted and implemented on an international basis. IMO is the forum at which this process takes place.

International shipping transports about 90 per cent of global trade to peoples and communities all over the world. Shipping is the most efficient and cost-effective method of international transportation for most goods; it provides a dependable, low-cost means of transporting goods globally, facilitating commerce and helping to create prosperity among nations and peoples.

The world relies on a safe, secure and efficient international shipping industry – and this is provided by the regulatory framework developed and maintained by IM;, it measures cover all aspects of international shipping – including ship design, construction, equipment, manning, operation and disposal – to ensure that this vital sector for remains safe, environmentally sound, energy efficient and secure.

Shipping is an essential component of any program for future sustainable economic growth. Through IMO, the Organization's Member States, civil society and the shipping industry are already working together to ensure a continued and strengthened contribution towards a green economy and growth in a sustainable manner. The promotion of sustainable shipping and sustainable maritime development is one of the major priorities of IMO in the coming years.

Energy efficiency, new technology and innovation, maritime education and training, maritime security, maritime traffic management and the development of the maritime infrastructure: the development and implementation, through IMO, of global standards covering these and other issues will underpin IMO's commitment to provide the institutional

---

[196] http://www.imo.org

framework necessary for a green - or rather blue - and sustainable global maritime transportation system.

It has always been recognized that the best way of improving safety at sea is by developing international regulations that are followed by all shipping nations and from the mid-19th century onwards a number of such treaties were adopted. Several countries proposed that a permanent international body should be established to promote maritime safety more effectively, but it was not until the establishment of the United Nations itself that these hopes were realized. In 1948 an international conference in Geneva adopted a convention formally establishing IMO (the original name was the Inter-Governmental Maritime Consultative Organization, or IMCO, but the name was changed in 1982 to IMO). The IMO Convention entered into force in 1958 and the new Organization met for the first time the following year.

The purposes of the Organization, as summarized by Article 1(a) of the Convention, are *"to provide machinery for cooperation among Governments in the field of governmental regulation and practices relating to technical matters of all kinds affecting shipping engaged in international trade; to encourage and facilitate the general adoption of the highest practicable standards in matters concerning maritime safety, efficiency of navigation and prevention and control of marine pollution from ships"*. The Organization is also empowered to deal with administrative and legal matters related to these purposes.

IMO's first task was to adopt a new version of the International Convention for the Safety of Life at Sea (SOLAS), the most important of all treaties dealing with maritime safety. This was achieved in 1960 and IMO then turned its attention to such matters as the facilitation of international maritime traffic, load lines and the carriage of dangerous goods, while the system of measuring the tonnage of ships was revised.

But although safety was and remains IMO's most important responsibility, a new problem began to emerge - pollution. The growth in the amount of oil being transported by sea and in the size of oil tankers was of particular concern and the Torrey Canyon disaster of 1967, in which 120,000 tons of oil was spilled, demonstrated the scale of the problem.

During the next few years IMO introduced a series of measures designed to prevent tanker accidents and to minimize their consequences. It also tackled the environmental threat caused by routine operations such as the cleaning of oil cargo tanks and the disposal of engine room wastes - in tonnage terms a bigger menace than accidental pollution.

The most important of all these measures was the International Convention for the Prevention of Pollution from Ships, 1973, as modified by the Protocol of 1978 relating thereto (MARPOL 73/78). It covers not only accidental and operational oil pollution but also pollution by chemicals, goods in packaged form, sewage, garbage and air pollution.

IMO was also given the task of establishing a system for providing compensation to those who had suffered financially as a result of pollution. Two treaties were adopted, in 1969 and 1971, which enabled victims of oil pollution to obtain compensation much more simply and quickly than had been possible before. Both treaties were amended in 1992, and again in 2000, to increase the limits of compensation payable to victims of pollution. A number of other legal conventions have been developed since, most of which concern liability and compensation issues.

Also in the 1970s a global search and rescue system was initiated, with the establishment of the International Mobile Satellite Organization (IMSO), which has greatly improved the provision of radio and other messages to ships.

The Global Maritime Distress and Safety System (GMDSS) was adopted in 1988 and began to be phased in from 1992. In February 1999, the GMDSS became fully operational, so that now a ship that is in distress anywhere in the world can be virtually guaranteed assistance, even if the ship's crew does not have time to radio for help, as the message will be transmitted automatically.

Two initiatives in the 1990s are especially important insofar as they relate to the human element in shipping. On 1. July 1998 the International Safety Management Code (IMSC) entered into force and became applicable to passenger ships, oil and chemical tankers, bulk carriers, gas carriers and cargo high speed craft of 500 gross tonnages and above. It became applicable to other cargo ships and mobile offshore drilling units of 500 gross tonnage and above from 1. July 2002.

On 1. February 1997, the 1995 amendments to the International Convention on Standards of Training, Certification and Watch-keeping for Seafarers, 1978 entered into force. They greatly improve seafarer standards and, for the first time, give IMO itself powers to check Government actions with Parties required to submit information to IMO regarding their compliance with the Convention. A major revision of the STCW Convention and Code was completed in 2010 with the adoption of the "*Manila amendments to the STCW Convention and Code*".

New conventions relating to the marine environment were adopted in the 2000s, including one on anti-fouling systems (AFS 2001), another on ballast water management to prevent the invasion of alien species (BWM 2004) and another on ship recycling (Hong Kong International Convention for the Safe and Environmentally Sound Recycling of Ships, 2009).

The 2000s also saw a focus on maritime security, with the entry into force in July 2004 of a new, comprehensive security regime for international shipping, including the International Ship and Port Facility Security (ISPS) Code, made mandatory under amendments to SOLAS adopted in 2002.

In 2005, IMO adopted amendments to the Convention for the Suppression of Unlawful Acts (SUA) Against the Safety of Maritime Navigation, 1988 and its related Protocol (the 2005 SUA Protocols), which amongst other things, introduce the right of a a State Party desires to board a ship flying the flag of another State Party when the requesting Party has reasonable grounds to suspect that the ship or a person on board the ship is, has been, or is about to be involved in, the commission of an offence under the Convention.

As IMO instruments have entered into force and been implemented, developments in technology and/or lessons learned from accidents have led to changes and amendments being adopted. The focus on implementation continues, with the technical co-operation program a key strand of IMO's work.

The IMO Member State Audit Scheme, which becomes mandatory under a number of key IMO instruments on 1. January 2016, will play a key role in supporting effective implementation by providing an audited Member State with a comprehensive and objective assessment of how effectively it administers and implements those mandatory IMO instruments which are covered by the Scheme.

IMO's mission statement:

"The mission of the International Maritime Organization (IMO) as a United Nations specialized agency is to promote safe, secure, environmentally sound, efficient and sustainable shipping through cooperation. This will be accomplished by adopting the highest practicable standards of maritime safety and security, efficiency of navigation and prevention and control of pollution from ships, as well as through consideration of the related legal matters and effective implementation of IMO's instruments with a view to their universal and uniform application."

IMO currently has today 171 Member States and three Associate Members. Non-governmental international organizations that have the capability to make a substantial contribution to the work of IMO may be granted consultative status by the Council with the approval of the Assembly.

Any organization seeking consultative status with IMO has to demonstrate considerable expertise as well as the capacity to contribute, within its field of competence, to the work of IMO. It must also show that it has no means of access to the work of IMO through other organizations already in consultative status and that it is truly international in its membership, namely that it has a range of members covering a broad geographical scope and, usually, more than one region.

The IMO Council considers applications for consultative status by non-governmental international organizations once a year, at its first session, which is usually held in June. Applications, including the letter, the questionnaire and any additional relevant documentation must have reached the IMO Secretariat by 31 March in order to be submitted to the Council session held in June that year; applications received after that date will be considered the following year. To date there are 77 international non-governmental organizations in consultative status with IMO.

IMO may enter into agreements of cooperation with other intergovernmental organizations on matters of common interest with a view to ensuring maximum coordination in respect of such matters. To date there are 65 intergovernmental organizations which have signed agreements of cooperation with IMO.

Automated Information System (AIS)

The International Maritime Organization's International Convention for the Safety of Life at Sea requires Automatic Identification System (AIS)[197] to be fitted aboard international voyaging ships with gross tonnage (GT) of 300 or more, and all passenger ships regardless of size.

The 2002 IMO SOLAS Agreement included a mandate that required most vessels over 300GT on international voyages to fit a Class A type AIS transceiver. This was the first mandate for the use of AIS equipment and affected approximately 100,000 vessels.

AIS as the name indicates is an automatic tracking system used on ships and by vessel traffic services (VTS) or Vessel Traffic (Management) System (VTM/VTMS) for identifying and locating vessels by electronically exchanging data with other nearby ships, AIS base stations, and satellites. When satellites are used to detect AIS signatures then the term Satellite-based AIS (S-AIS) is used. AIS information supplements marine radar, which continues to be the primary method of collision avoidance for water transport.

Information provided by AIS equipment, such as unique identification, position, course, and speed, can be displayed on a screen or an ECDIS. AIS is intended to assist a vessel's watch standing officers and allow maritime authorities to track and monitor vessel movements. AI integrates standardized VHF transceivers with a positioning system such as a GPS or LORAN-C receiver, with other electronic navigation sensors, such as a gyrocompass or rate of turn indicator. Vessels fitted with AIS transceivers can be tracked by AIS base stations located along coast lines or, when out of range of terrestrial networks, through a growing number of satellites that are fitted with special AIS receivers which are capable of de-conflicting a large number of signatures.

In 2006, the AIS standards committee published the Class B type AIS transceiver specification, designed to enable a simpler and lower cost AIS device. Low cost Class B transceivers became available in the same year triggering mandate adoptions by numerous countries and making large scale installation of AIS devices on vessels of all sizes commercially available.

Since 2006, the AIS technical standard committees have continued to evolve the AIS standard and product types to cover a wide range of applications from the largest vessel to small fishing vessels and life boats. In parallel, governments and authorities have instigated projects to fit varying classes of vessels with an AIS device to improve safety and security.

Most mandates are focused on commercial vessels, with leisure vessels selectively choosing to fit. In 2010 most commercial vessels operating on the European Inland Waterways were required to fit an Inland waterway certified Class A, all EU fishing boats over 16m will

---

[197] https://en.wikipedia.org/wiki/Automatic_Identification_System

have to have a Class A by May 2014, and the US has a long pending extension to their existing AIS fit rules which is expected to come into force during 2013.

It was estimated that as of 2012, approximately 250,000 vessels had fitted an AIS transceiver of some type, with a further 1 million required to do so in the near future; even larger projects under consideration.

Correlating optical and radar imagery with AIS and Satellite-based AIS (S-AIS) signatures enables the end-user to rapidly identify all types of vessel. A great strength of AIS and S-AIS is the ease with which it can be correlated with additional information from other sources such as radar, optical, ESM, and more SAR related tools such as GMDSS SARSAT and AMVER. Satellite-based radar and other sources can contribute to maritime surveillance by detecting all vessels in specific maritime areas of interest, a particularly useful attribute when trying to co-ordinate a long-range rescue effort or when dealing with VTS issues.

AIS transceivers automatically broadcast information, such as their position, speed, and navigational status, at regular intervals via a VHF transmitter built into the transceiver. The information originates from the ship's navigational sensors, typically its global navigation satellite system (GNSS) receiver and gyrocompass. Other information, such as the vessel name and VHF call sign, is programmed when installing the equipment and is also transmitted regularly. The signals are received by AIS transceivers fitted on other ships or on land based systems, such as VTS systems.

The received information can be displayed on a screen or chart plotter, showing the other vessels' positions in much the same manner as a radar display. Data is transmitted via a tracking system which makes use of a Self-Organized Time Division Multiple Access (SOTDMA) datalink designed by Swedish inventor Håkan Lans.

As technical specifications AIS uses the globally allocated Marine Band Channels on the High Side of the duplex from two VHF radio "channels" (87B) and (88B) while Frequency Modulation is not used:

- Channel A 161.975 MHz (87B)
- Channel B 162.025 MHz (88B)

The simplex channels 87A and 88A use a lower frequency so they are not even affected by this allocation and can still be used as designated for the maritime mobile frequency plan.

AIS uses Gaussian Minimum-Shift Keying (GMSK) modulation. Ordinary VHF receivers can receive AIS with the filtering disabled (the filtering destroys the GMSK data).

AIS uses the National Marine Electronics Association 0183 (NMEA 0183)[198] standard, which

---

[198] https://en.wikipedia.org/wiki/NMEA_0183

is a combined electrical and data specification for communication between marine electronics such as echo sounder, sonars, anemometer, gyrocompass, autopilot, GPS receivers and many other types of instruments. It has been defined by, and is controlled by, the US-American National Marine Electronics Association. It replaces the earlier NMEA 0180 and NMEA 0182 standards. In marine applications, it is slowly being phased out in favor of the newer NMEA 2000 standard.

The electrical standard that is used is EIA-422, although most hardware with NMEA-0183 outputs is also able to drive a single EIA-232 port. Although the standard calls for isolated inputs and outputs, there are various series of hardware that do not adhere to this requirement.

The NMEA 0183 standard uses a simple ASCII, serial communications protocol that defines how data are transmitted in a "sentence" from one "talker" to multiple "listeners" at a time. Through the use of intermediate expanders, a talker can have a unidirectional conversation with a nearly unlimited number of listeners, and using multiplexers, multiple sensors can talk to a single computer port.

At the application layer, the standard also defines the contents of each sentence (message) type, so that all listeners can parse messages accurately. The NMEA standard uses two primary sentences to for AIS data

- !AIVDM (Received Data from other vessels)
- !AIVDO (Your own vessels information)

National Marine Electronics Association 2000 (NMEA 2000)[199], abbreviated to NMEA2k or N2K and standardized as IEC 61162-3, is a plug-and-play communications standard used for connecting marine sensors and display units within ships and boats. Communication runs at 250 kilobits-per-second and allows any sensor to talk to any display unit or other device compatible with NMEA 2000 protocols. Electrically NMEA 2000 is compatible with the Controller Area Network (CAN Bus) used on road vehicles and fuel engines. The higher-level protocol format is based on SAE J1939, with specific messages for the marine environment.

Raymarine SeaTalk 2, Raymarine SeaTalkNG, Simrad Simnet, Furuno CAN are rebranded implementations of NMEA 2000, though may use physical connectors different from the standardized DeviceNet Micro-C M12 5-pin screw connector, all of which are electrically compatible and can be directly connected.

The protocol is used to create a network of electronic devices - chiefly marine instruments - on a boat. Various instruments that meet the NMEA 2000 standard are connected to one central cable, known as a backbone. The backbone powers each instrument and relays data among all of the instruments on the network. This allows one display unit to show

---

[199] https://en.wikipedia.org/wiki/NMEA_2000

many different types of information. It also allows the instruments to work together, since they share data. NMEA 2000 is meant to be plug-and-play to allow devices made by different manufacturers to communicate with each other.

Examples of marine electronic devices to include in a network are GPS receivers, auto pilots, wind instruments, depth sounders, navigation instruments, engine instruments, and nautical chart plotters. The interconnectivity among instruments in the network allows, for example, the GPS receiver to correct the course that the autopilot is steering.

AIS is intended, primarily, to allow ships to view marine traffic in their area and to be seen by that traffic. This requires a dedicated VHF AIS transceiver that allows local traffic to be viewed on an AIS enabled chart plotter or computer monitor while transmitting information about the ship itself to other AIS receivers. Port authorities or other shore-based facilities may be equipped with receivers only, so that they can view the local traffic without the need to transmit their own location. All AIS transceivers equipped traffic can be viewed this way very reliably but is limited to the VHF range, about 10–20 nautical miles.

If a suitable chart plotter is not available, local area AIS transceiver signals may be viewed via a computer using one of several computer applications such as a ShipPlotter, which displays the complete information about ships that are within VHF range of the own ships position using the Universal Automatic Identification System (AIS), and GNUAIS[200], a LINUX based Automatic Identification System. These demodulate the signal from a modified marine VHF radiotelephone tuned to the AIS frequencies and convert into a digital format that the computer can read and display on a monitor; this data may then be shared via a local or wide area network via TCP or UDP protocols but will still be limited to the collective range of the radio receivers used in the network.

Because computer AIS monitoring applications and normal VHF radio transceivers do not possess AIS transceivers, they may be used by shore-based facilities that have no need to transmit or as an inexpensive alternative to a dedicated AIS device for smaller vessels to view local traffic but, of course, the user will remain unseen by other traffic on the network.

A secondary, unplanned and emerging use for AIS data is to make it viewable publicly, on the internet, without the need for an AIS receiver. Global AIS transceiver data collected from both satellite and internet-connected shore-based stations are aggregated and made available on the internet through a number of service providers. Data aggregated this way can be viewed on any internet-capable device to provide near global, real-time position data from anywhere in the world. Typical data includes vessel name, details, location, and speed and heading on a map, is searchable, has potentially unlimited, global range and the history is archived.

Most of this data is free of charge but satellite data and special services such as searching the archives are usually supplied at a cost. The data is a read-only view and the users will

---

[200] http://gnuais.sourceforge.net

not be seen on the AIS network itself. Shore-based AIS receivers contributing to the internet are mostly run by a large number of volunteers. AIS mobile apps are also readily available for use with Android, Windows and iOS devices.

While AIS is a backbone of today's civilian and military ship monitoring, even the ship owners and cargo dispatchers use these services to find and track vessels and cargo GPS to track their transported cargoes. However, private ship owners and marine enthusiasts use the system to add time, date and other data to their photograph collections. The original purpose of AIS was solely collision avoidance but many other applications have since developed and continue to be developed. AIS is currently used for:

Collision Avoidance

AIS was developed by the IMO technical committees as a technology to avoid collisions among large vessels at sea that are not within range of shore-based systems. The technology identifies every vessel individually, along with its specific position and movements, enabling a virtual picture to be created in real time. The AIS standards include a variety of automatic calculations based on these position reports such as Closest Point of Approach (CPA) and collision alarms.

As AIS is not used by all vessels, AIS is usually used in conjunction with radar. When a ship is navigating at sea, information about the movement and identity of other ships in the vicinity is critical for navigators to make decisions to avoid collision with other ships and dangers (shoal or rocks). Visual observation (e.g., unaided, binoculars, and night vision), audio exchanges (i.e., whistle, horns, and VHF radio), and radar or Automatic Radar Plotting Aid are historically used for this purpose. These preventative mechanisms, however, sometimes fail due to time delays, radar limitations, miscalculations, and display malfunctions and can result in a collision. While requirements of AIS are to display only very basic text information, the data obtained can be integrated with a graphical electronic chart or a radar display, providing consolidated navigational information on a single display.

Fishing fleet monitoring and control

AIS is widely used by national authorities to track and monitor the activities of their national fishing fleets. AIS enables authorities to reliably and cost effectively monitor fishing vessel activities along their coast line, typically out to a range of 100 km (60 mi), depending on location and quality of coast based receivers/base stations with supplementary data from satellite based networks.

Maritime Security/Maritime Situational Awareness

AIS enables authorities to identify specific vessels and their activity within or near a nation's Exclusive Economic Zone. When AIS data is fused with existing radar systems, authorities are able to differentiate between vessels more easily. AIS data can be automatically processed to create normalized activity patterns for individual vessels, which when breached, create an alert, thus highlighting potential threats for more efficient use of security assets. AIS improves maritime domain awareness and allows for heightened security and control.

Additionally AIS enriches the traditional national and organizational RMPs worldwide, and since AIS can be applied to freshwater river systems and lakes it enlarges the operational coverage with AIS-Streaming.

Aids to Navigation

The AIS aids to navigation (AtoN) product standard was developed with the ability to broadcast the positions and names of objects other than vessels, such as navigational aid and marker positions and dynamic data reflecting the marker's environment (i.e., currents and climatic conditions). These aids can be located on shore, such as in a lighthouse, or on water, platforms, or buoys. The U.S. Coast Guard has suggested that AIS might replace racon (radar beacons) currently used for electronic navigation aids.

AtoNs enable authorities to remotely monitor the status of a buoy, such as the status of the lantern, as well as transmit live data from sensors (such as weather and sea state) located on the buoy back to vessels fitted with AIS transceivers or local authorities. An AtoN will broadcast its position and Identity along with all the other information. The AtoN standard also permits the transmit of 'Virtual AtoN' positions whereby a single device may transmit messages with a 'false' position such that an AtoN marker appears on electronic charts, although a physical AtoN may not be present at that location.

Search And Rescue (SAR)

For coordinating on-scene resources of a marine search and rescue (SAR) operation, it is imperative to have data on the position and navigation status of other ships in the vicinity. In such cases, AIS can provide additional information and enhance awareness of available resources, even if the AIS range is limited to VHF radio range. The AIS standard also envisioned the possible use on SAR aircraft, and included a message (AIS Message 9) for aircraft to report their position.

To aid SAR vessels and aircraft in locating people in distress, the specification (IEC 61097-14 Ed 1.0) for an AIS-based SAR transmitter (AIS-SART) was developed by the IEC's TC80 AIS work group. AIS-SART was added to Global Maritime Distress Safety System regulations effective January 1, 2010. AIS-SARTs have been available on the market since at least 2009. Recent regulations have mandated the installation of AIS systems on all Safety Of Life At Sea (SOLAS) vessels and vessels over 300 tons.

Accident Investigation

AIS information received by VTS is important for accident investigation since it provides accurate historical data on time, identity, GPS-based position, compass heading, course over ground, speed (by log/SOG), and rates of turn, rather than the less accurate information provided by radar. A more complete picture of the events could be obtained by Voyage Data Recorder (VDR) data if available and maintained on board for details of the movement of the ship, voice communication and radar pictures during the accidents. However, VDR data are not maintained due to the limited twelve hours storage by IMO requirement.

Ocean Currents Estimates

Ocean surface current estimates based on the analysis of AIS data have been available from French company, e-Odyn, since December 2015.

Infrastructure Protection

AIS information can be used by owners of marine seabed infrastructure, such as cables or pipelines, to monitor the activities of vessels close to their assets in close to real time. This information can then be used to trigger alerts to inform the owner and potentially avoid an incident where damage to the asset might occur.

Fleet and Cargo Tracking

Internet disseminated AIS can be used by fleet or ship managers to keep track of the global location of their ships. Cargo dispatchers, or the owners of goods in transit can track the progress of cargo and anticipate arrival times in port.

Satellite-based AIS (S-AIS)

In the 1990s AIS was developed as a high intensity, short-range identification and tracking network and, at the time, it was not anticipated to be detectable from space. Some smart people like however imagined the limited VHF range of about 10–20 nautical miles not only the surface of the earth but also in space available.

The Satellite-based Automatic Identification System (AIS)[201], also referred to as the Space-based AIS, was created and developed as a global ship surveillance system that uses small low orbit satellites carrying the AIS transponders to receive the ship's AIS information from space and then relay them to the ground station. Consequently similar to other satellite communication and navigation systems, the satellite-based AIS system consists of five components, i.e. small low orbit satellites in space, shipborne AIS equipment, ground station, user and communication link.

While the ship's information is automatically exchanged between AIS-equipped ships via VHF communication link, the satellite on which the AIS transponder is installed running on the low earth orbit at the same time is able to receive the VHF signal transmitted for the ship's AIS equipment since the VHF radio wave with the significant signal strength has been proven to be able reach the altitude up to 1000km from the earth ground. The satellite transfers the received VHF signal to the ground station in charge of controlling the whole system.

The ground station therefore can distribute the ship's information transferred by the AIS satellite to the authorized user. The communication links between the satellite and the ground as well as between the ground and the user are bi-directional whereas the communication link from the ship to the satellite is uni-directional. Consequently, the satellite-based AIS is capable of globally monitoring the ship's movement in real time if the number of the satellite and the ground station is satisfied.

Since 2005, various entities have been experimenting with detecting AIS transmissions using satellite-based receivers and, since 2008, companies such as exactEarth, ORBCOMM, Spacequest and also government programs have deployed AIS receivers on satellites. The TDMA radio access scheme used by the AIS system creates significant technical issues for the reliable reception of AIS messages from all types of transceivers: Class A, Class B, Identifier, AtoN and SART. However, the industry is seeking to address these issues through the development of new technologies and over the coming years the current restriction of satellite AIS systems to Class A messages is likely to dramatically improve with the addition of Class B and Identifier messages.

The fundamental challenge for AIS satellite operators is the ability to receive very large numbers of AIS messages simultaneously from a satellite's large reception footprint. There is an inherent issue within the AIS standard; the TDMA radio access scheme defined in the

---

[201] https://en.wikipedia.org/wiki/Automatic_identification_system

AIS standard creates 4,500 available time-slots in each minute but this can be easily overwhelmed by the large satellite reception footprints and the increasing numbers of AIS transceivers, resulting in message collisions, which the satellite receiver cannot process. Companies such as exactEarth are developing new technologies such as ABSEA, that will be embedded within terrestrial and satellite-based transceivers, which will assist the reliable detection of Class B messages from space without affecting the performance of terrestrial AIS.

The addition of satellite-based Class A and B messages could enable truly global AIS coverage but, because the satellite-based TDMA limitations will never match the reception performance of the terrestrial-based network, satellites will augment rather than replace the terrestrial system.

Shipboard AIS transceivers have a horizontal range that is highly variable, but typically only up to about 74 kilometers (46 mi). They reach much further vertically – up to the 400 km orbit of the International Space Station (ISS).

In November 2009, the STS-129 space shuttle mission attached two antennas—an AIS VHF antenna, and an Amateur Radio antenna to the Columbus module of the ISS. Both antennas were built in cooperation between ESA and the ARISS team (Amateur Radio on ISS). Starting from May 2010 the European Space Agency is testing an AIS receiver from Kongsberg Seatex (Norway) in a consortium led by the Norwegian Defence Research Establishment in the frame of technology demonstration for space-based ship monitoring. This is a first step towards a satellite-based AIS-monitoring service.

In 2008, ORBCOMM launched AIS enabled satellites in conjunction with a US Coast Guard contract to demonstrate the ability to collect AIS messages from space. In 2009, Luxspace, a Luxembourg-based company, launched the RUBIN-9.1 satellite (AIS Pathfinder 2). The satellite is operated in cooperation with SES and REDU Space Services. In late 2011 and early 2012, ORBCOMM and Luxspace launched the Vesselsat AIS microsatellites, one in an equatorial orbit and the other in a polar orbit. (VesselSat-2 and VesselSat-1).

In 2007, the U.S. tested space-based AIS tracking with the TacSat-2 satellite. However, the received signals were corrupted because of the simultaneous receipt of many signals from the satellite footprint.

In July 2009, SpaceQuest launched AprizeSat-3 and AprizeSat-4 with AIS receivers. These receivers were successfully able to receive the U.S. Coast Guard's SART test beacons off of Hawaii in 2010.[9] In July 2010, SpaceQuest and exactEarth of Canada announced an arrangement whereby data from AprizeSat-3 and AprizeSat-4 would be incorporated into the exactEarth system and made available worldwide as part of their exactAIS(TM)service.

On July 12, 2010, The Norwegian AISSat-1 satellite was successfully launched into polar orbit. The purpose of the satellite is to improve surveillance of maritime activities in the High North. AISSat-1 is a nano-satellite, measuring only 20x20x20 cm, with an AIS receiver made by Kongsberg Seatex. It weighs six kilograms and is shaped like a cube.

On 20 April 2011, Indian Space Research Organization launched Resourcesat-2 containing a S-AIS payload for monitoring maritime traffic in the Indian Ocean Search & Rescue (SAR) zone. AIS data is processed at National Remote Sensing Centre and archived at Indian Space Science Data Centre.

On 25. February 2013 - after one year launch delay - Aalborg University did launch AAUSAT3. It is a 1U cubesat, weights 800 grams, solely developed by students from Department of Electronic Systems. It carries two AIS receivers, a traditional and a SDR-based receiver. The project was proposed and sponsored by the Danish Safety Maritime Organization. It has been a huge success and has in the first 100 days downloaded more than 800,000 AIS messages and several 1 MHz raw samples of radio signal. It receives both AIS channels simultaneously and has received class A as well as class B messages. Cost including launch was less than €200,000.

Today, Canadian based exactEarth operates the largest AIS satellite network, providing global coverage using 8 satellites. This network will be significantly expanded with the announcement of a partnership with Harris Corp to utilize 58 hosted payloads on the Iridium NEXT constellation. Additionally exactEarth is involved in the development of ABSEA technology which will enable its network to reliably detect a high proportion of Class B type messages, as well as Class A.

ORBCOMM will be launching 17 additional satellites, as part of its OG2 (ORBCOMM Generation 2) satellite replenishment, that will all carry AIS receivers, and will download at ORBCOMM's 16 existing earth stations around the globe.

On 14. July 2014, ORBCOMM launched the first 6 of its 17 OG2 satellites aboard a Spacex Falcon 9 rocket from Cape Canaveral, Florida. Each OG2 satellite carries an AIS receiver payload. All 6 OG2 satellites were successfully deployed into orbit and started sending telemetry to ORBCOMM soon after launch. Successful commissioning of these satellites has provided ORBCOMM with a constellation of 7 AIS-equipped satellites. Additionally, Spacex has given ORBCOMM a launch window of December 2015 to launch its additional 11 OG2 satellites, all equipped with AIS receivers.

Electronic Chart Display and Information System (ECDIS)

The Electronic Chart Display and Information System (ECDIS)[202] is a development in the navigational chart system used in naval vessels and ships. With the use of the electronic chart system, it has become easier for a ship's navigating crew to pinpoint locations and attain directions.

ECDIS complies with IMO Regulation V/19 & V/27 of SOLAS convention as amended, by displaying selected information from a System Electronic Navigational Chart (SENC). ECDIS equipment complying with SOLAS requirements can be used as an alternative to paper charts. Besides enhancing navigational safety, ECDIS greatly eases the navigator's workload with its automatic capabilities such as route planning, route monitoring, automatic ETA computation and ENC updating. In addition, ECDIS provides many other sophisticated navigation and safety features, including continuous data recording for later analysis. IMO refers to similar systems not meeting the regulations as Electronic Chart Systems (ECS) as define by the International Hydrographic Organization (IHO).

An ECDIS system displays the information from electronic navigational charts (ENC) or Digital Nautical Charts (DNC) and integrates position information from position, heading and speed through water reference systems and optionally other navigational sensors. Other sensors which could interface with an ECDIS are radar, Navtex, automatic identification systems (AIS), and depth sounders. There are two most widely used electronic chart formats.

a) Electronic Navigational Charts (ENC) are vector charts that conform to the requirements for the chart databases for ECDIS, with standardized content, structure and format, issued for use with ECDIS on the authority of government authorized hydrographic offices. ENCs are vector charts that also conform to International Hydrographic Organization (IHO) specifications stated in IHO Publication S-57.
ENCs contain all the chart information necessary for safe navigation, and may contain supplementary information in addition to that contained in the paper chart (e.g., Sailing Directions). These supplementary information may be considered necessary for safe navigation and can be displayed together as a seamless chart. Systems using ENC charts can be programmed to give warning of impending danger in relation to the vessel's position and movement. Chart systems certified according to marine regulations are required to show these dangers.

b) Raster navigational charts (RNC) are raster graphics charts that conform to IHO specifications and are produced by converting paper charts to digital image by scanner. The image is similar to digital camera pictures, which could be zoomed in for more detailed information as it does in ENCs. IHO Publication S-61 provides

---

[202] https://en.wikipedia.org/wiki/Electronic_Chart_Display_and_Information_System

guidelines for the production of raster data. IMO Resolution MSC.86(70) permits ECDIS equipment to operate in a Raster Chart Display System (RCDS) mode in the absence of ENC.

Raster Navigational Charts (RNC) are created by the National Oceanic and Atmospheric Administration (NOAA) of the United States Government. Each original chart is scanned at high resolution with color separate overlays. The raster file also contains data that is Georeferencing; enabling computer based navigation attached to a GPS to locate and display the chart. The charts are stored in BSB format, proprietary standard of the NOAA. Image manipulation tools such as GDAL can read the image information, but there also is georeferenced data in the navigational charts.

An electronic navigational chart is an official database created by a national hydrographic office for use with an Electronic Chart Display and Information System (ECDIS). An electronic chart must conform to standards stated in the International Hydrographic Organization (IHO) Publication S-57 before it can be certified as an ENC. Only ENCs can be used within ECDIS to meet the International Maritime Organization (IMO) performance standard for ECDIS.

ENCs are available for wholesale distribution to chart agents and resellers from Regional Electronic Navigational Chart Centers (RENCs). The RENCs are not-for-profit organizations made up of ENC-producer countries. RENCs independently check each ENC submitted by the contributing countries to ensure that they conform to the relevant IHO standards. The RENCs also act collectively as one-stop wholesalers of most of the world's ENCs.

IHO Publication S-63 developed by the IHO Data Protection Scheme Working Group is used to encrypt and digitally sign ENC data. Chart data is captured based on standards stated in IHO Publication S-57, and is displayed according to a display standard set out in IHO Publication S-52 to ensure consistency of data rendering between different systems.

ECDIS (as defined by IHO Publications S-52 and S-57) is an approved marine navigational chart and information system, which is accepted as complying with the conventional paper charts required by Regulation V/19 of the 1974 IMO SOLAS Convention as amended. The performance requirements for ECDIS are defined by IMO and the consequent test standards have been developed by the International Electrotechnical Commission (IEC) in International Standard IEC 61174.

In the future, the ENC will be part of a product specification family which is based on the "*IHO Universal Hydrographic Data Model*", known as S-100. The product specification number S-101 has been assigned to the ENC.

ECDIS plays a role in Maritime Surveillance/MSA as an onboard system connected to the Automatic Identification System (AIS) and is an important part of Navigation and Security.

IMO adopted compulsory carriage of ECDIS and the ENC on new high speed craft from 1. July 2010 and progressively for other craft from 2012 to 2018.

Global Integrated Shipping Information System (GISIS)

The Global Integrated Shipping Information System (GISIS)[203] is a web site provided by IMO with access to Maritime Security Information from various resources:

- Ship and Company Particulars: Search the world fleet of ships by IMO Number and look up company particulars by IMO Company Number.
- Maritime Security: Information communicated under the provisions of SOLAS regulation XI-2/13 (SOLAS chapter X1-2 and the ISPS Code).
- Contact Points: Contact lists of competent authorities and authorized organizations relating to IMO matters.
- Recognized Organizations: Information submitted by Member States under MSC/Circ.1010-MEPC/Circ.382.
- Marine Casualties and Incidents: Data on marine casualties and incidents, as defined by circulars MSC-MEPC.3/Circ.3.
- Port Reception Facilities: Data on the available port reception facilities for the reception of ship-generated waste.
- Pollution Prevention Equipment and Anti-fouling Systems: Equipment required by MARPOL 73/78 and the BWM Convention, and anti-fouling systems compliant with the AFS Convention.
- Status of Treaties: Status of ratification of IMO conventions.
- Piracy and Armed Robbery: Reported incidents of piracy and armed robbery.
- Facilitation of International Maritime Traffic: Information on stowaway incidents, E-Addresses of Governmental Authorities and notifications pursuant to article VIII of the FAL Convention.
- Non-mandatory Instruments: Comprehensive list of non-mandatory IMO instruments.
- Simulators: Information on simulators available for use in maritime training.
- Global SAR Plan: Information on the availability of Search and Rescue (SAR) Services.
- Condition Assessment Scheme: Electronic database for the implementation of the Condition Assessment Scheme - Resolution MEPC.94 (46), as amended.
- Cargoes: Information received from IMO members relating to containers, grain and solid bulk cargoes and dangerous goods in packaged form.
- GMDSS: Global Maritime Distress and Safety System (GMDSS)
- Ship Fuel Oil Consumption: Mandatory reporting of fuel oil consumption by ships.
- MARPOL Annex VI: Notifications communicated under the provisions of MARPOL Annex VI (Regulations for the Prevention of Air Pollution from Ships).

---

[203] http://gisis.imo.org

- Evaluation of Hooks: Reports on evaluation of hooks according to the Guidelines for evaluation and replacement of lifeboat release and retrieval systems (MSC.1/Circ.1392).
- Survey and Certification: Specimen certificate and e-certificate, exemptions and equivalents, and Voluntary early implementation.
- Member State Audits: Information on audits under the IMO Member State Audit Scheme.
- Ballast Water Chemicals: GESAMP-BWWG Database of chemicals most commonly associated with treated ballast water.
- Ballast Water Management: Information on exemptions granted to ships, designated ballast water exchange areas, additional measures and warnings concerning ballast water uptakes.
- Inter-agency platform: Inter-agency platform for information sharing on migrant smuggling by sea

Global Maritime Distress and Safety System (GMDSS)

The Global Maritime Distress and Safety System (GMDSS)[204] is an internationally agreed-upon set of safety procedures, types of equipment, and communication protocols used to increase safety and make it easier to rescue distressed ships, boats and aircraft.

GMDSS consists of several systems, some of which are new, but many of which have been in operation for many years. The system is intended to perform the following functions: alerting (including position determination of the unit in distress), search and rescue coordination, locating (homing), maritime safety information broadcasts, general communications, and bridge-to-bridge communications. Specific radio carriage requirements depend upon the ship's area of operation, rather than its tonnage. The system also provides redundant means of distress alerting, and emergency sources of power.

Recreational vessels do not need to comply with GMDSS radio carriage requirements, but will increasingly use the Digital Selective Calling (DSC) VHF radios. Offshore vessels may elect to equip themselves further. Vessels under 300 Gross tonnages (GT) are not subject to GMDSS requirements.

Since the invention of radio at the end of the 19th century, ships at sea have relied on Morse code, invented by Samuel Morse and first used in 1844, for distress and safety telecommunications. The need for ship and coast radio stations to have and use radiotelegraph equipment, and to listen to a common radio frequency for Morse encoded distress calls, was recognized after the sinking of the liner RMS Titanic in the North Atlantic in 1912.

The U.S. Congress enacted legislation soon after, requiring U.S. ships to use Morse code radiotelegraph equipment for distress calls. The International Telecommunications Union (ITU), now a United Nations agency, followed suit for ships of all nations. Morse encoded distress calling has saved thousands of lives since its inception almost a century ago, but its use requires skilled radio operators spending many hours listening to the radio distress frequency. Its range on the medium frequency (MF) distress band (500 kHz) is limited, and the amount of traffic Morse signals can carry is also limited.

Not all ship-to-shore radio communications were short range. Some radio stations provided long-range radiotelephony services, such as radio telegrams and radio telex calls, on the HF bands (3–30 MHz) enabling worldwide communications with ships. For example, Portishead Radio, which was the world's busiest radiotelephony station, provided HF long-range services. In 1974, it had 154 radio operators who handled over 20 million words per year. Such large radiotelephony stations employed large numbers of people and were expensive to operate. By the end of the 1980s, satellite services had started to take an increasingly large share of the market for ship-to-shore communications.

For these reasons, the International Maritime Organization (IMO), a United Nations agency

---

[204] https://en.wikipedia.org/wiki/Global_Maritime_Distress_and_Safety_System

specializing in safety of shipping and preventing ships from polluting the seas, began looking at ways of improving maritime distress and safety communications. In 1979, a group of experts drafted the International Convention on Maritime Search and Rescue, which called for development of a global search and rescue plan. This group also passed a resolution calling for development by IMO of a Global Maritime Distress and Safety System (GMDSS) to provide the communication support needed to implement the search and rescue plan.

GMDSS is based upon a combination of satellite and terrestrial radio services, and has changed international distress communications from being primarily ship-to-ship based to ship-to-shore (Rescue Coordination Center) based. It spelled the end of Morse code communications for all but a few users, such as amateur radio operators. The GMDSS provides for automatic distress alerting and locating in cases where a radio operator doesn't have time to send an SOS or MAYDAY call, and, for the first time, requires ships to receive broadcasts of maritime safety information which could prevent a distress from happening in the first place.

In 1988, IMO amended the Safety of Life at Sea (SOLAS) Convention, requiring ships subject to it fit GMDSS equipment. Such ships were required to carry NAVTEX and satellite EPIRBs by August 1, 1993, and had to fit all other GMDSS equipment by February 1, 1999. US ships were allowed to fit GMDSS in replacement of Morse telegraphy equipment by the Telecommunications Act of 1996.

A GMDSS system may include High Frequency (HF) radiotelephone and RadioTelex (narrow-band direct printing) equipment, with calls initiated by digital selective calling (DSC). Worldwide broadcasts of maritime safety information can also make transmissions on HF narrow-band direct printing channels. The main types of equipment used in GMDSS are COSPAS-SARSAT, Navtex (Navigational Telex), and INMARSAT.

The GMDSS installation on ships include one (two on vessels over 500 GT) Search and Rescue Locating device(s) called Search and Rescue Radar Transponders (SART) which are used to locate survival craft or distressed vessels by creating a series of twelve dots on a rescuing ship's 3 cm radar display. The detection range between these devices and ships, dependent upon the height of the ship's radar mast and the height of the Search and Rescue Locating device, is normally about 15 km (8 nautical miles). Once detected by radar, the Search and Rescue Locating device will produce a visual and aural indication to the persons in distress.

The Digital Selective Calling (DSC)[205] is a standard for sending pre-defined digital messages via the medium frequency (MF), high frequency (HF) and very high frequency (VHF) maritime radio systems. DSC is a core part of the Global Maritime Distress Safety System (GMDSS).

---

[205] https://en.wikipedia.org/wiki/Digital_selective_calling

DSC was introduced by the IMO on MF, HF and VHF maritime radios as part of the GMDSS system. DSC is primarily intended to initiate ship-to-ship, ship-to-shore and shore-to-ship radiotelephone and MF/HF radio-telex calls. DSC calls can also be made to individual stations, groups of stations, or "all stations" in one's radio range. Each DSC-equipped ship, shore station and group is assigned a unique 9-digit Maritime Mobile Service Identity.

DSC distress alerts, which consist of a preformatted distress message, are used to initiate emergency communications with ships and rescue coordination centers. DSC was intended to eliminate the need for persons on a ship's bridge or on shore to continuously guard radio receivers on voice radio channels, including VHF channel 16 (156.8 MHz) and 2182 kHz now used for distress, safety and calling. A listening watch aboard GMDSS-equipped ships on 2182 kHz ended on 1. February 1999. In May 2002, IMO decided to postpone cessation of a VHF listening watch aboard ships. That watch-keeping requirement had been scheduled to end on 1. February 2005.

IMO and ITU both require that the DSC-equipped MF/HF and VHF radios be externally connected to a satellite navigation receiver (GPS). That connection will ensure accurate location information is sent to a rescue coordination center if a distress alert is transmitted. The FCC requires that all new VHF and MF/HF maritime radiotelephones type accepted after June 1999 have at least a basic DSC capability.

VHF digital selective calling also has other capabilities beyond those required for the GMDSS. The Coast Guard uses this system to track vessels in Prince William Sound, Alaska, Vessel Traffic Service. IMO and the USCG also plan to require ships carry a Universal Shipborne automatic identification system, which will be DSC-compatible. Countries having a GMDSS A1 Area should be able to identify and track AIS-equipped vessels in its waters without any additional radio equipment. A DSC-equipped radio cannot be interrogated and tracked unless that option was included by the manufacturer, and unless the user configures it to allow tracking.

GMDSS telecommunications equipment should not be reserved for emergency use only. The International Maritime Organization encourages mariners to use GMDSS equipment for routine as well as safety telecommunications.

DSC was developed to replace a call in older procedures. Because a DSC signal uses a stable signal with a narrow bandwidth and the receiver has no squelch, it has a slightly longer range than analog signals, with up 25 percent longer range and significantly faster. DSC senders are programmed with the ship's Maritime Mobile Service Identity (MMSI) and may be connected to the ship's Global Positioning System (GPS), which allows the apparatus to know who it is, what time it is and where it is. This allows a distress signal to be sent very quickly.

Often, ships use separate VHF DSC and MF/HF DSC controllers. For VHF, DSC has its own dedicated Receiver for monitoring Channel 70, but uses the main VHF transceiver for transmission. However, for the user, the controller is often a single unit.[1] MF/HF DSC devices monitor multiple distress, urgency and safety bands in the 2, 4, 6, 8, 12 and 16 MHz bands.

At minimum, controllers will monitor 2187.5 kHz and 8414.5 kHz and one more band. When sending a distress signal, the DSC device will at minimum include the ship's MMSI number. It will also include the coordinates if available and, if necessary, the channel for the following Radiotelephony or RadioTelex messages.

The distress can be sent either as a single-frequency or multi-frequency attempt. In the former, a distress signal is sent on one band and the system will wait up to four minutes for a DSC acknowledgment from a coast station. If none is received, it will repeat the distress alert up to five times. In a multi-frequency attempt, the distress signal is sent on the MF and all the HF distress frequencies in turn. As this requires retuning the antenna for each sending, without waiting for an acknowledgment, a multi-frequency attempt should only be done if there are only a few minutes until the ship's batteries are under water. As the distress message can only be sent on one of the bands, many ships and coast stations may be listening to a band without the message, and will after five minutes relay the distress signal to a coast station.

Distress calls can be both non-designated and designated. The latter allows one of ten pre-defined designations to be sent along with the distress signal. These are "*abandoning ship*", "*fire or explosion*", "*flooding*", "*collision*", "*grounding*", "*listing*", "*sinking*", "*disabled and adrift*", "*piracy or attack*" and "*man overboard*".

To avoid false distress alerts, distress buttons normally have protective covers, often with a spring-loaded cover so two hands need to be used simultaneously. Alternatively, some devices have two-button systems. Operators are required to cancel falsely sent distress alerts with a transmission on the channel designated by the distress signal.

A coast station which receives a DSC distress alert will immediately send an acknowledgment. The sending device will then both stop repeating the alert, and tune to the designated channel for the distress message to be sent. Ships receiving a distress alert who are outside coast station range or do not receive an acknowledgment, are required to relay the distress alert by any means to land.

Class A devices, used on commercial ships, have the ability to send distress, distress relay, all ships urgency, all ships safety, individual, group, geographic area and telephone alerts. Class D devices, used for most leisure vessels, can send distress, all ships urgency, all ships safety and individual on channels 06, 08, 72 and 77. The latter is only required to have one antenna and is thus not required to watch Channel 70 when in use. For routine alerts, which are used to establish communication with another station on a working channel, the receiver acknowledges to confirm that communication can be done on the appropriate channel.

The DSC is a synchronous system using characters composed from a ten-bit error detecting code. The bits are encoded using frequency shift keying. For High Frequency and Medium Frequency two tones 170 Hz apart either side of the allocated frequency with 100 Baud symbol rate are used. For VHF the two tones used are 1300 and 2100 Hz with a symbol rate of 1200 Baud. Each character is transmitted twice with a time delay. The detailed specification is published in the International Telecommunications Union recommendation ITU-R M.493, revision 14 being the most recent.

INMARSAT and Iridium frequency bands interfere at 1626.5 MHz thus each satcom radio has the ability to interfere with the other. Usually the far more powerful INMARSAT radio disrupts the Iridium radio up to 10–800 meters away.

Gulf of Finland Reporting (GOFREP)

The Gulf of Finland Reporting (GOFREP) is a Mandatory Ship Reporting System under SOLAS Regulation V/11 (adapted by IMO on 1. July 2004 under MSC.139). Shore-based facilities at TALLINN TRAFFIC, HELSINKI TRAFFIC and St. PETERSBURG TRAFFIC are able to monitor shipping movements and provide advice and information about navigational hazards and weather conditions. Estonia, Finland and Russia believe that safe navigation in the Gulf of Finland is enhanced by the GOFREP system.

The aims are to contribute to the safety of navigation through and across the GOFREP area; to increase the protection of the marine environment; to monitor compliance with the International regulations for preventing collisions at sea (COLREGS). The GOFREP area covers the international waters in the Gulf of Finland east of the Western Reporting Line.

Estonia and Finland have implemented additional mandatory ship reporting systems in their territorial waters outside their VTS areas providing the same services and impose the same requirements on shipping as the system operating in the international waters.

International Ship and Port Facility Security Code (ISPS Code)

The International Maritime Organization (IMO) states that the International Ship and Port Facility Security Code (ISPS Code)[206] is "*a comprehensive set of measures to enhance the security of ships and port facilities, developed in response to the perceived threats to ships and port facilities in the wake of the 9/11 attacks in the United States*".

Development and implementation were sped up drastically in reaction to the 11. September 2001 attacks and the bombing of the French oil tanker Limburg. The U.S. Coast Guard, as the lead agency in the United States delegation to the IMO, advocated for the measure. The Code was agreed at a meeting of the 108 signatories to the SOLAS convention in London in December 2002. The measures agreed under the Code were brought into force on 1. July 2004.

The International Ship and Port Facility Security (ISPS) Code is an amendment to the Safety of Life at Sea (SOLAS) Convention (1974/1988) on minimum security arrangements for ships, ports and government agencies. Having come into force in 2004, it prescribes responsibilities to governments, shipping companies, shipboard personnel, and port/facility personnel to "*detect security threats and take preventative measures against security incidents affecting ships or port facilities used in international trade*".

Europe has enacted the International regulations with EC Regulation (EC) No 725/2004 of the European Parliament and of the Council of 31. March 2004, on enhancing ship and port facility security, while GBR has enacted The Ship and Port Facility (Security) Regulations 2004, these bring the EU regulation 725/2004 into UK law.

The United States has issued regulations to enact the provisions of the Maritime Transportation Security Act of 2002 and to align domestic regulations with the maritime security standards of SOLAS and the ISPS Code. These regulations are found in Title 33 of the Code of Federal Regulations, Parts 101 through 107. Part 104 contains vessel security regulations, including some provisions that apply to foreign ships in U.S. waters.

The ISPS Code is implemented through chapter XI-2 Special measures to enhance maritime security in the International Convention for the Safety of Life at Sea (SOLAS). The Code is a two-part document describing minimum requirements for security of ships and ports. Part A provides mandatory requirements. Part B provides guidance for implementation.

The Code does not specify specific measures that each port and ship must take to ensure the safety of the facility against terrorism because of the many different types and sizes of these facilities. Instead it outlines "*a standardized, consistent framework for evaluating risk, enabling governments to offset changes in threat with changes in vulnerability for ships and port facilities.*"

The ISPS Code applies to ships on international voyages (including passenger ships, cargo

---

[206] https://en.wikipedia.org/wiki/International_Ship_and_Port_Facility_Security_Code

ships of 500 GT and upwards, and mobile offshore drilling units) and the port facilities serving such ships.

- The main objectives of the ISPS Code are:
- To detect security threats and implement security measures
- To establish roles and responsibilities concerning maritime security for governments, local administrations, ship and port industries at the national and international level
- To collate and promulgate security-related information
- To provide a methodology for security assessments so as to have in place plans and procedures to react to changing security levels

For ships the framework includes requirements for:

- Ship security plans
- Ship security officers
- Company security officers
- Certain onboard equipment

For port facilities, the requirements include:

- Port facility security plans
- Port facility security officers
- Certain security equipment

In addition the requirements for ships and for port facilities include:

- Monitoring and controlling access
- Monitoring the activities of people and cargo
- Ensuring security communications are readily available

Navigational Telex (Navtex)

Navigational Telex (Navtex)[207] is an international, automated system for instantly distributing Maritime Safety Information (MSI) which includes navigational warnings, weather forecasts and weather warnings, search and rescue notices and similar information to ships.

Navtex is an international automated medium frequency direct-printing service for delivery of navigational and meteorological warnings and forecasts, as well as urgent maritime safety information to ships. The small, low-cost and self-contained printing radio receiver is installed on the bridge, or the place from where the ship is navigated, and checks each incoming message to see if it has been received during an earlier transmission, or if it is of a category of no interest to the ship's master.

The frequency of transmission of these messages is 518 kHz in English, while 490 kHz is sometime used to broadcast in a local language. The messages are coded with a header code identified by the using single letters of the alphabet to represent broadcasting stations, type of messages, and followed by two figures indicating the serial number of the message. For example: FA56 where F is the ID of the transmitting station, A indicates the message category navigational warning, and 56 is the consecutive message number.

Navtex was developed to provide a low-cost, simple, and automated means of receiving this information aboard ships at sea within approximately 370 km (200 nautical miles) off shore.

There are no user fees associated with receiving Navtex broadcasts, as the transmissions are typically transmitted from the National Weather Authority (Italy) or Navy or Coast Guard (as in the US) or national navigation authority (Canada). Where the messages contain weather forecasts, an abbreviated format very similar to the shipping forecast is used.

Navtex is also a component of the International Maritime Organization/International Hydrographic Organization Worldwide Navigation Warning Service (WWNWS). Navtex is also a major element of the Global Maritime Distress Safety System (GMDSS). International Convention for the Safety of Life at Sea (SOLAS) mandated certain classes of vessels must carry Navtex, since 1. August 1993.

---

[207] https://en.wikipedia.org/wiki/Navtex

Long Range Identification and Tracking (LRIT)

The Long Range Identification and Tracking (LRIT)[208] of ships was established as an international system on 19 May 2006 by the International Maritime Organization (IMO) as resolution MSC.202(81), mainly due to the shortcomings of the Automated Information System (AIS). Long Range Identification and Tracking (LRIT) was proposed by the United States Coast Guard (USCG) at the International Maritime Organization (IMO) in London during the aftermath of the 11. September 2001 attacks to track the approximately 50,000 large ships around the world.

Under Regulation 19-1, para. 5, ships are required to automatically transmit information as to their identity, position, date and time. This applies, pursuant to para. 2.1, to passenger and merchant ships engaged in international voyages and of 300 gross tonnage and upwards. According to para. 8.1, this information serves 'security and other purposes as agreed' by the IMO. The resolution amends chapter V of the International Convention for the Safety of Life at Sea (SOLAS), regulation 19-1 and binds all governments which have contracted to the IMO.

LRIT was designed to collect and disseminate vessel position information received from ships that are subject to the SOLAS convention and provides for a long-range tracking of all vessels within 1000nm. Being based on HF transmission and satellites, LRIT can provide world-wide coverage, on the other hand with less detailed information than from the Automatic Identification System (AIS).

The LRIT regulation will apply to the following ship types engaged on international voyages:

- All passenger ships including high-speed craft,
- Cargo ships, including high-speed craft of 300 gross tonnages and above, and
- Mobile offshore drilling units.

These ships must report their position to their flag administration at least four times a day. Most vessels set their existing satellite communications systems to automatically make these reports. Other contracting governments may request information about vessels in which they have a legitimate interest under the regulation.

The LRIT system consists of the already installed (generally) shipborne satellite communications equipment, communications service providers (CSPs), application service providers (ASPs), LRIT data centers, the LRIT data distribution plan and the International LRIT data exchange. Certain aspects of the performance of the LRIT system are reviewed or audited by the LRIT coordinator acting on behalf of the IMO and its contracting governments.

The functions of LRIT are different from that of AIS which operates in the VHF radio band, with a range only slightly greater than line-of-sight. While AIS was originally designed for

---

[208] https://en.wikipedia.org/wiki/Long-range_identification_and_tracking_(ships)

short-range operation as a collision avoidance and navigational aid, it is now possible to receive Satellite-based AIS (S-AIS) signals in many, but not all, parts of the world. S-AIS and AIS are completely different from LRIT. The only similarity is that AIS is also collected from space for determining location of vessels, but requires no action from the vessels themselves except they must have their AIS system turned on.

LRIT requires the active, willing participation of the vessel involved, which is, in and of itself, a very useful indication as to whether the vessel in question is a lawful actor. Thus the information collected from the two systems, S-AIS and LRIT, are mutually complementary, and S-AIS clearly does not make LRIT superfluous in any manner. Indeed, because of co-channel interference near densely populated or congested sea areas satellites are having a difficult time in detecting AIS from space in those areas. Fixes are under development by several organizations, but how effective they will be remains to be seen.

Marshall Islands, one of the largest ship registries in the world, established one of the first prototype Data Centers, using Pole Star Space Applications. Liberia, the second largest ship registry in the world has established a LRIT Data Centre in 2008. The Recognized LRIT provider is Pole Star Space Applications.

The Flag State has to determine to which LRIT Data Centre a vessel must report. The selected Data Centre collects the information and must ensure that the LRIT information is only transmitted to those entitled to receive it. However, if a vessel is not providing information to a coastal State by virtue of a decision of the flag State of that vessel, no action should be taken by the coastal State to treat that vessel as acting in a suspicious manner and interfere with its passage. In addition the system and equipment providing LRIT information may be switched off or otherwise cease providing information, if international regulations provide for the protection of navigational information or when the operation is considered by the master to compromise the safety and security of the ship or crew.

The national LRIT Data Centers are:

- In January 2009 Canada become one of the first SOLAS contracting governments to implement a national data center and comply with the LRIT regulation.
- Liberia, the second largest ship registry in the world has established a LRIT Data Centre in 2008. The recognized LRIT provider is Pole Star Space Applications.
- Marshall Islands, one of the largest ship registries in the world, established one of the first prototype Data Centers, using Pole Star Space Applications.
- The Panama Flag Registry appointed Absolute Maritime Tracking Services, Inc. (AMTS) as the sole LRIT Application Service Provider (ASP) and National Data Center (NDC) provider for all Panama flagged vessels.
- Singapore established its LRIT National Data Centre with Pole Star Space Applications as its LRIT Recognized ASP.

- In January 2009, the United States became one of the first SOLAS contracting governments to implement a National Data Centre and comply with the LRIT regulation. Currently the US Authorized Application Service Provider (ASP) is CLS America.
- In January 2009 Brazil implemented a National Data Centre and was one of the first SOLAS contracting governments to become compliant with the LRIT regulation. In August 11th 2010 implemented the Regional LRIT Data Centre Brazil, providing services for Brazil and Uruguay. In 2014 the RDC BRAZIL providing services also for Namibia.
- The Venezuelan flag registry appointed Fulcrum Maritime Systems as the sole LRIT application service provider (ASP) and national data center (NDC) provider for all Venezuelan flagged vessels.
- The Chilean flag registry appointed collected localization satellites (CLS) as the sole LRIT application service provider (ASP). This Data Center provides services for all Chilean and Mexican flagged vessels.
- The Republic of Ecuador entered in LRIT production environment at April 15 of 2010. Ecuador owns a National LRIT Data Center (NDC) and recognizes their Maritime Authority as Application Service Provider (ASP).
- The Vanuatu flag registry appointed collected localization satellites (CLS) as the sole LRIT application service provider (ASP) and national data center (RDC) provider for all Vanuatu flagged vessels.
- Honduras Flag appointed Fulcrum Maritime Systems as Testing ASP.
- Several African states have formed a LRIT Cooperative Data Centre. South Africa National Data Centre provides services to a number of African states, including Ghana and the Gambia.

Following the EU Council Resolution of 2 October 2007, EU Member States (MS) decided to establish an EU LRIT Data Centre (EU LRIT DC). According to the Council Resolution, the Commission is in charge of managing the EU LRIT DC, in cooperation with Member States, through the European Maritime Safety Agency (EMSA).

EMSA is in particular in charge of the technical development, operation and maintenance of the EU LRIT DC. It also *"stresses that the objective of the EU LRIT DC should include maritime security, Search and Rescue (SAR), maritime safety and protection of the marine environment, taking into consideration respective developments within the IMO context."*

The 100 Series Rules

The 100 Series Rules[209] are an international model standard and example benchmark of best practice for the use of force in the maritime security and anti-piracy fields for application by privately contracted armed security personnel (PCASP) and private maritime security companies (PMSCs) on board ships.

The Rules are set out for the benefit of the Master, Ship owner, charterer, insurer, underwriters, PMSCs, PCASP and interested third parties, providing guidance on lawful graduated response measures and lawful use of force, including lethal force, in accordance with the right of self-defense in the context of maritime piracy, armed robbery or hijacking. The Rules aim to provide for transparency of rules, clarity in use and accountability of actions in those situations, and hope to fill gaps in these areas often lamented by the stakeholders of maritime industry and maritime security.

The 100 Series Rules have been developed for the benefit of the entire maritime industry and under-pinned by a thorough public international and criminal law legal review of what is *"reasonable and necessary"* when force is used, as a lawful last resort, in self-defense and have become a de-facto UN specification.

---

[209] http://www.100seriesrules.com

West European Tanker Reporting System (WETREP)

The West European Tanker Reporting System (WETREP) is a mandatory ship reporting system for all oil tankers over 600 tons DWT carrying heavy types of oils and entering the Western European Particularly Sensitive Sea Area (PSSA). (Government / military vessels are exempt.) It entered into force on 1. July 2005 and is part of the IMO SOLAS convention 7. WETREP reports must be sent when:

- Entering the PSSA
- Leaving a port within the PSSA
- Deviate from declared route
- Exiting the PSSA

The objective of the system is to provide advance information to authorities responsible for pollution prevention and search & rescue, in order that they can react quickly in case of an accident. The report must include: the ship's name, call sign, IMO number, MMSI number, date, time, position, course, speed, last and next port of call with ETA, type and quantity of oil or other hazardous substances, number of persons on board, and information on defects, damage, deficiencies etc.

The report must be sent to the nearest co-ordination center of a responsible authority of the Coastal State participating in the system, which can be a Vessel Traffic Service, RCC, or coast radio station. There is a list of these authorities, all MRCC or MRSC: 3 in IRL, 9 in GBR, 1 in BEL, 2 in FRA, 3 in ESP, 1 in PRT. Reports may be sent by any modern communication form, including Inmarsat-C, fax and e-mail as appropriate.

WETREP is for the exchange of information only and does not provide any additional authority for mandating changes in the vessel's operations. According to the IMO Resolution, this reporting system will be implemented consistent with UNCLOS9, SOLAS7 and other relevant international instruments so that the reporting system will not provide the basis to impinge on a transiting vessel's passage through the reporting area. Proprietary information obtained as a requirement of the mandatory ship reporting system WETREP will be protected under this system consistent with the

Guidelines and Criteria for Ship Reporting Systems, as amended (IMO Resolution A.851(20)). Failure to submit a report will result in information being passed to the flag State Authorities for investigation and possible prosecution. Full details of the cargoes and the geographical area and format of the WETREP are available in the IMO Circular SN/Circ.242 and the web site Maritime Knowledge on WETREP.

United States of America (USA)

The United States of America (USA) by geographical position, coastal area, economic and military capability one of the largest Maritime Entities; its maritime forces are still the most powerful global player next to Russia and China.

The website Global Firepower (GFP)[210] provides since 2006 analytical display of data concerning over 135 modern military powers with the ranking based on each nation's potential war-making capability across land, sea and air fought with conventional weapons. The results incorporate values related to resources, finances, and geography with over 55 different factors ultimately making up the final rankings. In 2018 a total of 136 countries were included in the Global Firepower database ranking the USA before Russia, China and India.

The USA are the world's third-largest country by size after Russia and Canada and by population after China and India, and are bordering both the North Atlantic Ocean and the North Pacific Ocean, between Canada and Mexico. The USA and China also represent next to the EU the largest economies and most international trade dependent nations/organization.

The USA is in international environmental agreements party to Air Pollution, Air Pollution-Nitrogen Oxides, Antarctic-Environmental Protocol, Antarctic-Marine Living Resources, Antarctic Seals, Antarctic Treaty, Climate Change, Desertification, Endangered Species, Environmental Modification, Marine Dumping, Marine Life Conservation, Ozone Layer Protection, Ship Pollution, Tropical Timber 83, Tropical Timber 94, Wetlands, Whaling. Signed, but not ratified are Air Pollution-Persistent Organic Pollutants, Air Pollution-Volatile Organic Compounds, Biodiversity, Climate Change-Kyoto Protocol, Hazardous Wastes.

Maritime Domain Awareness (MDA) is a very important part of United States' maritime law enforcement campaign pertaining to strengthening marine security – both internally as well as internationally. The Maritime Domain Awareness is a very wide and encompassing subject and therefore merits a special mention while maritime security of nations is taken into account. The Secretary of the Navy is the DoD Executive Agent for maritime domain awareness.

The USA developed an extensive National Maritime Domain Awareness Architecture Plan. The main motto and the objective of the Maritime Domain Awareness is to collect the maximum information and intelligence about any ship or vessel in the country's waters. With the collected data, a complete inference can be drawn about all those marine areas that could cause potential damage with respect to safety, eco-system and the economic system. This process is known as 'actionable intelligence.'

---

[210] https://www.globalfirepower.com

African Partnership Station (APS)

African Partnership Station (APS)[211] is an international initiative developed by United States Naval Forces Europe-Africa, which works cooperatively with USA and international partners to improve maritime safety and security in Africa as part of US Africa Command's Security Cooperation program.

APS is a strategic program designed to build the skills, expertise and professionalism of African militaries, coast guards and mariners. APS is not limited to one ship or platform nor is it delivered only at certain times. The program is delivered in many forms including ship visits, aircraft visits, training teams, and Seabees[212] construction projects throughout most of the year. APS is part of a long-term commitment on the part of all participating nations and organizations from Africa, the United States, Europe, and South America.

APS activities consist of joint exercises, port visits, hands-on practical courses, professional training and community outreach with the coastal nations of Africa. The focus is on building maritime capacity of the nations and increasing the level of cooperation between them to improve maritime safety and security. The goal is to improve the ability of the nations involved to extend the rule of law within their territorial waters and exclusive economic zones and better combat illegal fishing, human smuggling, drug trafficking, oil theft and piracy. APS also works to increase maritime safety by teaching skills that enhance a nation's ability to respond to mariners in distress.

The first APS deployment was from November 2007 to April 2008. Countries visited included Senegal, Togo, Ghana, São Tomé and Príncipe, Cameroon, Liberia, Gabon, and Equatorial Guinea and included USS FORT McHENRY (LSD-43) and HSV SWIFT, with an international staff embarked on FORT McHENRY. The time in between major deployments was covered by mobile training team visits, maritime patrol aircraft exercises and port visits by individual naval vessels.

During the summer and fall of 2008 two ships began what was at the time called a LEDET, or Law Enforcement Detachment. These ships were US Coast Guard Cutter DALLAS (USCGC DALLAS (WHEC-716)) and USS LEYTE GULF (CG-55). These missions were designed to bring African law enforcement officials onboard US ships, in concert with Coast Guard personnel, in order to conduct the first real-time operations, building upon the many skills and capabilities acquired on previous training visits.

USS NASHVILLE (LPD-13) was the second large amphibious ship to deploy to Africa under Africa Partnership Station; it deployed from February 2009 to May 2009. The Nashville was the largest ship to perform the APS mission in 2009. APS Nashville visited Senegal, Ghana, Gabon, Cameroon, and Nigeria, spending one to two weeks in each port. APS Nashville's

---

[211] https://en.wikipedia.org/wiki/Africa_Partnership_Station
[212] United States Naval Construction Battalions, better known as the Seabees, form the Naval Construction Force (NCF) of the United States Navy.

embarked staff had a larger international flavor with military members from Nigeria, Cameroon, Senegal, Ghana, Gabon, Italy, Portugal, Cape Verde, Sierra Leone, Togo, Equatorial Guinea, Kenya, United Kingdom, France, Germany, Spain, Denmark, Malta and Brazil.

In February 2009 APS expanded to South and East Africa when the USS ROBERT G. BRADLEY (FFG-49) visited Mozambique, Tanzania and Kenya. Over the summer of 2009 (after USS NASHVILLE completed her APS mission) other ships continued the initiative. USS ARLEIGH BURKE (DDG-51) visited East Africa, bringing APS once again to the East coast of Africa and expanding the range of cooperative training. At the same time HSV SWIFT and the US Coast Guard Cutter LEGARE (USCGC LEGARE) continued the APS mission in West and Central Africa.

While APS Swift conducted a series of training, humanitarian and outreach missions in the west, LEGARE led and participated in the first African Maritime Law Enforcement Partnership or AMLEP (the name used to replace the LEDETs mentioned above). While on the mission with members of the Sierra Leone Maritime Wing APS LEGARE made a significant impact when the joint mission boarded a Taiwanese vessel illegally fishing with thousands of dollars of fish. In April 2015 the USS SPEARHEAD and its embarked detachment of the U.S. Navy, civil service mariners and U.S., Spanish and British Marines conducted a portion of USS SPEARHEAD's support to Africa Partnership Station while in Port Gentil, Gabon.

In the Fall of 2009 the first APS mission led by a non-US country commenced when the Dutch Rotterdam class amphibious transport dock JOHAN DE WITT conducted the mission with US, Portuguese and Belgian Sailors and Marines embarked as training teams, and with Seabees and other subject matter experts. The ship conducted port visits to Senegal, Liberia, Sierra Leone and Ghana.

In winter and spring of 2010 APS again made banner deployments. This time USS GUNSTON HALL (LSD-44) visited West and Central Africa and a 2-ship flotilla visited East Africa at virtually the same time. The East Africa mission included USS NICHOLAS (FFG-47) and HSV SWIFT.

Customs-Trade Partnership against Terrorism (C-TPAT)

After the U.S. Customs Service became aware of the incomplete knowledge of the supply chain, it responded by launching the strictly voluntary Customs-Trade Partnership against Terrorism (C-TPAT)[213] in 2001. The Customs and Border Protection (CBP) established the Commercial Operations Advisory Committee (COAC), which initially consisted of 50 international trade industry experts. COAC gave CBP an expanded view of security in the context of international trade.

Customs Trade Partnership Against Terrorism (CTPAT) is but one layer in U.S. Customs and Border Protection's (CBP) multi-layered cargo enforcement strategy. Through this program, CBP works with the trade community to strengthen international supply chains and improve United States border security. CTPAT is a voluntary public-private sector partnership program which recognizes that CBP can provide the highest level of cargo security only through close cooperation with the principle stakeholders of the international supply chain such as importers, carriers, consolidators, licensed customs brokers, and manufacturers. The Security and Accountability for Every Port Act of 2006 provided a statutory framework for the CTPAT program and imposed strict program oversight requirements.

From its inception in November 2001, CTPAT continued to grow. Today, more than 11,400 certified partners spanning the gamut of the trade community, have been accepted into the program. The partners include U.S. importers/exporters, U.S./Canada highway carriers; U.S./Mexico highway carriers; rail and sea carriers; licensed U.S. Customs brokers; U.S. marine port authority/terminal operators; U.S. freight consolidators; ocean transportation intermediaries and non-operating common carriers; Mexican and Canadian manufacturers; and Mexican long-haul carriers, all of whom account for over 52 percent (by value) of cargo imported into the U.S.

When an entity joins CTPAT, an agreement is made to work with CBP to protect the supply chain, identify security gaps, and implement specific security measures and best practices. Applicants must address a broad range of security topics and present security profiles that list action plans to align security throughout the supply chain. CTPAT members are considered to be of low risk, and are therefore less likely to be examined at a U.S. port of entry.

Participation in CTPAT is voluntary and there are no costs associated with joining the program. Moreover, a company does not need an intermediary in order to apply to the program and work with CBP; the application process is easy and it is done online. The first step is for the company to review the CTPAT Minimum Security Criteria for their business entity to determine eligibility for the program. The second step is for the company to submit a basic application via the CTPAT Portal system and to agree to voluntarily participate. The third step is for the company to complete a supply chain security profile. The security profile

---

[213] https://www.cbp.gov/border-security/ports-entry/cargo-security/ctpat

explains how the company is meeting CTPAT's minimum security criteria. In order to do this, the company should have already conducted a risk assessment. Upon satisfactory completion of the application and supply chain security profile, the applicant company is assigned a CTPAT Supply Chain Security Specialist to review the submitted materials and to provide program guidance on an on-going basis. The CTPAT program will then have up to 90 days to certify the company into the program or to reject the application.

The idea behind C-TPAT is that trade members provide Customs and Border Protection (CBP) with information about the security of their supply chain. Shipments from shippers which do not participate in C-TPAT may require further scrutiny. CBP also provides importers an incentive to correct errors in importation paperwork through the Importer Self-Assessment (ISA) program. If the importer notifies CBP of the error prior to CBP becoming aware of it, the importer's penalty may be reduced.

CTPAT Partners enjoy a variety of benefits, including taking an active role in working closer with the U.S. Government in its war against terrorism. As they do this, Partners are able to better identify their own security vulnerabilities and take corrective actions to mitigate risks. Some of the benefits of the program include:

- Reduced number of CBP examinations
- Front of the line inspections
- Possible exemption from Stratified Exams
- Shorter wait times at the border
- Assignment of a Supply Chain Security Specialist to the company
- Access to the Free and Secure Trade (FAST) Lanes at the land borders
- Access to the CTPAT web-based Portal system and a library of training materials
- Possibility of enjoying additional benefits by being recognized as a trusted trade Partner by foreign Customs administrations that have signed Mutual Recognition with the United States
- Eligibility for other U.S. Government pilot programs, such as the Food and Drug Administration's Secure Supply Chain program
- Business resumption priority following a natural disaster or terrorist attack
- Importer eligibility to participate in the Importer Self-Assessment Program (ISA)
- Priority consideration at CBP's industry-focused Centers of Excellence and Expertise

As stated in a 2008 study, the annual cost to importers participating in the C-TPAT program was $30,000. However, the benefits of C-TPAT to these participants was not being realized uniformly, only 33% reported that the benefits outweighed the costs. Many trade professionals view these programs as a way to shift some of the burden of ensuring security of the supply chain onto the trade industry and not as a mutually beneficial information sharing environment.

In addition to programs like C-TPAT and ISA, other programs affect cargo-related information sharing and associated safety and security. CBP instituted the 24-hour advance vessel manifest rule in 2003. The rule's goal is to identify high risk containers before they arrive at a U.S. port of entry. The cargo manifest is the primary 17 DHS/CBP/PIA-006(c) input into the Automated Targeting System (ATS), and no other Government Accountability Office information is mandatory for the containers risk assignment.

Global Command and Control System (GCCS)

The US Global Command and Control System (GCCS)[214] is a collection of Service-oriented architecture (SOA) systems and applications which was developed to replace the US Worldwide Military Command and Control System[215]. in the US and serve as main core element for the NATO MCCIS.

GCCS systems contain web services which are used by many applications supporting combat operations, intelligence analysis and production, targeting, ground weapons and radar analysis, as well as terrain and weather analysis. Some next generation applications designed for GCCS may support collaboration using chat systems, newsgroups and email.

Three major GCCS versions exist to address the DoD and NATO service component operational requirements for battlespace management:

1. Global Command and Control System - Maritime (GCCS-M, for US Navy & Marine Corps)
2. Global Command and Control System - Army (GCCS-A, for US Army and Air Force)
3. Global Command and Control System - Joint (GCCS-J, for interoperability with NATO partners)

GCCS may use NIPRNet, SIPRNet, JWICS, or other IP-based networks for connectivity. In some installations, GCCS aggregates over 94 different sources of data. Each US-American forces use variants like the GCCS-A in the Army and Air Force or Global Command and Control System - Maritime (GCCS-M) for the US Navy & Marine Corps (Maritime).

Global Command and Control System - Army (GCCS-A)

Global Command and Control System (GCCS) - Army is the maritime version out of the collection of Service-oriented architecture (SOA) systems and applications developed to replace the Worldwide Military Command and Control System.

Global Command and Control System - Joint (GCCS-J)

Global Command and Control System (GCCS) - Joint is the joint version out of the collection Service-oriented architecture (SOA) systems and applications developed to replace

---

[214] http://en.wikipedia.org/wiki/Global_Command_and_Control_System
[215] http://en.wikipedia.org/wiki/Worldwide_Military_Command_and_Control_System

the Worldwide Military Command and Control System.

The Global Command and Control System - Joint (GCCS-J) of the US MOD is used to correlate and distribute situational awareness, Force Protection, and the battlespace; based on allied and coalition interoperability. This allows each area of responsibility view to be tailored to their command role (e.g., International Security Assistance Force (ISAF)) to view the Common Operational Picture (COP). The COP is a key tool for commanders in planning, conducting operations, monitoring, and coordinating the mission execution to achieve success.

COP is used to execute operational directives within the Joint Task Force and individual units for various efforts including Joint Operational Targets/Fires. The Joint Operational Targeting and Fires is used to facilitate the synchronization of air operations through the coordination of air support and the exchange of operational and intelligence information.

Global Command and Control System – Maritime (GCCS-M)

Global Command and Control System (GCCS) - Maritime is the maritime version out of the collection of Service-oriented architecture (SOA) systems and applications developed to replace the Worldwide Military Command and Control System as Family of Systems (FoS). GCCS-M is also the original core of the NATO Maritime Command and Control System (MCCIS).

GCCS-M is the Navy's primary fielded Command and Control System. The objective of the GCCS-M program is to satisfy Fleet C4I requirements through the rapid and efficient development and fielding of C4I capability. GCCS-M enhances the operational commander's warfighting capability and aids in the decision-making process by receiving, retrieving and displaying information relative to the current tactical situation.

GCCS-M receives, processes, displays, and manages data on the readiness of neutral, friendly and hostile forces in order to execute the full range of Navy missions (e.g., strategic deterrence, sea control, power projection, etc.) in near-real-time via external communication channels, local area networks and direct interfaces with other C2 systems.

GCCS-M supports decision making at all echelons of command with a single, integrated, scalable C4I system that fuses, correlates, filters, maintains and displays location and attribute information on friendly, hostile and neutral land, sea and air forces, integrated with available intelligence and environmental information. It operates in near real-time and constantly updates unit positions and other situational awareness data.

GCCS-M also records data in appropriate databases and maintains a history of changes to those records. System users can then use the data to construct relevant tactical pictures using maps, charts, topography overlays, oceanographic overlays, meteorological overlays, imagery, and all-source intelligence information coordinated into a Common Operational Picture that can be shared locally and with other sites. Navy commanders review and evaluate the general tactical situation, plan actions and operations, direct forces, synchronize tactical movements, and integrate force maneuver with firepower. The system operates in a variety of environments and supports joint, coalition, and allied forces.

GCCS-M is implemented Afloat and at Ashore fixed command centers and is replaced by the Maritime Tactical Command and Control (MTC2) upgrades of the Tactical Command System by the US Navy's Command, Control, Computer and Intelligence (C3I) systems. MTC2 processes C3I information for all warfare mission areas including planning, direction and reconstruction of missions for peacetime, wartime and times of crises and combine the different GCCS versions into an harmonized information system for the US Forces.

Global Maritime Partnership (GMP)

USN-Admiral Michael Mullen, unveiled the concept of the "*Thousand-Ship Navy*" as a new taxonomy for international naval cooperation. Embraced by the George W. Bush administration and renamed Global Maritime Partnership (GMP) the Initiative and concept was rapidly embraced by the community of nations as a way to secure the global commons. In the ensuing years this concept has become a new international norm and the sine qua non for international naval cooperation.

As international navies have gained experience operating together across a wide spectrum of operations from conflict to humanitarian efforts, they have also found that the networking challenges have been daunting and these C4ISR challenges have impeded effective maritime partnering.

GMP will also describe how lessons learned from past networking and coalition efforts can inform global security efforts today. GMP shares the results of a tests among the five AUSCANNZUKUS nations that provides one example of how to address these C4ISR challenges by harmonizing international naval C4ISR acquisition efforts.

While the United States Navy will continue to be the predominant power in the Asian Pacific for the foreseeable future, China and India will continue to build up their military capabilities. In recent years, the Asia pacific navies have increased their capacity and willingness to participate and cooperate in maritime security efforts within the region and globally. a noteworthy example of the GMP in action is the ongoing counter-piracy operations in the Gulf of Aden, where navies from the Asian Pacific countries such as Australia, Malaysia, Singapore and south Korea work together with the United States and other coalition partners as part of the combined task force CTF 151. Also contributing to the counter-piracy mission are the Chinese, Indian, and Japanese maritime forces under the ambition of respective national tasking.

It is noteworthy that the role of commander CTF 151 has been assumed by non-us commanders, including an Admiral from Turkey in 2009 and one from the Republic of Singapore in 2010 and 2011. This demonstrates a similar potential for the GMP leadership role to be played by other nations. Overall, the experiences and interoperability that the Asian Pacific maritime forces have cultivated through various multilateral and bilateral operations and exercises provide a good foundation in building the "*1,000-Ship Navy.*"

The way the GMP will take shape in the Asia Pacific in the future will depend not only on the US maritime security strategy but also on the evolving regional security environment. In order to maintain its leadership role within the Asia Pacific region, the US needs to recognize the emergence of China and India, and adopt a GMP policy that is inclusive of these two countries by leveraging on existing multilateral frameworks of cooperation.

In an article by LTC Joshua Toh[216] the shift and momentum in the balance of power in favor of China and India has been thoroughly outlined. Based on the December 2009 economic forecasting conducted by Goldman Sachs, china may overtake the US to become the largest economy in the world by 2027, while India may overtake Japan to become the third largest economy the same year.

These economic forecasts suggest that India and China will become regional maritime powers by the middle of the 21st century, on the premise that their military modernization programs will match their future economic prowess.

The US National Intelligence Council estimates that by 2025, China will be the world's second largest economy and will be a leading military power. According to Frost and Sullivan estimates, China is currently the fourth largest military spender globally with a defense budget of $70.2 billion, which is equal to that of France, although below both the United Kingdom and the USA. Frost and Sullivan estimate that the Chinese may procure two aircraft carriers by 2015 (which are likely to be complemented by SUKHOI SU-33 Fighter Aircraft), five Type 094 submarines, and transport aircraft (similar in size to c-17 Globemaster III). These military capabilities will enable China to project force on a sustained basis beyond its coastal periphery within the next 10-20 years.

China has enhanced its military exchanges and cooperation with various counterparts and is an active participant in multilateral activities such as the ASEAN Regional Forum (ARF) and the Western Pacific Naval Symposium (WPNS). China actively participates in the ARF because it sees the potential to further its interests with the ASEAN countries. China actively participates in the WPNS because it views the WPNS as a truly multilateral maritime arrangement rather than one that is US-centric.

Over the next 15-20 years, the US National Intelligence Council predicts that Indian leaders will strive for a multi-polar international system in which New Delhi will be one of the poles. India's impressive economic growth has spurred its military modernization, its ability to influence the Indian Ocean Region (IOR), and extended its strategic interests beyond the Gulf of Arabia and IOR to the Malacca Strait and beyond. Frost and Sullivan reported that India's defense budget reached $27.5 billion in 2008, a 10 percent increase over the previous year, and made it the ninth largest defense spender in the world. It is noteworthy that the Indian defense expenditure has almost doubled in five years. The impressive economic growth is translating into rapid military modernization, and continues to boost India's confidence as an emerging power in the world.

With its growing international confidence, the Indian navy held a two-day Indian Ocean Naval Symposium (IONS) in New Delhi in January 2008 in an effort to enhance trust between the naval leaders of the region. During exercise MILAN in February 2010, the Indian navy hosted 12 nations in the multilateral exercise,17 with ships from Australia, Bangladesh,

---

[216] Realising the Global Maritime Partnership: Strategies for enhancing cooperation within the ASIA Pacific Region by by LTC Joshua Toh: https://www.mindef.gov.sg/oms/content/dam/imindef_media_library

Indonesia, Malaysia, Myanmar, Singapore, Sri Lanka, and Thailand and representatives from navies of Brunei, Darussalam, Philippines, Vietnam and new Zealand. The implication of an emerging India is that the Indian navy will be a key player in the GMP initiative in the region.

However, the US National Intelligence Council postulates that India's new multilateral outreach program may be aimed at maximizing India's autonomy, and not at aligning India with any country or international coalition. One thing for certain, though, is that India and China will desire to play a bigger role in the Asia Pacific.

In any form of coalition or cooperation, navies "*need to practice and to acquire the technology, operational and logistic procedures and doctrinal approaches that are needed for effective interoperability.*" For the GMP to work, regional maritime services have to address gaps in maritime information sharing and interoperability at sea.

First, there is a lack of concerted information sharing at the regional level in Maritime Domain Awareness (MDA). to catalyze the collaboration, the Republic of Singapore navy in 2009 established a Military Information Fusion Center (IFC) which works closely with the maritime industry's RECAAP Information Sharing enter (RECAAP ISC). The IFC functions as a regional information-sharing hub, and currently links to the operational centers of 15 different countries through an internet-based multilateral information sharing system called REMIX. The Indian navy also has a Maritime Shipping Information System (MSIS) for the Indian Ocean region.

The Maritime Safety and Security Information System (MSSIS) of the United States Department of Transportation (DoT) is already linked to Singapore's IFC system, and it is in the process of being linked to India's MSIS. The Royal Malaysian navy is exploring the idea of a Maritime Information Region (MIR) in the Sulawesi and Sulu region between Malaysia, Indonesia and the Philippines. despite these efforts, regional countries are slow to join in the MDA movement. Stronger participation by regional governments and industries is necessary to generate greater comprehensive Maritime Domain Awareness.

Second, an interoperability gap exists amongst various Asia Pacific navies, which must be narrowed in order to realize the GMP. The current state of interoperability is largely based on the derived interoperability facilitated by the us geographic combatant commander's Theatre Security Cooperation Plans (TSCPs) and other standardization and interoperability efforts.

Some regional navies have enhanced their interoperability through bilateral exercises and the conduct of regional maritime cooperation, such as the Malacca Strait Patrols. However, a gap in interoperability clearly exists between those navies that have significant interactions with the US Navy (such as Australia, India, Indonesia, Japan, Malaysia, Philippines, South Korea, Singapore, and Thailand) and those with little or no US military exchanges (such as Cambodia, China, Myanmar, Pakistan, and Vietnam). Hence, the interoperability gap and limited willingness to share information will impede the effectiveness of the GMP.

While the foundation for GMP may exist, implementing the GMP will require overcoming key challenges such as power dynamics, regional geo-political sensitivities, and interoperability issues to counterbalance these developments. The two strategies for enhancing the GMP within the Asia Pacific region include:

1. engaging china and India in the GMP, and
2. leveraging on the Western Pacific Naval Symposium (WPNS) and the ASEAN Regional Forum (ARF) frameworks to realize the GMP.

In order to establish the Global Maritime Partnership (GMP) the USA also rely on the International Sea Power Symposia (ISS).

Global Positioning System (GPS)

The Global Positioning System (GPS)[217], originally named Navstar GPS, is a satellite-based radio-navigation system owned by the United States government and operated by the United States Air Force. It is a global navigation satellite system that provides geolocation and time information to a GPS receiver anywhere on or near the Earth where there is an unobstructed line of sight to four or more GPS satellites. Obstacles such as mountains and buildings block the relatively weak GPS signals.

At least four different people have been acknowledged to be clearly associated with the invention of this revolutionary technology which was ultimately developed by the US Department of Defense to assist the military forces. Roger L. Easton was former head of Naval Research Laboratory's space applications branch was the brain behind several engineering applications and technologies that enabled the development of the GPS. Ivan Getting is credited for pressing forward *"the concept of using an advanced system of satellites to allow the calculation of exquisitely precise positioning data for rapidly moving vehicles, ranging from cars to missiles"* and for the *"concept and development of the GPS"*. Bradford Parkinson was at the forefront of the NAVSTAR GPS Joint Program Office from 1972 to 1978 and *"as the program's first manager, he has been the chief architect of GPS throughout the system's conception, engineering development, and implementation"*. This has also earned Parkinson the title the 'Father of GPS'. Dr. Gladys West contribution was in programming an IBM 7030 'Stretch' computer to deliver *"increasingly refined calculations for an extremely accurate geodetic Earth model, a geoid, optimized for what ultimately became the Global Positioning System (GPS) orbit"*.

There are however many more like i.e. James Julius Spilker Jr. and his team analyzed the worst-case cross-correlation between codes for the L band GPS signals with their +/- 5 KHz Doppler offsets and clearly show that the codes with Doppler offset indeed were the worst case, and he recommended the 1023-period codes even though with no Doppler they were no better than the 511-period codes. Those 1023-period codes are the C/A codes now supporting more than 2 billion users.

The GPS does not require the user to transmit any data, and it operates independently of any telephonic or internet reception, though these technologies can enhance the usefulness of the GPS positioning information. The GPS provides critical positioning capabilities to military, civil, and commercial users around the world. The United States government created the system, maintains it, and makes it freely accessible to anyone with a GPS receiver.

The GPS project was launched by the U.S. Department of Defense in 1973 for use by the United States military and became fully operational in 1995. It was allowed for civilian use in the 1980s. Advances in technology and new demands on the existing system have now

---

[217] https://en.wikipedia.org/wiki/Global_Positioning_System

led to efforts to modernize the GPS and implement the next generation of GPS Block IIIA satellites and Next Generation Operational Control System (OCX). Announcements from Vice President Al Gore and the White House in 1998 initiated these changes. In 2000, the U.S. Congress authorized the modernization effort, GPS III. During the 1990s, GPS quality was degraded by the United States government in a program called "*Selective Availability*"; this was discontinued in May 2000 by a law signed by President Bill Clinton.

The GPS system is provided by the United States government, which can selectively deny access to the system, as happened to the Indian military in 1999 during the Kargil War, or degrade the service at any time. As a result, several countries have developed or are in the process of setting up other global or regional satellite navigation systems. The Russian Global Navigation Satellite System (GLONASS) was developed contemporaneously with GPS, but suffered from incomplete coverage of the globe until the mid-2000s. GLONASS can be added to GPS devices, making more satellites available and enabling positions to be fixed more quickly and accurately, to within two meters (6.6 ft). China's BeiDou Navigation Satellite System is due to achieve global reach in 2020. There are also the European Union Galileo positioning system, and India's NAVIC. Japan's Quasi-Zenith Satellite System, scheduled to commence in November 2018, will be a GPS satellite-based augmentation system to enhance GPS's accuracy.

When selective availability was lifted in 2000, GPS had about a five-meter (16 ft) accuracy. The latest stage of accuracy enhancement uses the L5 band and is now fully deployed. GPS receivers released in 2018 that use the L5 band can have much higher accuracy, pinpointing to within 30 centimeters or 11.8 inches. Today many civilian applications use one or more of GPS's three basic components: absolute location, relative movement, and time transfer.

However, The U.S. government controls the export of some civilian receivers. All GPS receivers capable of functioning above 18 km (59,000 ft) above sea level and 515 m/s (1,000 kn; 2,000 km/h; 1,000 mph), or designed or modified for use with unmanned air vehicles like, i.e., ballistic or cruise missile systems, are classified as munitions/weapons - which means they require State Department export licenses. This rule applies even to otherwise purely civilian units that only receive the L1 frequency and the C/A (Coarse/Acquisition) code.

Disabling operation above these limits exempts the receiver from classification as a munition. Vendor interpretations differ. The rule refers to operation at both the target altitude and speed, but some receivers stop operating even when stationary. This has caused problems with some amateur radio balloon launches that regularly reach 30 km (100,000 feet).

These limits only apply to units or components exported from the United States. A growing trade in various components exists, including GPS units from other countries. These are expressly sold as ITAR-free. The navigational signals transmitted by GPS satellites encode a variety of information including satellite positions, the state of the internal clocks, and the health of the network. These signals are transmitted on two separate carrier frequencies that are common to all satellites in the network. Two different encodings are used: a public encoding that enables lower resolution navigation, and an encrypted encoding used by the U.S. military.

GPS type navigation was first used in war in the 1991 Persian Gulf War, before GPS was fully developed in 1995, to assist Coalition Forces to navigate and perform maneuvers in the war. The war also demonstrated the vulnerability of GPS to being jammed, when Iraqi forces installed jamming devices on likely targets that emitted radio noise, disrupting reception of the weak GPS signal.

Billions of people and a growing number of autonomous vehicles rely today on mobile navigation services from Google, Uber, and others to provide real-time driving directions based on GPS. However, new proof-of-concepts demonstrated how hackers could inconspicuously steer a targeted object to the wrong destination or, worse, endanger passengers by sending them down the wrong way of a one-way road or shipping lane.

The attack starts with an inexpensive piece of hardware that's planted in or on the targeted object to spoof the radio signals used by civilian GPS services. It then uses algorithms to plot a fake "*ghost route*" that mimics the turn-by-turn navigation directions contained in the original route. Depending on the hackers' ultimate motivations, the attack can be used to divert an emergency or rescue team or a specific passenger to an unintended location or to follow an unsafe route. The attack works best in areas the navigator doesn't know well, and it assumes hackers have a general idea of the object's intended destination.

Examination of global ship tracking data for the past years has shown several instances of multiple vessels reporting their locations as being on land at airports far from where the ships were operating off shore. About 20 vessels in the Black Sea where affected in June 2017. Most interestingly, all three analyzed locations involve airports, the Gelendzhik Airport and the Sochi International Airport near the Black Sea, and the St. Petersburg Airport near the North Sea. Some of the vessels that mistakenly appeared in Sochi Airport were really located near Gelendzhik, about 200 kilometers (124 miles) away, two vessels appeared at Sochi Airport, 20 kilometers (12 miles) from their actual positions near the Sochi harbor. These incidents dovetail with reports that people in downtown Moscow often find their GPS receivers placing them at Moscow's Domodedovo airport, about 25 miles away.

International Seapower Symposium (ISS)

The International Seapower Symposia (ISS)[218] are a series of biennial meetings of the world's chiefs of navy that has met at the United States Naval War College since 1969. The proceedings of these, symposia since ISS XVI in 2003, are published on the Internet.

With the objective of promoting mutual understanding among the several leaders of the world's maritime nations, the First International Sea Power Symposium (ISS I) was convened at the Naval War College in Newport, Rhode Island on 17.–20. November 1969. This four-day symposium was conceived by the President of the Naval War College, Vice Admiral Richard G. Colbert, U.S. Navy, who served as host for the first ISS. As a result of the success of this symposium, plans were made to continue these discussions as a biennial event. All International Seapower Symposia, since, have been conducted at the Naval War College, and hosted by the incumbent Chief of Naval Operations with only two exceptions.

---

[218] https://en.wikipedia.org/wiki/International_Seapower_Symposium

Joint Interagency Task Force (JIATF)

A Joint Task Force is a "*joint*" (multi-service) ad hoc military formation. The task force concept originated with the United States Navy in the 1920s and 1930s. "Combined" is the British-American military term for multi-national formations.

- CTF - Commander Task Force, sometimes Combined Task Force
- CCTF - Commander Combined Task Force
- CJTF - Combined Joint Task Force

There are between 25 – 30 Joint task forces of the United States Armed Forces, while the Joint Interagency Task Forces aim towards a better cooperation with in the US MDA concept.

The Joint Interagency Task Force (JIATF) South[219] mission within the U.S. Southern Command AOR is to plan, conduct, and direct interagency detection, monitoring, and sorting operations of air and maritime drug smuggling activities; plan and conduct flexible operations to detect, monitor, disrupt and deter the cultivation, production and transportation of illicit narcotics; utilize and integrate C4I systems to efficiently coordinate operations and intelligence information with other counterdrug centers, law enforcement agencies, and domestic and international counterdrug partners; and collect, fuse, and disseminate counterdrug information from all participating agencies to the detection and monitoring forces for tactical action.

Joint Interagency Task Force (JIATF) East was created as a result of Presidential Decision Directive 14 which ordered a review of the nation's command and control and intelligence centers involved in international counter-narcotics operations. On 7. April 1994, Dr. Lee Brown, Director of the Office of National Drug Control Policy, signed the National Interdiction Command and Control Plan which directed establishment of three national interagency task forces (JIATF East in Key West, Florida; JIATF South in Panama; and JIATF West in Alameda, California) and the Domestic Air Interdiction Coordination Center at March Air Force Base in Riverside, California.

JIATF East was formed by integrating additional law enforcement personnel into the former Joint Task Force Four (JTF-4) organization which was officially formed 22. February 1989 under the direction of the U. S. Commander in Chief Atlantic. On 1. June 1997, the Commander in Chief U. S. Southern Command expanded his area of responsibility to include the Caribbean and the waters bordering South America, and assumed command and control of JIATF East. In compliance with the 1979 Panama Canal Treaty and the necessity to complete the military drawdown in Panama by the end of 1999, the decision was made

---

[219] http://www.globalsecurity.org/military/agency/dod/jitf.htm

to merge JIATF South and JIATF East into one organization. Transfer of the JIATF South mission to the merged JIATF was completed 1. May 1999.

The interagency concept of the task force is illustrated by the leadership composed primarily of representatives from the Department of Defense, Department of Transportation (U. S. Coast Guard) and the Department of the Treasury (U. S. Customs Service). Other assigned agencies include Drug Enforcement Administration; Federal Bureau of Investigation; Defense Intelligence Agency; Naval Criminal Investigative Service; and the National Security Agency. Great Britain, France and the Netherlands provide ships, aircraft, and liaison officers to the task force; and the Flag Officer of the Netherlands Forces Caribbean commands one task group in the task force. Since the merger in 1999, the countries of Argentina, Brazil, Colombia, Ecuador, Peru and Venezuela also have assigned liaison officers to JIATF East. The result is a fully integrated, international task force organized to capitalize on the force multiplier effect of the various agencies and countries involved.

The focus of the command is a Joint Operations Command Center where intelligence and operations functions are fused in a state-of-the-art command, control, communications, computers, and intelligence facility. The task force coordinates the employment of USN and USCG ships and aircraft, USAF and USCS aircraft, and aircraft and ships from allied nations and law enforcement agencies -- a complete integration of sophisticated multi-agency forces committed to the cause of interdicting the flow of illicit drugs. JIATF East is located at Naval Air Station Key West, Truman Annex, Key West, Florida, in facilities that originally housed the Fleet Sonar School and former JTF-4.

Joint Interagency Task Force West (JIATF-W or JIATF West)[220] is a standing United States military joint task force with the mission of combating drug-related transnational organized crime in the Indo-Asia-Pacific. JIATF West's area of responsibility (AOR) is that of United States Indo-Pacific Command (USINDOPACOM). JIATF West is one of two Joint Interagency Task Forces with a counter-narcotics mission. The other is Joint Interagency Task Force South. The task force is run as USPACOM's *executive agent* for counterdrug activities providing support to partner nation law enforcement. Approximately 166 active duty and reserve U.S. military forces; Department of Defense civilian employees; contractors; and U.S. and foreign law enforcement agency personnel are members of the task force.

The Bureau for International Narcotics and Law Enforcement Affairs of the United States Department of State describes the task force's mission as to *"in cooperation with U.S. interagency and foreign partners, conduct activities to detect, disrupt, and dismantle drug-related transnational threats in Asia and the Pacific in order to protect U.S. security interests at home and abroad."*

JIATF West lists its *"task force partners"* as including the U.S. Army, Navy, Marines, Air Force, and Coast Guard; the Drug Enforcement Administration (DEA), Defense Intelligence Agency (DIA), Federal Bureau of Investigation (FBI), National Geospatial Intelligence

---

[220] https://en.wikipedia.org/wiki/Joint_Interagency_Task_Force_West

Agency (NGA), Naval Criminal Investigative Service (NCIS), United States Customs and Border Protection (CBP), United States Immigration and Customs Enforcement (ICE); and the Australian Customs Service, Australian Federal Police, and New Zealand Police. The current Director of the Task Force as of 2017 is Rear Admiral Donna L. Cottrell, USCG.

<u>Joint Interagency Task Force South (JIATF)</u>[221] is a United States multi-service commanded by a Coast Guard Flag Officer, multiagency task force based at Naval Air Station Key West (Truman Annex), Key West, Florida, and is subordinate command to United States Southern Command.

JIATF plans and conducts flexible operations to detect, monitor, disrupt and deter the cultivation, production and transportation of illicit narcotics, utilize and integrate C4I systems to efficiently coordinate operations and intelligence information with other counterdrug centers, law enforcement agencies, and domestic and international counterdrug partners; and collect, fuse, and disseminate counterdrug information from all participating agencies to the detection and monitoring forces for tactical action.

It conducts counter illicit trafficking operations, intelligence fusion and multi-sensor correlation to detect, monitor, and handoff suspected illicit trafficking targets; promotes security cooperation and coordinates country team and partner nation initiatives in order to defeat the flow of illicit traffic.

In response to a need for unified command and control of drug interdiction activities, the FY 1989 National Defense Authorization Act designated the Department of Defense as the lead agency for the detection and monitoring program targeted against the aerial and maritime traffic attempting to bring drugs into the United States. Commander Joint Task Force FOUR (CJTF-4) in Key West, Commander, Joint Task Force FIVE in Alameda, California and Commander, Joint Task Force 6 in El Paso, Texas were established to direct the anti-drug surveillance efforts in the Atlantic/Caribbean, Pacific, and Mexico border areas respectively. The Joint Task Forces have been operating since October 1989.

The Joint Task Force 4 operations center received radar data from the AN/FPS-118 Over-the-horizon radar located at Moscow Air Force Station, Maine, until the system was turned off and placed in *"warm storage"* after the end of the Cold War.

On 1. June 1997, the Commander in Chief of the U.S. Southern Command expanded his area of responsibility to include the Caribbean and the waters bordering South America, and assumed command and control of JIATF East. In compliance with the 1979 Panama Canal Treaty and the necessity to complete the military drawdown in Panama by the end of 1999, the decision was made to merge JIATF South and JIATF East into one organization. Transfer of the JIATF South mission to the merged JIATF was completed on 1. May 1999.

Due to the previous history of the command, Task Groups 4.1, 4.2, 4.3, and 4.4, and others, are in use controlling U.S. and allied assets assigned to JIATF South. In February 2007, a

---

[221] https://en.wikipedia.org/wiki/Joint_Interagency_Task_Force_South

Dutch magazine described the relationships as follows: under the command of the Director JIATF South, the U.S. Tactical Commander held the position of Commander Task Group 4.1, United States Air Force forces CTG 4.2, US Navy forces CTG 4.3, the Director of the Dutch Caribbean Coast Guard (DCCG), who is always the commander of the Dutch Navy in the Caribbean area (CZMCARIB), Commander Task Group 4.4 (CTG 4.4), and US Customs force CTG 4.5. Since 2008 an additional Task Group known as CTG 4.6 has been commanded by the French Navy Commander (Antilles).

Linked Operations-Intelligence Centers Europe (LOCE)

The Linked Operations-Intelligence Centers Europe (LOCE, initially Limited Operational Capability Europe), a United States European Command (USEUCOM) system that provides U.S. forces, NATO forces, and other national allied military organizations with near-real-time, correlated situation and order of battle (OB) information. It supports threat analysis, target recommendations, indications and warning, and cueing of collection assets. The LOCE system brings a fused, all-source intelligence focus to current crisis situations with the capacity to support future operations and exercises. The system is structured to present a complete picture of the threat environment with the ability to highlight specific critical nodes. LOCE's intelligence data exchange capabilities and secure voice communications, LOCE facilitates multi-national operations and integrated allied participation in the intelligence and operations planning cycles. LOCE assists in maintaining the intelligence infrastructure by distributed database management, rationalized intelligence databases supporting multi-source OB, artificial intelligence applications in correlation, and enhanced targeting applications.

LOCE is a US-developed system utilized at several hundred NATO locations, but with limited bandwidth capability in the late 1990s. There had been a significant expansion in the number of LOCE sites over the 1990s, with most of these terminals deployed in support of theatre operations, including Operation Joint Guard in the former Yugoslavia. LOCE started as a US system and was adopted by US allies because it was further along in development and offered better capability than what was available to them at the time. The system consists of web-enabled PCs (clients), a centralized set of servers, and dedicated communications circuits, providing multimedia E-mail, bulletin board, TACELINT, secondary imagery, order of battle databases, network services, and a secure voice capability.

The LOCE system supports combined intelligence operations by connecting users at all echelons, from the national ministry of defense to the tactical level. LOCE is a SECRET REL NATO intelligence system that serves as USEUCOM's intelligence system for coalition warfare. It is also the declared US gateway to NATO's Battlefield Information Collection and Exploitation System (BICES). LOCE provides users with connectivity among NATO and US operational units and decision-makers in the form of gateways such as BICES, SHAPE's CRONOS network and SACLANT's Maritime Command and Control Information System (MCCIS). This gives each user access to near-real-time (NRT), all-source, correlated air, ground and naval intelligence analysis and products. It supports I&W, current intelligence, collection management, and most aspects of the targeting cycle including nominations, air tasking orders, and battle damage assessments. It also provides the TBM data architecture supporting shared early warning among NATO and theater components. LOCE is the designated US injection system for the sharing of theater missile defense data under the Shared Early Warning Program.

USEUCOM has the responsibility for maintaining LOCE. Originally designed as a NATO-only information dissemination system, LOCE has greatly aided the dissemination of US intelligence to coalition partners in the Balkans. The LOCE Correlation Center, located at the Joint Analysis Center, RAF Molesworth, UK, functions as the US gateway for exchange of operations intelligence with NATO. LOCE remains a top priority for continued support and funding for needed enhancements as NATO expands and combined operations increase.

LOCE is the backbone intelligence system for NATO, and remains a top priority for continued support and funding for needed enhancements as NATO expands and coalition missions increase in intensity and frequency. While NATO coalition partners now are fully using LOCE and sharing information among themselves, there are many cases where US forces are not taking full advantage of the information in LOCE. Electronic transfer of LOCE information to US systems, e.g., Warlord, is not currently available. And LOCE utility is still limited in many locations because of severely constricted bandwidth and/or air gaps. LOCE is definitely not a user-friendly system, as it has difficult operating instructions.

Initially, information entered into LOCE was "*fat-fingered in*", meaning that an operator physically typed the relevant information into the LOCE system which created an opportunity for error in translations. The later process was to copy the US information to disk or some other media and then transfer it to LOCE, which still requires manual intervention but with less errors and risk.

The 1996 Defense Science Board Task Force concluded that use of an electronic gateway with the appropriate "*guard*" technology would not significantly increase the risk over today's methodology and may even improve it. The cycle time for transfer would be faster because less effort is required on the part of the operator(s), who would simply "*send*" it to LOCE rather than first copying it. The net benefit to the warfighter would come from being able to better operate inside the enemy's information cycle.

The Task Force concluded that the philosophy that "*We must maintain an air gap*" is not the right solution to security concerns in this environment. DIA has several less than perfect, multilevel security solutions available now, and the Task Force recommended that DIA pick one, install it at the Joint Analysis Center (JAC) as well as other locations operating to the same constraints and not wait for the 100% solution, given that the benefits of doing so appear to outweigh the risk in this circumstance.

While significant progress had been made in strengthening the LOCE system by extending the a range of information that can be carried on it and encouraging allies and coalition partners to make their own contributions of information, by the turn of the century US forces were not exploiting LOCE as they could. Limitations with the LOCE system that contributed to its underutilization by US forces included: LOCE bandwidth is far too low at major nodes (only 19.2 kbps and often is less than that depending on how a site is configured) and does not allow for effective information push to the brigade level; US forces cannot easily move between LOCE and US databases, the ACE is reacting rather than pushing information, and there is no electronic connectivity between the Army's Warlord system and LOCE.

Integrating LOCE with the Joint Broadcast System (JBS) delivery system would allow LOCE users much faster access to larger product files on a routine basis and free up some of its very limited bandwidth for other important uses. At the same time, the LOCE concept could be migrated into the DISN architecture to provide a seamless flow of information into and out of LOCE and US systems consistent with the security guidelines and the previous discussion relative to electronic interfaces. In addition, LOCE utility would be increased if it were made compatible with 5D and Netscape, thereby allowing the use of standard web browsers for access to information and accelerating LOCE's compliance with the standard architecture for Intelink.

Efforts were funded under the Joint Military Intelligence Program to use the best functionality from LOCE and develop the system into the Joint Deployable Intelligence Support Systems (JDISS) common intelligence baseline, thereby eliminating different systems with near duplicate functionality and centering on JDISS as the DoD common intelligence workstation baseline. RDT&E funding was used to work on the development of LOCE functionality onto JDISS, develop LOCE tools as a model of intelligence services for a JDISS coalition system, develop all functionality to the Defense Information Infrastructure (DII), development of JDISS segments in the Global Command and Control System and the serviced systems Command, Control, Communications, Computers and Intelligence (C4I) systems, and adopt new technology as it becomes available into the JDISS intelligence environment.

The JDISS/LOCE development ensured interoperability with the Navy's Joint Maritime Command Information System, Army's All Source Analysis System Warlord system, Air Force's Combat Information System, and USMC's Interactive Analysis System, while all systems continue to evolve to a common DII.

Upgrades were tested to allow sharing of intelligence data between NATO and US systems. New software developed and successfully tested at USAFE to interface the Combat Intelligence System (CIS) and the LOCE system was loaded and successfully tested at the 1996 Fort Franklin Battle Lab demonstration. Previously to this fix only 83% of the data being sent from LOCE to CIS was successfully passed, and it took about five hours to complete. This upgrade produced a 100% successful download and is took only 15 minutes to load the LOCE database into the CIS workstation and updated the CIS database with this information.

During Operation ALLIED FORCE LOCE's primary benefit was that it provided access to the NATO Air Tasking Order. LOCE was deployed to Rhein Main Air Base, Germany; JTF-SH at Einsiedlerhof Air Station, Germany; and Mont de Marsan, France. Once a long haul circuit connection was made to the appropriate LOCE Remote Communications Server, the user required several pieces of equipment. There was some confusion over the responsibility for the implementation of LOCE requirements, and therefore many responsibilities were never addressed during the air campaign. One source of confusion was the transfer of office of primary responsibility early on in the operation.

As of early 2000, the LOCE network had expanded to approximately 400 remote sites, and was expected to grow to 500 by the end of 2000. Various communications technology upgrades underway were FOC by the end of FY2000.

By 2005 LOCE was a technological leader in meeting the goals of the US government in the areas of coalition warfare and streamlining and increasing the use of intelligence and responsiveness to customer's needs. Its ability to integrate and correlate data and rapidly disseminate the data based on consumer demands, as well as its flexibility in serving a wide variety of coalition partners, allows LOCE to stand out among intelligence support systems.

LOCE uses integrated commercial off-the-shelf (COTS) and government off-the-shelf (GOTS) software packages. By 2005 LOCE supported a user population of 1,200 with the ability to handle 300 concurrent users. LOCE is accessed by U.S., NATO, and allied forces through more than 300 Sun Sparc and NT workstations geographically located throughout the U.S., Canada, and Europe. LOCE is installed at approximately 130 sites, including ships and bases.

Basic LOCE functions include database and repository services, automatic sensor report correlation, electronic mail, imagery dissemination, graphical situation displays, and secure voice communications. Databases are maintained at the Correlation Center (CORCEN) located at the USEUCOM Joint Analysis Center (JAC), RAF Molesworth, UK. The intelligence reports database (IRDB) presents the results of near-real-time sensor reports and intelligence reports received from units *"on the ground"*.

The Combined Orders of Battle (COB) hold the results of OB analysis by the designated OB Manager. The imagery database consists of a storage and retrieval system to provide U.S. and allied imagery to consumers. This imagery forms the basis for operational planning, battle damage assessment (BDA), and confirmation of OB analysis results. The LOCE Bulletin Board, and HTML servers at DIA, the JAC, and the U.S. Army Intelligence Readiness Facility (USAIRF) provide a database for unstructured text reporting for items of general interest. Together, these databases form the basis for a common and consistent view of the area of concern. Using a LOCE workstation (LWS), an analyst can query against one or more system databases.

The LOCE system architecture consists of an automated data processing (ADP) component, a communications component, and interfaces with external NATO domains. Primary ADP components include a fixed CORCEN, deployable workstation capability, browser-accessed web servers, and the LWS. Communications are implemented via integrated voice and data over dedicated and dial-up circuits. These circuits are generally carried over leased land lines, military communications, and indigenous satellite communications. External system interfaces include near-real-time sensor reporting, such as track updating, other intelligence products, like the U.S. NATO releasable integrated database (IDB), and input from foreign intelligence dissemination systems and sources.

Anteon has played a major role in helping LOCE become one of the most widely used C4I information systems within the intelligence community. Anteon has been directly involved in the expansion of the LOCE program, which has grown from 2000 to 2005 from 60 to more than 300 LOCE workstations located throughout Europe, Canada, and the US. Anteon is the operations, training, and support contractor for LOCE. As such, we provide systems engineering and technical assistance, software operations and maintenance, configuration management, communications engineering support, inventory management, and user training.

Anteon engineers and maintains the complex LOCE communications network by providing full-service site surveys, configuration, and installation support for all systems on the LOCE network. Anteon has employees working to support the LOCE system in Naples and Vicenza, Italy, RAF Molesworth, Ramstein, AFB and Stuttgart, Germany, and Alexandria and Norfolk, VA, USA. Anteon maintains formal LOCE training centers in Vicenza, Ramstein, and Molesworth. In addition, Anteon deploys mobile trainers throughout Europe. Dedicated employees ensure rapid response and quality support are provided to all LOCE users located throughout Europe and North America.

Maritime Domain Awareness Executive Steering Committee (MDA ESC)

The National MDA Stakeholders Board Executive Steering Committee of the United States of America is comprised of the two offices charged with coordinating interagency MDA efforts, the Office of Global Maritime Situational Awareness (OGMSA) and GMAII, who co-chair the ESC, and the MDA Executive Agents from the U.S. agencies with the most active roles in developing maritime domain awareness in support of the National Strategy for Maritime Security.

Currently, the departments of Transportation, Homeland Security, and Defense have named executive agents who sit on the ESC. The ESC meets more frequently than the Security Homeland Board (SHB) and streamlines interactions between key MDA players. The ESC has been authorized by the SHB and Maritime Security Interagency Policy Committee (MSIPC) of the U.S. National Security Council (NSC) to decide on courses of action for all recommendations put forth by the full Board; establish and review progress of sub-committees as directed by the co-chairs; provide the MSIPC with final recommendations to update strategic-level guidance and revise policy as approved.

Maritime Information Sharing Environment (MISE)

The <u>Maritime Information Sharing Environment (MISE)</u>[222] if the approach of the government of the United States of America aiming at a global common understanding through common standards, low cost, and implementable solutions for maritime information sharing while providing mechanisms to mitigate associated legal and policy concerns. The Maritime National Information Sharing Environment (MISE) is therefore a similar approach as the Common Information Sharing Environment (CISE) in the European Union.

The MISE method of achieving a common understanding starts with first describing an operational use-case and second identifying the essential data elements supporting the use-case. The cargo operations model satisfies these two MISE needs for the particular use-case. The information dictionary produced from our model consists of the essential data elements and the work to which they are critical.

As a participant in the MISE, data providers and consumers manage and share maritime information through common data definitions and security attributes, resulting in an internet accessible, unclassified information sharing capability. Current versions of related documents as well as an online implementation guide have been posted to facilitate new MISE partners.

A current US federal initiative to standardize the format of information exchange is the development of the National Information Exchange Model (NIEM), which serves as a data model and reference vocabulary. NIEM provides the common language as a standard way of defining the contents of messages being exchanged. The MISE Implementation provides practitioners with developer guidance and specific examples for interfacing with MISE. Specifically, it shows how to create messages that conform to the National Information Exchange Model (NIEM) Maritime IEPD formats, how to implement security to successfully access the environment, and how to interface with the services to publish and consume messages from the environment.

The MISE Implementation Guide is a walkthrough with code snippets demonstrating how to publish and consume data from the MISE. The MISE implementation guide provides practitioners with all of the technical details, guidance and specific examples for interfacing with the Maritime Information Sharing Environment (MISE) defined by the National MDA Architecture. The MISE services provide publish, search, and retrieve for all information products defined in the National MDA Architecture. The sections cover the separate steps necessary to interface with the MISE, from data modeling to security to the HTTP REST services.

The MISE Toolkits are libraries and sample code. The Java client toolkit provides the necessary libraries to connect to the MISE. All of the code examples shown in the implementation

---

[222] https://www.niem.gov

guide are included in the toolkit. It can be provided in the compiled (JAR) toolkit, the source code, or both.

The NIEM-Maritime Information Exchange Packages (IEPDs) list a set of maritime exchange specifications defined using the National Information Exchange Model (NIEM) Maritime Domain. The foundation for maritime information sharing via the MISE is therefore a set of NIEM-Maritime Information Exchange Packages (IEPDs) that defined by the National MDA Architecture. These are fully NIEM-conformant exchange packages. Each IEPD has the IEPD package and associated exchange model document containing the logical models and detailed descriptions of each IEPD.

- Position (POS) - Exchange Model Description, a geospatial position, course, heading, speed, and status of a vessel at a given time. A series of position reports can be combined to produce track information.
- Indicators and Notifications (IAN) - Exchange Model Description. Indicators are information used to inform or contribute to an analytical process. Notifications include warnings of a possible event and alerts about the execution of an event.
- Notice of Arrival (NOA) - Exchange Model Description. A 96-hour advance notice that all vessels inbound to US ports are required to submit, which lists vessel, crew, passenger, and cargo information.
- Vessel Information (VINFO) - Exchange Model Description. Static vessel characteristics information, such as vessel tombstone data.
- Consolidated Vessel Information and Security Reporting (CVISR) - Exchange Model Description. Defined and standardized levels characterizing how much is known about a vessel (and associated people, cargo, and infrastructure) at a given time. CVISR Is generally an assembled product consisting of essential elements from the previous four IEPDs.

The MISE Specifications list HTTP-level request & response documentation for all functionality in the MISE services. As defined in the National MDA Architecture Plan, the technical specifications below define the interface for security and information exchange with the MISE. These specifications provide specific response codes, parameters, and other specific details. These specifications should be considered the authoritative source for questions related to the MISE.

- MISE Attribute Specification
- MISE Interface Security Specification
- MISE Publish Specification
- MISE Search/Retrieve Specification

The MISE Security provides details about the security implementation, trust fabric, authentication over SSL, and SAML for secure interface with the MISE Security Services. All interactions to publish and consume data within the MISE are secured interactions over SSL between trusted systems with X.509 Certificates and the registration of Trusted System in the MISE Trust Fabric. The following documents are provided for an understanding of the security implementation:

- National MDA Architecture Plan for an overview of the MISE security approach.
- MISE Interface Security Specification for the details of how trusted systems securely connect to the ISI.
- MISE Attribute Specification for an explanation of the common attributes used for entitlement management.

Maritime Liaison Office (MARLO) Bahrain

For 25 years, the United States Maritime Liaison Office (MARLO) Bahrain has facilitated the exchange of information between the US Navy, the 25-Member Nation Combined Maritime Forces, and maritime interests in US Central Command's area of responsibility. Our liaisons travel regularly throughout the Gulf & East Africa to foster communication, stay apprised of changes in local port infrastructure and management, proactively address concerns and questions. They remain, first and foremost, an active advocate for commercial mariners, and dedicated to providing accurate and timely information on safety of shipping and marine navigation.

A quarter century ago, MARLO was created (during the Iran-Iraq *Tanker Wars*) to promote cooperation between the U.S. Navy and maritime interests. Since then, MARLO has diversified its offerings and broadened its influence worldwide. These days, MARLO sister units (known as Maritime Liaison Units, or simply MARLUs) operate in Naples, Miami, Japan, Singapore and points between. Yet despite the many changes, our mission remains steadfast—to serve as a conduit of information to and from the maritime industry.

Maritime Tactical Command and Control (MTC2)

The Tactical Command System upgrades the US Navy's Command, Control, Computer and Intelligence (C3I) systems and processes C3I information for all warfare mission areas including planning, direction and reconstruction of missions for peacetime, wartime and times of crises.

Maritime Tactical Command and Control (MTC2) will provide software to perform maritime tactical Command and Control amongst naval platforms and between platforms and their superior and subordinate Commanders and interchange C2 information seamlessly with allied navies of the United States, Britain, Canada, Australia and New Zealand. MTC2 will provide necessary hardware upgrades i.e. to the Canadian Naval Information Systems (NAVIS) to support the new software demands.

MTC2 provides Navies with the ability to deliver maritime domain-unique tactical Command and Control (C2) capabilities from Maritime Operations Centers (MOC) down to the lowest tactical unit of operations and align to the Navy Tactical Cloud (NTC). MTC2 supports alignment and provides interoperability of US Navy C2 with the Department of Defense (DoD) joint C2 (jC2) way-forward. The program also aligns to the jC2 data and service exposure and consumption goals, architectures, and Net-Centric Enterprise Service efforts.

These resources support the evolutionary acquisition, materiel solution analysis, technology development, engineering and software development of these capabilities. Global Force Management - Data Initiative (GFM-DI) is the Department-wide enterprise solution that enables visibility/accessibility/sharing of data applicable to the entire DoD force structure. GFM-DI is the enterprise solution for force structure representation and MTC2 will be the data source for the Navy's force structure representation. The Maritime Tactical Command and Control (MTC2) will replace the Global Command and Control System - Maritime (GCCS-M).

National Information Exchange Model (NIEM)

The National Information Exchange Model (NIEM)[223] is an XML-based information exchange framework from the United States. NIEM represents a collaborative partnership of agencies and organizations across all levels of government (federal, state, tribal, and local) and with private industry. The purpose of this partnership is to effectively and efficiently share critical information at key decision points throughout the whole of the justice, public safety, emergency and disaster management, intelligence, and homeland security enterprise.

NIEM is designed to develop, disseminate, and support enterprise-wide information exchange standards and processes that will enable jurisdictions to automate information sharing. NIEM is an outgrowth of the United States Department of Justice's Global Justice XML Data Model (GJXDM) project. NIEM is now being expanded to include other federal and state agencies such as the Office of the Director of National Intelligence, Federal Bureau of Investigation, Texas, Florida, New York, Pennsylvania, and others, like the National Information Exchange Model-Maritime (NIEM-M).

National Information Exchange Model-Maritime (NIEM-M)

The National Information Exchange Model-Maritime (NIEM-M) is the maritime part of the National Information Exchange Model (NIEM) and became the de facto US standard for the national data exchange with the publication *Agency Information Sharing Functional Specification*", dated march 4, 2010, of the Office of Management and Budget:

"All agencies are required to evaluate the adoption and use of the National Information Exchange Model (NIEM) as the basis for developing reference information exchange package descriptions ...".

NIEM and NIEM-M are a network-centric or network-enabled approach, can therefore be compared to the Common Information Sharing (CISE) Initiative in the EU, in some way also to the Federated Mission Network (FMN) aim. The Global Justice XML Data Model (GJXDM) is therefore a comparable effort as the CISE Model. A harmonization and exchange between the efforts of US-American national, NATO and European Standardization would be most beneficial for the international exchange of information in Maritime Situational Awareness.

---

[223] https://en.wikipedia.org/wiki/National_Information_Exchange_Model

National Maritime Intelligence-Integration Office (NMIO)

The National Maritime Intelligence-Integration Office (NMIO)[224] is a United States Navy entity located in the National Maritime Intelligence Center (NMIC) Facility in Suitland in Prince George's County, Maryland, southeast of Washington, DC. It is a part of the Suitland Federal Center.

As National Intelligence Manager (NIM) for Maritime, the National Maritime Intelligence-Integration Office (NMIO)[225] serves as the principal advisor to the Director of National Intelligence (DNI) on maritime issues and is the unified maritime voice of the United States Intelligence Community. In December 2016, NMIO was formally designated the NIM for Maritime by the DNI. NMIO collaborates with the Global Maritime Community of Interest (GMCOI) consisting of federal, state, territorial, tribal, industry, and academia to provide whole of government solutions to maritime information sharing challenges.

NMIO is not an intelligence production center. NMIO works at the national and international level to facilitate the integration of maritime information and intelligence collection and analysis in support of national policy and decision makers, Maritime Domain Awareness objectives, and interagency operations, at all levels of the U.S. Government. It breaks down barriers to information sharing and creates the framework for the GMCOI to share data. NMIO enables maritime stakeholders to proactively identify, locate, track, and defeat threats to the United States and its global partners.

The mission statement of NMIO is to advance maritime intelligence integration, information sharing, and domain awareness to foster unity of effort for decision advantage that protects the United States, its allies, and partners against threats in or emanating from the global maritime domain.

The NMIO was established as a result of guidance from the 9/11 Commission, The Intelligence Reform and Terror Prevention Act (IRTPA), Presidential directives, and maritime security plans. Its goal is to support national policymakers and decision-makers on maritime issues and perform actions as directed by the Office of the Director of National Intelligence (ODNI) Strategic Guidance to create unity of effort and position leaders for decision advantage in a decisive and efficient manner. The office effort focuses on collaboration between Federal, state, local, tribal and territorial governments, acting in concert with international partners as well as representatives from the private sector and academia.

NMIO is co-located with the Office of Naval Intelligence (ONI), the Naval Information Warfare Activity (NIWA), and the Coast Guard Intelligence Coordination Center (ICC).

The National Maritime Intelligence Integration Office (NMIO) is first of all as the name tells a national organization of the United States of America and directly located under the US

---

[224] https://en.wikipedia.org/wiki/National_Maritime_Intelligence-Integration_Office
[225] http://nmio.ise.gov

Precedential Office.

To advance maritime intelligence integration, information sharing and situational awareness within the US, to build unity of effort and to enable better decision-making to protect the United States, its allies, and international partners against threats in or emanating from the global maritime domain.

NMIO promotes maritime domain intelligence and information sharing by creating a Global Maritime Community of Interest (GMCOI), to unify and synchronize efforts. The challenges facing information sharing are:

- Sharing across technology platforms and systems
- Sharing protected information across all parties with a need to know
- Access to databases
- Enabling sharing with Federal, state, local and international partners

The NMIO International Partner Roundtable provides a forum for briefings and discussion of maritime topics of common concern and aims to achieve greater collaboration on ongoing NMIO and interagency projects with participants from Australia, Canada, Finland, France, Italy, New Zealand, Romania and the United Kingdom.

National Maritime Law Enforcement Academy (NMLEA)

The National Maritime Law Enforcement Academy (NMLEA)[226] is a non-profit member-driven organization, that provides assessment, evaluation, training, research, education and consulting services to the professionals that patrol, protect and preserve our nation's rivers, bays, harbors, lakes, ports and coastlines formed in 2000 under the guidance of Admiral Owen Siler, the USCG's 15th Commandant.

The National Maritime Law Enforcement Academy (NMLEA) offices are in Washington, DC and Tallahassee, Florida. Through customized exportable maritime tactical, rescue, and operational courses based on international best practices; an online Port Training Program exclusive to America's maritime security, various port vulnerability and risk assessments, training and exercise development and evaluation, and multiple product, technology, process and security consultation services, the Academy staff has provided over 15,000 officers and maritime professionals with tools that help to enhance the safety and security of their Nation's waterways, and to maritime organizations around the world.

---

[226] https://www.nmlea.org

North American Aerospace Defense Command (NORAD)

The <u>North American Aerospace Defense Command (NORAD)</u>[227], known until March 1981 as the North American Air Defense Command, is a combined organization of the United States and Canada that provides aerospace warning, air sovereignty, and defense for Northern America.

Headquarters for NORAD and the NORAD/United States Northern Command (USNORTH-COM) center are located at Peterson Air Force Base in El Paso County, near Colorado Springs, Colorado. The nearby Cheyenne Mountain Complex has the Alternate Command Center. The NORAD commander and deputy commander (CINCNORAD) are, respectively, a United States four-star general or equivalent and a Canadian three-star general or equivalent.

CINCNORAD maintains the NORAD headquarters at Peterson Air Force Base near Colorado Springs, Colorado. The NORAD and USNORTHCOM Command Center at Peterson AFB serves as a central collection and coordination facility for a worldwide system of sensors designed to provide the commander and the leadership of Canada and the U.S. with an accurate picture of any aerospace or maritime threat. NORAD has administratively divided the North American landmass into three regions: the Alaska NORAD (ANR) Region, under Eleventh Air Force (11 AF); the Canadian NORAD (CANR) Region, under 1 Canadian Air Division, and the Continental U.S. (CONR) Region, under 1 AF/CONR-AFNORTH. Both the CONR and CANR regions are divided into eastern and western sectors.

The Alaska NORAD Region (ANR) maintains continuous capability to detect, validate and warn off any atmospheric threat in its area of operations from its Regional Operations Control Center (ROCC) at Joint Base Elmendorf–Richardson, Alaska (which is an amalgamation of the United States Air Force's Elmendorf Air Force Base and the United States Army's Fort Richardson, which were merged in 2010).

ANR also maintains the readiness to conduct a continuum of aerospace control missions, which include daily air sovereignty in peacetime, contingency and/or deterrence in time of tension, and active air defense against manned and unmanned air-breathing atmospheric vehicles in times of crisis.

ANR is supported by both active duty and reserve units. Active duty forces are provided by 11 AF and the Canadian Armed Forces (CAF), and reserve forces provided by the Alaska Air National Guard. Both 11 AF and the CAF provide active duty personnel to the ROCC to maintain continuous surveillance of Alaskan airspace.

The Canadian Air Division/Canadian NORAD Region Headquarters is at CFB Winnipeg, Manitoba. It was established on 22. April 1983. It is responsible for providing surveillance

---

[227] http://www.norad.mil/

and control of Canadian airspace. The Royal Canadian Air Force provides alert assets to NORAD. CANR is divided into two sectors, which are designated as the Canada East Sector and Canada West Sector. Both Sector Operations Control Centers (SOCCs) are co-located at CFB North Bay Ontario. The routine operation of the SOCCs includes reporting track data, sensor status and aircraft alert status to NORAD headquarters.

Canadian air defense forces assigned to NORAD include 409 Tactical Fighter Squadron at CFB Cold Lake, Alberta and 425 Tactical Fighter Squadron at CFB Bagotville, Quebec. All squadrons fly the McDonnell Douglas CF-18 Hornet fighter aircraft.

To monitor for drug trafficking, in cooperation with the Royal Canadian Mounted Police and the United States drug law enforcement agencies, the Canadian NORAD Region monitors all air traffic approaching the coast of Canada. Any aircraft that has not filed a flight plan may be directed to land and be inspected by RCMP and Canada Border Services Agency.

The Continental NORAD Region (CONR) is the component of NORAD that provides airspace surveillance and control and directs air sovereignty activities for the Contiguous United States (CONUS).

CONR is the NORAD designation of the United States Air Force First Air Force/AFNORTH. Its headquarters is located at Tyndall Air Force Base, Florida. The First Air Force (1 AF) became responsible for the USAF air defense mission on 30 September 1990. AFNORTH is the United States Air Force component of United States Northern Command (NORTHCOM).

AF/CONR-AFNORTH comprises State Air National Guard Fighter Wings assigned an air defense mission to 1 AF/CONR-AFNORTH, made up primarily of citizen Airmen. The primary weapons systems are the McDonnell Douglas F-15 Eagle and General Dynamics F-16 Fighting Falcon aircraft.

It plans, conducts, controls, coordinates and ensures air sovereignty and provides for the unilateral defense of the United States. It is organized with a combined First Air Force command post at Tyndall Air Force Base and two Sector Operations Control Centers (SOCC) at Rome, New York for the US East ROCC (Eastern Air Defense Sector) and McChord Field, Washington for the US West ROCC (Western Air Defense Sector) manned by active duty personnel to maintain continuous surveillance of CONUS airspace.

In its role as the CONUS NORAD Region, AF/CONR-AFNORTH also performs counter-drug surveillance operations. As in 1989 NORAD operations expanded to cover counter-drug operations, e.g., tracking of small aircraft entering and operating within the United States and Canada. DEW line sites were replaced between 1986 and 1995 by the North Warning System. The Cheyenne Mountain site was also upgraded, but none of the proposed OTH-B radars are currently in operation.

After the 11. September 2001 attacks, the NORAD Air Warning Center's mission expanded to include the interior airspace of North America in relation to aerospace warning and aerospace control. Aerospace warning includes the detection, validation, and warning of attack against North America whether by aircraft, missiles, or space vehicles, through mutual support arrangements with other commands.

The Aerospace control includes ensuring air sovereignty and air defense of the airspace of Canada and the United States. The renewal of the NORAD Agreement in May 2006 added a maritime warning mission, which entails a shared awareness and understanding of the activities conducted in U.S. and Canadian maritime approaches, maritime areas and internal waterways which formed NORAD into a Maritime Entity.

Office of Global Maritime Situational Awareness (OGMSA)

The Office of Global Maritime Situational Awareness (OGMSA)[228] is a United States interagency office with a global reach tasked with enhancing global maritime domain awareness. The focus of OGMSA is interagency coordination within U.S. federal, state and local governments, increased international cooperation and improved commercial cooperation – all to facilitate seamless maritime information sharing. OGMSA identifies the data owners and the information they can contribute, demonstrates how participation will be beneficial, resolves barriers to participation, and fosters development of the architecture, protocols and policies to make maritime information available and searchable.

Global Maritime Situational Awareness (GMSA)[229] is defined in the U.S. National Concept of Operations for Maritime Domain Awareness, December 2007, as *"the comprehensive fusion of data from every agency and by every nation to improve knowledge of the maritime domain"*. It is an integral element of Maritime Domain Awareness (MDA).

Essentially, no one country, department, or agency holds all of the authorities and capabilities to have effective Maritime Domain Awareness on its own. However, by combining separate pieces of information from agencies at the federal, state, local, and tribal level around the world with information from the maritime industry and other non-governmental organizations, it is possible to keep track of the status of every ocean-bound and sea-bound vessel. GMSA results from combining intelligence given by other regions of the world into a complete picture for identifying trends and detecting anomalies.

In the United States, the director for GMSA is responsible for managing data critical to building the situational awareness component of global MDA. The director also develops and recommends policy guidance for coordinated collection, fusion, analysis, and dissemination of GMSA information and products, as well as information integration policies, protocols and standards. The director also recommends improvements to situational awareness-related activities supporting maritime information collection, fusion, analysis and dissemination. The director co-chairs the U.S. National MDA Stakeholder Board, sits on the MDA Stakeholder Board Executive Steering Committee, and is a member of the U.S. Maritime Security Policy Coordinating Committee (MSPCC).

The GMSA staff consists of and is supported by dedicated subject matter experts from across the federal government as selected by the Director from departmental nominees from the Department of Homeland Security, Department of Defense, Director of National Intelligence, Department of Justice, Department of Transportation, Department of Commerce, Department of State, Department of the Treasury, and Department of Energy's National Nuclear Security Administration. The GMSA Director and staff form the Office of

---

[228] https://en.wikipedia.org/wiki/Office_of_Global_Maritime_Situational_Awareness#External_links
[229] https://en.wikipedia.org/wiki/Global_Maritime_Situational_Awareness

Global Maritime Situational Awareness.

OGMSA acts as a catalyst for and among all entities in the public and private sectors with maritime interests to develop an information sharing environment. Within this environment, partners can embrace and achieve the common objective of obtaining and sharing information as a mechanism to increase safety, security and economic prosperity in the maritime domain and have the supporting architecture to do so. The goal is not for OGMSA to own the data or control the processes.

Rather, OGMSA facilitates the identification and resolution of information sharing barriers. The focus of OGMSA is interagency coordination within U.S. federal, state and local governments, increased international cooperation and improved commercial cooperation – all to facilitate seamless maritime information sharing.

OGMSA identifies the data owners and the information they can contribute, demonstrates how participation will be beneficial, resolves barriers to participation, and fosters development of the architecture, protocols and policies to make maritime information available and searchable.

Major initiatives:

- Global Maritime Information Sharing Symposium (GMISS),
- Maritime Safety and Security Information System (MSSIS),
- Maritime Domain Awareness Enterprise Hubs,
- Inter-Agency Investment Strategy Monitoring, Reporting and Coordination.

The Global Maritime Information Sharing Symposium (GMISS) is an annual event hosted by the National MDA Coordination Office (NMCO) to align US Government outreach to the maritime industry and improve and increase industry-government maritime information sharing partnerships.

The symposium supports the U.S. Department of Transportation Maritime Administration (MARAD) in developing a coherent federal outreach and coordination effort. Leading up to this symposium, the U.S. Navy (USN) and U.S. Coast Guard (USCG) had expressed the need to expand their relationship with industry as set forth in the Cooperative Strategy for a 21st Century Seapower.

As government agencies have increasingly realized the benefits to be gained by partnering with the maritime industry however, there has been a flurry of "commercial outreach" and a deluge of information requests that are overwhelming industry and rapidly eroding its eagerness and ability to respond. The maritime community, both directly and through its various associations, has expressed frustration and confusion over the seemingly uncoordinated efforts in support of and demands being place on maritime trade and security by government entities.

The GMISS was undertaken as an effort to harness this government energy for engagement and establish a means to coordinate a more cohesive dialogue between government and maritime industry representatives. GMISS is an opportunity to demonstrate that a USN/USCG/MARAD partnership can open the doors of communication to further advance the maritime strategy, and equally, begin to provide the coordination necessary to present a unified voice to the maritime community. The long range (multi-year) objectives of the GMISS include:

- Coordinate the U.S. Government's maritime commercial outreach
- Implement industry/government working groups
- Highlight specific government relationships
- Promote regional involvement
- Engage a diverse range of stakeholders

Office of Naval Intelligence (ONI)

Established in 1882, the Office of Naval Intelligence (ONI) is America's longest continuously operating intelligence service, and provides maritime intelligence to the U.S. Navy and joint warfighting forces, as well national decision makers and other consumers in the Intelligence Community. ONI specializes in the collection, analysis and production of scientific, technical, geopolitical, military and maritime intelligence. ONI employs more than 3,000 military and civilian Intelligence Professionals including active and reserve officers and enlisted Sailors and Marines and contracted personnel at the modern National Maritime Intelligence Center facility in Washington, D.C., and at other strategic locations around the world.

The Nimitz Operational Intelligence Center executes ONI's responsibility for Maritime Domain Awareness (MDA) and Global Maritime Intelligence Integration (GMII). The Nimitz Center ensures the production of intelligence to all fleet elements, including Maritime Operational Centers. On their website, ONI provides weekly unclassified Maritime Operational Intelligence Reports on Piracy in the Horn of Africa. The report includes a piracy analysis and piracy warnings. The report is a review of recent Somali piracy incidents and an outlook on threats around the Horn of Africa.

The following terms have been adopted by ONI to describe the range of criminal anti-shipping activity and impediments to safe navigation in our worldwide reporting and analysis:

- Attempted boarding (Close approach or hull-to-hull contact with report that boarding paraphernalia were employed or visible in the approaching boat)
- Blocking (hampering safe navigation, docking, or undocking, of a vessel as a means of protest)
- Boarding (Unauthorized boarding of a vessel by persons not part of its complement without successfully taking control of the vessel)
- Firing upon (weapons discharged at or toward a vessel)
- Hijacking (Unauthorized seizure and retention of a vessel by persons not part of its complement)
- Kidnapping (Unauthorized forcible removal of persons belonging to the vessel from it)
- Robbery (theft from a vessel or from persons aboard the vessel)
- Suspicious approach (all other unexplained close proximity of an unknown vessel)

Next to these categories, there are differences in interpretations as well. As an example, when a person is shooting on board of a vessel, it can be a warning signal from a fishing vessel for the vessel approaching his fishing area or it also could be a pirate attack. Some organizations may report this incident as a pirate attack, while other organizations do not.

Surface Picture (SURPIC)

Surface Picture (SURPIC)[230] is an application of the US Coast Guard Rescue Coordination Centers (RCCs) used to determine the best Amver asset to utilize in the event of a distress at sea. The SURPIC II application accesses the Amver database and dead reckons all the vessels within the distress area using the last know good position in Amver. The information is then displayed for the operator at the RCC via the Amver Vessel Summary dialog box. Vessels are listed by call sign, name, predicted position, hours to intercept, distance to intercept, course variance, flag of registry, SAR(Q) medical, reported medical and satellite number. This information can be sorted by call sign, name, hours to intercept, distance to intercept, flag of registry, reported medical and satellite.

The SURPIC application also allows the RCC operator to research Amver, Lloyds and owner information for a selected vessel. RCC operators can review a vessel's dimensions, medical capabilities and owner contact information within seconds. Operators can review report and vessel voyage information for a selected vessel. All this information is there to assist the operator in utilizing the best search and rescue (SAR) resource available. In addition to this information, operators are able to "*plot*" the SURPIC information. This gives the RCC's a visual representation of the Amver Vessel Summary dialog box. Visualization of the SAR scenario can assist the operator in making a more accurate decision when selecting the best asset to utilize. On the right is what the RCCs would see when they "plot" the data. For this example, the call sign tag that accompanies each vessel icon has been removed, to keep the propriety of the information intact. The cross hairs represent the actual distress position. Operators have the ability to select any vessel from this display and research its vessel information with a simple click of the mouse.

Requests for SURPIC's from RCC's outside the United States are processed by the nearest Coast Guard RCC, and then forwarded by the most expeditious means (telephone, fax, telex, e-mail) to assist in that nation's response to an emergency within its area of responsibility under international agreements. The SURPIC II application is only distributed to U.S. Coast Guard RCCs to ensure Amver data is used only for SAR purposes.

The genesis behind the SURPIC improvements was the need to fully utilize the new information available through Amver's Lloyd's Register - Fairplay data feed. With the new data feed, some information SURPIC used would no longer be provided by Lloyd's. This allowed Amver to refine the SURPIC II application, incorporating new information and reconfiguring some of the dialog boxes to be more user friendly.

Minor changes were made to the Vessel Summary and Amver Data dialog boxes. The Vessel Summary dialog was configured to allow 3-digit country codes, and the WX column was removed. The WX column had not been utilized for some time, as new processes had replaced this method of ship identification. The Amver Data dialog box was reconfigured,

---

[230] https://vos.noaa.gov/MWL/april_05/amver.shtml

providing the same information in a more concise format.

The more robust changes are seen in the Lloyds Vessel Data and Owner Data dialogs. In the old Lloyds Vessel Data dialog, information was spread out, and five data fields were no longer going to be provided. This dialog is shown below with the data fields that are no longer being provided highlighted. Please note that, on all of the following dialog boxes, sensitive vessel information was removed.

The new Lloyds Vessel Data dialog will take the new information along with the current information and provide it in a concise easy to read format. The new dialog box is shown above right.

The changes to the Owner Information dialog boxes are similar in nature to the Lloyds Vessel Data changes, except that all information is still being provided as before in addition to the new data fields. On the bottom left is what the old Owner Information dialog looked like.

United States Department of Transportation (US DOT)

The United States Department of Transportation (US DOT)[231] was established by an act of Congress on October 15, 1966. The Department's first official day of operation was 1. April 1967.

The top priorities at DOT are to keep the traveling public safe and secure, increase their mobility, and have our transportation system contribute to the nation's economic growth. DOT employs almost 55,000 people across the country, in the Office of the Secretary of Transportation (OST) and its operating administrations and bureaus, each with its own management and organizational structure.

- Office of the Secretary (OST)
- National Highway Traffic Safety Administration (NHTSA)
- Federal Aviation Administration (FAA)
- Office of Inspector General (OIG)
- Federal Highway Administration (FHWA)
- Pipeline and Hazardous Materials Safety Administration (PHMSA)
- Federal Motor Carrier Safety Administration (FMCSA)
- Federal Railroad Administration (FRA)
- Saint Lawrence Seaway Development Corporation (SLSDC)
- Federal Transit Administration (FTA)
- Maritime Administration (MARAD)

Leadership of the DOT is provided by the Secretary of Transportation, who is the principal adviser to the President in all matters relating to federal transportation programs. The Secretary is assisted by the Deputy Secretary in this role. The Office of the Secretary (OST) oversees the formulation of national transportation policy and promotes intermodal transportation. Other responsibilities range from negotiation and implementation of international transportation agreements, assuring the fitness of US airlines, enforcing airline consumer protection regulations, issuance of regulations to prevent alcohol and illegal drug misuse in transportation systems and preparing transportation legislation.

The Federal Aviation Administration (FAA) oversees the safety of civil aviation. The safety mission of the FAA is first and foremost and includes the issuance and enforcement of regulations and standards related to the manufacture, operation, certification and maintenance of aircraft. The agency is responsible for the rating and certification of airmen and for certification of airports serving air carriers. It also regulates a program to protect the security of civil aviation, and enforces regulations under the Hazardous Materials Transportation Act for shipments by air.

---

[231] https://www.transportation.gov

The FAA, which operates a network of airport towers, air route traffic control centers, and flight service stations, develops air traffic rules, allocates the use of airspace, and provides for the security control of air traffic to meet national defense requirements. Other responsibilities include the construction or installation of visual and electronic aids to air navigation and promotion of aviation safety internationally. The FAA, which regulates and encourages the U.S. commercial space transportation industry, also licenses commercial space launch facilities and private sector launches.

The Federal Highway Administration (FHWA) coordinates highway transportation programs in cooperation with states and other partners to enhance the country's safety, economic vitality, quality of life, and the environment. Major program areas include the Federal-Aid Highway Program, which provides federal financial assistance to the States to construct and improve the National Highway System, urban and rural roads, and bridges. This program provides funds for general improvements and development of safe highways and roads.

The Federal Lands Highway Program provides access to and within national forests, national parks, Indian reservations and other public lands by preparing plans and contracts, supervising construction facilities, and conducting bridge inspections and surveys. The FHWA also manages a comprehensive research, development, and technology program.

The Federal Motor Carrier Safety Administration was established within the Department of Transportation on January 1, 2000, pursuant to the Motor Carrier Safety Improvement Act of 1999 (Public Law No. 106-159, 113 Stat. 1748 (December 9, 1999)). Formerly a part of the Federal Highway Administration, the Federal Motor Carrier Safety Administration's primary mission is to prevent commercial motor vehicle-related fatalities and injuries. Administration activities contribute to:

- Ensuring safety in motor carrier operations through strong enforcement of safety regulations, targeting high-risk carriers and commercial motor vehicle drivers.
- Improving safety information systems and commercial motor vehicle technologies; strengthening commercial motor vehicle equipment and operating standards
- Increasing safety awareness.

To accomplish these activities, the Administration works with Federal, state, and local enforcement agencies, the motor carrier industry, labor safety interest groups, and others.

The Federal Railroad Administration (FRA) promotes safe and environmentally sound rail transportation. With the responsibility of ensuring railroad safety throughout the nation, the FRA employs safety inspectors to monitor railroad compliance with federally mandated safety standards including track maintenance, inspection standards and operating practices.

The FRA conducts research and development tests to evaluate projects in support of its safety mission and to enhance the railroad system as a national transportation resource. Public education campaigns on highway-rail grade crossing safety and the danger of trespassing on rail property are also administered by FRA.

The Federal Transit Administration (FTA) assists in developing improved mass transportation systems for cities and communities nationwide. Through its grant programs, FTA helps plan, build, and operate transit systems with convenience, cost and accessibility in mind. While buses and rail vehicles are the most common type of public transportation, other kinds include commuter ferryboats, trolleys, inclined railways, subways, and people movers. In providing financial, technical and planning assistance, the agency provides leadership and resources for safe and technologically advanced local transit systems while assisting in the development of local and regional traffic reduction.

The FTA maintains the National Transit library (NTL), a repository of reports, documents, and data generated by professionals and others from around the country. The NTL is designed to facilitate document sharing among people interested in transit and transit related topics.

The Maritime Administration (MARAD) promotes development and maintenance of an adequate, well-balanced, United States merchant marine, sufficient to carry the Nation's domestic waterborne commerce and a substantial portion of its waterborne foreign commerce, and capable of serving as a naval and military auxiliary in time of war or national emergency. MARAD also seeks to ensure that the United States enjoys adequate shipbuilding and repair service, efficient ports, effective intermodal water and land transportation systems, and reserve shipping capacity in time of national emergency.

The National Highway Traffic Safety Administration (NHTSA) is responsible for reducing deaths, injuries and economic losses resulting from motor vehicle crashes. NHTSA sets and enforces safety performance standards for motor vehicles and equipment, and through grants to state and local governments enables them to conduct effective local highway safety programs. NHTSA investigates safety defects in motor vehicles, sets and enforces fuel economy standards, helps states and local communities reduce the threat of drunk drivers, promotes the use of safety belts, child safety seats and air bags, investigates odometer fraud, establishes and enforces vehicle anti-theft regulations and provides consumer information on motor vehicle safety topics.

Research on driver behavior and traffic safety is conducted by NHTSA to develop the most efficient and effective means of bringing about safety improvements. A toll-free Auto Safety Hotline, 1-888-DASH-2-DOT, furnishes consumers with a wide range of auto safety information. Callers also can help identify safety problems in motor vehicles, tires and automotive equipment such as child safety seats.

The Pipeline and Hazardous Materials Safety Administration (PHMSA) oversees the safety of more than 800,000 daily shipments of hazardous materials in the United States and 64 percent of the nation's energy that is transported by pipelines. PHMSA is dedicated solely to safety by working toward the elimination of transportation-related deaths and injuries in hazardous materials and pipeline transportation, and by promoting transportation solutions that enhance communities and protect the natural environment.

The Saint Lawrence Seaway Development Corporation (SLSDC)[232] operates and maintains a safe, reliable and efficient waterway for commercial and noncommercial vessels between the Great Lakes and the Atlantic Ocean. The SLSDC, in tandem with the Saint Lawrence Seaway Authority of Canada, oversees operations safety, vessel inspections, traffic control, and navigation aids on the Great Lakes and the Saint Lawrence Seaway.

Important to the economic development of the Great Lakes region, SLSDC works to develop trade opportunities to benefit port communities, shippers and receivers and related industries in the area.

---

[232] https://www.seaway.dot.gov

Maritime Safety and Security System (MSSIS)

The <u>Maritime Safety and Security Information System (MSSIS)</u>[233] and SeaVision are networks and portals owned and operated by the United States Department of Transportation (US DOT), and are part of the US Department of Homeland Security.

The US DOT computer systems, including all related equipment, are provided for the processing of official US Federal Government information. This means all the information viewed in the network and via the client software is considered US property. This disclaimer governs the use of the Maritime Safety and Security Information System (MSSIS) and SeaVision.

Both networks and software are for the use with US organizations and agencies and about 70-80 participating nations for:

- Military Organizations, to include Air, Land and Maritime forces
- Government Commerce and Fishery regulation agencies
- Law Enforcement agencies
- Border Security agencies
- Government Port Operations and Security agencies

The MSSIS is a partly freely-shared, unclassified, near real-time data collection and distribution network. Its member countries share data from Automatic Identification Systems (AIS). MSSIS is intended to promote multilateral collaboration and data-sharing among international participants, with a primary goal of increasing maritime security and safety.

Data sources may range from a single sensor to an entire national vessel tracking network. MSSIS is perfectly suitable as a one-stop source for streaming global maritime data.

Because the data distributed by MSSIS maintains its original, internationally recognized format and is delivered to users in near real time, member organizations are able to utilize the feed to meet their specific mission requirements.

---

[233] https://mssis.volpe.dot.gov

SeaVision

SeaVision[234] and the Maritime Safety and Security Information System (MSSIS) are both networks and portals owned and operated by the United States Department of Transportation (US DOT) and are part of the US Department of Homeland Security.

The US DOT computer systems, including all related equipment, are provided for the processing of official US Federal Government information. This means all the information viewed in the network and via the client software is considered US property. This disclaimer governs the use of the Maritime Safety and Security Information System (MSSIS) and SeaVision. Both networks and software are for the use with US organizations and agencies, NATO and about 70-80 participating nations for:

- Military Organizations, to include Air, Land and Maritime forces
- Government Commerce and Fishery regulation agencies
- Law Enforcement agencies
- Border Security agencies
- Government Port Operations and Security agencies

While MSSIS is accessible for governmental organizations and the information partly shared, SeaVision is mainly restricted to US- governmental use. Both systems contain information on sea and air traffic and other relevant issues which is not shared with any partners in the networks.

SeaVision was developed specifically for use by the U.S. Navy and government agencies of international partner nations. The U.S. Navy may grant permission to Other Government Agency (OGA) users to access SeaVision. This access may be modified or revoked at the discretion of the U.S. Navy. OGA user are accessing and using SeaVision as-is, with its present system capabilities and all future releases.

The U.S. Navy reserves the right to modify or enhance the existing system features and/or data accessibility of the SeaVision system at any time. This includes, but is not limited to, the following functions: data visualization, analytics, and extraction/export services, as well as users' assignments to SeaVision personas and communities.

---

[234] https://seavision.volpe.dot.gov

Western Pacific Naval Symposium (WPNS)

At the International Sea Power Symposium in 1987 agreement was reached "to establish a forum where leaders of regional navies could meet to discuss cooperative initiatives". The first meeting was held in 1988. The Western Pacific Naval Symposium (WPNS)[235] consists of a series of biennial meetings of the Pacific nations to discuss naval matters held on even numbered years. A WPNS workshop is held on odd numbered years in between the symposiums.

The WPNS has 24 countries participating: China, Australia, Brunei, Cambodia, Canada, Chile, France, Indonesia, Japan, Malaysia, NZ, Papua New Guinea, Peru, Philippines, South Korea, Russia, Singapore, Thailand, Tonga, USA, Vietnam, Bangladesh, India, and Mexico.

At the 2014 WPNS, agreement was reached on the Code for Unplanned Encounters at Sea[236]. Under the auspices of the WPNS, the Royal Singapore Navy developed an introduced the Regional Maritime Information Exchange (ReMIX) System, the secure and internet based website of the Royal Singapore Navy in the Information Fusion Center (IFC), with access to maritime reports and news and collaboration tools.

---

[235] http://wpns.mod.gov.cn/
[236] https://en.wikipedia.org/wiki/Code_for_Unplanned_Encounters_at_Sea

World Port Source (WPS)

The World Port Source (WPS)[237] is a private owned Internet website of publicly accessible seaport information.

Due to the complex interrelation of ports, cargo carriers, cruise ship lines, shipping agencies, terminal operators and dozens of other transportation related businesses, government agencies and public authorities the completion of this website will be a journey of discovery rather than a destination with closure.

The first objective in developing this site is to provide contact information and satellite images of ports and harbors throughout the world. Over time, this foundation of world ports will be cross referenced with the people and companies who make their livelihood servicing the world's largest and most valuable transportation network.

World Port Source provides interactive satellite images, maps and contact information for 4,764 ports in 196 countries around the world. Quickly find any port using our regional map of the world.

---

[237] http://www.worldportsource.com

SYSTEMATOGENESIS

Concept, Development & Experimentation (CD&E) to Production

The term Technology[238] ("*science of craft*", from Greek τέχνη, techne, "*art, skill, cunning of hand*"; and -λογία, -logia) stands for the collection of techniques, skills, methods, and processes used in the production of goods or services or in the accomplishment of objectives, such as scientific investigation. Technology can be the knowledge of techniques, processes, and the like, or it can be embedded in machines to allow for operation without detailed knowledge of their workings.

Today the system of practical knowledge and the study of doctrine, as in ancient Greek, 600 BC. Plato (428-348 BC) viewed "*techne*" (technology) and knowledge as being closely related. Aristotle (384-322 BC) went a step further by asserting that technology was the systematic use of knowledge for intelligent human action. Today technologies are often exemplified in a material product, for example an object can be termed state of the art.

Situational Awareness is largely the responsibility of different national and international entities in a domestic setting with different business cultures, processes, and technical systems supporting the work environment.

"*Successful information sharing activities are the result of operational, information and technological understanding achieved through a well-defined and routinely implemented process.*"[239]

International governance architectures often form in response to a need to coordinate behavior among countries around an issue such as maritime security involving organizations and technology. In the absence of an overarching governance organization, countries have to align and coordinate activities and form new forms of cooperation, a beneficial environment; a new "*system*".

These system shall be a regularly interacting or interdependent group of units of stakeholders forming an integrated whole. Every system is delineated by its spatial and temporal

---

[238] https://en.wikipedia.org/wiki/Technology
[239] Frank, S. and Radlinksi, T. Maritime Information Sharing Environment: An Information Exchange. Brief to Mr. DeVries, Office of the DOD CIO. August 13, 2013.

boundaries, surrounded and influenced by its environment, described by its structure and purpose and expressed in its functioning. These stakeholder systems requirements might then lead to new technologies, new technical systems.

Given the plethora of initiatives under Maritime Situational Awareness rapidly evolving in many nations and regions around the world, the challenge of linking the best of these diverse processes to form an effective international framework for maritime security cooperation and awareness is compounded by differing cultural norms, political agendas and technical standards.

Situational Awareness begins in general with a framework of domestic law enforcement and defense activities coordinated amongst various national agencies. This collaborative framework extends individual mandates and collective interests to facilitate broader co-operation for geographic regional activities, and can culminate in intra-regional global initiatives. Cooperation promotes a convergence of national strategic and security interests. International institutions help facilitate alignment of interests and provide a forum for establishing a reputation for implementing the efficient employment of complex strategies.

Whether operating at a domestic level, or a global level, any initiative for MSA must tackle the difficult legal and policy issues which surround agency responsibility and co-ordination with other domestic or international groups. Sovereign governments must confirm that it is in their national interest to work with neighboring states and commercial or non-governmental agencies to improve safety & security.

Timely and effective information sharing must remain a central priority for the stakeholders who have a responsibility to inform and update information in line with desired operational outcomes. These networks must link constabulary with military and civilian networks where it makes good operational and economic sense to do so. There are practical concerns on how such information is to be structured electronically and securely managed; the holistic MSA data design.

In many ways, new surveillance and communication technology is making MSA more feasible, but such improvements carry with them the challenges of expense and increased complexity for establishing the requisite infrastructure that not all national partners can afford. The growing number of national and regional initiatives creates independent systems and processes which detract from common standards and often duplicate national, organizational or individual efforts with systems and networks already existent.

Only the alignment of best practices, implementation of these enablers can eventually provide a framework for MSA on a global scale. Selecting the coordinating body for such a large undertaking must be made through consensus formed by all Maritime Entities affiliates.

The international community of interest needs to determine activities such as surveillance and information requirements, as well as, establish standards for the tools and applications needed to share and conduct collaborative risk assessments on a global scale. The key objective here is aligning and linking the broad spectrum of maritime capabilities which will enable the relevant stakeholders to rapidly sort through volumes of data to quickly assess the appropriate level of response to deal with a specific threat.

The development of global standards for safe navigation, communication, and operation of maritime shipping has long been an area of international cooperation, with the United Nations, representing the most international body with a reputation for impartiality among many participants, is an obvious choice for facilitating, providing legitimacy to, but not necessarily executing, i.e. a global MSA interoperability profile.

The International Maritime Organization (IMO) Maritime Security Directorate has a mandate which fits closely with collecting the requirements for a global MSA Interoperability Profile. It is a logical candidate for consideration as a facilitator for aligning processes and coordinating standards that enable regions to build on existing international cooperative frameworks to achieve global MSA. However IMO organization does and shall not reach down to the operational user level and daily business processes.

The MSA Interoperability Profile does include any Reginal Maritime Situational Awareness cooperation and effort being part of a global common MSA, which is a patch work of national and international initiatives and networks in need of a common authority in order to enable the information exchange between all singular entities and their implementations in various different multinational and omni-organizational structures for Standardization.

Pragmatism will drive all stakeholders to be more closely integrated with key partners across governments, industry, academia and principal allies in inter-agency and multinational engagements for an increased complex problem solving. Enhanced innovation and creativity through diversity of thought and collaboration for new and advanced capabilities are key factors in achieving a better awareness.

The ability to exploit new technologies and approaches requires access to people with appropriate skills, knowledge and experience. In a competitive and changing work force formed through the industrial markets, securing access to expertise in important sectors such as science, technology, engineering and mathematics, especially in information and communications, intelligence, cyber, space, nuclear, computing and simulation, medical and legal areas, requires a pragmatic collaborative approach.

Though not a panacea, the Research & Development (R&D) to develop technology will remain an essential element of gaining advantage via Situational Awareness in the future operating environment and an important driver of military change. The tempo of technological change will accelerate and human interaction with more and more complex technology will increase significantly. Combinations of civil and military technologies will allow state and non-state actors to access sophisticated capabilities that were once the preserve of just a few states.

Technological advantages across key capability areas significantly erode through globalization in combination with constrained resources, because people are at the heart of our ability to innovate and adapt.

The most significant changes are likely to come from the rapid development of information technologies, new sensors and novel weapons, developments in artificial intelligence, biological and material sciences and a rapid growth of remote and automated systems. Accessing and developing the knowledge, skills and experience to recognize and respond quickly to transformative ideas and technologies, many of which will be driven by the private sector, will be a primary challenge.

The vast amounts of information, in any given situation, for use in any entity, need today methodological, scientific and technical support to combine and compile the data in systems and networks and in order to produce a specified, analyzed and validated picture for awareness.

Key element for success in establishing knowledge is the data design from the raw source to validated information. Data design is structuring data so that it can be used successfully and in a wide range of places. Data design focuses on the end-user through a design process to discover what works and what does not and provide a visual communication towards the end user based on a technical analysis.

Technical analysis was an arcane workmanship before the internet blasted online tools out in the world wide web. Chartists perform technical analysis in their secret rooms with data that was carefully gathered from professional sources. Those were the times when stock prices and data did not have a medium through which to be promptly accessible to general society and be going through openly accessible software to create the charts that are accessible today.

With internet taking over almost every household, the technical analysis turned into a something anybody could do. Complex charts, technical indicators, and analysis that were previously the sole space of a couple of generously compensated Wall Street analysts are presently accessible to any individual who wants it, regularly for nothing.

At the same instance the beneficial internet usage brought new threats to MSA by opening the so far exclusive club of international and national organizations to the possibilities of attacks from outsiders.

The roundabout 50,000 ships sailing the sea at any one time have joined an ever-expanding list of objects that can be hacked, running often outdated and unsupported operating systems. Ship crews are highly dynamic, the users of these maritime computer systems are constantly in flux. The linkage between onboard and terrestrial systems as well as the navigational aids are prone to cyber-attacks. The MSA Cloud is a vulnerable space.

Managing large quantities of structured and unstructured data is a primary function of information systems. Data *"given"* (or sent) about an incidence has no indication about the authority transmitting, sharing or receiving. It is the operational task that determines the lead and to whom data has "*to be given*" (or send) to, as described in the concept of "Neutrality of Data[240]".

A Data Model is the concept of data being made basically available - in the sense of being stored – in a data base for operational use:

- In software engineering a data model is an abstract model on how data is stored and accessed. It explicitly determines the structure of data for storage.
- The main aim of data models is to support the development of information systems (by providing the definition and format of the data[241]).
- Designing a data model is the complex task of software engineers.

Designing a data model has several challenges:

- Business rules[242] specify the work flow in a particular place or organisation and are often fixed in the structure of a data model. Small changes in the way business is conducted lead to large changes in the data model, computer systems and interfaces.
- Entity types not identified or incorrectly identified lead to duplication of data and therefore additional costs for system development and maintenance.
- Data models for different systems are arbitrarily different, resulting in complex interfaces needed to enable the sharing of data. Data structure and meaning of data has to been standardised for electronic and automatic data sharing.

Data design is a new term, first introduced in the 1970's by Richard Saul Wurman and Metadata and classification are an integral part of data design. Data design is tied in with making content easy for individuals to understand. Individuals have designed data since they first started using attracted and composed marks to impart. In any case, the discipline of data design experienced childhood in the 1930s with its roots in the Isotype Institute's transformational way to deal with data presentation.

Wayfinding is a specific specialization of data design that moves far from customary designing to engineering and urban arranging. The designers who make graphics for signage can also compose the underlying wayfinding research about how individuals explore and

---

[240] http://beckh.wordpress.com/2011/07/05/the-neutrality-of-data

[241] "… if this is done consistently across systems then compatibility of data can be achieved. If the same data structures are used to store and access data then different applications can share data. … However, systems and interfaces often cost more than they should, to build, operate, and maintain. They may also constrain the business rather than support it. A major cause is that the quality of the data models implemented in systems and interfaces is poor" by West and Fowler (1999).

[242] Business rules describe the operations, definitions and constraints that apply to an organization, they can apply to people, processes, corporate behavior and computing systems. The common approaches to visualize Business Rules are flow charts, to define them for electronic exchange the eXtended Markup Language (XML) is used.

appreciate spaces. Wayfinding and signage systems are accordingly critical.

Many time data engineers fall flat at technical analysis simply because they don't have the necessary basic information, no operational end-user background, to understand how to translate technical indications appropriately in the first place. Data Engineering is the specialty of building a structure for overseeing data using Data design for the best visual presentation to the user and is therefore critical for the visualization outcome and end product.

Data Visualization is the process of presenting data to others in a rich illustration and plays a vital role in every field. Data visualization tools and Data visualization are particularly important to influence processes in situational awareness. Data visualization has helped a considerable measure in presenting of data genuinely among the masses so that the general population can get the reasonable thought regarding what the data is attempting to say or endeavoring to give data.

Data visualization is advancing from the customary charts, graphs, warm maps, histograms and scatter plots used to represent numerical values that are then measured against at least one dimensions. There are numerous advantages of utilizing infographics in advertising and on websites that prove the benefits for the use in situational awareness:

- Give Facts and Figures: Individuals cherish perusing facts, numbers, statistics and other such data. Infographics can give this data in a structured, engaging manner. Infographics are substantially more visually appealing than content which can increase interest which will give your site or blog additionally drawing power.
- Greater Engagement: Visual graphics and images increase engagement. Using infographics on your blog could raise the level of visitor engagement by an extremely substantial sum.
- Backlinks: Infographics can be an exceptionally successful method for building your backlinks. Urge individuals to share your infographics with their partners in cooperation and provide a connection back to published own sources.
- Easier to Understand: Besides infographics being visually engaging they are easier to understand than content data. The human cerebrum can process and recall visual data superior to content. This will take individuals back to the source for gathering more data.
- Infographics are Search Providers: The searches for infographics in the internet have increased by over 800% in two years. Adding infographics in published data will therefore support communication activities between partners/stakeholders.

Information must no longer flow in vertical stovepipes, but instead be widely published and available on multiple networks. Agile Command, Control, Communications, Computers, Combat Systems, Intelligence, Surveillance, and Reconnaissance (C5ISR) and user oriented visualization are the best response to today's challanges, which are now almost invariably characterized as complex, with few simple cause and effect relationships. There are practical constraints, principally security, physical connectivity and volume of data that can be handled at information choke points, but processes are shaped by user needs.

There are three important aspects of a more agile C5ISR, the extent and integration of networks, increased and improved collaboration, and a break-up of traditional command hierarchies towards flatter C5ISR networks and greater delegations of authority, albeit still guided by command intent.

Technology in general is not limited to any field, standard, specification and products used in interfaces and/or systems need to be interoperable in order to enable a Global Situational Awareness. Therefore a technology may only at first glance be attributed to a maritime, land, air, space or other specified environment, in practice, and due to the subject of situational awareness, there can never be a cemented line to distinguish where each specific area and duty might begin or end, as the situation might change within a split second and require a transfer of responsibility and leadership.

Technology is an important tool in preventing, responding, managing and mitigating potential threats. However, the effectiveness and legitimacy of technology depends on the human activity that is associated with its use. An exploration of the non-technical enabling factors are instrumental in delivering the hoped-for technological results, e.g. standards, legislation, human factors, partnerships, etc., as the overall system is only as robust as its weakest part, and in meeting security needs, human and organizational aspects have proven themselves, on frequent occasion, to be the weakest links.

The identification and authentication of individuals is the most common function required for the protection of critical infrastructures. Almost all proposed solutions to address the requirements of this function are based on the use of biometric technology. Implementation of biometric technology has often raised concerns, and its use is controlled by several special regulations. Important points to be considered are the most effective manner for authorities to communicate in the communities before, and during, a crisis as well as the key role to be played by the media. The understanding of organizational structures and cultures of users' has direct relevance to an effective and efficient crisis management response.

Standardized and interoperable equipment for first responders is essential. As assessment of the current situational status, in particular via integrated wired and wireless communication systems is mission critical. Communication systems are of vital importance to first responders which must be able to seamlessly and dynamically interconnect multiple agency users, who have multiple functions, and multiple information and communications technology systems. Sensor information and accurate global positioning are necessary ingredients in building situational awareness for timely and effective command and control, and for the monitoring of resources and personnel and communication relay.

Advantage should be taken of the networking and transparency mechanisms with the aim to exchange best practice, standardize operating procedures and equipment and developed poles of excellence for key crisis management activities and necessary efforts as well as avoidable workloads. There is e.g. no need to define common standardized data formats to ensure information coding permits data exchange between people and systems. The definition of common data formats would further be rather challenging since more than 80 % of the world's database content is in unstructured largely textual format.

Data formats are time consuming and costly artifacts when necessary changes have to be implemented in systems. It is impractical and therefore unlikely, that large amounts of data would be retrospectively codified against a common standard for interoperable representation of the information (e.g. common formats, common data model) or even against a common language with the difficulties of achieving a common understanding of nomenclatures.

Fortunately today technology offers a more adaptable way than data formats, with XML, Schemas and Name Space the translation processes can be defined and ensured. A more practical approach would be therefore to adopt common interchange standards for data (e.g. XML Name Spaces), either bilaterally between stakeholders or preferably for the stakeholders as a whole (UN/NATO/EU Standardization). Gateways and/or translator are identified to convert formats and protocols as required. In terms of interoperability, the ability to rely on interoperable communication mechanisms is a key basic enabler across all missions.

Across all mission areas, is the importance of access control to facilities, areas, systems and information. At the heart of access control is an inherent trust between the parties. Robust access control and authentication models are required for the sharing and exchange of information, particularly those relating to sensitive data. Controlling access requires the identification of the accessing entity and thereafter positive confirmation that the claimed identity is correct (i.e. authentication).

Speed is the driving factor for authentication and information management techniques, database design and high speed communication bandwidths will all figure strongly in final solutions. Authentication speed is also affected by the chosen method of identification which varies considerably depending on the entity requesting access - be that an individual, system or application. Each will have different requirements affecting their technological solutions and consequently affect interoperability and design parameters. Biometrics is viewed as the solution of choice for the identification and authentication for individuals. For widespread deployment of biometric-based solutions for identification and authentication interoperability between biometric data readers and their corresponding authentication databases is of critical importance.

While historically grown sensor environments list a wide range of classical systems for signal and object detecting and ranging, technical developments have brought a vast amount of new possible information sources to be harvested for a better situational awareness.

The technology for Space-based Side Aperture Radar (SAR) is for example very effective in ship detection, especially for the detection of "*dark targets*" (non-self-reporting and/or non-emitting vessels), however has shown its potential at the same time in land missions or air warfare. In most use cases of tracking or targeting the revisit rates of SAR unfortunately do not provide persistency.

SAR is being operationally utilized by nations, but research challenges remain including detection of smaller vessels, ship feature extraction (size, direction, classification), and false alarm reduction. Space-based sensors including Automatic Identification System (AIS), Synthetic Aperture Radar (SAR), and Electro-Optical (EO) are routinely providing data for MSA.

Space-based AIS is providing the capability for non-coastal regions with a detection performance and persistence that depend on satellite revisit rates, ship traffic density, and other interference sources. S-AIS is both commercially and nationally available, while Space-based EO is an emerging capability for ship detection. One newer solution might be a high-number of brick-sized Nano-Satellites combined in providing a maritime picture.

Radio-controlled satellite communication is limited by the range of the long wavelengths, the available frequency bands are scarce and the radio signals can be easily intercepted. The laser technology can transmit 10,000 times more information, the number of channels is almost unlimited and the communication is relatively secure.

Weather influences and disturbs the transmission of information. The University of Geneva has developed an ultra-powerful laser that creates a temporary hole in the cloud by heating the air up to 1.500° C. The shock wave pushes the water droplets sideways which creates a hole of a few centimeters through the entire cloud cover big enough for a laser beam. Thus, a laser beam containing the information can be transmitted at the same time. The new technology could reach usable by 2025.

The fleet of recreational ships and most modern commercial vessel carry today equipment allowing passengers and crew to use their smartphones onboard. This usage can be tracked and used in tracking the maritime traffic.

The University of California, Santa Barbara, demonstrated the possibility of spying on the interior of a home with a smartphone and a new software by intercepting the WLAN radio signals. The slightest changes in the signal strength caused by moving inside the apartment is calculated by the new software. The use of special wireless antennas or access to the wireless router is no longer needed, the smartphone and the software is enough. Although the process provides only rough information about the location, the movements can be recognized, when they get up, open doors or leave the place. The receiving phone needs to be walked around the apartment outside several times to precisely locate the wireless router and determine the different signal strengths. Then there are hardly any measurements that can be taken, except that the WLAN signal is provided with a noise or the walls are covered with aluminum foil.

The same principle is used since the detection of radio waves and the invention of radio receivers in order to locate ships and triangular their position and movements based on their radio transmissions. Passive detection by radio, light or heat emissions via satellites, as well as active counter measurements and interference, are getting more sophisticated,

Sensors, platforms, miniaturized and interconnected technology (Internet of Things, IOT) can be linked into large networks of Nano-satellites, evaluated in Big Data analytics, and integrated into decision support tools. Multi-sensor systems[243] are an important enabler reducing data fusion challenges downstream by decreasing the time difference and uncertainty between observations.

Single sources of information, within a sensor system or any other network, are today no longer sufficient for analyzing a situation because, too many variables can affect the output when trying to establish a knowledge-based awareness, and to many pieces of information are distributed in different political, sectorial, geographical areas, in order to provide them in a common graphical view. A picture can consist from a single to thousands or millions of tracks with any other data collected during a defined period of time accumulated in a Case File (CF). The tracks can later be evaluated and data confirmed by a human operator producing the Recognized Picture (RP).

Processing that enables cross cueing between sensors might be a future technology trend and while a vast virtual constellation of sensors currently exists in principle, none of them is being singular exploited or operationally limited for ship detection. AIS track can be found thousands of miles inside African rivers and play a role for Land Awareness. Any plane out in the ocean is part of a maritime picture on a military vessel for Air Awareness and Air Defense, and in Subsurface Awareness a submarine carrying Intercontinental Ballistic Missiles may fall in the interest of combined-joint global Ballistic Missile Defense (BMD). Sensors are universal in their importance. The responsibility and need to share cannot be defined by a border line on a map or the letter in a law. The importance of a single track lays within the eye of the individual observer.

Networks need to be distinguished into hierarchical and centralized network architectures. Existing sources must be linked to new systems forming a network. Various networks then again will require to be linked in order to enable information sharing in a federated environment. This holistic concept is found for example in the Maritime Surveillance Networking (MARSUR), Sea Surveillance Cooperation Baltic Sea (SUCBAS), the Common Information Sharing Environment (CISE), or, in their original root source of concepts history, the Network Enabled Capability (NEC) from NATO, later transferred and used in the EU/EDA as EUNEC. NATO has initiated the Federated Mission Network (FMN) in order to enable the connection and sharing of various possible partners, affiliates and their individual systems.

Based on a Concept comes the decision to conduct a Study. The study points at a possible experimentation, which in turn produces the core knowledge for a project tender. The

---

[243] i.e. the RADARSAT Constellation Mission that will carry both SAR and AIS payloads or the TerraSAR-X1 (also referred to as TSX or TSX-1) follow-on TSX-NG (TerraSAR-X Next Generation)

project might result in a first prototype. The first prototype might get stabilized foundation by conduction more studies and experiments, sometimes multiple follow-on projects and advanced prototypes; but only few of all concepts, studies, experiments, projects and prototypes will make it to a final system.

The world of technical systems is like a "*systematogenesis*"[244], mainly the first one in its field will make it to the fusion of gametes, to initiate the development of a new individual organism, or here a new technical system. Oogenesis[245], ovogenesis, or oogenesis, in relation to concept, development and experimentation lead to the differentiation of the ovum (here the prototype) into a cell competent to further development. However, only when the value is recognized and politically fertilized, it will continue to develop from a primary system oocyte when maturation is ensured. From the combination of these observations in Concept, Development & Experimentation the term "*systematogenesis*" was derived.

---

[244] compare to spermatogenesis: https://en.wikipedia.org/wiki/Oogenesis
[245] https://en.wikipedia.org/wiki/Oogenesis

Basic Knowledge on Terms, Abbreviations and Definitions

Terms, Abbreviations and their definitions provide the required jargon specific and basic knowledge to a subject matter presented in a strategy or concept, as overall guidance mainly copied down the document hierarchy. Each operational and technical document will define the suitable own abbreviation directory with operational and technical terms and definitions.

In an easy reading approach some of the core terms and their definitions are nevertheless listed along the text for a general understanding, not aiming at completeness. In the long run, terms, abbreviations, and definitions, help any reader or project manager and they needs to be reflected.

Terms are words and compound words or multi-word expressions that in specific contexts are given specific meanings—these may deviate from the meanings the same words have in other contexts and in everyday language. Terminology is a discipline that studies, among other things, the development of such terms and their interrelationships within a specialized domain.

Terminology is the study of terms and their use. Terminology differs from lexicography, as it involves the study of concepts, conceptual systems and their labels (*terms*), whereas lexicography studies words and their meanings.

Terminology is a discipline that systematically studies the "*labelling or designating of concepts*" particular to one or more subject fields or domains of human activity. It does this through the research and analysis of terms in context for the purpose of documenting and promoting consistent usage. Terminology can be limited to one or more languages (for example, "*multilingual terminology*" and "*bilingual terminology*"), or may have an interdisciplinary focus on the use of terms in different fields.

Definitions are important, for example when talking about Security, like securing a building, there exist different understandings on what security and securing means, especially in the military world:

- Army Forces will put guards on every door and window, will not let anybody out without an access badge including their own commander.
- Marine Infantry will blow-up the doors and windows, clean the house from any resistance taking no prisoners and host the national flag on the rooftop.
- The Air Force will send a remote controlled UAV and level the entire quarter via joystick and hope it hit the right target.
- A Navy guy will simply turn of all the lights, lock the doors, and throw the key underneath the doormat before going to the next bar.

An abbreviation (from Latin *brevis*, meaning *short*) is a abridged form of a word or phrase. It consists of a group of letters taken from the word or phrase. For example, the word "abbreviation" can itself be represented by the abbreviation *abbr.*, *abbrv.*, or *abbrev*.

Bottom line is, that the definition of any term is depending on the originator and target audience. A definition is a statement of the meaning of a term (a word, phrase, or other set of symbols) and can be associated with two large categories:

- Intentional definitions (which try to give the essence of a term)
- and extensional definitions (which proceed by listing the objects that a term describes).
- Another important category of definitions is the class of ostensive definitions, which convey the meaning of a term by pointing out examples.

A term may have many different senses and multiple meanings depending on the context, and thus require multiple definitions. In Situational Awareness the same term, with the identical abbreviation, might have various, slightly to absolute different, meanings.

When it comes to Terms, Abbreviations and Definitions in MSA or any other field, instead of one there are constantly additional new stones rolling down the hill. Not even Sisyphus is able to roll them all up for an appropriate Tyros status-quo and update.

Information can be stored in IT systems, but it can only be useful and become knowledge when it has been used. The term "*to be used*" means that the information is taken from the system, is contextualized and thereby achieves a specific meaning for the human being, the knowledge.

There are many levels of information, from a lower level to a high or higher level and the other way around, residing at senior management level considered at a high level, and the junior management level can be seen as knowledge by the lower level. But between knowledge and wisdom there is a bridge called "*knowing*". Knowledge is not the same as knowing - knowing is deeper, and it comes when knowledge has been reflected and internalized in steps:

1. Step: Define and construct "*The Question*"
2. Step: Locate the required Information
3. Step: Select / Analyze the Information
4. Step: Organize / Synthesize the Information
5. Step: Present the Data
6. Step: Evaluate "*The Answer*"

Gaining knowledge like a spiral. The more we know, the more we will unconsciously consider some of our knowledge to be information because it has become common knowledge to us. Unfortunately, since it is an unconscious process, we automatically assume that everybody should know about "*our*" common knowledge. When the knowledge value is *"degraded"* or *"down-graded"* it will be seen as simply information again, and when it is passed down without the background knowledge, it becomes information without a context and nearly useless.

When knowledge has been taken out of the context it is nearly impossible for any newcomer to learn and understand. In awareness it is therefore essential to have any information in relation to the overall context and situation.

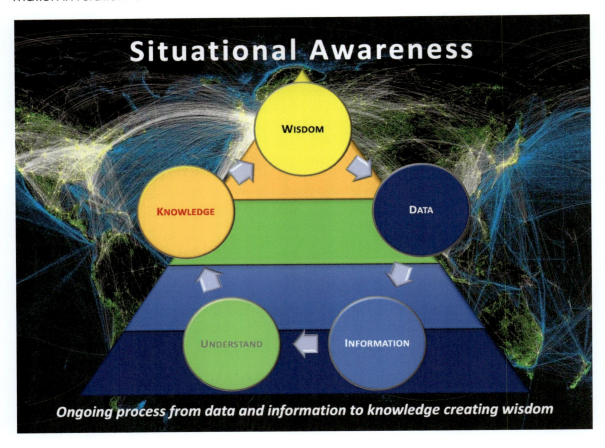

Architecture

Architecting is a practice for conducting enterprise analysis, design, planning, and implementation, using a holistic engineering approach at all times, for the implementation of strategy. Architecting applies principles and practices to guide organizations through the business/mission, information, application and technology changes necessary to implement their strategies. Good architecture practices include the usage of architectural artefacts to describe, assess, evaluate and document relevant aspects of an architecture.

The NATO Architecture Framework (NAF) provides a standardized way to develop architecture artefacts, by defining:

- Methodology – how to develop architectures and run an architecture project (Chapter 2),
- Viewpoints – conventions for the construction, interpretation and use of architecture views for communicating the enterprise architecture to different stakeholders (Chapter 3),
- Meta-Model – the application of commercial meta-models identified as compliant with NATO policy (Chapter 4), and
- a Glossary, References and Bibliography (Chapter 5).

ISO/IEC/IEEE 42010 describes architecture as: *"The fundamental concepts or properties of a system in its environment embodied in its elements, relationships, and in the principles of its design and evolution"*. In the case of an Architecture Framework, a system is anything that can be considered with a systemic approach, such as a:

- Product,
- Service,
- Information System,
- System of Systems, or
- Enterprise

Architectures are developed for many purposes and their development can be described as both a process and a discipline. Architectures aid the development of systems that deliver solutions that can meet an organization's needs in order to achieve its mission.

An architecture framework is a specification of how to organize and present an enterprise through architecture descriptions. ISO/IEC/IEEE 42010 describes an architecture framework as: *"The conventions, principles and practices for the description of architectures established within a specific domain of application and/or community of stakeholders"*.

An Enterprise Architecture (EA) is a way of formalizing stakeholder concerns and presenting them in the context of the enterprise. For example EA can encompass both business and technical concepts to emphasize the dependencies between them. This approach enables change to proceed with a clearer understanding of the touch-points and problem areas. EA takes a holistic approach in order to manage problems associated with the system-of-interest to show the interaction of technology and business processes. The purpose of EA is to optimize across the enterprise, the often fragmented legacy of processes (both manual and automated) and systems, into an integrated environment that is responsive to change and supports the delivery of the business strategy. The purpose of EA is not to model the entire enterprise.

The purpose of the Architecture Definition process is to generate system architecture alternatives, to select one or more alternative(s) that frame stakeholder concerns and meet system requirements, and to express this in a set of consistent views.

Iteration of the Architecture Definition process with the Business or Mission Analysis process, System Requirements Definition process, Design Definition process, and Stakeholder Needs and Requirements Definition process is often employed so that there is a negotiated understanding of the problem to be solved and a satisfactory solution is identified. The results of the Architecture Definition process are widely used across the life cycle processes. Architecture definition may be applied at many levels of abstraction, highlighting the relevant detail that is necessary for the decisions at that level.

An architecture may be used to provide a complete expression of any part of the system in an enterprise context. The meta-model defines the essential modelling elements that can be used to describe the system in an enterprise context and its environment. However care must be taken to have a clear purpose in mind for developing any architecture. Architecture Frameworks may define a common language-independent and tool-independent formalism for architecture representation, and it provides the means to help achieve better communication between architects as well as between architects and stakeholders.

The use of standardized viewpoints serves as a lingua franca as it provides a unified way of describing complex real world objects. It is important both to architects and stakeholders that those involved in an architecture process are aware of this fact and use it to their common interest. This common language will also help to establish a common arena for discussing architectures and consequences across communities of interest to Maritime Entities across Nations and organizations.

Enterprise Level architectures, particularly federated architectures, are used at the enterprise level to make decisions that improve:

- human resource utilization,
- deployment of assets,
- investments,
- identification of the enterprise boundary (external interfaces) and assignment of functional responsibility, and
- structuring the functional activities in terms of projects.

Architectures are developed to support strategic planning, transformation, and various types of analyses (i.e. gap, impact, risk) and the decisions made during each of those processes. Additional uses include identifying capability needs, relating needs to systems development and integration, attaining interoperability and supportability, and managing investments.

Project Level architectures are used at the project level to identify capability requirements and operational resource needs that meet business objectives. Project architectures may then be integrated to support decision making at the enterprise level. Architectures facilitate decision making by conveying the necessary information. Setting architectures within the enterprise context ensures complete, actionable information for more reliable decisions. The following describes architecture data usage for different types of decisions:

- Portfolio management identifies objectives and goals to be satisfied with regards to owned assets (capabilities and systems) and processes to be governed.
- Capability and Interoperability Readiness assesses capabilities and their implementation (systems, platforms, services and aggregated solutions) against needs and their net-readiness to identify gaps in interoperable features.
- Operational Concept Planning examines how various mission participants, processes, roles, responsibilities, and information need to work together, to recognize potential problems that may be encountered, and to identify quick fixes that may be available to accomplish a mission.
- Acquisition Program Management and System Development expresses the plan and management activities to acquire and develop system concepts, design, and implementation (as they mature over time), which enable and support operational requirements and provide traceability to those requirements. This process must be compliant with the Enterprise objective and operational requirements. It refines operational analysis, performs system analysis, and improves both materiel and non-materiel solution analysis.
- Modelling and Simulation deal with modelling and simulation techniques which can be used in order to assess the business and mission analysis. For example, in the military context through the implementation of mission threads and scenarios, thus providing an environment for thorough testing of identified use cases.

Architectures must be produced as a means to achieve higher level enterprise objectives. Architecture related processes should be seen as a technique for managing complexity rather than activities to produce models. A common set of architecture processes is judged to be the best way of achieving success in the formation of a federation of systems approach. This concept is valid for organizations themselves, but also between their various partners.

Concept

A <u>CONCEPT</u>[246] is a generalization or abstraction from experience or the result of a transformation of existing ideas. More understandable described is a concept basically any idea, something that is conceived in the human mind. Concepts derived from high strategies were listed under Maritime Entities when they formed a strong maritime community.

Demonstration

A <u>DEMONSTRATION</u> can be a conclusive proof, a method of proven by example rather than simple explanation, an incomplete version of product to showcase idea, performance, method or features of the product, a scientific experiment to illustrate principles, a repository of computer based educational demonstrations.

The <u>DEMONSTRATOR</u>[247] is the device or object used to perform the demonstration, may be adapted for use in a experimentation, and can be the basis for a prototype.

Experiment

An <u>EXPERIMENT</u>[248] is a procedure carried out to verify, refute, or validate a hypothesis. Experiments provide insight into cause-and-effect by demonstrating what outcome occurs when a particular factor is manipulated. Experiments vary greatly in goal and scale, but always rely on repeatable procedure and logical analysis of the results.

<u>Concept Development and Experimentation (CD&E)</u>[249] is the application of the structure and methods of experimental science to the challenge of developing future military capability; in the context of Maritime Security the future Maritime Capability.

CD&E is a forward-looking process for developing and evaluating new concepts, before committing extensive resources. CDE is a process to identify the best solution not only from a technical perspective, but also for possible solutions for challenges involving doctrine, organization, training, and material to achieve significant advances in future operations.

---

[246] https://en.wikipedia.org/wiki/Concept
[247] https://en.wikipedia.org/wiki/Demonstrator
[248] https://en.wikipedia.org/wiki/Experiment
[249] https://en.wikipedia.org/wiki/Concept_development_and_experimentation

CD&E is a way of thinking your way through the future before spending money. Developing and identifying future-oriented concepts allows one to:

- test their validity/feasibility
- take advantage of other studies/experiments conducted and
- save resources and avoid duplication.

The potential impacts on interoperability and increased capabilities by emerging concepts must be captured by some process and exploited.

Concept development gives broad and sometimes ill-defined ideas a chance to be examined by groups of experts in a logical process. These ideas can come from different sources: e.g. Ministry of Defense, industry, servicemen organizations or partners. They can be generated by staff processes, operational experience, formal analytical work, or published proposals. There need be no boundaries on the types of ideas that enter the concept development process, although some simpler ones that modify techniques or procedures might be 'fast-tracked' into practice because they are intuitively sound.

Typically, promising ideas with a broader scope are explored and refined through workshops and larger seminars to the point where more mature concepts are formed. These concepts are further debated in committees or working groups and, if accepted, are submitted to the experimental process. This may then lead to a project definition based on the prior collected results from studies and experimentations.

Project

In contemporary business and science a project[250] is a collaborative enterprise, involving research or design, which is carefully planned to achieve a particular aim, a set of interrelated tasks to be executed over a fixed period and within certain cost and other limitations. The defined time frame has a set start and end date.

Projects can be further defined as temporary rather than permanent social systems or work systems that are constituted by teams within or across organizations to accomplish particular tasks under time constraints.

An ongoing project is usually transforms or evolves into a program[251]. The term program can describe anything in business and management, media, art and entertainment, an organization, medicine, science and philosophy and technology. The main difference to the term project is that a program has an initial start point but not an in a time line defined end state. The program has objectives to be reached within a roughly estimated duration without the definition of termination.

Prototype

All concepts, demonstrations and experimentations aim in providing the background for a prototype[252] as an early sample, model, or release of a product built to test the given concept or process or to act as a thing to be replicated or learned from. The term prototype is used in a variety of contexts, including semantics, design, electronics, and software programming. A prototype is designed to test and try a new design to enhance precision by system analysts and users.

Prototyping serves to provide specifications for a real, working system rather than a theoretical one. In some workflow models, creating a prototype (a process sometimes called materialization) is the step between the formalization and the evaluation of an idea.

---

[250] https://en.wikipedia.org/wiki/Project
[251] https://en.wikipedia.org/wiki/Program
[252] https://en.wikipedia.org/wiki/Prototype

Standardization

The International Maritime Organization (IMO), the sponsorship of the G7++, G20, the European Union, NATO and most coastal countries nationally, either initiated or support a vast number of concepts, demonstration and experimentations, projects and programs under Maritime Security and Maritime Situational Awareness (MSA).

One key element is Standardization. In projects, programs, systems and networking it needs to be divided into operational standards and technical standards.

The operational standards include all documents that define the business rules, legal aspects, etc. pp. for the governing organization and/or the operational community.

The technical standards can be divided into:

- International Standards
- National or Organizational Standards
- Military Standards
- Open Standards
- Proprietary Standards[253]

Each country and international institution thrives to standardize development to ensure an easy and secure guidance throughout any process. The United States Department of Defense – as just one example many follow – uses variations of:

- Doctrine, Organization, Training, Material, Leadership and Education, Personnel and Facilities (DOTMLPF)
- or Doctrine, Organization, Training, Material, Logistic & Medical, Personnel, Legal & Finance (DOTMLMPLF)

DOTMLPF is defined in the Joint Capabilities Integration Development System, or JCIDS Process. The JCIDS process provides a solution space that considers solutions involving any combination of doctrine, organization, training, materiel, leadership and education, personnel and facilities (DOTMLPF).

The interesting question in every new concept is, what was traded in at the end to gain what new feature for the future. Despite all efforts to reach Standardization, we can observe multiple siblings, sometimes identical twins, when politics or business rules interfere with the natural selection of the best system. When there is plenty of funds available and

---

[253] A proprietary standard is a actually more a specification or standard owned and used by a company, organization, or individual and is in most cases no bringing the same benefit as the other standards. However, sometimes a proprietary standard can become one of the above due to usage or a monopole.

only a limited mindset or an over-complex approach, paired with a slight idea on the subject, then are you are in most cases in the middle of a program.

Since combatant commanders define requirements in consultation with the Office of the Secretary of Defense (OSD), they are able to consider gaps in the context of strategic direction for the total US military force and influence the direction of requirements earlier in the acquisition process. It also serves as a mnemonic for staff planners to consider certain issues prior to undertaking a new effort.

The idea at that time was to fix the capability gap, and CJCSI 3170.01G – Joint Capabilities Integration and Development System, 1. March 2009, is the one governing instruction that encompasses both materiel (requiring new defense acquisition programs) and non-materiel (not requiring new defense acquisition program) solutions.

NATO adopted it in Doctrine, Organization, Training, Material, Leadership and Education, Personnel, Facilities and Interoperability (DOTMLPFI) which includes the areas:

- Doctrine (ACO lead)
- Legal (ACT lead)
- Organization (ACO lead)
- Training (ACT lead)
- Material & Technology (ACT lead)
- Facilities (ACO lead)
- and Interoperability (ACT lead)

All Study Areas to include: background, scope & objectives, current assets, activities, deliverables, timeline, resources & organizations (NATO & non-NATO) involvement, including interdependencies between studies.

Communications and Information Systems (CIS) to C5ISR

A system[254] can refer to anything interacting with infrastructure, organization, personnel, and components. See here Information System (IS). System is a "*whole compounded of several parts or members* ", literary "*composition*", a set of interacting or interdependent components forming an integrated whole. A system is a set of elements and relationships which are different from relationships of the set or its elements to other elements or sets.

Fields that study the general properties of systems include systems theory, cybernetics, dynamical systems, thermodynamics and complex systems. They investigate the abstract properties of systems' matter and organization, looking for concepts and principles that are independent of domain, substance, type, or temporal scale. The term system may also refer to a set of rules that governs structure and/or behavior.

Most systems share common characteristics, including:

- Systems have structure, defined by components and their composition;
- Systems have behavior, which involves inputs, processing and outputs of material, energy, information, or data;
- Systems have interconnectivity: the various parts of a system have functional as well as structural relationships to each other.
- Systems may have some functions or groups of functions

Network theory[255] is the study of graphs as a representation of either symmetric relations or asymmetric relations between discrete objects. In computer science and network science, network theory is a part of graph theory: a network can be defined as a graph in which nodes and/or edges have attributes (i.e. names).

Network theory has applications in many disciplines including statistical physics, particle physics, computer science, electrical engineering, biology, economics, finance, operations research, climatology, ecology and sociology. Applications of network theory include logistical networks, the World Wide Web, Internet, gene regulatory networks, metabolic networks, social networks, epistemological networks, etc..

Maritime Entities and their systems are connected in networks. Humans practice social networking using digital media allowing communication and information exchange between separated nodes via technical networks interconnecting electronically relevant components.

At the beginning of the Internet, networking was largely either government-sponsored (ARPANET, CYCLADES) or vendor-developed and proprietary, the latter effort consisting of

---

[254] https://en.wikipedia.org/wiki/System
[255] https://en.wikipedia.org/wiki/Network_theory

protocol standards such as SNA, Appletalk, NetWare and DECnet. It was common for large networks to support multiple network protocol suites, with many devices unable to interoperate with other devices because of a lack of common protocols. However, while OSI developed its networking standards, TCP/IP came into widespread use on multivendor networks for inter-networking, while on the local network level both Ethernet and token ring gained prominence.

Open Systems Interconnection (OSI) is an effort to standardize networking that was started in 1977 by the International Organization for Standardization (ISO), along with the ITU-T and industry, attempting to get industry participants to agree on common network standards to provide multi-vendor interoperability.

The Open Systems Interconnection model (OSI model)[256] is a conceptual model that characterizes and standardizes the communication functions of a telecommunication or computing system without regard to its underlying internal structure and technology. Its goal is the interoperability of diverse communication systems with standard protocols. The model partitions a communication system into abstraction layers. The original version of the model defined seven layers.

The OSI reference model was a major advance in the teaching of network concepts. It promoted the idea of a consistent model of protocol layers, defining interoperability between network devices and software. The OSI model was established by ISO in 1984.

Networks are using different topologies. The idea of topology and the foundation of graph theory was prefigured by the negative resolution of Leonhard Euler in 1736 for the historically notable problem in mathematics, called "The Seven Bridges of Königsberg[257]".

The city of Königsberg in Prussia (now Kaliningrad, Russia) was set on both sides of the Pregel River, and included two large islands - Kneiphof and Lomse - which were connected to each other, or to the two mainland portions of the city, by seven bridges. The problem was to devise a walk through the city that would cross each of those bridges once and only once. By way of specifying the logical task unambiguously, solutions involving either

- reaching an island or mainland bank other than via one of the bridges, or
- accessing any bridge without crossing to its other end

are explicitly unacceptable.

Euler proved that the problem has no solution. The difficulty he faced was the development of a suitable technique of analysis, and of subsequent tests that established this assertion with mathematical rigor. Euler's solution of the Seven Bridges of Königsberg problem is considered to be the first true proof in the theory of networks.

---

[256] https://en.wikipedia.org/wiki/Open_Systems_Interconnection
[257] https://en.wikipedia.org/wiki/Seven_Bridges_of_Königsberg

System-of-Systems (SOS)

System-of-Systems, Federated or Decentralized Systems concepts aim to preserve financial resources and avoid duplication in a federation of system through an interfacing approach. System-of-Systems (SOS) is a collection of task-oriented or dedicated networks of systems that pool their resources and capabilities together to obtain a new, more complex, 'meta-system' which offers more functionality and performance than simply the sum of the constituent systems. System-of-Systems was in modified by the GBR MOD from the US DOD definition stating, *"a SOS is a set or arrangement of systems that result when independent and useful systems are combined into a larger system that delivers capabilities not deliverable by any individual system."*

System-of-Systems (SOS) is a critical concept and research discipline for which frames of reference, thought processes, quantitative analysis, tools, and design methods are often incomplete. The methodology for defining, abstracting, modeling, and analyzing system of systems problems is typically referred to as system of systems engineering. Modern systems that comprise System-of-Systems (SOS) are large-scale concurrent and distributed networks with challenges that are not monolithic; rather they have five common characteristics:

- operational independence of the individual systems
- managerial independence of the systems
- geographical distribution
- emergent behavior and evolutionary development
- description in the field of evolutionary acquisition of complex adaptive systems

Taken together, all these descriptions of System-of-Systems (SOS) suggest that a complete system of systems engineering framework is needed to improve decision support for system of systems problems. Specifically, an effective system of systems engineering framework is needed to help decision makers to determine whether related infrastructure, policy, and/or technology considerations as an interrelated whole are good, bad, or neutral over time. The need to solve System-of-Systems (SOS) problems is urgent not only because of the growing complexity of today's grand challenges, but also because such problems require large monetary and resource investments with multi-generational consequences. While the individual systems constituting a System-of-Systems (SOS) can be very different and operate independently, their interactions typically expose and deliver important emergent properties. These emergent patterns have an evolving nature that stakeholders for these problems must recognize, analyze, and understand.

The System-of-Systems (SOS) approach does not advocate particular tools, methods, or practices; instead, it promotes a new way of thinking for solving grand challenges where the interactions of technology, policy, and economics are the primary drivers System-of-Systems (SOS) study is related to the general study of designing, complexity and systems engineering, but also brings to the forefront the additional challenge of design. System-of-

Systems (SOS) typically exhibit the behaviors of complex systems. But not all complex problems fall in the realm of SOS. Inherent to System-of-Systems (SOS) problems are several combinations of traits, not all of which are exhibited by every such problem. System-of-Systems (SOS) is used in concepts as Network Enabled Capabilities (NEC) in NATO or EU and in co-operations like e.g. SUCBAS, MARSUR and many more.

The System-of-Systems Approach (SOSA) is delivering enhanced capability through achieving commonality, re-use and the interoperability of independent procured systems. Projects, Systems & Programs for a System-of-Systems approach need to understand their Enterprise Constraints in the SOS context. Projects and program development should not make decisions against local criteria in isolation of enterprise level considerations, since local decisions often have global significance.

System Of Systems Approach (SOSA) and Systems Engineering / Systems Thinking needs to be recognized as part of the solution to enable delivery of successful projects. A SOSA approach needs to be applied systematically and logically as part of the systems engineering lifecycle of a project. Projects & Programs need to apply to own SOSA Principles, Operating Model and Rules.

System Software is computer software designed to operate the computer hardware and to provide and maintain a platform for running application software.

Types of system software[258]:

- The computer BIOS and device firmware, which provide basic functionality to operate and control the hardware connected to or built into the computer.
- The operating system (prominent examples being Microsoft Windows, Mac OS X and Linux), which allows the parts of a computer to work together by performing tasks like transferring data between memory and disks or rendering output onto a display device. It also provides a platform to run high-level system software and application software.
- Utility software, which helps to analyze, configure, optimize and maintain the computer.

In some publications, the term system software is also used to designate software development tools (like a compiler, linker or debugger). System software is usually not what a user would buy a computer for - instead, it can be seen as the basics of a computer which come built-in or pre-installed. In contrast to system software, software that allows users to do things like create text documents, play games, listen to music, or surf the web is called application software.

A Status is a state, condition, or situation, therefore the System Status is the state, condition,

---

[258] http://en.wikipedia.org/wiki/System_software

or situation of any system under the defined terminology, like i.e. under "*operational*", and most of the times it is connected with a System Status Message.

Communication (from Latin "*communicare*", meaning "*to share*") is the act of conveying meanings from one entity or group to another through the use of mutually understood signs, symbols, and semiotic rules. Communication is the process by which information, meanings and feelings are shared by people through an (intended or unintended) exchange of verbal and non-verbal messages. The main steps inherent to all communication are:

The formation of communicative motivation or reason.

- Message composition (further internal or technical elaboration on what exactly to express).
- Message encoding (for example, into digital data, written text, speech, pictures, gestures and so on).
- Transmission of the encoded message as a sequence of signals using a specific channel or medium.
- Noise sources such as natural forces and in some cases human activity (both intentional and accidental) begin influencing the quality of signals propagating from the sender to one or more receivers.
- Reception of signals and reassembling of the encoded message from a sequence of received signals.
- Decoding of the reassembled encoded message.
- Interpretation and making sense of the presumed original message.

Communications (COMMS) is the science and practice of the conveyance of information of any kind from one person or place to another except by direct unassisted conversation or correspondence. In most organizations the information transfer is regulated under agreed conventions.

A Communication System (CS) is in general any assembly of equipment, methods and procedures, and if necessary personnel, organized to accomplish information transfer functions. A communication system provides communication between its users and may embrace transmission systems, switching systems and user systems. A communication system may also include storage or processing functions in support of information transfer.

Communications and Information Systems (CIS) is a collective term extended up to Command, Control and Communications, Computers, Combat Systems, Intelligence, Surveillance and Reconnaissance (C5ISR). Another collection of CIS are Intelligence, Surveillance, Target, Acquisition & Reconnaissance (ISTAR)[259], which in its macroscopic

---

[259] https://en.wikipedia.org/wiki/Intelligence,_surveillance,_target_acquisition,_and_reconnaissance

sense is a practice that links several battlefield functions together to assist a combat force in employing its sensors and managing the information they gather. Command and Control (C2)[260] has been coupled in the past with Communication/Communications, (Military) Intelligence, Information/Information Systems, Computers/Computing, Surveillance, Target Acquisition, Reconnaissance, Interoperability, Collaboration, Electronic Warfare and some of the more common defined terminologies and abbreviations variations include:

- C2 (Command and Control)
- C2I (Command, Control & Intelligence)
- C2I (Command, Control & Information, a less common usage)
- C2IS (Command and Control Information Systems)
- C2ISR (C2I plus Surveillance and Reconnaissance)
- C2ISTAR (C2 plus ISTAR (Intelligence, Surveillance, Target Acquisition, and Reconnaissance))
- C3 (Command, Control and Communications, in human activity focus)
- C3 (Command, Control & Communications, in technology focus)
- C3 (Consultation, Command, and Control, used in NATO)
- C3I (4 combination possibilities; the most common is Command, Control, Communications and Intelligence)
- C3ISTAR (C3 plus ISTAR)
- C3ISREW (C2ISR plus Communications plus Electronic Warfare, in technology focus)
- C4 (Command, Control, Communications, Computers as C4, C4I, C4ISR, C4ISTAR, C4ISREW, C4ISTAREW – plus Computers in technology focus or Computing in human activity focus)
- C4I2 (Command, Control, Communications, Computers, Intelligence, and Interoperability)
- C3ISR (Command, Control, Communications Systems, Intelligence, Surveillance and Reconnaissance)
- C4ISR (Command, Control, Communications, Computers, Intelligence, Surveillance and Reconnaissance)
- C5I (Command, Control, Communications, Computers, Collaboration and Intelligence)
- C5ISR (Command, Control and Communications, Computers, Combat Systems, Intelligence, Surveillance and Reconnaissance)
- NC2 (Nuclear command and control)
- NC3 (Nuclear command and control and communications)

---

[260] https://en.wikipedia.org/wiki/Command_and_control

Other abbreviations are possible, however their terminology and definition depend on the individual use case of the relevant Maritime Entity.

Communications (COMMS) transform through the information provided. When interests collide, policies matter, but leadership and management count. Mistakes dismay friends and provide enemies and opponents with unintentional assistance. Communication is not the problem, but it is a problem. To focus the processes and efforts one has to understand and engage the target audience to create, strengthen or preserve conditions favorable to advance interests and objectives through the use of coordinated information, themes, plans, programs, and actions synchronized with other elements of power. COMMS can also just relate to any communication equipment.

There is an abundance of systems and networks in use, most are in constant transformation to adapt to the fast changing requirements, new evolve to deal with new requirements. The listed system and networks in *"Maritime Situational Awareness for Tyrians"* are some selected of the most relevant, however a list in the SA/MSA field can be never complete, is always a status quo an basic reference to start an own research project.

Situational Awareness is never static. *"Maritime Situational Awareness for Tyrians"* is a brief introduction on MSA, Maritime Entities and related Technologies, a first step into a very exiting story of constant change and transformation. The information provided is an account of personal work experience, hopefully avoiding the phenomenon of „*expertise induced amnesia*". However, for a dynamic topic like MSA a comprehensive coverage is hardly achievable, therefore submissions of correction or additional information are most welcome.

Some organizations, agencies, sometimes even companies, are not the main active entity, they maybe use a system for data collection, but aim mainly at the task as information provider. Such are in some way neither an entity not are systems or network but possible simply a source for data collection and information gathering, however some are also collected below.

In the end the dynamic field will require a constant awareness about the changing world of awareness, performing online research, participation, reading and listening, enhancing the own skills in the art of communication.

"Situational Awareness is less the technology and more the art of communication of the individual important in the jungle of information, however the supporting tools are the most complex and relevant.[261]"

---

[261] Joachim Beckh

Fusion

Fusion[262], as in Data Fusion, Information Integration, Information Fusion, Deduplication and referential integrity is the merging of information from disparate sources with differing conceptual, contextual and typographical representations. It is used in data mining and consolidation of data from unstructured or semi-structured resources. Typically, information integration refers to textual representations of knowledge but is sometimes applied to rich-media content. The technologies available to integrate information include string metrics which allow the detection of similar text in different data sources by fuzzy matching.

In Situational Awareness Track Fusion enables the access to all position information from various sources about one track in order to investigate the source, target and situation related information separately. Track Correlation[263] will keep fire-control on a single track position established from various sources. The source with highest trust in position (i.e. radar, laser directing) will be used for the correlated track.

Correlations are useful because they can indicate a predictive relationship that can be exploited in practice. The narrower term "correlation" refers to a linear relationship between two quantities, whereas the broader term "association" refers broadly to any such relationship between two measured quantities that renders them statistically dependent. To proper analyze information a correlation engine and a fusion engine is required to achieve the best result for an operator to make a final decision on ambiguities and anomalies; see also Pedigree and Fusion & Correlation Models.

The term Fusion is described in NATO with: "*In intelligence usage, the blending of intelligence and/or information from multiple sources or agencies into a coherent picture. The origin of the initial individual items should then no longer be apparent*".

Data fusion, is generally defined as the use of techniques that combine data from multiple sources and gather that information into discrete, actionable items in order to achieve inferences, which will be more efficient and narrowly tailored than if they were achieved by means of disparate sources. Data fusion processes are often categorized as low, intermediate or high, depending on the processing stage at which fusion takes place. Low level fusion, (Data fusion) combines several sources of raw data to produce new raw data. The expectation is that fused data is more informative and synthetic than the original inputs. For example, sensor fusion is also known as (multi-sensor) data fusion and is a subset of information fusion.

---

[262] https://en.wikipedia.org/wiki/Fusion
[263] https://en.wikipedia.org/wiki/Correlation_(disambiguation)

Correlation

Track Correlation will keep fire-control on a single track position established from various sources. The source with highest trust in position (i.e. radar, laser directing) will be used for the correlated track. In NATO correlation is described as: "... *the determination that an ... (track) ... appearing on a detection or display device or visually, is the same as that on which information is being received from another source*".

In Situational Awareness Track Fusion enables the access to all position information from various sources about one track in order to investigate the source, target and situation related information separately.

Correlation refers to any of a broad class of statistical relationships involving dependence and dependence refers to any statistical relationship between two random variables or two sets of data. Correlations are useful because they can indicate a predictive relationship that can be exploited in practice. The narrower term "*correlation*" refers to a linear relationship between two quantities, whereas the broader term "*association*" refers broadly to any such relationship between two measured quantities that renders them statistically dependent. To proper analyze information a correlation engine and a fusion engine is required to achieve the best result for an operator to make a final decision on ambiguities and anomalies.

Familiar examples of dependent phenomena include the correlation between the physical statures of parents and their offspring, and the correlation between the demand for a product and its price. Correlations are useful because they can indicate a predictive relationship that can be exploited in practice. For example, an electrical utility may produce less power on a mild day based on the correlation between electricity demand and weather. In this example there is a causal relationship, because extreme weather causes people to use more electricity for heating or cooling; however, statistical dependence is not sufficient to demonstrate the presence of such a causal relationship.

Formally, dependence refers to any situation in which random variables do not satisfy a mathematical condition of probabilistic independence. In loose usage, correlation can refer to any departure of two or more random variables from independence, but technically it refers to any of several more specialized types of relationship between mean values. There are several correlation coefficients measuring the degree of correlation. The most common of these is the Pearson correlation coefficient[264], which is sensitive only to a linear relationship between two variables (which may exist even if one is a nonlinear function of the other). Other correlation coefficients have been developed to be more robust than the Pearson correlation — that is, more sensitive to nonlinear relationships.

---

[264] http://en.wikipedia.org/wiki/Pearson_product-moment_correlation_coefficient

Fusion & Correlation Models

In the past different models have been developed to describe the complex subjects of Fusion and Correlation. Among other models, such as the *"Waterfall Model"*, the *"Intelligence Cycle"* or the *"Dasarathy Model"* are four models which might be considered:

- Joint Directors of Laboratories Data Fusion Working Group Model (JDL Model), with roots in the U.S. *"data fusion community"* of the U.S. Armed Forces;
- Boyd Control Loop with the division into the four phases of *"Observe"*, *"Orient"*, *"Decide"* and *"Act"* also referred as the *"OODA Loop"*;
- Omnibus Model on the basis of the *"Boyd Control Loop"* (OOAD) model combines several older models and
- Extended OOAD Loop developed by Lockheed Martin with some aspects of the previous models and the mechanism for multiple parallel, concurrent and interactive processes of data fusion.

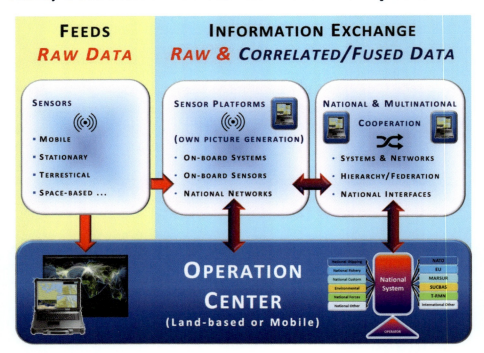

Raw, correlated and fused data for land-based or mobile Operation Centers

Interoperability

Interoperability[265] is a property referring to the ability of diverse systems and organizations to work together (inter-operate). The term is often used in a technical systems engineering sense, or alternatively in a broad sense, taking into account social, political, and organizational factors that impact system to system performance. Interoperability can therefore be seen under the fields of Operational Interoperability, Technical Interoperability, Procedural Interoperability and possible other dependencies.

Operational, Technical, Procedural Interoperability and possible other dependencies

Security organizations increasingly face technical, operational, and human interoperability issues at their geographical and organizational borders. A vigorous political will to share assets and standards will empower us all in jointly handling the security issues posed by a progressively more interlinked world.

Interoperability has many facets, the most important are Concepts (Concepts are establishing Security aims and tasks, define its place and role in the structure, and sets linkages

---

[265] http://en.wikipedia.org/wiki/Interoperability

with National Resources), Doctrine (Common doctrine as the guiding element of all activities), Tactics (Common tactics would further support), Logistics, Communication, Material, Training and Standardization.

The ISO-IEC provides the following definitions of the levels of standardization:

- Commonality (highest level) "The state achieved when the same doctrine, procedures or equipment are used".
- Interchangeability (middle level) "The ability of one product, process or service to be used in place of another to fulfill the same requirements"
- Compatibility (lowest level) "The suitability of products, processes, or services for use together under specific conditions to fulfill relevant requirements without causing unacceptable interactions".

Technical Interoperability

Technical Interoperability refers to the technical exchange in the operational use of information in relation to the meaning and content. There is the need to identify architecture and products, define Information Exchange Requirements (IER) the business process (Procedural Interoperability) as well as technical and human interface (Operational Interoperability) requirements together with technical and human interface standards.

This includes an interoperability profile, a comprehensive framework, an interoperability matrix capturing IER compliance of participating systems, and interoperability study addressing key aspects of the Concept Development Plan.

Operational Interoperability

Operational Interoperability relates to the meaning, content and the operational use of information to be exchanged. Technical Interoperability refers only to the technical exchange in the operational use of information in relation to the meaning and content. There is the need to identify the business process (Procedural Interoperability) as well as technical and human interface (Operational Interoperability) requirements together with technical and human interface standards. This includes an interoperability profile, a comprehensive framework, an interoperability matrix capturing IER compliance of participating systems, and interoperability study addressing key aspects of the Concept Development Plan.

The effectiveness of forces in peace time, crisis or in conflict, depends on the ability of the forces provided to operate together coherently, effectively and efficiently. Joint operations should be prepared for, planned and conducted in a manner that makes the best use of the relative strengths and capabilities of the forces which members offer for an operation. Interoperability of formations and units of a joint and multinational unit has three dimensions, technical (i.e. hardware, systems,) procedural (i.e. doctrines, procedures) and human (i.e. language, terminology, and training). Forces commit to information sharing through the lessons learned process, in particular about interoperability shortfalls.. At the operational level, emphasis should be placed on the integration of the contributing nations' forces and the synergy that can be attained. The success of the process will determine the ability of a joint force to achieve its commander's objectives.

Procedural Interoperability

While technical Interoperability refers only to the technical exchange in the operational use of information in relation to the meaning and content, the Procedural Interoperability refers to the procedures in the operational use and technical exchange of information in relation to the meaning and content.

There is the need to identify the business process (Procedural Interoperability) as well as technical and human interface (Operational Interoperability) requirements together with technical and human interface standards. This includes an interoperability profile, a comprehensive framework, an interoperability matrix capturing IER compliance of participating systems, and interoperability study addressing key aspects of the Concept Development Plan.

Information Exchange Requirement (IER)

Information Exchange between partners in a cooperation aims in enhancing the Maritime Security Environment, bringing benefits to the Maritime Safety, Security, Environmental and Economic matters by sharing the Information between the relevant authorities of the participating nations. Based on developed concepts with follow-on technical solution and procedures based on common standards and distribution principles with a design to have low impact on the autonomous national systems by sharing and consuming information through services depending on bilateral agreements, with priority given to locally and nationally produced track history as well as notifications from well managed and sustainable data bases.

An Information Exchange Requirement (IER) is the description, in terms of characteristics, of the requirement to transfer information between two or more end users. The characteristics described include source, recipients, contents, size, timeliness, security and trigger.

Information Exchange Requirement (IER) is the identified and detailed described process of information exchange between technical (system) or operational (human) nodes that can be envisaged in the Operational Information Requirements sub-view and as graphical view in an Information Exchange Requirement Matrix (IER Matrix). An IER is a statement of need to exchange information between co-operating forces or HQs specifying the information to be exchanged in a standardized manner within the context of the mission, key tasks, required degree of inter-operability and the parameters of CIS involved.

As Information and its exchange are quite dynamic so are the defined requirements. The static IER definitions linked with the tender of a project will need to be constantly mapped with the status quo of the user community in order to provide the best product suited for operational use.

Network Topology

Network topology[266] is the arrangement of the elements (links, nodes, etc.) of a communication network. Network topology can be used to define or describe the arrangement of various types of telecommunication networks, including command and control radio networks, industrial field busses, and computer networks.

Network topology is the topological structure of a network and may be depicted physically or logically. It is an application of graph theory wherein communicating devices are modeled as nodes and the connections between the devices are modeled as links or lines between the nodes. Physical topology is the placement of the various components of a network (e.g., device location and cable installation), while logical topology illustrates how data flows within a network. Distances between nodes, physical interconnections, transmission rates, or signal types may differ between two different networks, yet their topologies may be identical. A network's physical topology is a particular concern of the physical layer of the OSI model.

Examples of network topologies are found in local area networks (LAN), a common computer network installation. Any given node in the LAN has one or more physical links to other devices in the network; graphically mapping these links results in a geometric shape that can be used to describe the physical topology of the network. A wide variety of physical topologies have been used in LANs, including ring, bus, mesh and star. Conversely, mapping the data flow between the components determines the logical topology of the network. In comparison, Controller Area Networks, common in vehicles, are primarily distributed control system networks of one or more controllers interconnected with sensors and actuators over, invariably, a physical bus topology. Two basic categories of network topologies exist, physical topologies and logical topologies.

The transmission medium layout used to link devices is the physical topology of the network. For conductive or fiber optical mediums, this refers to the layout of cabling, the locations of nodes, and the links between the nodes and the cabling. The physical topology of a network is determined by the capabilities of the network access devices and media, the level of control or fault tolerance desired, and the cost associated with cabling or telecommunications circuits.

In contrast, logical topology is the way that the signals act on the network media, or the way that the data passes through the network from one device to the next without regard to the physical interconnection of the devices. A network's logical topology is not necessarily the same as its physical topology. For example, the original twisted pair Ethernet using repeater hubs was a logical bus topology carried on a physical star topology. Token ring is a logical ring topology, but is wired as a physical star from the media access unit.

---

[266] https://en.wikipedia.org/wiki/Network_topology

Logical topologies are often closely associated with media access control methods and protocols. Some networks are able to dynamically change their logical topology through configuration changes to their routers and switches.

In Maritime Situational Awareness or SA in general, the Maritime Entities either have already own technical system in place and only want to link those to enhance the information exchange or they have no technical system in place an rely solely on the data provided by the other partners, meaning they require a graphical user Interface (GUI) to view and process the information. Here basically there are two types of network topologies for Maritime Entities in a cooperation:

- Hierarchical Topology
- Full Mesh Network Topology
- Hierarchical Topology and Full Mesh Network Topology might both provide a graphical user Interface (GUI) to view and process the available information.

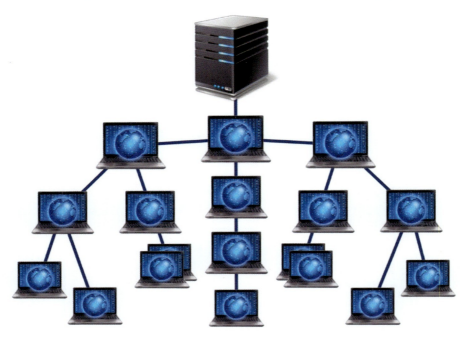

Centralized Server and Services for all participants

The Hierarchical Topology requires a new central server, system – and most of the time a centralized agency/organization - to link all Entities and ensure the distribution and exchange of the information from the connected systems of all stakeholders. Most existing systems are in an hierarchical or mixed topology, like in a ring, star, tree, line, bus or mesh.

Mesh Network[267] is a type of network wherein each node may act as an independent router, regardless of whether it is connected to another network or not. It allows for continuous connections and reconfiguration around broken or blocked paths by "*hopping*" from node to node until the destination is reached. A mesh network whose nodes are all connected to each other is a fully connected network[268].

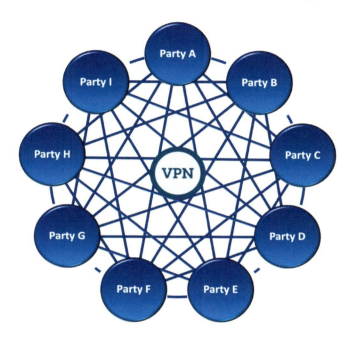

Full Mesh Network Topology

Number of connections (c) linking all nodes (n) in a full mesh network $c = n(n-1)/2$

Mesh networks differ from other networks in that the component parts can all connect to each other via multiple hops, and they generally are not mobile. Mesh networks can be seen as one type of ad hoc network. Mobile ad hoc networks (MANET) and mesh networks are therefore closely related, but MANET also have to deal with the problems introduced by the mobility of the nodes. Mesh networks are self-healing: the network can still operate when one node breaks down or a connection goes bad.

As a result, the network may typically be very reliable, as there is often more than one path between a source and a destination in the network. Although mostly used in wireless scenarios, this concept is also applicable to wired networks and software interaction. The animation at the right illustrates how wireless mesh networks can self-form and self-heal. For

---

[267] http://en.wikipedia.org/wiki/Mesh_networking
[268] http://en.wikipedia.org/wiki/Fully_connected_network

more animations see History of Wireless Mesh Networking[269].

Wireless Mesh Networks (WMN) were originally developed for military applications and are typical of mesh architectures. Over the past decade the size, cost, and power requirements of radios has declined, enabling more radios to be included within each device acting as a mesh node. The additional radios within each node enable it to support multiple functions such as client access, backhaul service, and scanning (required for high speed handover in mobile applications). Additionally, the reduction in radio size, cost, and power has enabled the mesh nodes to become more modular - one node or device now can contain multiple radio cards or modules, allowing the nodes to be customized to handle a unique set of functions and frequency bands.

The Full Mesh Network Topology does not require any new central server, system and a centralized agency/organization and links each Maritime Entity directly with each and every other participating stakeholder. The Full Mesh Network Topology is more complex and requires, with a growing number of participants, most of the time a centralized agency/organization in order to manage the exchange system, which can be either implemented straight into the system to be adapted or as additional hardware with the external system implementation which then connects to the system to be linked.

The ideal Full Mesh Network Topology has all participants connected on the same level within a certain time period. In reality due to the different business processes, requirements, budgets, procurement will and many more factors, the complete Full Mesh Network is hard to achieve, and same accounts for any other technical network, where the partner will connect at different times.

Full Mesh Topology occurs when every node has a circuit connecting it to every other node in a network. Full mesh is very expensive to implement but yields the greatest amount of redundancy, so in the event that one of those nodes fails, network traffic can be directed to any of the other nodes. Full mesh is usually reserved for backbone networks.

Partial mesh topology is less expensive to implement and yields less redundancy than full mesh topology. With partial mesh, some nodes are organized in a full mesh scheme but others are only connected to one or two in the network. Partial mesh topology is commonly found in peripheral networks connected to a full meshed backbone.

Full Mesh Topology is more complex to establish and maintain than a hierarchical network, however it is also more robust against failures or cyber-attacks as there is no central node[270] as target and main point of failure.

---

[269] http://en.wikipedia.org/wiki/History_of_wireless_mesh_networking
[270] Except for a central dedicated time server required to synchronize all nodes no matter which topology. I case of loss of the time server operation of the network may be affected, but information exchange will be in general still possible.

Examples are the few Full Mesh Networks in existence today, the Sea Surveillance Cooperation Baltic Sea (SUCBAS) and the Maritime Surveillance (MARSUR) Networking. While SUCBAS with today 9 participants was able to establish the network links within a few years, MARSUR has more than double the number of members and has still not all links to all nodes (participants) established. One reason is shown in the complex MARSUR Network Topology and the required links between the nodes of the 22 members (status 2019), which amounts to 231 Virtual Private Network (VPN) connections.

MARSUR Full Mesh Network Topology

Each of the 22 nodes connects to 21 other nodes (c = n(n−1)/2): c = 231 links

Based on the early specification of the Federated Mission Network (FMN) from the NATO TIDE MARSUR and SUCBAS are today still the only operational federated network in military use. Based on the development of the Afghanistan Mission Network (AMN) the NATO Military Committee (MC) tasked Allied Command transformation together with the Allied Command Operations (ACO) in the development of the Future Mission Network (FMN). Pursuant the development of the level of ambition, and of governance and management principles, NATO later adopted the term "*Federated Mission Networking*" (FMN):

"The interaction of people, processes and technology to exchange information and/or services among federated mission participants including but not limited to the use of a set of interconnected autonomous computer networks for the conduct of coalition operations and exercises".

Note that in SUCBAS, as well as in MARSUR, all nodes are connected to either a system or a Graphical User Interface (GUI, here the SUCBAS Stand-Alone-Client), while in MARSUR either a system is connected or the MARSUR User Interface (MARSUR-UI), however not all nodes are linked with each other.

SUCBAS Network State-of-Play 2018

Each Maritime Entity has an individual implementation process and therefore an independent time line on when services can and will be made available to other partners. The type of network topology is not determine the overall time line for system connectivity, however the complexity of the network topology chosen does often have an impact on the projects and the implementation.

In a decentralized or federated architecture the finalization of the standardized interface is just the start for the individual projects of each participant, which will have to evaluate the cost, provide the required funding and the appropriate time line.

Any project requires a funding according to its complexity. The funding needs to be requested in the financial planning for the follow-on annual household of an organization. Defining the requirements, cost estimate and funding are set to a minimum of two years. Depending on the complexity of the project the duration will be between in short-term within a year or mid- to long-term, meaning up to 5 or 10 years, until the product will be delivered.

At least two participants need to connect to establish a first network link and declare an Initial Operational Capability (IOC) and after all services are operational declare a Full Operational Capability (FOC). Other participant will follow according to their individual time lines and will join the network to provide further links and possible information. However, only when the last participant of the community has finished the implementation and is connected to the full mesh network, a Full Operational Capability (FOC) of the whole community is achieved.

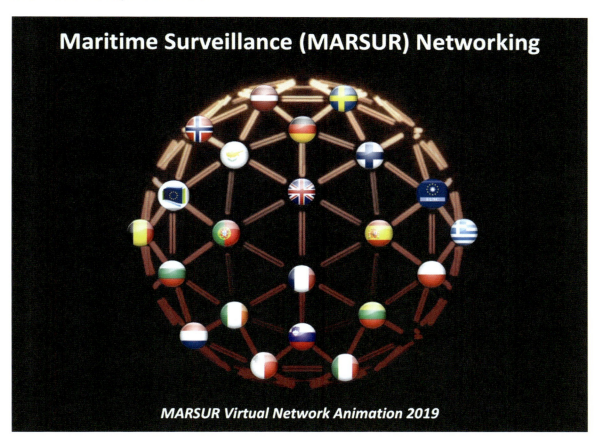

The more members join a community, the slower the processes in the groups will be and the slower the implementation of the product. While the 8 members in the SUCBAS cooperation have reached a full meshed network and connected every participant, the 19 members in MARSUR (status 2018) had reached only a partly implementation.

Some nation had national systems and in addition even a MARSUR User Interface connected, some had only the national system or a MARSUR User Interface, while others were still not even connected to the VPN. This is due to the higher numbers of participants, their time of joining the cooperation and the structure of the community. In 2019 MARSUR had already 21 members and an application of the European Defence Agency (EDA) as the 22. member. The nation will most likely all join with a own system or use the MARSUR User Interface, while EDA as a European Agency is no operational Headquarter, does not lead any military operations and will therefore probably never have either one.

One reason for slow speed in implementation of a mesh network are the different time schedules of each participant. When an EDA/EU project has finished and delivered a prototype or product, the implementation cycle in each nation starts; however their respective ending will hardly ever be at the same time. Below is a diagram explaining the complex implementation cycle in the Maritime Surveillance (MARSUR) Full Mesh Network.

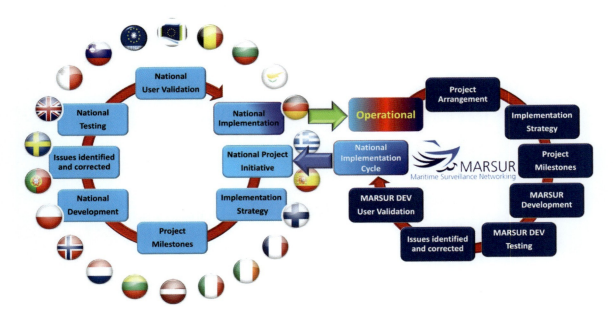

22 participants with an individual implementation process to be considered

Operating Capability

IOC (Initial Operating Capability or Initial Operational Capability)

IOC is sometimes referred as Initial Operating Capability or Initial Operational Capability, just as FOC has different national definitions.

Initial Operating Capability or Initial Operational Capability (IOC) is the state achieved when a capability is available in its minimum usefully deployable form. The term is often used in government or military procurement and the U.S. Department of Defence chose to use the term Initial Operational Capability (versus initial "operating" capability) when referring to IOC. The date at which IOC is achieved often defines the in-service date (ISD) for an associated system.

Declaration of an initial operating capability may imply that the capability will be developed in the future, for example by modifications or adjustments to improve the system's performance, deployment of greater numbers of systems (perhaps of different types), or testing and training that permit wider application of the capability. Once the capability is fully developed, Full Operational Capability (FOC) may be declared. For example, the capability may be fielded to a limited number of users with plans to roll out to all users incrementally over a period of time (possibly incorporating changes along the way).

The point at which the first users begin using the capability is IOC, with FOC achieved when all intended users (by agreement between the developer and the user) have the capability. This does not preclude additional users from obtaining the capability after FOC. The in-service date (ISD) can occur when the capability defined in the User Requirement Definition (URD) is assessed as available for operational use - in its minimum usefully deployable form. It is the date on which an Initial Operating Capability (IOC) is achieved.

Keep in mind that *"assessed as available for operational use"* can only be defined by national OPERATORS, who have to evaluate the system services according to a predefined test schedule and based on their judgment the individual national authority will then issue an official system release with certificate.

- UK AOF: Initial Operating Capability (IOC). The state achieved when Military Capability is available in its minimum usefully deployable form. The date of achievement is the ISD.
- US JCIDS: In general, attained when some units and/or organizations in the force structure scheduled to receive a system have received it and have the ability to employ and maintain it. The specifics for any particular system IOC are defined in that system's Capability Development Document (CDD) and Capability Production Document (CPD).

FOC (Full Operational Capability)

Once the capability is fully developed and the step of the Initial Operating Capability or Initial Operational Capability (IOC) surpassed or skipped, Full Operational Capability (FOC) may be declared. For example, the capability may be fielded to a limited number of users with plans to roll out to all users incrementally over a period of time (possibly incorporating changes along the way).

FOC is achieved when all intended users (by agreement between the developer and the user) have the capability in use. This does not preclude additional users from obtaining the capability after FOC. The in-service date (ISD) can occur when the capability defined in the User Requirement Definition (URD) is assessed as available for operational use - in its minimum usefully deployable form.

It is the date on which an Initial Operating Capability (IOC) was achieved. Keep in mind that *"assessed as available for operational use"* can only be defined by the national OPERATORS, who have to evaluate the system services according to a predefined test schedule and based on their judgement the individual national authority will then issue an official system release with certificate.

Policy

A policy is a principle or rule to guide decision and achieve a rational outcome. Policy heads the Doctrine, where for the policy in most multinational environments a consensus is required, while a doctrine can be finalized by the majority of the members using it.

A Policy is developed in response to changing circumstances in the political-military strategic environment, agreed political guidance, practical lessons learned or new technology and is essentially prescriptive. Among other factors which influence the development of doctrine, it primarily evolves in response to changes in policy, war fighting capabilities and/or force employment considerations. Thus it is recognized that policy, as agreed by the highest national authorities, normally leads and directs doctrine.

Security Policies are clear instructions that provide the guidelines for safeguarding information, and are fundamental building blocks in developing effective controls to counter potential security threats. These policies are even more significant when it comes to preventing and detecting social engineering attacks.

Effective security controls are implemented by training employees with well-documented policies and procedures. The risk assessment has the primary goal to prioritize which information assets are in need of immediate safeguarding, and whether instituting safeguards will be cost-effective based on a cost-benefit analysis.

The person assigned to draft information security policies needs to understand that the policies should be written in a style free of technical jargon and readily understood by nontechnical addressees. The policies will need to be modified and/or supplemented. A process for a regular review and updating should be put in place. The policies are in relation with classifications.

In case of a security incident a reporting mechanism and measurements required have to be already in place. This includes the nominated receiver and distributor of alerts concerning possible security incidents.

Procedure

A Procedure is a specified series of actions or operations which have to be executed in the same manner in order to always obtain the same result under the same circumstances (i.e. emergency procedures). In telecommunications, this is the premise under which a SOP (Standard Operating Procedure) is generated.

Standard Operating Procedure (SOP)

A Standard Operating Procedure (SOP)[271] is a set of step-by-step instructions compiled by an organization to help workers carry out complex routine operations. SOPs aim to achieve efficiency, quality output and uniformity of performance, while reducing miscommunication and failure to comply with industry regulations.

The military (i.e. in the U.S. and UK) sometimes uses the term standing (rather than standard) operating procedure because a military SOP refers to a unit's unique procedures, which are not necessarily standard to another unit. The word "standard" can imply that only one (standard) procedure is to be used across all units.

A Standard Operating Procedure (SOP) is specifically designed to describe and guide multiple iterations of the same operational procedure over a broad number of locations, on multiple occasions, and over an open period of time. A procedure change usually induces a change in the daily routines of the operation centers. Such SOP is continuously updated and adapted to the operational environment, or discontinued.

Less precisely speaking, a SOP can indicate a sequence of activities, tasks, steps, decisions, calculations and processes, that when undertaken in the sequence laid down produces the described result, product or outcome.

---

[271] https://en.wikipedia.org/wiki/Standard_operating_procedure

Safety

Safety[272] is the condition of being protected against harmful conditions or events, or the control of hazards to reduce risk.

Safety is the state of being "*safe*", the condition of being protected against physical, social, spiritual, financial, political, emotional, occupational, psychological, educational or other types or consequences of failure, damage, error, accidents, harm or any other event which could be considered non-desirable. This can take the form of being protected from the event or from exposure to something that causes health or economical losses. It can include protection of people or of possessions.

In distributed computing, safety properties informally require that "*something bad will never happen*" in a distributed system or distributed algorithm. Unlike liveness properties, safety properties can be violated by a finite execution of a distributed system. In a database system, a promise to never return data with null fields is an example of a safety guarantee. All properties can be expressed as the intersection of safety and liveness properties.

Note that most languages that descend from Latin and the language of ancient Rome like Italian, French, Portuguese, Rhaeto-Romance, Romanian, Spanish, Catalan, Sardinian, Corsican, Lombard, Occitan, Gascon, Aromanian, Sardinian, Sicilian, Venetian, Galician, Neapolitan, Friulan and German, do not distinguish between Safety and Security, use for both an equivalent term, and can therefore be ambiguous.

---

[272] http://en.wikipedia.org/wiki/Safety

Security

Security is the degree of protection against danger, damage, loss, and criminal activity. Security[273] in form of physical protection measurements are i.e. structures and processes that provide or improve security as a condition. Security as a national condition was defined in a United Nations study (1986), so that countries can develop and progress freely. Security has related concepts:

- Safety,
- Continuity,
- Reliability.

The key difference between security and reliability is that security must take into account the actions of people attempting to cause destruction. Each country and organization has a own definition of Security, i.e. National Security.

Security is inseparable from the social, cultural and political values that distinguish social life in all its diversity. Security research and innovation must address the long-term vulnerability of these values via economic, cultural, political, and technological systems. Humans are at the core of security processes: They endure and respond to natural disasters. They perpetrate or are victimized by organized crime, trafficking and terrorism.

Because security is inextricably bound to a society's daily political, economic and cultural values, technological innovation cannot fully contribute to security unless it focuses on the human being.

Two old meanings of the word secure are to be free from apprehension and free from danger. Feeling secure is a fleeting sentiment guaranteeing very little, while being secure suggests something substantial and enduring. However, confusing the two tends to leave people feeling quite insecure. The things and events allegedly empowered with the ability to have people being secure only have them temporarily feeling secure.

Security from a social perspective has three major characteristics:

- It is about people – both as the source and the object of insecurity
- It is about society – in the knowledge that some threats will target people's identity, culture, and way of life
- It is about values – and which proactive and reactive measures can protect citizens while reflecting their values and way of life

Research and innovation in security demands a framework of legal and ethical guidelines,

---

[273] http://en.wikipedia.org/wiki/Security

a "*legitimacy perimeter*", to ensure social acceptance and trust, alongside effective political leadership and communication. These will open markets for trusted new solutions.

The cohesion of society will depend heavily on the strength of its convictions and commitment to its institutions, culture and identity. In times of crisis this requires that individuals work together, based on joint preparation and mutual trust, confidence and support. Such interaction is crucial to societal robustness and resilience, but it is complex and needs to be understood.

Trust refers to overall judgement about what can be expected from both people and technologies. It is a core component of security. Security implies nurturing trust among people, institutions and technologies. Under conditions of threat trust enhances transparency and social inclusion. It plays a decisive role at the interface between citizens and governments, social services and institutions, information agencies, technological systems, and local and global markets. Yet trust is not a "*given*", but must be earned.

It flows from a determined combination of direct human contact, informal transmission of knowledge, experience and tradition, culture, reputation, solidarity, expertise and communication. It rests on transparency, fairness and justness, but also enhances efficiency. Fortunately, many concepts and challenges are similar in the security and defense areas, and across various security disciplines (e.g. police forces and private security services).

National Security

National Security[274] is the Security requirement to maintain the survival of the nation/state through the use of economic, military and political power and the exercise of diplomacy.

Originally conceived as protection against military attack, national security is now widely understood to include non-military dimensions, including the security from terrorism, crime, economic security, energy security, environmental security, food security, cyber security etc. Similarly, national security risks include, in addition to the actions of other nation states, action by violent non-state actors, narcotic cartels, and multinational corporations, and also the effects of natural disasters.

Governments rely on a range of measures, including political, economic, and military power, as well as diplomacy to enforce national security. They may also act to build the conditions of security regionally and internationally by reducing transnational causes of insecurity, such as climate change, economic inequality, political exclusion, and nuclear proliferation.

Governments depend on international cooperation and organizations such as the United

---

[274] http://en.wikipedia.org/wiki/National_security

Nations, NATO, EU, OSZE, any many others, in order to ensure National Security.

Measures taken to ensure national security include:

- using diplomacy to rally allies and isolate threats;
- marshalling economic power to facilitate or compel cooperation;
- maintaining effective armed forces;
- implementing civil defense and emergency preparedness measures (including anti-terrorism legislation);
- ensuring the resilience and redundancy of critical infrastructure;
- using intelligence services to detect and defeat or avoid threats and espionage, and to protect classified information;
- using counter-intelligence services or secret police to protect the nation from internal threats.

Security Modes

Security Modes[275] refer to information systems security modes of operations used in mandatory access control (MAC) systems. Often, these systems contain information at various levels of security classification. Type of Security Modes are:

- Dedicated Security Mode
- System High Security Mode
- Compartmented Security Mode
- Multilevel Security Mode

In Dedicated Security Mode of operation, all users must have:

1. Signed Non-Disclosure Agreement (NDA) for ALL information on the system;
2. Proper clearance for ALL information on the system;
3. Formal access approval for ALL information on the system
4. A valid need to know for ALL information on the system.

All users can access ALL data.

In System High Security Mode of operation, all users must have:

1. Signed Non-Disclosure Agreement (NDA) for ALL information on the system;
2. Proper clearance for ALL information on the system;
3. Formal access approval for ALL information on the system;

---

[275] http://en.wikipedia.org/wiki/Security_modes

4. A valid need to know for SOME information on the system.

All users can access SOME data, based on their need to know.

In Compartmented Security Mode of operation, all users must have:

1. Signed Non-Disclosure Agreement (NDA) for ALL information on the system. Proper clearance for ALL information on the system;
2. Formal access approval for SOME information they will access on the system;
3. A valid need to know for SOME information on the system. All users can access SOME data, based on their need to know and formal access approval.

In Multilevel Security Mode of operation, all users must have:

1. Signed Non-Disclosure Agreement (NDA) for ALL information on the system;
2. Proper clearance for SOME information on the system;
3. Formal access approval for SOME information on the system;
4. A valid need to know for SOME information on the system.

All users can access SOME data, based on their need to know, clearance and formal access approval.

Compartmented Information

Compartmented Information is an addition to the general risk-based Classified Information levels a constraint on access, such as (in the U.S.) Special Intelligence (SI) which protects intelligence sources and methods with labels like: "*No Foreign dissemination*" (NOFORN), which restricts dissemination to U.S. nationals, or "*Originator Controlled dissemination*" (ORCON), which ensures that the originator can track possessors of the information. Documents in some compartments are marked with specific "*code words*" in addition to the classification level; famous for example is "*ULTRA*" in the II WW for decrypted messages from the German cipher machine ENIGMA.

Information Security (INFOSEC) / Information Technology Security (IT Security)

Information Security (INFOSEC) or Information Technology Security (IT Security) relates to the identification and application of security measures to protect information processed, stored or transmitted in communication, information and other electronic systems against loss of confidentiality, integrity or availability, whether accidental or intentional. Adequate countermeasures shall be taken in order prevent access to EU information by unauthorized users, to prevent the denial of access to EU information authorized users, and to prevent corruption or unauthorized modification or deletion of EU information (SECURITY

REGULATIONS OF THE COUNCIL OF THE EUROPEAN UNION[276]).

Information Security (IS) is part of Operations Security (OPSEC) and has the goal of Information Security (INFOSEC) is to protect information (stored, processed or transmitted), as well as the host systems, against a loss of confidentiality, integrity and availability through a variety of procedural, technical and administrative controls.

INFOSEC includes a range of measures that are applied on a routine basis under the auspices of security policy to protect information. However, it is driven by the generic classification of data, not the criticality of information to a particular activity. Info Ops provides guidance on the application, or waiving, of INFOSEC measures in order to protect friendly decision-makers from adversarial Info Ops, to deny access to EEFI and to support influence or deception. This must be balanced with the need to maintain friendly decision tempo.

INFOSEC is an integral element of all military operations and encompasses Communications Security (COMSEC), Computer Security (COMPUSEC), Computer Network Defense (CND), an integral part of Computer Network Operations (CNO), and together with personnel, document, physical and procedural security, it must be considered at the earliest conceptual stages and throughout the planning of an operation.

Information Security

Information Security (sometimes shortened to InfoSec) is the practice of defending information from unauthorized access, use, disclosure, disruption, modification, perusal, inspection, recording or destruction. It is a general term that can be used regardless of the form the data may take (electronic, physical, etc. pp.). Some typical terms when dealing with information security are Information Technology Security (IT-Security) and Information Assurance (IA).

Information Technology Security (IT-Security)

IT-Security is sometimes referred to as computer security[277], IT Security is information security when applied to technology (most often some form of computer system). It is worthwhile to note that a computer does not necessarily mean a home desktop. A computer is any device with a processor and some memory (even a calculator).

IT security specialists are almost always found in any major enterprise/establishment due to the nature and value of the data within larger businesses. They are responsible for keeping all of the technology within the company secure from malicious cyber-attacks that

---

[276] http://eur-lex.europa.eu/LexUriServ/LexUriServ.do?uri=OJ:L:2001:101:0001:0001:EN:PDF
[277] http://en.wikipedia.org/wiki/Computer_security

often attempt to breach into critical private information or gain control of the internal systems.

Information Assurance (IA)

Information Assurance is the act of ensuring that data is not lost when critical issues arise. These issues include but are not limited to; natural disasters, computer/server malfunction, physical theft, or any other instance where data has the potential of being lost. Since most information is stored on computers in our modern era, information assurance is typically dealt with by IT security specialists. One of the most common methods of providing information assurance is to have an off-site backup of the data in case one of the mentioned issues arise.

Governments, military, corporations, financial institutions, hospitals, and private businesses amass a great deal of confidential information about their employees, customers, products, research and financial status. Most of this information is now collected, processed and stored on electronic computers and transmitted across networks to other computers.

Printed in Poland
by Amazon Fulfillment
Poland Sp. z o.o., Wrocław